Graduate Texts in Mathematics 15

Managing Editors: P. R. Halmos
C. C. Moore

Sterling K. Berberian

Lectures in Functional Analysis and Analysis and Operator Theory 1974.

Springer-Verlag New York · Heidelberg · Berlin

Sterling K. Berberian

The University of Texas at Austin
Department of Mathematics
Austin, Texas 78712

QA
320
B398

Managing Editors

P. R. Halmos

Indiana University
Department of Mathematics
Swain Hall East
Bloomington, Indiana 47401

C. C. Moore

University of California
at Berkeley
Department of Mathematics
Berkeley, Ca. 94720

AMS Subject Classification (1973)
Primary: 43A20, 43A65, 46A05, 46C05, 46H05, 46J05, 46K10,
        46L05, 46L10, 47A25, 47B15

Library of Congress Cataloging in Publication Data

Berberian, S. K.   1926–
  Lectures in functional analysis and operator theory
(Graduate texts in mathematics v. 15)
74-9754

ISBN 0-387-90080-2 Springer-Verlag New York Heidelberg Berlin
ISBN 3-540-90080-2 Springer-Verlag Berlin Heidelberg New York

Preface

An 'analyst' is a mathematician who is seen habitually in the company of the real or complex numbers; a 'functional analyst' is an analyst who is not squeamish about using Zorn's lemma, definitely relishes the use of topology, and does not stand in the way of the internal algebraic impulses of the subject.

The term 'functional analysis' dates from the early years of the subject, when abstract methods were novel and the principal motivation came from spaces whose elements are functions. Like other generic terms it is a catchall, encompassing a large part of the mathematical analysis of the last four decades. No portable textbook can hope to survey the subject. The present book is offered as a sampler. It is limited by my narrow glimpse of a vast domain, and biased in favor of operator theory. Yet the subject is young enough that there is a wide consensus on a central core (Hahn-Banach theorem, uniform boundedness principle, Kreĭn-Mil'man theorem, Gel'fand representation theorem, etc.); the emphasis of the book is on this core (Chapters 1–6).

The prerequisites for reading the book are as follows.

Algebra: Elementary concepts of groups, rings and vector spaces.

Topology: Chapters I and II of N. Bourbaki's *Topologie générale*, plus the Weierstrass-Stone approximation theorem (not needed until Section 59).

Complex analysis: Liouville's theorem is cited in the proof of the Gel'fand-Mazur theorem (51.8); the algebraic completeness of the complex field is used in the proof of the polynomial spectral mapping theorem (53.3); the Taylor expansion of an analytic function figures in the proof of Gel'fand's formula for the spectral radius (55.1).

Measure theory: Needed only for understanding some of the applications (Sections 15, 39, 69, 70).

The prerequisites are respected in the text but are sometimes violated in the exercises. In particular, exercises marked with a * are not really exercises—they are generally too difficult—but are reports of advanced topics, relevant to the text, offered as signposts to deeper water.

v

The book is based on lecture notes for a one-year course, addressed to general mathematics students at the second year graduate level, which I have offered on three occasions. {Twice at the University of Iowa (1961–62, 1965–66), once at the University of Texas (1969–70)}. I helped myself to generous portions of expositions I have admired. These include a course of lectures by Paul R. Halmos, entitled "Topological algebra" (University of Chicago, 1952); a mimeographed set of lecture notes, with the same title, by Irving Kaplansky (University of Chicago, 1952); Arthur H. Kruse's personal notes on Kaplansky's course; and Chapters I and II of N. Bourbaki's *Espaces vectoriels topologiques.*

My deepest debt is to Paul R. Halmos. His superbly organized and polished lectures kindled my interest in the subject, and his inspiration and encouragement have helped to sustain it. This book is gratefully dedicated to him.

Sterling K. Berberian

Austin, Texas
February, 1973

Contents

Starred sections may be omitted without jeopardy to the unstarred ones.

Chapter 4. Normed Spaces, Banach Spaces, Hilbert Spaces

Chapter 5. Category

Chapter 6. Banach Algebras

Chapter 7. C^*-Algebras

Chapter 8. Miscellaneous Applications

*0. Apéritif (Wiener's Theorem)

We write $\mathbb{Z}$, $\mathbb{R}$, and $\mathbb{C}$, respectively, for the ring of integers, the field of real numbers and the field of complex numbers.

Our objective in this section is to formulate in functional-analytic terms—and almost prove—the following classical result of N. Wiener: *The reciprocal of a nonvanishing absolutely convergent trigonometric series is itself an absolutely convergent trigonometric series.* In other words:

Wiener's theorem. *If $(a_n)_{n \in \mathbb{Z}}$ is a sequence of complex numbers such that*

$$\sum_{n=-\infty}^{\infty} |a_n| < \infty,$$

and if the (continuous, periodic) function $f : \mathbb{R} \to \mathbb{C}$ defined by

$$f(t) = \sum_{n=-\infty}^{\infty} a_n e^{int} \qquad (t \in \mathbb{R})$$

never vanishes, then there exists a sequence $(b_n)_{n \in \mathbb{Z}}$ of complex numbers, with

$$\sum_{n=-\infty}^{\infty} |b_n| < \infty,$$

such that

$$\frac{1}{f(t)} = \sum_{n=-\infty}^{\infty} b_n e^{int} \qquad (t \in \mathbb{R}).$$

The proof to be described here was discovered by I. M. Gel'fand [53]; an early triumph of the functional-analytic point of view, Gel'fand's proof still provides a highly instructive and appealing point of entry into the subject.

The statement of Wiener's theorem immediately draws attention to the sequence of coefficients (a_n) (strictly speaking, we should call it a 'bilateral sequence'): do the sequence and the function determine each other uniquely? The affirmative answer depends on being able to solve for the a_n in terms of the function f:

Remark 1. *If $(a_n)_{n \in \mathbb{Z}}$ is a complex sequence such that*

$$\sum_{n=-\infty}^{\infty} |a_n| < \infty,$$

then the formula

(1)
$$f(t) = \sum_{n=-\infty}^{\infty} a_n e^{int} \qquad (t \in \mathbb{R})$$

---

* Starred sections may be omitted without jeopardy to the unstarred ones.

defines a continuous function $f: \mathbb{R} \to \mathbb{C}$ that is periodic of period 2π, the series being uniformly and absolutely convergent on $\mathbb{R}$. Moreover, the coefficients a_n may be expressed in terms of f:

$$(2) \qquad a_n = \frac{1}{2\pi} \int_0^{2\pi} f(t) e^{-int} \, dt \qquad (n \in \mathbb{Z}).$$

Proof. If $h: \mathbb{R} \to \mathbb{C}$ is any bounded, continuous function—say $|h(t)| \leq K$ for all $t \in \mathbb{R}$—then the series

$$\sum_{n=-\infty}^{\infty} a_n e^{int} h(t)$$

converges uniformly and absolutely on $\mathbb{R}$; this follows from the validity of the inequality

$$\sum_{n=-\infty}^{\infty} |a_n e^{int} h(t)| \leq K \sum_{n=-\infty}^{\infty} |a_n|$$

for all $t \in \mathbb{R}$. (We are applying the 'Weierstrass M-test' with $M_n = K|a_n|$.) For $h(t) \equiv 1$ this yields the assertion concerning (1). For general h we then have

$$f(t)h(t) = \sum_{n=-\infty}^{\infty} a_n e^{int} h(t) \qquad (t \in \mathbb{R});$$

integrating term by term (as we may by the uniform convergence) we have

$$(*) \qquad \int_0^{2\pi} f(t)h(t) \, dt = \sum_{n=-\infty}^{\infty} a_n \int_0^{2\pi} e^{int} h(t) \, dt.$$

Fix an integer m and apply (*) to the function $h(t) \equiv e^{-imt}$; by elementary calculus,

$$\int_0^{2\pi} e^{int} e^{-imt} \, dt = 2\pi \delta_{nm}$$

(where $\delta_{nm} = 0$ or 1 according as $n \neq m$ or $n = m$), thus (*) yields

$$\int_0^{2\pi} f(t) e^{-imt} \, dt = 2\pi a_m. \qquad \blacksquare$$

Let us denote by $\mathscr{A}$ the set of all functions $f: \mathbb{R} \to \mathbb{C}$ that are produced by Remark 1—one for each absolutely summable sequence $(a_n)_{n \in \mathbb{Z}}$. Obviously included in $\mathscr{A}$ are the functions u_n $(n \in \mathbb{Z})$ defined by the formulas

$$(3) \qquad u_n(t) = e^{int} \qquad (t \in \mathbb{R}).$$

Also included are the finite linear combinations of the u_n, that is, the functions of the form

$$(4) \qquad \sum_{n \in J} a_n u_n,$$

where J is a finite subset of $\mathbb{Z}$ (such functions are called *trigonometric poly-nomials*). In particular, $\mathscr{A}$ contains the constant function $u_0(t) \equiv 1$, and, more generally, the constant functions cu_0 $(c \in \mathbb{C})$. Note that, for all $t \in \mathbb{R}$,

$$(5) \qquad\qquad u_{m+n}(t) = u_m(t)u_n(t),$$

$$(6) \qquad\qquad u_{-n}(t) = \frac{1}{u_n(t)} = (u_n(t))^*;$$

briefly, $u_{m+n} = u_m u_n$, $u_{-n} = 1/u_n = u_n^*$ (here the * denotes complex conjugation).

The message of Wiener's theorem is that if $f \in \mathscr{A}$ and f never vanishes, then $1/f \in \mathscr{A}$. In the course of the proof, it is shown that $\mathscr{A}$ is an algebraic system of functions, closed under the usual pointwise operations of addition and multiplication. Before getting down to the details, it is instructive to note that $\mathscr{A}$ consists of the functions that are continuous, periodic of period 2π, and have absolutely summable Fourier coefficients:

Remark 2. *Suppose $f \colon \mathbb{R} \to \mathbb{C}$ is a continuous, periodic function of period 2π, define*

$$a_n = \frac{1}{2\pi} \int_0^{2\pi} f(t)e^{-int}dt \qquad (n \in \mathbb{Z}),$$

and suppose that

$$\sum_{n=-\infty}^{\infty} |a_n| < \infty.$$

Then

$$f(t) = \sum_{n=-\infty}^{\infty} a_n e^{int} \qquad (t \in \mathbb{R}).$$

Proof. By Remark 1, the formula

$$g(t) = \sum_{n=-\infty}^{\infty} a_n e^{int}$$

defines a function $g \colon \mathbb{R} \to \mathbb{C}$ that is continuous and periodic of period 2π, such that

$$a_n = \frac{1}{2\pi} \int_0^{2\pi} g(t)e^{-int}\,dt \qquad (n \in \mathbb{Z}).$$

Thus

$$\frac{1}{2\pi} \int_0^{2\pi} [g(t) - f(t)]u_{-n}(t)\,dt = a_n - a_n = 0$$

for all $n \in \mathbb{Z}$, therefore

$$(*) \qquad\qquad \int_0^{2\pi} [g(t) - f(t)]h(t)\,dt = 0$$

for every trigonometric polynomial h (cf. (4)). A classical theorem of Weierstrass asserts that every continuous, periodic function $k : \mathbb{R} \to \mathbb{C}$ of period 2π is the uniform limit of a sequence of trigonometric polynomials; in view of (*), we infer that

$$\int_0^{2\pi} [g(t) - f(t)]k(t)\, dt = 0$$

for every such function k. Applying this to the function

$$k(t) = [g(t) - f(t)]^*$$

we see that

$$\int_0^{2\pi} |g(t) - f(t)|^2\, dt = 0;$$

since the integrand is continuous, we conclude that $g - f = 0$. ∎

In studying the algebraic properties of the function set $\mathscr{A}$, it will be expedient to impose an algebraic structure on the set of absolutely summable sequences $(a_n)_{n \in \mathbb{Z}}$.

Definition. We write $l^1(\mathbb{Z})$ for the set of all sequences $(a_n)_{n \in \mathbb{Z}}$ with $a_n \in \mathbb{C}$ and $\sum_{-\infty}^{\infty} |a_n| < \infty$; such a sequence may be regarded as a function $x : \mathbb{Z} \to \mathbb{C}$ with $x(n) = a_n$ $(n \in \mathbb{Z})$, thus $l^1(\mathbb{Z})$ may be regarded as the set of all functions $x : \mathbb{Z} \to \mathbb{C}$ such that

$$\sum_{n=-\infty}^{\infty} |x(n)| < \infty.$$

Then $x, y \in l^1(\mathbb{Z})$ are equal iff $x(n) = y(n)$ for all $n \in \mathbb{Z}$—viewed as sequences, x and y are equal term by term. If $x \in l^1(\mathbb{Z})$ we write

$$(7) \qquad \|x\|_1 = \sum_{n=-\infty}^{\infty} |x(n)|,$$

called the *norm* (or l^1-norm) of x; thus

$$\|x\|_1 = \sup_J \sum_{n \in J} |x(n)|,$$

where J varies over all nonempty finite subsets of $\mathbb{Z}$. {The notation $l^1(\mathbb{Z})$ is a traditional one, with roots in the theory of Lebesgue integration (cf. Section 39).}

Remark 3. $l^1(\mathbb{Z})$ *is a complex vector space of functions on* $\mathbb{Z}$, *i.e., if* $x, y \in l^1(\mathbb{Z})$ *and* $c \in \mathbb{C}$, *then* $x + y$, $cx \in l^1(\mathbb{Z})$. *Moreover,*

$$(8) \qquad \|x + y\|_1 \leq \|x\|_1 + \|y\|_1,$$

$$(9) \qquad \|cx\|_1 = |c|\, \|x\|_1.$$

The formula

(10)
$$d(x, y) = \|x - y\|_1$$

defines a metric d on $l^1(\mathbb{Z})$.

Proof. The functions $x + y$ and cx are defined by the formulas

$$(x + y)(n) = x(n) + y(n), \qquad (cx)(n) = cx(n),$$

i.e., these are the 'pointwise' (or 'term-by-term') operations. For every finite subset J of $\mathbb{Z}$, we have

$$\sum_{n \in J} |(x + y)(n)| \le \sum_{n \in J} |x(n)| + \sum_{n \in J} |y(n)| \le \|x\|_1 + \|y\|_1,$$

and

$$\sum_{n \in J} |(cx)(n)| = |c| \sum_{n \in J} |x(n)|;$$

varying J, we infer that $x + y$ and cx are also in $l^1(\mathbb{Z})$ and that (8) and (9) hold.

The fact that d satisfies the triangle inequality

$$d(x, z) \le d(x, y) + d(y, z)$$

results from applying (8) to the relation

$$x - z = (x - y) + (y - z).$$

The properties $d(x, y) = d(y, x)$, $d(x, y) \ge 0$, and $d(x, y) = 0$ iff $x = y$, are obvious. ∎

Definition. If $x \in l^1(\mathbb{Z})$ it is useful to have a notation and a name for the function $\mathbb{R} \to \mathbb{C}$ defined in Remark 1: we write

(11)
$$\hat{x}(t) = \sum_{n=-\infty}^{\infty} x(n)e^{int} \qquad (t \in \mathbb{R})$$

and call $\hat{x}$ the *Fourier transform* of x. {This is not quite accurate. In the usual formula for Fourier transform, e^{-int} appears instead of e^{int}. Also, the Fourier transform of x is often regarded as a function $\hat{x}$ on the unit circle $\mathbb{T} = \{\lambda \in \mathbb{C} : |\lambda| = 1\}$, with $\hat{x}(\lambda) = \sum_{n=-\infty}^{\infty} x(n)\lambda^{-n}$. These conventions are explained in Section 70 (in a more general setting).}

Remark 4. *$\mathscr{A}$ is a vector space with respect to the pointwise operations, and the mapping $x \mapsto \hat{x}$ is a vector space isomorphism of $l^1(\mathbb{Z})$ onto $\mathscr{A}$.*

Proof. It is clear from Remark 1 that $x \mapsto \hat{x}$ is a bijective mapping of $l^1(\mathbb{Z})$ onto $\mathscr{A}$. Thus, assuming $x, y \in l^1(\mathbb{Z})$ and $c \in \mathbb{C}$, it will clearly suffice to show that $(x + y)^\wedge = \hat{x} + \hat{y}$ and $(cx)^\wedge = c\hat{x}$; but these relations follow at once from the elementary term-by-term operations on infinite series. ∎

To show that $\mathscr{A}$ is closed under pointwise products (hence is an algebra of functions), it is convenient to impose a suitable multiplication $(x, y) \mapsto xy$ on $l^1(\mathbb{Z})$ and then show that $\hat{x}\hat{y} = (xy)^\wedge$. To guess the correct definition of product in $l^1(\mathbb{Z})$, we look at the Fourier transform of certain special elements of $l^1(\mathbb{Z})$:

Definition.   For each $n \in \mathbb{Z}$ we define $e_n \in l^1(\mathbb{Z})$ by the formula

(12) $$e_n(m) = \delta_{mn} \qquad (m \in \mathbb{Z}).$$

It is clear from the definition of Fourier transform that $\hat{e}_n = u_n$ $(n \in \mathbb{Z})$. The linear subspace spanned by the e_n is denoted $c_{00}(\mathbb{Z})$; its elements are the functions $x : \mathbb{Z} \to \mathbb{C}$ with finite support, i.e., such that the set

$$\{n \in \mathbb{Z} : x(n) \neq 0\}$$

is finite. An element $x \in c_{00}(\mathbb{Z})$ may be written as a sum

$$x = \sum_{n \in \mathbb{Z}} x(n)e_n,$$

where $x(n) = 0$ for all but finitely many n; the Fourier transform of such a function,

$$\hat{x} = \sum_{n \in \mathbb{Z}} x(n)u_n,$$

is the typical trigonometric polynomial. We write $\mathscr{T}$ for the set of all trigonometric polynomials. Thus $\mathscr{T}$ is a linear subspace of $\mathscr{A}$, and $x \mapsto \hat{x}$ is a vector space isomorphism of $c_{00}(\mathbb{Z})$ onto $\mathscr{T}$. The linear subspace $c_{00}(\mathbb{Z})$ is dense in $l^1(\mathbb{Z})$—in fact, the e_n are a 'basis' of $l^1(\mathbb{Z})$ in the following sense:

Remark 5.   *If $x \in l^1(\mathbb{Z})$ then*

(13) $$x = \sum_{n=-\infty}^{\infty} x(n)e_n,$$

in the sense that the net of finite subsums converges to x in the metric of $l^1(\mathbb{Z})$.

Proof.   Fix $x \in l^1(\mathbb{Z})$ and, for every nonempty finite subset J of $\mathbb{Z}$, write

$$x_J = \sum_{n \in J} x(n)e_n;$$

it is to be shown that $x_J \to x$ in the sense that, for every $\varepsilon > 0$, there exists a finite subset K of $\mathbb{Z}$ such that

$$J \supset K \Rightarrow \|x - x_J\|_1 \leq \varepsilon.$$

Indeed, choose K so that

$$\sum_{n \in \complement K} |x(n)| \leq \varepsilon.$$

If $J \supset K$ then $x - x_J$ vanishes on J and agrees with x on $\complement J$; since $\complement J \subset \complement K$, we have

$$\|x - x_J\|_1 = \sum_{n \in \complement J} |x(n)| \leq \sum_{n \in \complement K} |x(n)| \leq \varepsilon. \qquad \blacksquare$$

As noted in (5), we have $u_m u_n = u_{m+n}$ $(m, n \in \mathbb{Z})$; it follows that the vector space $\mathscr{T}$ of trigonometric polynomials is in fact an algebra of functions. The vector space isomorphism $\hat{x} \mapsto x$ $(\hat{x} \in \mathscr{T})$ therefore induces an algebra structure on $c_{00}(\mathbb{Z})$. What does this product $(x, y) \mapsto xy$ on $c_{00}(\mathbb{Z})$ look like? The answer is as follows:

Remark 6. *If $x, y \in c_{00}(\mathbb{Z})$ then*

$$(14) \qquad (xy)(m) = \sum_{n=-\infty}^{\infty} x(m-n)y(n) \qquad (m \in \mathbb{Z}).$$

Proof. (Note that for each $m \in \mathbb{Z}$, all but finitely many terms on the right side of (14) are 0.) First let us calculate the products $e_m e_n$: by definition, $e_m e_n$ is the element of $c_{00}(\mathbb{Z})$ whose Fourier transform is $\hat{e}_m \hat{e}_n$, that is,

$$(e_m e_n)^{\wedge} = \hat{e}_m \hat{e}_n = u_m u_n = u_{m+n} = \hat{e}_{m+n};$$

therefore

$$(15) \qquad e_m e_n = e_{m+n} \qquad (m, n \in \mathbb{Z}).$$

By definition, the mapping $(x, y) \mapsto xy$ $(x, y \in c_{00}(\mathbb{Z}))$ is bilinear; it follows from (15) that if $x, y \in c_{00}(\mathbb{Z})$ then

$$(16) \qquad xy = \left(\sum_{k=-\infty}^{\infty} x(k)e_k \right) \left(\sum_{n=-\infty}^{\infty} y(n)e_n \right)$$

$$= \sum_{k=-\infty}^{\infty} \sum_{n=-\infty}^{\infty} x(k)y(n)e_{k+n}.$$

For each $m \in \mathbb{Z}$, $(xy)(m)$ is the total coefficient of e_m on the right side of (16), thus

$$(xy)(m) = \sum_{k+n=m} x(k)y(n) = \sum_{n=-\infty}^{\infty} x(m-n)y(n). \qquad \blacksquare$$

Formula (14) suggests a plausible definition of product in $l^1(\mathbb{Z})$; we now check that such a definition is feasible:

Remark 7. *If $x, y \in l^1(\mathbb{Z})$ then, for each $m \in \mathbb{Z}$, the series*

$$\sum_{n=-\infty}^{\infty} x(m-n)y(n)$$

converges absolutely; denoting the sum of the series by $z(m)$, we have $z \in l^1(\mathbb{Z})$ and $\|z\|_1 \leq \|x\|_1 \|y\|_1$.

Proof. For each $n \in \mathbb{Z}$ we have

$$\sum_{m=-\infty}^{\infty} |x(m-n)y(n)| = \left(\sum_{m=-\infty}^{\infty} |x(m-n)| \right) |y(n)| = \|x\|_1 |y(n)|,$$

therefore

$$\sum_{n=-\infty}^{\infty} \sum_{m=-\infty}^{\infty} |x(m-n)y(n)| = \|x\|_1 \sum_{m=-\infty}^{\infty} |y(n)| = \|x\|_1 \|y\|_1.$$

Reversing the order of summation (as we may),

$$(*) \qquad \sum_{m=-\infty}^{\infty} \sum_{n=-\infty}^{\infty} |x(m-n)y(n)| = \|x\|_1 \|y\|_1 < \infty.$$

It follows that for each m, the series indicated in the statement of the proposition is absolutely convergent, and, denoting its sum by $z(m)$,

$$|z(m)| \leq \sum_{n=-\infty}^{\infty} |x(m-n)y(n)|;$$

summing on m, we have

$$\sum_{m=-\infty}^{\infty} |z(m)| \leq \sum_{m=-\infty}^{\infty} \sum_{n=-\infty}^{\infty} |x(m-n)y(n)| = \|x\|_1 \|y\|_1$$

by (*), thus $z \in l^1(\mathbb{Z})$ and $\|z\|_1 \leq \|x\|_1 \|y\|_1$. ∎

Definition. If $x, y \in l^1(\mathbb{Z})$ as in Remark 7, we define $xy: \mathbb{Z} \to \mathbb{C}$ by the formula

$$(17) \qquad (xy)(m) = \sum_{n=-\infty}^{\infty} x(m-n)y(n) \qquad (m \in \mathbb{Z}).$$

Thus $xy \in l^1(\mathbb{Z})$ and

$$(18) \qquad \|xy\|_1 \leq \|x\|_1 \|y\|_1;$$

xy is called the *convolution* of x and y.

With convolution in hand, the proof that $\mathscr{A}$ is closed under pointwise products is straightforward:

Remark 8. *If $x, y \in l^1(\mathbb{Z})$ then $\hat{x}\hat{y} = (xy)^\smallfrown$.*

Proof. For all $t \in \mathbb{R}$,

$$(xy)^\smallfrown(t) = \sum_{m=-\infty}^{\infty} (xy)(m)e^{imt}$$

$$= \sum_{m=-\infty}^{\infty} \left(\sum_{n=-\infty}^{\infty} x(m-n)y(n) \right) e^{imt}$$

$$= \sum_{m=-\infty}^{\infty} \left(\sum_{n=-\infty}^{\infty} x(m-n)e^{i(m-n)t}y(n)e^{int} \right)$$

$$= \sum_{n=-\infty}^{\infty} \left(\sum_{m=-\infty}^{\infty} x(m-n)e^{i(m-n)t} \right) y(n)e^{int}$$

$$= \sum_{n=-\infty}^{\infty} \hat{x}(t)y(n)e^{int} = \hat{x}(t) \sum_{n=-\infty}^{\infty} y(n)e^{int} = \hat{x}(t)\hat{y}(t),$$

the interchange of the order of summation being permissible because the doubly indexed family

$$|x(m-n)e^{i(m-n)t}y(n)e^{int}| = |x(m-n)y(n)|$$

is summable (see the proof of Remark 7). ∎

The algebraic properties of convolution now come free of charge:

Remark 9. *With convolution as product, $l^1(\mathbb{Z})$ is a commutative, associative algebra with unity element e_0.*

Proof. By Remark 8, $\mathscr{A}$ is closed under pointwise products; it is already known to be a vector space of functions, thus it is an algebra of functions. Moreover, Remark 8 shows that the vector space isomorphism $\hat{x} \mapsto x$ of $\mathscr{A}$ onto $l^1(\mathbb{Z})$ preserves products, therefore the convolution product in $l^1(\mathbb{Z})$ enjoys the same algebraic properties as the pointwise product does in $\mathscr{A}$. Since $\hat{e}_0(t) \equiv 1$, e_0 serves as unity element for the convolution product (this is also easy to see directly). ∎

Let us peruse again the statement of Wiener's theorem. Assuming that $f \in \mathscr{A}$ and that f never vanishes, it is to be shown that $1/f \in \mathscr{A}$. This amounts to showing that the principal ideal $f\mathscr{A}$ is equal to $\mathscr{A}$. Suppose to the contrary that $f\mathscr{A}$ is a proper ideal of $\mathscr{A}$. By Zorn's lemma, there exists a maximal ideal $\mathscr{M}$ of $\mathscr{A}$ such that $f\mathscr{A} \subset \mathscr{M}$.

Thus, supposing to the contrary that $1/f \notin \mathscr{A}$, we have argued that f belongs to some maximal ideal $\mathscr{M}$. How is the contradiction to be obtained?

The answer is contained in another question: What do the maximal ideals of $\mathscr{A}$ look like? There are the obvious maximal ideals: fix a point $t \in \mathbb{R}$ and consider the algebra epimorphism $\mathscr{A} \to \mathbb{C}$ defined by $g \mapsto g(t)$; the kernel

$$\mathscr{M}_t = \{g \in \mathscr{A} : g(t) = 0\}$$

is a maximal ideal of $\mathscr{A}$ (because the quotient ring $\mathscr{A}/\mathscr{M}_t$ is isomorphic to the simple ring $\mathbb{C}$). The original hypothesis is that f belongs to no $\mathscr{M}_t$. But f belongs to $\mathscr{M}$. The desired contradiction results from the following:

Remark 10. *Every maximal ideal $\mathscr{M}$ of $\mathscr{A}$ has the form $\mathscr{M} = \mathscr{M}_t$ for suitable $t \in \mathbb{R}$.*

The proof of this lies beyond the elementary techniques at hand; we defer it until Section 63, where the powerful machinery of the general Gel'fand theory (Section 52) reduces it to an effortless bagatelle. {A forced march on Remark 10 is possible at this stage, but would cover much the same ground as the general Gel'fand theory.} For the present, let us examine more closely what is at issue. Suppose $\mathscr{M}$ is a maximal ideal of $\mathscr{A}$. Since $\mathscr{A}$ is a commutative ring with unity, it results from elementary ring theory that the quotient ring $\mathscr{A}/\mathscr{M}$ is a field. Moreover, $\mathscr{M}$ is a linear subspace of $\mathscr{A}$ (because $\mathscr{A}$ contains the constant functions), therefore $\mathscr{A}/\mathscr{M}$ is an algebra over $\mathbb{C}$; thus $\mathscr{A}/\mathscr{M}$ is a field extension of $\mathbb{C}$. The heart of the matter is to prove that the extension collapses, i.e., is one-dimensional—so to speak, $\mathscr{A}/\mathscr{M} = \mathbb{C}$. {It is not surprising that the proof belongs to a circle of ideas that includes the 'fundamental theorem of algebra.'} The technique of the proof will entail reformulating this as a problem in the isomorphic algebra $l^1(\mathbb{Z})$; one is given the maximal ideal

$$M = \{x \in l^1(\mathbb{Z}) : \hat{x} \in \mathscr{M}\}$$

of $l^1(\mathbb{Z})$, and the problem is to show that the quotient algebra $l^1(\mathbb{Z})/M$ is one-dimensional over $\mathbb{C}$. The key new ingredient in these considerations is the fact (not needed for Remarks 1–9) that $l^1(\mathbb{Z})$ is *complete* for the metric $d(x, y) = \|x - y\|_1$; the proof of completeness is an elementary exercise that belies the remarkable consequences that flow from it. A first consequence of completeness is that a maximal ideal M of $l^1(\mathbb{Z})$ is automatically a closed subset of $l^1(\mathbb{Z})$; this ensures that the field $l^1(\mathbb{Z})/M$ has enough additional structure (topological/geometrical) to force it into coincidence with $\mathbb{C}$. Another consequence of this is that the natural algebra epimorphism $\varphi : l^1(\mathbb{Z}) \to l^1(\mathbb{Z})/M = \mathbb{C}$ is continuous. Combining this with Remark 5 and formula (15), it is easy to see that the complex number $\mu = \varphi(e_0)$ satisfies $|\mu| = 1$ and that

$$\varphi(x) = \sum_{n=-\infty}^{\infty} x(n)\mu^n \qquad (x \in l^1(\mathbb{Z}));$$

writing $\mu = e^{it}$ for suitable $t \in \mathbb{R}$, this means that

$$\varphi(x) = \hat{x}(t) \qquad (x \in l^1(\mathbb{Z})),$$

and the proof of Remark 10 is completed by the observation that M is the kernel of φ.

The foregoing informal discussion is nothing like a proof, but the implication is plain: a key role is to be played by the interplay between algebra, topology, and classical analysis. A marriage of algebra and topology, held together by analysis—a capsule definition of functional analysis that describes reasonably well the viewpoint of this book.

Exercises

(0.1) $l^1(\mathbb{Z})$ is a complete metric space with respect to the metric $d(x, y) = \|x - y\|_1$.

(0.2)   If $x, y \in l^1(\mathbb{Z})$ and if $x \geq 0$, $y \geq 0$ (as functions on $\mathbb{Z}$), then $\|xy\|_1 = \|x\|_1\|y\|_1$.

(0.3) $\mathcal{M}_s = \mathcal{M}_t$ if and only if $s - t$ is a multiple of 2π.

Chapter 1

Topological Groups

1. Topological algebraic structures. The concept of a topological algebraic structure is a major unifying theme in functional analysis. The concept is best explained through representative examples of such structures.

(1.1) Let X be a monoid, i.e., a set X with an internal law of composition $X \times X \to X$, denoted, say, $(x, y) \mapsto xy$, that is associative and admits a neutral element. A topology on X is said to be compatible with the monoid structure if the mapping $(x, y) \mapsto xy$ is continuous, where it is understood that $X \times X$ bears the product topology. A *topological monoid* is a pair (X, τ), where X is a monoid and τ is a topology on X compatible with the monoid structure.

(1.2) Let G be a group (notated, say, multiplicatively). A topology on G is said to be compatible with the group structure if the mappings $(x, y) \mapsto xy$ and $x \mapsto x^{-1}$ are continuous. A *topological group* is a pair (G, τ), where G is a group and τ is a topology on G compatible with the group structure.

(1.3) Let A be a ring (we shall deal exclusively with associative rings). A topology on A is said to be compatible with the ring structure if the mappings $(x, y) \mapsto x + y$, $x \mapsto -x$ and $(x, y) \mapsto xy$ are continuous. A *topological ring* is a pair (A, τ), where A is a ring and τ is a topology on A compatible with the ring structure. Such a topology is, in particular, compatible with the additive group structure, thus the additive group of a topological ring is an abelian topological group. {However, assuming A has a unity element, the multiplicative group G of invertible elements of A need not be a topological group, since the mapping $x \mapsto x^{-1}$ $(x \in G)$ may fail to be continuous.}

(1.4) Let D be a division ring. A topology on D is said to be compatible with the division ring structure if the mappings $(x, y) \mapsto x + y$, $x \mapsto -x$, $(x, y) \mapsto xy$ and $x \mapsto x^{-1}$ $(x \neq 0)$ are continuous. A *topological division ring* is a pair (D, τ), where D is a division ring and τ is a topology on D compatible with the division ring structure; in other words, (D, τ) is a topological ring such that τ is also compatible with the multiplicative group structure of $D - \{0\}$. If, moreover, multiplication is commutative, then (D, τ) is called a *topological field*.

(1.5) Let (A, τ) be a topological ring and let M be an A-module (say a left module). A topology on M is said to be compatible with the A-module structure if the module operations $(x, y) \mapsto x + y$, $x \mapsto -x$ and $(a, x) \mapsto ax$ are continuous. A *topological module* (over A) is a quadruple $(A, \tau_A; M, \tau_M)$, where (A, τ_A) is a topological ring, M is an A-module, and τ_M is a topology

11

on M compatible with the module structure. {In particular, (M, τ_M) is an abelian topological group under addition.}

(**1.6**) Let (D, τ) be a topological division ring and let E be a vector space over D. A topology on E is said to be compatible with the vector space structure if the mappings $(x, y) \mapsto x + y$ and $(\lambda, x) \mapsto \overset{\centerdot}{\lambda} x$ are continuous. A *topological vector space* (briefly, TVS) over D is a quadruple $(D, \tau_D; E, \tau_E)$, where (D, τ_D) is a topological division ring, E is a vector space over D, and τ_E is a topology on E compatible with the vector space structure. Since $-x = (-1)x$, it follows that E is a topological D-module. {We shall deal exclusively with topological vector spaces over the real field $\mathbb{R}$ or the complex field $\mathbb{C}$.}

(**1.7**) Let A be an algebra (we deal exclusively with associative algebras) over a topological field F. A topology on A is said to be compatible with the algebra structure if the mappings $(x, y) \mapsto x + y$, $(\lambda, x) \mapsto \lambda x$ and $(x, y) \mapsto xy$ are continuous, in other words, the topology makes A simultaneously a topological vector space over F and a topological ring. A *topological algebra* is a quadruple $(F, \tau_F; A, \tau_A)$, where (F, τ_F) is a topological field, A is an algebra over F, and τ_A is a topology on A compatible with the algebra structure.

(**1.8**) With the above examples in view, the reader is invited to roll his own definition of 'topological algebraic structure.' The area of mathematics comprising topological algebraic structures is called 'topological algebra,' a minor conflict with the terminology of the preceding paragraph.

It is usually unnecessary to have explicit notations for the topology (or topologies) involved in a topological algebraic structure; for example 'G is a topological group' is permissible in place of '(G, τ) is a topological group,' provided that only one compatible topology on G is under discussion.

The emphasis in this book is on three particular structures: topological groups (especially abelian, metrizable, complete topological groups), topological vector spaces (especially locally convex spaces, Banach spaces and Hilbert spaces), and topological algebras (specifically, Banach algebras).

2. Topological groups. See the preceding section for the definition of a topological group (1.2). It is useful to begin with a modest repertory of examples:

(**2.1**) *G* any group, with the discrete topology.

(**2.2**) If G is a topological group and H is a subgroup of G, then H is also a topological group for the relative topology. {Proof: The restrictions of the continuous mappings $(x, y) \mapsto xy$ and $x \mapsto x^{-1}$ to $H \times H$ and H, respectively, are continuous for the relative topology on H.}

(**2.3**) The additive group of the complex field $\mathbb{C}$ is a topological group with the topology defined by the usual metric $d(\lambda, \mu) = |\lambda - \mu|$. {Compatibility means that $\lambda_n \to \lambda$ and $\mu_n \to \mu$ imply $\lambda_n + \mu_n \to \lambda + \mu$ and $-\lambda_n \to -\lambda.$}

(2.4) The additive groups of the real numbers $\mathbb{R}$, the rational numbers $\mathbb{Q}$, and the integers $\mathbb{Z}$ are topological groups for the usual metric topologies (cf. (2.2), (2.3)). Of course $\mathbb{Z}$ is also covered by (2.1).

(2.5) The multiplicative group $\mathbb{C} - \{0\}$ of nonzero complex numbers, with the usual topology, is a topological group: $\lambda_n \to \lambda$ and $\mu_n \to \mu$ imply $\lambda_n \mu_n \to \lambda \mu$ and $1/\lambda_n \to 1/\lambda$.

(2.6) The multiplicative groups $\mathbb{R} - \{0\}$ and $\mathbb{Q} - \{0\}$, with their usual topologies, are topological groups (cf. (2.2), (2.5)).

(2.7) The multiplicative group $\mathbb{T} = \{\lambda \in \mathbb{C} : |\lambda| = 1\}$, with the usual topology, is a topological group (cf. (2.2), (2.5)); $\mathbb{T}$ is called the *circle group*. Note that $\lambda^{-1} = \lambda^*$ (the complex conjugate of λ) for λ in $\mathbb{T}$.

(2.8) Theorem. *If a and b are fixed elements of a topological group G, then each of the following mappings is a homeomorphism of G:*

(1) $x \mapsto x^{-1}$,
(2) $x \mapsto ax$,
(3) $x \mapsto xb$,
(4) $x \mapsto axb$,
(5) $x \mapsto axa^{-1}$.

For fixed $a \in G$, the mapping

(6) $x \mapsto xax^{-1}$

is continuous.

Proof. (1) The mapping $x \mapsto x^{-1}$ is continuous and self-inverse.
(2), (3) $x \mapsto (a, x) \mapsto ax$ is the composition of two continuous mappings; the inverse mapping is $x \mapsto a^{-1}x$. Similarly for $x \mapsto xb$.
(4), (5) $x \mapsto ax \mapsto (ax)b$ is the composition of two homeomorphisms.
(6) $x \mapsto (xa, x^{-1}) \mapsto (xa)x^{-1}$ is the composition of two continuous mappings, the first mapping being continuous because its compositions with the coordinate projections (namely, $x \mapsto xa$ and $x \mapsto x^{-1}$) are continuous. ∎

(2.9) Corollary. *If G is a topological group and $\mathscr{B}$ is a fundamental system of neighborhoods of the neutral element e, then, for each $a \in G$, the classes $\{aV : V \in \mathscr{B}\}$ and $\{Va : V \in \mathscr{B}\}$ are fundamental systems of neighborhoods of a. Also, $\{V^{-1} : V \in \mathscr{B}\}$ is a fundamental system of neighborhoods of e.*

Proof. The mapping $x \mapsto ax$ is a homeomorphism of G (2.8), sending e to a, and therefore transforming $\mathscr{B}$ into a fundamental system of neighborhoods of a. Similarly, $x \mapsto xa$ (resp. $x \mapsto x^{-1}$) transforms $\mathscr{B}$ into a fundamental system of neighborhoods of $ea = a$ (resp. $e^{-1} = e$). ∎

(2.10) Corollary. *If A is a subset of a topological group G and if U is an open subset of G, then the following subsets are also open: U^{-1}, AU, UA.*

Proof. {For subsets S, T of G, one writes $ST = \{st : s \in S, t \in T\}$, with the convention that $ST = \varnothing$ if either $S = \varnothing$ or $T = \varnothing$. Similarly, $S^{-1} = \{s^{-1} : s \in S\}$.}

We can suppose that A and U are nonempty. Writing $Jx = x^{-1}$, $U^{-1} = J(U)$ is open because J is a homeomorphism (2.8). Similarly, aU and Ua are open for each $a \in A$, therefore so are the sets

$$AU = \bigcup_{a \in A} aU, \qquad UA = \bigcup_{a \in A} Ua. \qquad \blacksquare$$

(2.11) The terminology for the mappings in (2.8) is as follows: $x \mapsto x^{-1}$ is the *inversion* mapping; $x \mapsto ax$ is *left translation* by a; $x \mapsto xb$ is *right translation* by b; $x \mapsto axa^{-1}$ is the *inner automorphism* of G induced by a.

The fact that translations are homeomorphisms means that a topological group is 'topologically homogeneous': for any pair of points a, $b \in G$, the mapping $x \mapsto ba^{-1}x$ is a homeomorphism of G that sends a to b. Consequently, topological behavior at a is reflected at b. For example, if one point of G has a countable neighborhood base then every point does (cf. 2.9); if one point has a compact neighborhood then every point does; etc.

(2.12) Theorem. *If G and H are topological groups then the product topology on $G \times H$ is compatible with the product group structure.*

Proof. Assuming G and H are notated multiplicatively, the product group structure on $G \times H$ is defined by

$$(x, y)(x', y') = (xx', yy'), \qquad (x, y)^{-1} = (x^{-1}, y^{-1});$$

in particular, the neutral element of $G \times H$ is (e, e).

Since $x \mapsto x^{-1}$ and $y \mapsto y^{-1}$ are continuous, so is $(x, y) \mapsto (x^{-1}, y^{-1})$; thus, inversion in $G \times H$ is continuous.

It remains to show that the mapping $((x, y), (x', y')) \mapsto (xx', yy')$, from $(G \times H) \times (G \times H)$ to $G \times H$, is continuous; since $(G \times H) \times (G \times H)$ is homeomorphic with $G \times H \times G \times H$ via the mapping $((x, y), (x', y')) \mapsto (x, y, x', y')$, it is equivalent to establish the continuity of $(x, y, x', y') \mapsto (xx', yy')$. Indeed,

$$(x, y, x', y') \mapsto (x, x') \mapsto xx'$$

is the composition of two continuous mappings, as is the mapping

$$(x, y, x', y') \mapsto (y, y') \mapsto yy',$$

consequently $(x, y, x', y') \mapsto (xx', yy')$ is also continuous. $\blacksquare$

(2.13) If G and H are topological groups then $G \times H$, with the product topology and product group structure, is a topological group (2.12). More generally, one can show (directly, or by induction on (2.12)) that if $G_1, \ldots, G_n$ are topological groups then the product topology on $G_1 \times \cdots \times G_n$ is compatible with the product group structure, thus $G_1 \times \cdots \times G_n$ is a topological group, called the *product topological group* of $G_1, \ldots, G_n$. This is a useful construct for generating new examples of topological groups:

(2.14) Let $\mathbb{C}$ be the additive group of complex numbers with the usual topology (2.3), $\mathbb{C}^n$ the additive group of *n*ples $x = (\lambda_k)$, $y = (\mu_k)$. Each of the following metrics on $\mathbb{C}^n$ yields the product topology:

$$d(x, y) = \max_k |\lambda_k - \mu_k|,$$

and

$$d(x, y) = \left(\sum_1^n |\lambda_k - \mu_k|^p \right)^{1/p},$$

where p is a fixed real number ≥ 1 (the most popular choices are $p = 1$ and $p = 2$).

(2.15) If $\mathbb{R}$ is the additive group of real numbers with the usual topology (2.4), the product topological group $\mathbb{R}^n$ is called the *n-dimensional vector group*.

(2.16) If $\mathbb{T}$ is the circle group with the usual topology (2.7), the product topological group $\mathbb{T}^n$ is called the *n-dimensional torus*.

Exercises

(2.17) Let G be a group, with neutral element e. A topology on G is compatible with the group structure if and only if (i) $(x, y) \mapsto xy$ is continuous at (e, e), (ii) for each fixed y, $x \mapsto xy$ is continuous, (iii) for each fixed x, $y \mapsto xy$ is continuous, and (iv) $x \mapsto x^{-1}$ is continuous at e.

(2.18) A topology on a group G is compatible with the group structure if and only if $(x, y) \mapsto x^{-1}y$ is a continuous mapping of $G \times G$ into G.

(2.19) If $(G_\iota)_{\iota \in I}$ is any family (not necessarily finite) of topological groups, then the product topology on $G = \prod_{\iota \in I} G_\iota$ is compatible with the product group structure. G is called the *product topological group* of the family $(G_\iota)_{\iota \in I}$.

(2.20) With notation as in (2.19), let N be the set of all $x = (x_\iota)_{\iota \in I}$ in G such that x_ι is the neutral element of G_ι for all but finitely many ι. Then N is a normal subgroup of G, and N is dense in G (i.e., $\overline{N} = G$).

*(2.21) (R. Ellis) If G is a group with a locally compact topology such that, for each $a \in G$, the mappings $x \mapsto ax$ and $x \mapsto xa$ are continuous, then the topology is compatible with the group structure.

(2.22) Let G be the group of all homeomorphisms of $[0, 1]$ (with composition product $(f \circ g)(t) = f(g(t))$), equipped with the topology defined by the metric $d(f, g) = \sup_t |f(t) - g(t)|$. Then (i) $d(f \circ h, g \circ h) = d(f, g)$, and (ii) G is a topological group.

3. Neighborhoods of *e*.

Due to the homogeneity of a topological group (2.11), the topology is completely determined by the system of neighborhoods of the neutral element e (see (3.2) below); the pertinent properties of this neighborhood system are as follows.

(3.1) Theorem. *If G is a topological group, then the class $\mathscr{V}$ of all neighborhoods of the neutral element e has the following properties:*
(1) $e \in V$ *for all* $V \in \mathscr{V}$;
(2) *if* $V, W \in \mathscr{V}$ *then* $V \cap W \in \mathscr{V}$;
(3) *if* $V \in \mathscr{V}$ *then there exists* $W \in \mathscr{V}$ *such that* $WW \subset V$;
(4) *if* $V \in \mathscr{V}$ *then* $V^{-1} \in \mathscr{V}$;
(5) *if* $V \in \mathscr{V}$ *and* $a \in G$ *then* $aVa^{-1} \in \mathscr{V}$;
(6) *if* $V \in \mathscr{V}$ *and* $W \supset V$ *then* $W \in \mathscr{V}$.

Proof. (1), (2), (6) are general properties of the filter of neighborhoods of a point of a topological space. {Note that the term 'neighborhood,' used in the Bourbaki sense, means any superset of an open superset.}

(3) The mapping $f(x, y) = xy$ is continuous at (e, e), and $f(e, e) = e$. Given any neighborhood V of e, choose a neighborhood A of (e, e) in $G \times G$ such that $f(A) \subset V$. One can suppose that $A = W \times W$ for some $W \in \mathscr{V}$ (such neighborhoods of (e, e) are basic). Thus, $WW = f(W \times W) \subset V$.

(4), (5) $x \mapsto x^{-1}$ and $x \mapsto axa^{-1}$ are homeomorphisms of G mapping e onto e (cf. 2.9). ∎

The properties listed in (3.1) are characteristic of compatibility of a topology with the group structure:

(3.2) Theorem. *If G is an abstract group, with neutral element e, and if $\mathscr{V}$ is a nonempty class of subsets of G satisfying (1)–(6) of (3.1), then there exists a unique topology on G such that* (i) *the topology is compatible with the group structure, and* (ii) $\mathscr{V}$ *is the system of all neighborhoods of e.*

Proof. Uniqueness: Clear from (2.9).

Existence. By (1), (2) and (6), $\mathscr{V}$ is a filter of sets containing e. For each $a \in G$, define $\mathscr{V}_a = \{aV : V \in \mathscr{V}\}$; since $x \mapsto ax$ is bijective and maps e onto a, $\mathscr{V}_a$ is a filter of sets containing a. In particular, $\mathscr{V}_e = \mathscr{V}$.

We assert that the family $(\mathscr{V}_a)_{a \in G}$ is the family of neighborhood filters for a topology on G. Given $N \in \mathscr{V}_x$, it is to be shown that there exists $M \in \mathscr{V}_x$ such that $y \in M$ implies $N \in \mathscr{V}_y$. Say $N = xV$, $V \in \mathscr{V}$. By (3) choose $W \in \mathscr{V}$ such that $WW \subset V$, and set $M = xW$. Then $y \in M$ implies $yW \subset xWW \subset xV = N$, thus $N \supset yW \in \mathscr{V}_y$ and therefore $N \in \mathscr{V}_y$.

It remains to show that the topology on G defined in the preceding paragraph is compatible with the group structure.

Fix $a, b \in G$. To see that $(x, y) \mapsto xy$ is continuous at (a, b), let a neighborhood C of ab be given, say $C = abV$, $V \in \mathscr{V}$; it will suffice to find $U \in \mathscr{V}$ such that $(aU)(bU) \subset abV$, i.e., $(b^{-1}Ub)U \subset V$. By (3) choose $W \in \mathscr{V}$ with $WW \subset V$; then $bWb^{-1} \in \mathscr{V}$ by (5), and, setting $U = (bWb^{-1}) \cap W$ one has $U \in \mathscr{V}$ and $(b^{-1}Ub)U \subset [b^{-1}(bWb^{-1})b]W = WW \subset V$.

Finally, fixing $a \in G$, it is to be shown that $x \mapsto x^{-1}$ is continuous at a. Given any $C \in \mathscr{V}_{a^{-1}}$, say $C = a^{-1}V$, $V \in \mathscr{V}$, we seek $W \in \mathscr{V}$ such that

$(aW)^{-1} \subset a^{-1}V$, i.e., $W^{-1} \subset a^{-1}Va$, i.e., $W \subset a^{-1}V^{-1}a$. Indeed, $a^{-1}V^{-1}a \in \mathscr{V}$ by (4) and (5), thus $W = a^{-1}V^{-1}a$ is suitable. $\blacksquare$

Recall that, in a topological space, '$\mathscr{B}$ is a fundamental system of neighborhoods of x' means that $\mathscr{B}$ is a base for the filter of neighborhoods of x; that is, (i) every set in $\mathscr{B}$ is a neighborhood of x, and (ii) every neighborhood of x contains some set in $\mathscr{B}$. The statement 'B is a basic neighborhood of x' means that B is a generic element of some fundamental system $\mathscr{B}$ of neighborhoods of x. Theorem (3.2) can be formulated in terms of basic neighborhoods of e:

(3.3) Theorem. *Let G be an abstract group, with neutral element e, and let $\mathscr{B}$ be a nonempty class of subsets of G satisfying the following properties (1)–(5):*

(1) *$e \in B$ for all $B \in \mathscr{B}$;*
(2) *if $B_1, B_2 \in \mathscr{B}$ then there exists $B_3 \in \mathscr{B}$ such that $B_3 \subset B_1 \cap B_2$;*
(3) *if $B \in \mathscr{B}$ then there exists $C \in \mathscr{B}$ such that $CC \subset B$;*
(4) *if $B \in \mathscr{B}$ then there exists $C \in \mathscr{B}$ such that $C \subset B^{-1}$;*
(5) *if $B \in \mathscr{B}$ and $a \in G$ then there exists $C \in \mathscr{B}$ such that $C \subset aBa^{-1}$.*

Then, there exists a unique topology on G such that (i) *the topology is compatible with the group structure, and* (ii) *$\mathscr{B}$ is a fundamental system of neighborhoods of e.*

Proof. $\mathscr{B}$ is the base for precisely one filter $\mathscr{V}$, namely,

$$\mathscr{V} = \{V : V \supset B \text{ for some } B \in \mathscr{B}\}.$$

Evidently $\mathscr{V}$ satisfies the hypotheses (1)–(6) of Theorem (3.2). $\blacksquare$

A topological space X is said to be *separated* if, for each pair of distinct points x, y of X, there exist neighborhoods U, V of x, y, respectively, such that $U \cap V = \varnothing$. {Synonyms: Hausdorff, T_2.}

(3.4) Theorem. *If G is a topological group and $\mathscr{B}$ is any fundamental system of neighborhoods of the neutral element e, then the following conditions are equivalent:*

(a) *G is separated;*
(b) *$\{e\}$ is a closed subset of G;*
(c) *$\bigcap_{B \in \mathscr{B}} B = \{e\}$.*

Proof. (a) implies (b): In a separated space every singleton is closed.

(b) implies (c): Assuming $x \neq e$ it is to be shown that some B in $\mathscr{B}$ excludes x. Since $\{x\}$ is also closed (2.8) and $e \notin \{x\}$, there exists a neighborhood V of e such that $V \cap \{x\} = \varnothing$, i.e., $x \notin V$. Choose $B \in \mathscr{B}$ with $B \subset V$.

(c) implies (a): Assuming $x \neq y$, we seek a neighborhood V of e such that $(xV) \cap (yV) = \varnothing$. Since $y^{-1}x \neq e$, by (c) there exists $B \in \mathscr{B}$ with $y^{-1}x \notin B$. Choose $C \in \mathscr{B}$ with $CC \subset B$ (cf. (3) of (3.3)). Then $V = C \cap C^{-1}$ is a neighborhood of e, and $(xV) \cap (yV) \neq \varnothing$ would imply $y^{-1}x \in VV^{-1} = VV \subset CC \subset B$, a contradiction. $\blacksquare$

Exercises

(3.5) A topological group is a T_0-space if and only if it is separated. {A T_0-*space* is a topological space such that for every pair of distinct points there exists a neighborhood of one of the points that excludes the other (but one cannot specify which point is excluded).}

(3.6) If E is a topological space and $\Delta = \{(x, x) : x \in E\}$ is the *diagonal* of $E \times E$, then E is separated if and only if Δ is closed in $E \times E$. If G is a topological group, then $\Delta = \{(x, y) : x^{-1}y = e\}$ is the inverse image of $\{e\}$ under the continuous mapping $(x, y) \mapsto x^{-1}y$; this yields a brief proof of '(b) implies (a)' in (3.4).

(3.7) Suppose E is a topological space in which every singleton is closed; if $x \in E$ and $\mathscr{B}$ is a fundamental system of neighborhoods of x, then $\bigcap_{B \in \mathscr{B}} B = \{x\}$. This property is characteristic of spaces in which every singleton is closed.

4. Subgroups and quotient groups. Subgroup, product group, quotient group—these constructs are part of the elementary grammar of group theory. In the theory of topological spaces, the parallel constructs are subspace, product space, quotient space. For topological groups, both sets of constructs are available; it is natural to ask whether the results of applying a pair of parallel constructs are compatible. If H is a subgroup of a topological group G, is the relative topology on H compatible with its group structure?; the obviously affirmative answer (2.2) allows one to say, briefly, that a subgroup of a topological group is a topological group. If G and H are topological groups, then the product topology on $G \times H$ is compatible with product group structure (2.12); briefly, the product of topological groups is a topological group (see also (2.19)). If N is a normal subgroup of the topological group G, there is a quotient group structure on the set of cosets G/N, and, via the canonical mapping $\pi : G \to G/N$, a quotient topology on G/N; the main result in this section is that the quotient topology on G/N is compatible with the quotient group structure, i.e., a quotient group of a topological group is a topological group. Also included are some odds and ends about subgroups, for which there is no better place.

(4.1) Theorem. *In a topological group, the closure of a subgroup is a subgroup, and the closure of a normal subgroup is a normal subgroup.*

Proof. Suppose H is a subgroup of the topological group G. Write $f: G \times G \to G, f(x, y) = x^{-1}y$. The subset H has the property $f(H \times H) \subset H$; the desired conclusion, $f(\overline{H} \times \overline{H}) \subset \overline{H}$, follows from the continuity of f and the fact that $\overline{H} \times \overline{H}$ is the closure of $H \times H$.

If, in addition, H is normal in G, then for each $a \in G$ the homeomorphism $x \mapsto axa^{-1}$ (2.8) maps H into H and therefore $\overline{H}$ into $\overline{H}$. ∎

The following corollary is of interest only for nonseparated topological groups (3.4):

(4.2) Corollary. *In a topological group G, with neutral element e, $\overline{\{e\}}$ is a closed normal subgroup of G.*

(4.3) Definition. If H is a subgroup of a group G, G/H denotes the set of all *left cosets* xH (i.e., the set of all equivalence classes for the equivalence relation $x \, R \, y$ defined by $x^{-1}y \in H$). The *canonical mapping* $\pi : G \to G/H$ is defined by $\pi(x) = xH$. Dually, $H \backslash G$ denotes the set of all *right cosets* Hx. When H is normal in G, i.e., when $xH = Hx$ for all $x \in G$, we need not distinguish between G/H and $H \backslash G$; for definiteness, we use the notation G/H.

(4.4) Lemma. *If H is a subgroup of a topological group G, G/H is the space of left cosets equipped with the quotient topology, and $\pi : G \to G/H$ is the canonical mapping, then* (i) *π is an open, continuous mapping, and* (ii) *for each $a \in G$, the neighborhoods of $\pi(a)$ are precisely the sets $\pi(aV)$, where V is a neighborhood of the neutral element e of G.*

Proof. By definition, the quotient topology on G/H is the final topology for the mapping π, thus the open sets of G/H are precisely the sets $A \subset G/H$ such that $\pi^{-1}(A)$ is open in G; in particular, π is continuous.

(i) If U is an open subset of G, then $\pi^{-1}(\pi(U)) = UH$ is open in G (2.10), therefore $\pi(U)$ is open in G/H.

(ii) Fix $a \in G$. If V is a neighborhood of e then aV is a neighborhood of a, therefore $\pi(aV)$ is a neighborhood of $\pi(a)$ by (i). Conversely, if A is a neighborhood of $\pi(a)$ then, by the continuity of π, $\pi^{-1}(A)$ is a neighborhood of a. Say $\pi^{-1}(A) = aV$, V a neighborhood of e; then $\pi(aV) = \pi(\pi^{-1}(A)) = A$ since π is surjective. ∎

(4.5) Theorem. *Let N be a normal subgroup of a topological group G, let $\pi : G \to G/N$ be the canonical mapping, and equip G/N with the quotient group structure and the quotient topology. Then:*

(i) *G/N is a topological group;*

(ii) *π is an open, continuous mapping;*

(iii) *for each $a \in G$, the neighborhoods of $\pi(a)$ for the quotient topology are precisely the sets $\pi(a)\pi(V)$, where V is a neighborhood of the neutral element e of G;*

(iv) *G/N is separated if and only if N is a closed subset of G.*

Proof. (ii) was proved in the lemma.

(iii) Since N is normal, π is a homomorphism, therefore $\pi(aV) = \pi(a)\pi(V)$; quote the lemma.

(i) Let $\mathscr{V}$ be the set of all neighborhoods of e. By (iii), $\pi(\mathscr{V}) = \{\pi(V) : V \in \mathscr{V}\}$ is the set of all neighborhoods of $\pi(e)$, the neutral element of G/N. Let us show that $\pi(\mathscr{V})$ satisfies the neighborhood axioms (3.2) for a topological group structure.

Of course $\pi(\mathscr{V})$ is a filter of sets containing $\pi(e)$, i.e., it satisfies conditions (1), (2), and (6) of (3.2); indeed, it is the filter of all neighborhoods of $\pi(e)$ for the quotient topology. Suppose $V \in \mathscr{V}$; choosing $W \in \mathscr{V}$ with $WW \subset V$

one has $\pi(W)\pi(W) = \pi(WW) \subset \pi(V)$, thus $\pi(\mathscr{V})$ satisfies condition (3) of (3.2). If $V \in \mathscr{V}$ then $(\pi(V))^{-1} = \pi(V^{-1}) \in \pi(\mathscr{V})$ shows that $\pi(\mathscr{V})$ satisfies (4) of (3.2). If $V \in \mathscr{V}$ and $a \in G$ then $\pi(a)\pi(V)\pi(a)^{-1} = \pi(aVa^{-1}) \in \pi(\mathscr{V})$ shows that $\pi(\mathscr{V})$ satisfies (5) of (3.2).

It follows from (3.2) that there exists a topology τ on G/H compatible with the group structure, for which $\pi(\mathscr{V})$ is the system of neighborhoods of $\pi(e)$; since the τ-neighborhoods of $\pi(a)$ are the sets $\pi(a)\pi(V)$ (2.9), it follows from (iii) that τ coincides with the quotient topology.

(iv) If G/N is separated then $\{\pi(e)\}$ is closed in G/N, therefore $N = \pi^{-1}(\{\pi(e)\})$ is closed in G by the continuity of π. Conversely, if N is closed then $\complement N$ is open, thus $\pi(\complement N)$ is open by (ii); since $\pi(\complement N) = G/N - \{\pi(e)\}$, $\{\pi(e)\}$ is closed and therefore G/N is separated (3.4). ∎

Exercises

(4.6) Let G be a group equipped with a topology such that (i) $x \mapsto x^{-1}$ is continuous, and (ii) for each $a \in G$, $x \mapsto ax$ is continuous. Then the closure of a subgroup is a subgroup, and the closure of a normal subgroup is a normal subgroup.

(4.7) In a separated topological group G, the centralizer $C(a) = \{x \in G : xa = ax\}$ of each $a \in G$ is a closed subgroup of G, and the center $Z = \bigcap_{a \in G} C(a)$ is a closed normal subgroup of G. In a nonseparated topological group, the center need not be closed.

(4.8) If G is any topological group and $N = \overline{\{e\}}$, then G/N is a separated topological group.

(4.9) An open subgroup of a topological group is necessarily closed.

(4.10) Let H be a subgroup of a topological group G and let G/H be the space of left cosets (4.4). Then (i) G/H is separated iff H is closed in G, and (ii) G/H is discrete iff H is open in G.

(4.11) If G is any topological group and C is the connected component of the neutral element e, then (i) C is a closed normal subgroup of G, (ii) for each $a \in G$ the connected component of a is $aC = Ca$, and (iii) the quotient topological group G/C is separated and totally disconnected (i.e., its connected components are all singletons).

(4.12) Let G and H be topological groups, let $f: G \to H$ be a continuous homomorphism of G onto H, and assume that G is compact and H is separated. Then, (i) H is compact, (ii) the kernel N of f is a closed normal subgroup of G, and (iii) the quotient mapping $xN \mapsto f(x)$ is a bicontinuous isomorphism of G/N onto H.

(4.13) If G and H are topological groups and if $f: G \to H$ is a homomorphism of G onto H with kernel N, then the following conditions are equivalent: (a) f is an open mapping; (b) for every neighborhood V of e_G, $f(V)$ is a neighborhood of e_H; (c) the natural isomorphism of G/N onto H is bicontinuous.

(4.14) If G is a connected topological group and V is any symmetric (i.e., $V = V^{-1}$) neighborhood of e, then every element of G can be written as a finite product of elements of V, i.e., $G = \bigcup_{n=1}^{\infty} V^n$.

5. Uniformity in topological groups. The neighborhoods V of the neutral element e play a central role in the description and determination of the topology of a topological group G. Specifically, for fixed $x \in G$ and variable V, xV (or Vx) varies over the set of all neighborhoods of x. A slight rearrangement of emphasis leads to the idea of uniformity: for fixed V, xV (or Vx) is a neighborhood of x for each $x \in G$, so to speak of 'uniform size V' independent of the particular point x.

The example (2.4) of the additive group of $\mathbb{R}$ with the usual metric $d(x, y) = |x - y|$ is instructive. The set of open intervals $V_\varepsilon = (-\varepsilon, \varepsilon)$, $\varepsilon > 0$, is a fundamental system of neighborhoods of 0. Each V_ε is a measure of nearness to the origin: $y \in V_\varepsilon$ means $d(0, y) < \varepsilon$. The points of $x + V_\varepsilon$ are the points 'within V_ε' (i.e., within ε) of x: $y \in x + V_\varepsilon$ means $y - x \in V_\varepsilon$, i.e., $d(x, y) < \varepsilon$. Moreover, for fixed $\varepsilon > 0$ the neighborhoods $x + V_\varepsilon$, x variable, are of uniform 'size' (diameter).

In a general topological group G, with neutral element e, a neighborhood V of e represents a degree of nearness to e. By homogeneity, this notion of degree of nearness can be translated throughout the group: xV is the set of points 'within V of x,' in the sense that $y \in xV$ precisely when $x^{-1}y \in V$. {In general $xV \neq Vx$; the $y \in xV$ are, so to speak, 'left-within V of x,' and the $y \in Vx$ are 'right-within V of x.'} Thus the V's play the role of ε's in the theory of topological groups.

The concept of uniformity is discernibly at work in condition (d) of the following theorem:

(5.1) Theorem. *If G and H are topological groups, with neutral elements e_G and e_H, and if $f: G \to H$ is a group homomorphism, then the following conditions are equivalent:*

(a) *f is a continuous mapping;*

(b) *f is continuous at some $a \in G$;*

(c) *f is continuous at e_G;*

(d) *given any neighborhood V of e_H, there exists a neighborhood U of e_G such that $x^{-1}y \in U$ implies $f(x)^{-1}f(y) \in V$.*

Proof. By hypothesis, $f(xy) = f(x)f(y)$ for all $x, y \in G$.

(a) implies (b) trivially.

(b) implies (c): Let V be a neighborhood of $f(e_G) = e_H$; we seek a neighborhood U of e_G such that $f(U) \subset V$. Since $f(a)V$ is a neighborhood of $f(a)$, by hypothesis there exists a neighborhood U of e_G such that $f(aU) \subset f(a)V$ (cf. 2.9), that is, $f(a)f(U) \subset f(a)V$ and therefore $f(U) \subset V$. Moreover, $x^{-1}y \in U$ implies $f(x)^{-1}f(y) = f(x^{-1}y) \in f(U) \subset V$, thus (c) implies (d).

(d) implies (a): Fix $a \in G$; let us show that f is continuous at a. Let B be a neighborhood of $f(a)$, say $B = f(a)V$, V a neighborhood of e_H. Choose a neighborhood U of e_G satisfying the condition in (d). For any $y \in aU$ one has $a^{-1}y \in U$ and therefore $f(a)^{-1}f(y) \in V$, i.e., $f(y) \in f(a)V = B$; thus $f(aU) \subset B$. ∎

(5.2) The continuity of the homomorphism f at every x, as implied by the relation between U and V in condition (d) of (5.1), can be expressed as follows: $y \in xU$ implies $f(y) \in f(x)V$, i.e., $f(xU) \subset f(x)V$. To paraphrase: given any 'ε' (namely V) there exists a 'δ' (namely U) that works at every x. This is the idea of 'uniform continuity,' to which we shall return shortly. Let us first give a definite meaning to 'uniformity' in topological groups by making connection with the theory of uniform structures, as follows.

(5.3) Definition. Let G be a topological group and let $\mathscr{V}$ be the class of all neighborhoods of the neutral element e. For each $V \in \mathscr{V}$ define a subset V_s of $G \times G$ by

$$V_s = \{(x, y) : x^{-1}y \in V\}.$$

The class of all such V_s is denoted $\mathscr{V}_s$. Each V_s 'surrounds' the diagonal $\Delta = \{(x, x) : x \in G\}$ of $G \times G$, in the sense that $V_s \supset \Delta$.

(5.4) We recall two notations from the theory of relations (here G can be any set): if A and B are subsets of $G \times G$ (i.e., A and B are relations in G) then $A \circ B$ (the composite of A and B) and A^{-1} (the reverse of A) are the subsets of $G \times G$ defined by

$$A \circ B = \{(x, z) : \exists y \ni (x, y) \in A \ \& \ (y, z) \in B\},$$

$$A^{-1} = \{(y, x) : (x, y) \in A\}.$$

A *uniformity* on G is a filter $\mathscr{U}$ of subsets of $G \times G$ such that (i) $\Delta \subset A$ for all $A \in \mathscr{U}$, (ii) $A \in \mathscr{U}$ implies $A^{-1} \in \mathscr{U}$, and (iii) $A \in \mathscr{U}$ implies there exists $B \in \mathscr{U}$ such that $B \circ B \subset A$. A *uniform structure* is a pair $(G, \mathscr{U})$, where G is a set and $\mathscr{U}$ is a uniformity on G; the sets $A \in \mathscr{U}$ are called the *entourages* for the uniform structure. A *base* for the uniformity is a base $\mathscr{B}$ for the filter $\mathscr{U}$ of entourages. If G is a set and $\mathscr{B}$ is the base for a filter $\mathscr{U}$ of subsets of $G \times G$, then $\mathscr{U}$ is a uniformity if and only if (i') $\Delta \subset A$ for all $A \in \mathscr{B}$, (ii') $A \in \mathscr{B}$ implies there exists $B \in \mathscr{B}$ such that $B \subset A^{-1}$, and (iii') $A \in \mathscr{B}$ implies there exists $B \in \mathscr{B}$ such that $B \circ B \subset A$.

In any topological group, the filter of neighbourhoods of the neutral element determines two uniform structures, in general distinct, called the left uniform structure and the right uniform structure. For clarity we will consider them one at a time.

(5.5) Theorem. *If G is a topological group, $\mathscr{V}$ is the system of neighborhoods of the neutral element e, and $\mathscr{V}_s$ is the class of subsets of $G \times G$ described in (5.3), then $\mathscr{V}_s$ is the base for a uniformity $\mathscr{U}_s$ on G, $\mathscr{U}_s = \{A : A \supset V_s \text{ for some } V \in \mathscr{V}\}$.*

Proof. Already noted is that the V_s are supersets of the diagonal Δ. If $U, V \in \mathscr{V}$, evidently $U \subset V$ implies $U_s \subset V_s$; consequently if $U, V \in \mathscr{V}$ are arbitrary then for the neighborhood $W = U \cap V$ one has $W_s \subset U_s \cap V_s$. Thus $\mathscr{V}_s$ is the base for precisely one filter $\mathscr{U}_s$, namely, $\mathscr{U}_s = \{A : A \supset V_s \text{ for some } V_s \in \mathscr{V}_s\}$. It remains to verify the conditions (ii'), (iii') of (5.4).

If $V \in \mathscr{V}$ then also $V^{-1} \in \mathscr{V}$, and $(V^{-1})_s = (V_s)^{-1}$ results from the chain of equivalent relations $(x, y) \in (V^{-1})_s$, $x^{-1}y \in V^{-1}$, $y^{-1}x \in V$, $(y, x) \in V_s$, $(x, y) \in (V_s)^{-1}$. Thus $\mathscr{V}_s$ satisfies (ii').

If $V \in \mathscr{V}$ and if $W \in \mathscr{V}$ is so chosen that $WW \subset V$, then $W_s \circ W_s \subset V_s$ results from the following chain of relations, each of which implies the next: $(x, z) \in W_s \circ W_s$, $(x, y) \in W_s$ and $(y, z) \in W_s$ for some y, $x^{-1}y \in W$ and $y^{-1}z \in W$ for some y, $(x^{-1}y)(y^{-1}z) \in WW \subset V$ for some y, $x^{-1}z \in V$, $(x, z) \in V_s$. ∎

(5.6) Definition. With notation as in (5.5), the uniform structure $(G, \mathscr{U}_s)$ is called the *left uniform structure* of the topological group G, and $\mathscr{U}_s$ the *left uniformity* of G.

(5.7) With notation as in (5.5), if $\mathscr{B}$ is a fundamental system of neighborhoods of e (i.e., if $\mathscr{B}$ is a base for the filter $\mathscr{V}$) and if $\mathscr{B}_s = \{U_s : U \in \mathscr{B}\}$, then $\mathscr{B}_s$ is a fundamental system of entourages for the left uniform structure of G (i.e., $\mathscr{B}_s$ is a base for the filter $\mathscr{U}_s$); this is immediate from (5.5) and the fact that $U \subset V$ implies $U_s \subset V_s$.

(5.8) We recall another notation from the theory of relations (here G can be any set): if $A \subset G \times G$ and $x \in G$, one defines $A(x) = \{y : (x, y) \in A\}$. Informally, $A(x)$ is the projection, onto the y-axis, of the vertical slice of A at x. In a topological group, one recaptures all neighborhoods by slicing the V_s:

(5.9) Lemma. *If V is a neighborhood of the neutral element of a topological group G, then $V_s(x) = xV$ for all $x \in G$.*

Proof. The following relations are equivalent: $y \in V_s(x)$, $(x, y) \in V_s$, $x^{-1}y \in V$, $y \in xV$. (The argument is purely group-theoretic; topology enters only since we have chosen to define the notation V_s only for neighborhoods V of the neutral element of a topological group.) ∎

(5.10) Theorem. *If G is a topological group, then the uniform topology deduced from its left uniform structure $(G, \mathscr{U}_s)$ coincides with the given topology on G. In particular, G is a uniformizable topological space.*

Proof. Let $x \in G$. As V varies over a fundamental system of neighborhoods of e in the given topology, V_s ranges over a fundamental system of entourages for the uniformity $\mathscr{U}_s$ (5.7) and therefore $V_s(x)$ varies over a fundamental system of neighborhoods of x for the uniform topology deduced from $\mathscr{U}_s$; since the sets $V_s(x) = xV$ are also a fundamental system of neighborhoods of x for the given topology (2.9), the two topologies coincide. ∎

(5.11) The rationale for the above terminology and notation is as follows. The neighborhoods $xV = V_s(x)$ are left translates of the neighborhoods V of the neutral element, hence $\mathscr{U}_s$ is termed the *left* uniformity on G (see also (5.20)). The subscript s abbreviates 'sinistral' (in preference to l for 'left,' which is too easily confused typographically with the numeral 1).

Theorem (5.1) can now be reformulated as follows:

(5.12) Theorem. *Let G and H be topological groups, each equipped with its left uniform structure, and let $f: G \to H$ be a group homomorphism. Then, f is continuous if and only if it is uniformly continuous.*

Proof. The condition on U, V in (d) of (5.1) is that $(x, y) \in U_s$ implies $(f(x), f(y)) \in V_s$. ∎

The closure operation in a topological group can be simply expressed in terms of the neighborhoods of e:

(5.13) Theorem. *If A is any subset of a topological group, then $\bar{A} = \bigcap AV = \bigcap VA$, where V runs over any fundamental system of neighborhoods of the neutral element.*

Proof. Since the intersections are unaffected by the particular choice of fundamental system of neighborhoods, we may assume that V runs over the class $\mathscr{V}$ of *all* neighborhoods of e. If $V \in \mathscr{V}$ one defines

$$V_s(A) = \{x : \exists a \in A \ni (a, x) \in V_s\};$$

thus

$$V_s(A) = \{x : \exists a \in A \ni x \in V_s(a)\} = \bigcup_{a \in A} V_s(a) = \bigcup_{a \in A} aV = AV.$$

But $\bar{A} = \bigcap_{V \in \mathscr{V}} V_s(A)$ by the theory of uniform spaces, thus $\bar{A} = \bigcap_{V \in \mathscr{V}} AV$. Since $x \mapsto x^{-1}$ is a homeomorphism, and $V \in \mathscr{V}$ iff $V^{-1} \in \mathscr{V}$, one has (letting V vary over $\mathscr{V}$)

$$\bigcap VA = \bigcap V^{-1}A = \bigcap (A^{-1}V)^{-1} = \left(\bigcap A^{-1}V\right)^{-1}$$
$$= ((A^{-1})^{-})^{-1} = \bar{A}. \qquad \blacksquare$$

Another important dividend of uniformizability (see also (5.22)):

(5.14) Corollary. *In any topological group, every point has a fundamental system of closed neighborhoods.*

Proof. By translation it is sufficient to consider neighborhoods of the neutral element e (2.9). If U is any neighborhood of e, choose a neighborhood V of e such that $VV \subset U$ (3.1). Citing (5.13), $\bar{V} \subset VV \subset U$. Thus each neighborhood U contains a closed neighborhood $\bar{V}$. ∎

(5.15) The right uniform structure of a topological group G can be derived briefly as follows. Suppose the binary operation of G is denoted xy; let G' be the *opposite* group, i.e., the set G equipped with the opposite binary operation $x \circ y = yx$. Then G' is also a topological group for the given topology. Consider the *left* uniform structure for the topological group G': a neighborhood V of e defines the entourage $\{(x, y) : x^{-1} \circ y \in V\} = \{(x, y) : yx^{-1} \in V\}$, and, writing

$$V_r = \{(x, y) : yx^{-1} \in V\},$$

it follows at one from (5.5) that the class $\mathscr{V}_r = \{V_r : V \in \mathscr{V}\}$ is the base for a uniformity $\mathscr{U}_r$ on the set $G' = G$.

(5.16) Definition. With notation as in (5.15), the uniform structure $(G, \mathscr{U}_r)$ is called the *right uniform structure* of the topological group G, and $\mathscr{U}_r$ the *right uniformity* of G. {In strict analogy with (5.11), the appropriate subscript would be d, for 'dextral'; however, we are reserving the notation $\mathscr{U}_d$ for the uniformity derived from a metric d.}

Paralleling (5.10) we have:

(5.17) Theorem. *If G is a topological group, then the uniform topology deduced from its right uniform structure $(G, \mathscr{U}_r)$ coincides with the given topology on G.*

Proof. Write G' as in (5.15). By (5.10) the topology derived from the uniformity $\mathscr{U}_r$ coincides with the given topology on G', that is, with the given topology on G. Incidentally, $V_r(x) = \{y : (x, y) \in V_r\} = \{y : yx^{-1} \in V\} = \{y : y \in Vx\} = Vx$. ∎

(5.18) For a topological group G the uniformities $\mathscr{U}_s$ and $\mathscr{U}_r$ may be different (5.25); nevertheless, in view of (5.10) and (5.17), they define the same topology, namely, the given topology on G.

The left and right uniform structures of a topological group are transformed into one another by the inversion mapping:

(5.19) Theorem. *If G is a topological group and $Jx = x^{-1}$ is the inversion mapping, then J induces an isomorphism between the uniform structures $(G, \mathscr{U}_s)$ and $(G, \mathscr{U}_r)$. Explicitly, $(x, y) \mapsto (Jx, Jy)$ is a bijective mapping on $G \times G$ that transforms $\mathscr{U}_s$ into $\mathscr{U}_r$. Equivalently, J is a uniformly continuous mapping of $(G, \mathscr{U}_s)$ onto $(G, \mathscr{U}_r)$, with uniformly continuous inverse $J^{-1} = J$.*

Proof. If V is a neighborhood of e, the following relations are equivalent: $(x, y) \in V_s$, $x^{-1}y \in V$, $y^{-1}x \in V^{-1}$, $(Jy)(Jx)^{-1} \in V^{-1}$, $(Jx, Jy) \in (V^{-1})_r$. Thus $(J \times J)(V_s) = (V^{-1})_r$. Since the filters $\mathscr{U}_s$, $\mathscr{U}_r$ are generated by the sets V_s, $(V^{-1})_r$, respectively, the theorem follows. Incidentally, another proof results from (5.12) and (5.15), since J is a bicontinuous group isomorphism of G onto G'. {Stripped of fancy notation and terminology, the theorem essentially reduces to the calculation $(xV)^{-1} = V^{-1}x^{-1}$.} ∎

Exercises

(5.20) Let G be a topological group, fix an element $a \in G$, and write $f(x) = ax$ for left translation by a. If $g = f \times f$ is the product mapping, $g(x, y) = (f(x), f(y))$, then for any neighborhood V of e one has $g(V_s) = V_s$. Thus, the left translations in G are uniformly continuous for the left uniform structure on G.

(5.21) If G and H are topological groups, with left uniformities $\mathscr{U}_s(G)$ and $\mathscr{U}_s(H)$, then the product uniformity on $G \times H$ coincides with the left uniformity $\mathscr{U}_s(G \times H)$ on the product topological group $G \times H$. What about infinite products?

(5.22) It follows from (5.14) that if A is a closed set in a topological group G and if $x \notin A$, then there exist disjoint open sets U and V such that $x \in U$ and

$A \subset V$. The theory of uniform structures yields even more: if N is any neighborhood of x, there exists a continuous function $f: G \to [0, 1]$ such that $f(x) = 1$ and $f = 0$ on $\complement N$. {This last property is characteristic of uniformizable spaces.} In particular, if G is a T_0-space (3.5) then it is automatically a $T_{3\frac{1}{2}}$-space, i.e., a completely regular space.

(5.23) The following conditions on a topological group G are equivalent: (a) G is separated; (b) $\bigcap V_s = \Delta$; (c) $\bigcap V_r = \Delta$. (Here V runs over any fundamental system of neighborhoods of e.)

(5.24) If G is a compact group, it follows from (5.10) and (5.17) that $\mathscr{U}_s = \mathscr{U}_r$. {Recall that in the Bourbaki terminology, compact spaces are separated.}

(5.25) There exists a locally compact group G for which $\mathscr{U}_s \neq \mathscr{U}_r$.

***(5.26)** Every locally compact topological group G is normal (i.e., is a T_4-space): if A and B are disjoint closed subsets of G, then there exist disjoint open sets U and V such that $A \subset U$ and $B \subset V$.

6. Metrizable topological groups.

A topological space X is called *metrizable* if there exists a metric d on X such that the topology derived from d coincides with the given topology on X; such a metric is said to be *compatible* with the given topology. A metrizable topological space is obviously separated and first countable (i.e., at each point there exists a fundamental sequence of neighborhoods).

A topological group G is said to be *metrizable* if it is metrizable as a topological space; then each point of G (equivalently, the neutral element of G) has a fundamental sequence of neighborhoods. A remarkable fact, proved independently by G. Birkhoff and S. Kakutani, is that the converse is true: *If the neutral element of a topological group G has a fundamental sequence of neighborhoods (equivalently, if G is first countable) and if G is separated, then G is metrizable.* The present section is devoted to a proof of this key result. Actually, the Birkhoff-Kakutani result includes significantly more, a point to which we will return after the following definitions.

(6.1) Definition. A metric d on an abstract (i.e., not necessarily topological) group G is said to be *left-invariant* if $d(ax, ay) = d(x, y)$ identically, and *right-invariant* if $d(xa, ya) = d(x, y)$ identically.

The full statement of the Birkhoff-Kakutani theorem is that *a first countable, separated topological group is metrizable via a left-invariant metric.* Incidentally, if a topological group G admits a left-invariant compatible metric d, then the metric $d'(x, y) = d(x^{-1}, y^{-1})$ is obviously right-invariant, and, since $x \mapsto x^{-1}$ is a homeomorphism of G, d' generates the same topology as d, i.e., d' is also compatible with the topology of G. Thus, in the Birkhoff-Kakutani theorem it is immaterial whether one says 'left' or 'right.' {However, a metrizable topological group need not admit a compatible metric which is both left- and right-invariant (cf. (10.4), (10.8)).} Part of the machinery of the proof can be separated out as an elementary combinatorial lemma:

(6.2) Lemma. *Let X be a set and suppose $f: X \times X \to \mathbb{R}$ is a function satisfying the following conditions:*

(1) $f(x, y) \geq 0$ for all x, y;

(2) $f(x, x) = 0$ for all x;

(3) if $\varepsilon > 0$, the relations $f(w, x) \leq \varepsilon$, $f(x, y) \leq \varepsilon$, $f(y, z) \leq \varepsilon$ imply $f(w, z) \leq 2\varepsilon$.

Define a function $d: X \times X \to \mathbb{R}$ as follows. If $(x, y) \in X \times X$ and $\mathfrak{p} = \{x = x_0, x_1, \ldots, x_n = y\}$ is any finite system of points in X that begins at x and ends at y, write

$$|\mathfrak{p}| = \sum_1^n f(x_{k-1}, x_k)$$

and define

$$d(x, y) = \inf |\mathfrak{p}|,$$

where $\mathfrak{p}$ varies over all such finite systems. Then, d has the following properties:

(4) $\frac{1}{2}f(x, y) \leq d(x, y) \leq f(x, y)$;

(5) $d(x, z) \leq d(x, y) + d(y, z)$;

(6) if $f(x, y) = f(y, x)$ for all x, y, then $d(x, y) = d(y, x)$ for all x, y.

If $f(x, y) = f(y, x)$ for all x, y, and if $f(x, y) > 0$ whenever $x \neq y$, then d is a metric on X.

Proof. It is obvious that (6) holds and that the final assertion is immediate from (4), (5), (6); it remains only to verify (4) and (5).

First, we note that for each $\varepsilon > 0$, f satisfies the following 'weak triangle law': if $f(x, y) \leq \varepsilon$ and $f(y, z) \leq \varepsilon$ then $f(x, z) \leq 2\varepsilon$ (take $w = x$ in (3)). In particular, if $f(x, y) = f(y, z) = 0$ then $0 \leq f(x, z) \leq 2\varepsilon$ for every $\varepsilon > 0$ and therefore $f(x, z) = 0$. It follows by induction that if $f(x_0, x_1) = f(x_1, x_2) = \cdots = f(x_{n-1}, x_n) = 0$ then $f(x_0, x_n) = 0$; in other words, if $\mathfrak{p} = \{x = x_0, x_1, \ldots, x_n = y\}$ is a system such that $|\mathfrak{p}| = 0$, then $f(x, y) = 0$.

(4) For the system $\mathfrak{q} = \{x = x_0, x_1 = y\}$ one has $d(x, y) \leq |\mathfrak{q}| = f(x, y)$. To prove that $\frac{1}{2}f(x, y) \leq d(x, y)$ one must show that $\frac{1}{2}f(x, y) \leq |\mathfrak{p}|$ for every finite system $\mathfrak{p} = \{x = x_0, x_1, \ldots, x_n = y\}$. The proof is by induction on n. If $n = 1$ then $\mathfrak{p} = \{x = x_0, x_1 = y\}$ and so $|\mathfrak{p}| = f(x, y) \geq \frac{1}{2}f(x, y)$ by (1). Suppose $n \geq 2$ and assume the assertion true for systems of length $< n$. We consider three cases.

Case 1: $f(x_0, x_1) \geq \frac{1}{2}|\mathfrak{p}|$. Then $\frac{1}{2}|\mathfrak{p}| = |\mathfrak{p}| - \frac{1}{2}|\mathfrak{p}| \geq |\mathfrak{p}| - f(x_0, x_1) = \sum_2^n f(x_{k-1}, x_k) \geq \frac{1}{2}f(x_1, x_n)$ by the induction hypothesis, that is, $f(x_1, x_n) \leq |\mathfrak{p}|$; obviously $f(x_0, x_1) \leq |\mathfrak{p}|$, therefore $f(x_0, x_n) \leq 2|\mathfrak{p}|$ by the remarks at the beginning of the proof (even if $|\mathfrak{p}| = 0$), i.e., $\frac{1}{2}f(x, y) \leq |\mathfrak{p}|$.

Case 2: $f(x_{n-1}, x_n) \geq \frac{1}{2}|\mathfrak{p}|$. The argument is similar to that for Case 1.

Case 3: $f(x_0, x_1) < \frac{1}{2}|\mathfrak{p}|$ and $f(x_{n-1}, x_n) < \frac{1}{2}|\mathfrak{p}|$. In particular, $|\mathfrak{p}| > 0$ and $n \geq 3$. Let r be the largest integer such that

(*) $$\sum_1^r f(x_{k-1}, x_k) \leq \frac{1}{2}|\mathfrak{p}|;$$

$r \geq 1$ because $f(x_0, x_1) < \frac{1}{2}|\mathfrak{p}|$. Since $f(x_{n-1}, x_n) < \frac{1}{2}|\mathfrak{p}|$,

$$\sum_1^{n-1} f(x_{k-1}, x_k) = |\mathfrak{p}| - f(x_{n-1}, x_n) > \frac{1}{2}|\mathfrak{p}|,$$

therefore $r < n - 1$. Thus $1 \leq r \leq n - 2$. By the maximality of r, $\sum_1^{r+1} f(x_{k-1}, x_k) > \frac{1}{2}|\mathfrak{p}|$, therefore

(**) $\sum_{r+2}^n f(x_{k-1}, x_k) < \frac{1}{2}|\mathfrak{p}|$.

By induction and (*), $\frac{1}{2}f(x_0, x_r) \leq \sum_1^r f(x_{k-1}, x_k) \leq \frac{1}{2}|\mathfrak{p}|$, thus

(i) $f(x_0, x_r) \leq |\mathfrak{p}|$;

trivially

(ii) $f(x_r, x_{r+1}) \leq |\mathfrak{p}|$;

by induction and (**), $\frac{1}{2}f(x_{r+1}, x_n) \leq \sum_{r+2}^n f(x_{k-1}, x_k) < \frac{1}{2}|\mathfrak{p}|$, thus

(iii) $f(x_{r+1}, x_n) < |\mathfrak{p}|$.

Citing (3), it follows from (i), (ii), (iii) that $f(x_0, x_n) \leq 2|\mathfrak{p}|$, that is, $\frac{1}{2}f(x, y) \leq |\mathfrak{p}|$.

(5) Fix $x, y, z \in X$. Given any pair of systems

$$\mathfrak{p} = \{x = x_0, x_1, \ldots, x_n = y\}, \qquad \mathfrak{q} = \{y = y_0, y_1, \ldots, y_m = z\},$$

let $\mathfrak{s}$ be the concatenation of the two systems,

$$\mathfrak{s} = \{x = x_0, x_1, \ldots, x_n = y = y_0, y_1, \ldots, y_m = z\}.$$

Obviously $d(x, z) \leq |\mathfrak{s}| = |\mathfrak{p}| + |\mathfrak{q}|$; varying $\mathfrak{p}$ and $\mathfrak{q}$ independently, $d(x, z) \leq d(x, y) + d(y, z)$. ∎

(6.3) **Theorem.** (Birkhoff-Kakutani) *A topological group G is metrizable if and only if it is separated and the neutral element e has a countable fundamental system of neighborhoods. A metrizable topological group admits a left-invariant (or a right-invariant) compatible metric.*

Proof. Only if: Any metrizable space is separated and first countable.

If: Let G be a separated topological group possessing a fundamental sequence of neighborhoods U_n ($n = 1, 2, 3, \ldots$) of e. {Since the sets $(U_n)_s$ are evidently a fundamental system of entourages for the left uniform structure of G (5.5), the proof of the metrizability of G could be based on a general theorem on the metrizability of uniform structures (6.7). However, this would exceed the topological prerequisites stated in the Preface; instead, we proceed with a direct proof (on which, incidentally, a proof of the general uniform space result can easily be modeled).}

The first step is to construct an 'improved' fundamental sequence of neighborhoods V_n. Replacing U_n by $U_n \cap U_n^{-1}$, one can assume that the U_n are symmetric ($U_n = U_n^{-1}$). Let $V_1 = U_1$. By (3.1) there exists U_k such that $U_k^3 \subset U_2 \cap V_1$; let V_2 be the (say) first such U_k. Inductively, let V_n be the first U_k such that $U_k^3 \subset U_n \cap V_{n-1}$. Since $V_n \subset U_n$ for all n, the sequence of neighborhoods V_n is also fundamental; also

(i) $\bigcap_1^\infty V_k = \{e\}$

because G is separated (3.4), and by construction

(ii) $$V_{k+1}^3 \subseteq V_k \qquad (k = 1, 2, 3, \ldots).$$

Set $V_0 = G$. From (ii) we have

(iii) $$G = V_0 \supset V_1 \supset V_2 \supset V_3 \supset \cdots;$$

thus, every $x \in G$ belongs to some V_k, and it follows from (i) and (iii) that if $x \neq e$ then V_k excludes x from some k onward, that is, x belongs to only finitely many V_k. (Incidentally, if G admits a *finite* fundamental system of neighborhoods of e, then G is obviously discrete; in this case the discrete metric, $d(x, x) = 0$ for all x and $d(x, y) = 1$ when $x \neq y$, is a left-invariant compatible metric. Having disposed of this case, let us assume G nondiscrete.)

Each V_k represents a degree of 'nearness' to the 'origin' e; alternatively, $x^{-1}y \in V_k$ is a measure of the nearness of x to y. The problem is to express such qualitative statements in terms of real numbers (eventually, in terms of a metric). Left-invariance will then follow from the fact that the germinal relation $x^{-1}y \in V_k$ is itself left-invariant, i.e., $(ax)^{-1}(ay) = x^{-1}y$.

Suppose $x \neq y$. A qualitative assertion is that $x^{-1}y \in V_k$ for some k (e.g., $k = 0$). A quantitative assertion is that there exists a largest such k; this permits the definition

$$f(x, y) = \min \{(\tfrac{1}{2})^k : x^{-1}y \in V_k\}.$$

On the other hand, if $x = y$ then $x^{-1}y = e \in V_k$ for all k; setting $f(x, x) = 0$, one has

(iv) $$f(x, y) = \inf \{(\tfrac{1}{2})^k : x^{-1}y \in V_k\}$$

for all $x, y \in G$. The desired metric d will be derived from f via the lemma; we now show that the hypotheses of the lemma are fulfilled.

Obviously $f(x, y) \geq 0$, and $f(x, y) = 0$ iff $x = y$. Also, $f(x, y) = f(y, x)$ since the V_n are symmetric. To apply the lemma, we need only verify the condition (3); the left-invariance of d will follow at once from the evident property $f(ax, ay) = f(x, y)$. Assuming $\varepsilon > 0$, $f(w, x) \leq \varepsilon$, $f(x, y) \leq \varepsilon$, $f(y, z) \leq \varepsilon$, it is to be shown that $f(w, z) \leq 2\varepsilon$. This is trivial if $\varepsilon \geq \tfrac{1}{2}$ (because $f \leq 1$); suppose $0 < \varepsilon < \tfrac{1}{2}$. According to (iv) there exist positive integers i, j, k such that

$$w^{-1}x \in V_i \quad \text{and} \quad (\tfrac{1}{2})^i \leq \varepsilon,$$

$$x^{-1}y \in V_j \quad \text{and} \quad (\tfrac{1}{2})^j \leq \varepsilon,$$

$$y^{-1}z \in V_k \quad \text{and} \quad (\tfrac{1}{2})^k \leq \varepsilon;$$

if $r = \min \{i, j, k\}$, then $(\tfrac{1}{2})^r \leq \varepsilon$ and it follows from (ii) that

$$w^{-1}z = (w^{-1}x)(x^{-1}y)(y^{-1}z) \in V_i V_j V_k \subseteq V_r^3 \subseteq V_{r-1},$$

therefore $f(w, z) \leq (\tfrac{1}{2})^{r-1} = 2(\tfrac{1}{2})^r \leq 2\varepsilon$.

The lemma is now applicable, and yields a left-invariant metric d; it remains to show that d generates the given topology.

For any $\varepsilon > 0$ and any $a \in G$, define

$$U_\varepsilon(a) = \{x : f(a, x) < \varepsilon\}.$$

We assert that the sets $U_\varepsilon(a)$, $\varepsilon > 0$, form a fundamental system of neighborhoods of a for the given topology. First, every $U_\varepsilon(a)$ is a neighborhood of a; for, if k is a positive integer such that $(\frac{1}{2})^k < \varepsilon$, then $aV_k \subset U_\varepsilon(a)$ results from the chain of implications $x \in aV_k \Rightarrow a^{-1}x \in V_k \Rightarrow f(a, x) \leq (\frac{1}{2})^k < \varepsilon$. On the other hand, if A is any neighborhood of a, let us show that $U_\varepsilon(a) \subset A$ for some $\varepsilon > 0$. Let k be a positive integer such that $aV_k \subset A$ ($a^{-1}A$ is a neighborhood of e, and the V_k are basic), and set $\varepsilon = (\frac{1}{2})^k$. If $x \notin aV_k$ then $a^{-1}x$ can belong to V_j only for $j < k$, therefore $f(a, x) > (\frac{1}{2})^k = \varepsilon$ (by the definition of f) and so $x \notin U_\varepsilon(a)$; thus $U_\varepsilon(a) \subset aV_k \subset A$. (Incidentally, $U_\varepsilon(a) = aU_\varepsilon(e)$ follows at once from the left-invariance of f.)

By (4) of the lemma, $\frac{1}{2}f(x, y) \leq d(x, y) \leq f(x, y)$; it follows that, for $\varepsilon > 0, f(x, y) < \varepsilon \Rightarrow d(x, y) < \varepsilon \Rightarrow \frac{1}{2}f(x, y) < \varepsilon$, thus

(v) $$U_\varepsilon(x) \subset \{y : d(x, y) < \varepsilon\} \subset U_{2\varepsilon}(x);$$

since the $U_\varepsilon(x)$ are a fundamental system of neighborhoods of x for the given topology, and the open balls $\{y : d(x, y) < \varepsilon\}$ are a fundamental system of neighborhoods of x for the topology derived from the metric d, it is immediate from (v) that the two topologies coincide.

Finally, if G is metrizable then G is separated and first countable by the 'only if' part of the proof, and therefore G possesses a left-invariant compatible metric by the 'if' part of the proof. For 'right-invariant', see the remarks following (6.1). ∎

In a metrizable topological group, every subgroup (with the relative topology) is also a metrizable topological group; this is an elementary remark valid for any subspace of a metrizable topological space. For quotient groups, the situation is deeper:

(6.4) Corollary. *If G is a metrizable topological group and if N is a closed normal subgroup of G, then the quotient topological group G/N is metrizable.*

Proof. Let $\pi : G \to G/N$ be the canonical mapping. By hypothesis, the neutral element e of G admits a fundamental sequence of neighborhoods V_n; since π is continuous and open, it follows that $\pi(V_n)$ is a fundamental sequence of neighborhoods of $\pi(e)$, therefore G/N is metrizable (6.3). {See (8.14) for an alternative approach to the proof.} ∎

The possibility of switching from an arbitrary compatible metric to an invariant one seems, at first glance, innocuous; actually, it is of great significance for the theory of completeness to be given in the next section.

Exercises

(6.5) If G_n $(n = 1, 2, 3, \ldots)$ is a sequence of metrizable topological groups, then the product topological group $G = \prod_{n=1}^{\infty} G_n$ (2.19) is metrizable.

(6.6) (V. L. Klee)   If a separated topological group G admits a fundamental sequence of neighborhoods V_n such that $aV_na^{-1} = V_n$ for all $a \in G$, then G admits a compatible metric that is both left- and right-invariant. (Cf. Section 10.)

(6.7) (A. Weil)   If $(X, \mathcal{U})$ is a uniform structure such that (i) the intersection of all entourages is the diagonal, and (ii) there exists a fundamental sequence of entourages (i.e., the filter $\mathcal{U}$ has a countable base), then there exists a metric d on X such that the sets $V_\varepsilon = \{(x, y) : d(x, y) < \varepsilon\}$, $\varepsilon > 0$, are a base for the uniformity $\mathcal{U}$.

(6.8)   Let G be a group, let d be a right-invariant metric on G, and equip G with the topology derived from d. If, for each $a \in G$, the left translation $x \mapsto ax$ is continuous, then $x \mapsto x^{-1}$ is continuous. Thus if $(x, y) \mapsto xy$ is continuous, then G is a topological group. {Cf. (2.22).}

(6.9)   Let G be a metrizable topological group, let d be any metric compatible with the topology of G, and let $\mathcal{U}_d$ be the uniformity derived from d, i.e., the uniformity for which the sets $\{(x, y) : d(x, y) < \varepsilon\}$, $\varepsilon > 0$, are basic entourages. As usual, $\mathcal{U}_s$ denotes the left uniformity of G (5.6). Then (i) $\mathcal{U}_d \subset \mathcal{U}_s$ iff given any $\varepsilon > 0$, there exists $\delta > 0$ such that $d(x^{-1}y, e) < \delta$ implies $d(x, y) < \varepsilon$; (ii) $\mathcal{U}_s \subset \mathcal{U}_d$ iff given any $\varepsilon > 0$, there exists $\delta > 0$ such that $d(x, y) < \delta$ implies $d(x^{-1}y, e) < \varepsilon$. {Cf. (10.9).}

7. Metrizable complete topological groups.

There is a notion of completeness for metric spaces (X, d) and for uniform structures $(X, \mathcal{U})$ (the definitions will be reviewed shortly). Any metric d on a set X defines a uniform structure $(X, \mathcal{U}_d)$, where the sets $\{(x, y) : d(x, y) < \varepsilon\}$, $\varepsilon > 0$, are a base for the uniformity $\mathcal{U}_d$; it is a pleasant fact that d is a complete metric if and only if $(X, \mathcal{U}_d)$ is a complete uniform structure (7.7).

Suppose G is a metrizable topological group. Like any topological group, G has the uniform structures $(G, \mathcal{U}_s)$ and $(G, \mathcal{U}_r)$, described in (5.5) and (5.16). If d is any metric compatible with the topology of G, one has also the uniform structure $(G, \mathcal{U}_d)$; however, $\mathcal{U}_d$ depends on the choice of a particular metric d, and, regrettably, $\mathcal{U}_d$ may differ from the 'natural' (or 'intrinsic') uniformities $\mathcal{U}_s$ and $\mathcal{U}_r$ (6.9). Thus, for a metrizable topological group, there are competing notions of completeness: the completeness of the uniform structure $(G, \mathcal{U}_s)$ (which turns out to be equivalent to the completeness of $(G, \mathcal{U}_r)$) and the completeness of the metric d (which is equivalent to the completeness of the uniform structure $(G, \mathcal{U}_d)$). It is an unpleasant surprise that the relationship between these notions of completeness is not transparent; it can happen that $(G, \mathcal{U}_s)$ is complete but d is not (7.11), or that d is complete but $(G, \mathcal{U}_s)$ is not (7.14), though the situation is repaired when d is left-invariant (7.8). The present section is devoted to clarifying this relationship, and to showing that if $(G, \mathcal{U}_s)$ is complete, then there does exist a compatible complete metric d (7.9). Section 8 is a technical preliminary to Section 9; in Section 9 we show that the appropriate notion of completeness for a metrizable topological group survives on passage to quotients modulo a closed normal subgroup.

(7.1) We review here the concepts from general topology, in particular the theory of uniform structures, that are needed for the discussion.

Let X be a topological space. For each $x \in X$ let $\mathscr{V}_x$ denote the filter of all neighborhoods of x. A filter $\mathscr{F}$ on X (i.e., a filter $\mathscr{F}$ of subsets of X) is said to be *convergent* if there exists a point $x \in X$ such that $\mathscr{F} \supset \mathscr{V}_x$; one then says that $\mathscr{F}$ *converges to* x, denoted $\mathscr{F} \to x$. If $\mathscr{B}$ is any base for the filter $\mathscr{F}$ and if $\mathscr{B}_x$ is any fundamental system of neighborhoods of x (i.e., $\mathscr{B}_x$ is a base for the filter $\mathscr{V}_x$), then the condition $\mathscr{F} \to x$ is equivalent to the following: if $V \in \mathscr{B}_x$ there exists $A \in \mathscr{B}$ such that $A \subset V$. When X is separated, such a point x is unique, i.e., if $\mathscr{F} \to x$ and $\mathscr{F} \to y$ then $x = y$.

Let $(X, \mathscr{U})$ be a uniform structure. Equip X with the topology derived from the uniformity $\mathscr{U}$; thus, for each $x \in X$, the sets $U(x) = \{y : (x, y) \in U\}$, $U \in \mathscr{U}$, form a base for the neighborhood filter $\mathscr{V}_x$. If $A \subset X$ and $U \in \mathscr{U}$, one says that A is *small of order U* provided $A \times A \subset U$. A base $\mathscr{B}$ for a filter $\mathscr{F}$ on X is said to be *Cauchy* (with respect to the uniformity $\mathscr{U}$) if, for every $U \in \mathscr{U}$, there exists a set $A \in \mathscr{B}$ such that $A \times A \subset U$ (so to speak, $\mathscr{B}$ contains sets of arbitrarily small order). Evidently $\mathscr{B}$ is Cauchy if and only if $\mathscr{F}$ is Cauchy. Every convergent filter is Cauchy. The uniform structure $(X, \mathscr{U})$ is said to be *complete* if every Cauchy filter on X is convergent.

(7.2) Let us specialize the foregoing concepts to the left and right uniform structures of a topological group G, $(G, \mathscr{U}_s)$ and $(G, \mathscr{U}_r)$, described in (5.5) and (5.16). Let $\mathscr{V}$ be the filter of neighborhoods of the neutral element e. Thus, the sets $V_s = \{(x, y) : x^{-1}y \in V\}$ and $V_r = \{(x, y) : yx^{-1} \in V\}$, $V \in \mathscr{V}$, are basic entourages for the respective uniform structures.

A filter $\mathscr{F}$ on G is convergent to $x \in G$ in case, for each $V \in \mathscr{V}$, there exists a set $A \in \mathscr{F}$ such that $A \subset xV$ (equivalently, for each $V \in \mathscr{V}$ there exists a set $A \in \mathscr{F}$ such that $A \subset Vx$). A filter $\mathscr{F}$ on G is Cauchy with respect to the uniformity $\mathscr{U}_s$ in case, for every $V \in \mathscr{V}$, $\mathscr{F}$ contains a set A that is small of order V_s (i.e., $A^{-1}A \subset V$); $\mathscr{F}$ is Cauchy with respect to $\mathscr{U}_r$ in case, for every $V \in \mathscr{V}$, $\mathscr{F}$ contains a set A that is small of order V_r (i.e., $AA^{-1} \subset V$).

By (5.19) the inversion mapping $Jx = x^{-1}$ induces an isomorphism of the uniform structures $(G, \mathscr{U}_s)$ and $(G, \mathscr{U}_r)$, consequently one of the structures is complete if and only if the other is; this justifies the parenthetical remark in the following definition.

(7.3) A topological group G is said to be *complete* if the uniform structure $(G, \mathscr{U}_s)$ is complete (equivalently (5.19), the uniform structure $(G, \mathscr{U}_r)$ is complete).

(7.4) Let us note explicitly the meaning of completeness, in the sense of (7.3), for a topological group G: if $\mathscr{F}$ is any filter on G that is Cauchy with respect to $\mathscr{U}_s$ (i.e., for each neighborhood V of e there exists a set $A \in \mathscr{F}$ such that $A^{-1}A \subset V$), then $\mathscr{F} \to x$ for some $x \in G$; equivalently, if $\mathscr{F}$ is any filter on G that is Cauchy with respect to $\mathscr{U}_r$ (i.e., for each neighborhood V of e there exists a set $A \in \mathscr{F}$ such that $AA^{-1} \subset V$), then $\mathscr{F} \to x$ for some $x \in G$. One can simply say that in a complete topological group, every Cauchy filter is convergent; it does not matter whether "Cauchy" refers to the left uniform structure or the right uniform structure (but see (7.15)).

(7.5) If a complete topological group is metrizable, we shall refer to it as a *metrizable complete topological group* (rather than a 'complete metrizable topological group'), to emphasize the fact that the notion of completeness defined in (7.3) is topological-algebraic (rather than metric). The gist of the next series of results is that the way to avoid confusion is to stick to invariant metrics.

(7.6) **Lemma.** *If G is a metrizable topological group and if d is a left-invariant compatible metric on G (6.3), then the uniform structure $(G, \mathcal{U}_d)$ derived from d coincides with the left uniform structure $(G, \mathcal{U}_s)$.*

Proof. The problem is to show that $\mathcal{U}_d = \mathcal{U}_s$; it suffices to show that the two filters have a common base. Let $\varepsilon > 0$. A basic neighborhood of e is $V = \{x : d(e, x) < \varepsilon\}$, therefore a basic entourage for $\mathcal{U}_s$ is $V_s = \{(x, y) : x^{-1}y \in V\} = \{(x, y) : d(e, x^{-1}y) < \varepsilon\} = \{(x, y) : d(x, y) < \varepsilon\}$ (because d is left-invariant), and this is also a basic entourage for $\mathcal{U}_d$. {There is a dual result with 'left' replaced by 'right' throughout. See also (6.9).} ∎

(7.7) **Lemma.** *Let (X, d) be a metric space and let $(X, \mathcal{U}_d)$ be the uniform structure derived from d. Then, (X, d) is a complete metric space if and only if $(X, \mathcal{U}_d)$ is a complete uniform structure.*

Proof. The sets $U_n = \{(x, y) : d(x, y) < 1/n\}$ ($n = 1, 2, 3, \ldots$) form a fundamental system of entourages for $\mathcal{U}_d$. {Incidentally,

$$U_n(x) = \{y : (x, y) \in U_n\} = \{y : d(x, y) < 1/n\}$$

is a basic neighborhood of x for the topology derived from d, thus the uniform topology derived from $\mathcal{U}_d$ coincides with the metric topology derived from d; this shows that a metrizable topological space is uniformizable, but the rub is that topologically equivalent metrics may give rise to different uniform structures.}

If: Assume $(X, \mathcal{U}_d)$ is a complete uniform structure. Given a sequence $x_n \in X$ with $d(x_m, x_n) \to 0$, the problem is to find an $x \in X$ such that $d(x_n, x) \to 0$. For each n, let $A_n = \{x_i : i \geq n\}$ and let $\mathcal{F}$ be the filter with base $\{A_n : n = 1, 2, 3, \ldots\}$; that is, let $\mathcal{F}$ be the elementary filter determined by the sequence x_n.

The filter $\mathcal{F}$ is Cauchy with respect to $\mathcal{U}_d$; for, given a basic entourage $U = \{(x, y) : d(x, y) < \varepsilon\}$, $\varepsilon > 0$, if n is so chosen that $d(x_i, x_j) < \varepsilon$ for all $i, j \geq n$, then $A_n \times A_n \subset U$, thus $\mathcal{F}$ contains a set that is small of order U.

By hypothesis, $\mathcal{F} \to x$ for a suitable $x \in X$; let us show that $d(x_n, x) \to 0$. Given any $\varepsilon > 0$, consider the neighborhood $N = \{y : d(x, y) < \varepsilon\}$ of x. Since $\mathcal{F} \to x$, there exists an index n such that $A_n \subset N$, thus $d(x_i, x) < \varepsilon$ for all $i \geq n$.

Only if: Assume (X, d) is a complete metric space. Given a filter $\mathcal{F}$ on X that is Cauchy with respect to $\mathcal{U}_d$, we seek an $x \in X$ such that $\mathcal{F} \to x$. The point x will be produced by constructing a decreasing sequence of nonempty closed sets B_n such that diam $B_n \to 0$; the intersection of the B_n will then reduce to the desired singleton $\{x\}$.

For each n, there exists, by hypothesis, a set $A_n \in \mathscr{F}$ that is small of order U_{2n}, i.e., $d(x, y) < 1/2n$ for all $x, y \in A_n$. Then for all $x, y \in \overline{A}_n$ one has $d(x, y) \leq 1/2n < 1/n$ by the continuity of d, thus $\overline{A}_n$ is small of order U_n. Let $B_n = \overline{A}_1 \cap \cdots \cap \overline{A}_n$. Since $\overline{A}_k \supset A_k \in \mathscr{F}$ and $\mathscr{F}$ is closed under finite intersection, one has $B_n \in \mathscr{F}$ and in particular $B_n \neq \varnothing$. Moreover, B_n is closed, $B_1 \supset B_2 \supset B_3 \supset \cdots$, and diam $B_n \leq 1/n$ because $B_n \subset \overline{A}_n$. By a standard characterization of complete metric spaces, the intersection of the B_n is a singleton $\{x\}$.

The proof will be concluded by showing that $\mathscr{F} \to x$. Given a basic neighborhood N of x, say $N = \{y : d(x, y) < \varepsilon\}$, $\varepsilon > 0$, we wish to show that N contains some set in $\mathscr{F}$. Choose n so that $1/n < \varepsilon$; we show that $B_n \subset N$. Let $y \in B_n$; since also $x \in B_n$, one has $(x, y) \in B_n \times B_n \subset \overline{A}_n \times \overline{A}_n \subset U_n$, thus $d(x, y) < 1/n < \varepsilon$ and therefore $y \in N$. $\blacksquare$

(7.8) Theorem. *Let G be a metrizable topological group and let d be any left-invariant (or right-invariant) compatible metric (6.3). Then, G is a complete topological group if and only if d is a complete metric.*

Proof. Say d is left-invariant; then $\mathscr{U}_s = \mathscr{U}_d$ by (7.6). According to the definition (7.3), G is a complete topological group if and only if the uniform structure $(G, \mathscr{U}_s) = (G, \mathscr{U}_d)$ is complete, that is, if and only if d is a complete metric (7.7). If, on the other hand, d is right-invariant, one considers instead the right uniform structure $(G, \mathscr{U}_r)$ (see the remark at the end of the proof of (7.6)). $\blacksquare$

Implicit in (7.8) is the following striking consequence of the Birkhoff-Kakutani metrization theorem:

(7.9) Theorem. *If G is a metrizable complete topological group, there exists a metric d on G such that (i) the metric topology coincides with the given topology, (ii) d is left-invariant, and (iii) the metric space (G, d) is complete.*

Proof. If d is any left-invariant compatible metric (6.3) then the condition (iii) holds automatically by (7.8). $\blacksquare$

Exercises

(7.10) If a topological group G admits a neighborhood V of e which is complete for the uniform structure induced by either the left or the right uniform structure of G, then G is complete. In particular, every locally compact topological group is complete.

(7.11) Let $\mathbb{R} - \{0\}$ be the multiplicative group of nonzero real numbers, topologized via the usual metric $d(x, y) = |x - y|$ (2.6). Then $\mathbb{R} - \{0\}$ is a complete topological group (7.10), although the compatible metric d is not complete (consider $x_n = 1/n$). A compatible complete metric is $D(x, y) = |x - y| + |x^{-1} - y^{-1}|$.

(7.12) (J. Dieudonné)   Let G be the topological group of homeomorphisms of $[0, 1]$, with the sup metric d, as described in (2.22). For each positive integer n, let $f_n \in G$ be the continuous, piecewise linear function whose graph is determined by the points $(0, 0)$, $(1/2, 1 - 1/n)$, $(1, 1)$. Then (i) f_n is a Cauchy sequence with respect to the metric d, but f_n^{-1} is not; (ii) d is not a complete metric on G; (iii) G is not a complete topological group; (iv) the left and right uniform structures of G are different, i.e., $\mathcal{U}_s \neq \mathcal{U}_r$.

(7.13)   Let G be a metrizable topological group, let d be any compatible metric, and define $D(x, y) = d(x, y) + d(x^{-1}, y^{-1})$. Then D is also a compatible metric.

(7.14)   Let G, d be as in (7.12) and define $D(f, g) = d(f, g) + d(f^{-1}, g^{-1})$ as in (7.13). Although G is not a complete topological group (7.12), D *is* a compatible complete metric (cf. 7.11).

(7.15)   Let G be a separated topological group. In order that G 'be' a dense subgroup of some complete topological group (i.e., in order that there exist a complete topological group G' and a group monomorphism $f: G \to G'$ such that $f(G)$ is a dense subset of G' and such that f maps G homeomorphically onto the subspace $f(G)$ of G'), it is necessary and sufficient that the following condition hold: (*) if $\mathscr{F}$ is any filter on G that is Cauchy with respect to the left uniform structure $(G, \mathcal{U}_s)$, then the image of $\mathscr{F}$ under the inversion mapping $x \mapsto x^{-1}$, namely the filter $\mathscr{F}^{-1} = \{A^{-1} : A \in \mathscr{F}\}$, is also Cauchy with respect to $(G, \mathcal{U}_s)$. In view of (5.19), an equivalent condition is that every filter on G that is Cauchy with respect to $(G, \mathcal{U}_s)$ is also Cauchy with respect to $(G, \mathcal{U}_r)$. The condition (*) is trivially verified if $\mathcal{U}_s = \mathcal{U}_r$, (5.19); such groups, which of course include the abelian ones, are called bi-uniform (see Section 10 for a discussion of bi-uniform groups). However, condition (*) does not imply bi-uniformity: there exist locally compact groups—necessarily complete, by (7.10)—for which $\mathcal{U}_s \neq \mathcal{U}_r$ (an example is given in (10.8)). The condition (*) does not hold for the group of (7.12), even though the group admits a compatible complete metric (7.14).

(7.16)   If $f: G \to H$ is a bicontinuous group isomorphism, where G and H are topological groups and G is complete, then H is also complete.

8. Valued groups.

Quotient groups of metrizable groups are the main concern of this section. In addition, we introduce the concept of a value on a group; this is nothing more than a reformulation of the concept of invariant metric, but it has the advantage of a very suggestive notation. The results and techniques of this section are applied in the next section to prove decisive results on quotient groups of metrizable complete groups.

(8.1) **Definition.**   Let G be a group (notated multiplicatively, with neutral element e). A *value* on G is a nonnegative real-valued function $x \mapsto |x|$ on G such that

(i) $|e| = 0$,
(ii) $|x| > 0$ when $x \neq e$,
(iii) $|x^{-1}| = |x|$ for all x,
(iv) $|xy| \leq |x| + |y|$ for all x, y.

When G is written additively (as is often the case when G is abelian), the axioms for a value take a more familiar form:

(i') $|0| = 0$,
(ii') $|x| > 0$ when $x \neq 0$,
(iii') $|-x| = |x|$ for all x,
(iv') $|x + y| \leq |x| + |y|$ for all x, y.

(8.2) Definition. A *valued group* is a pair $(G, |x|)$, where G is a group and $x \mapsto |x|$ is a value on G.

In the next two theorems, we establish a one-to-one correspondence between the values on a group and the left-invariant metrics on the group.

(8.3) Theorem. *If $(G, |x|)$ is a valued group, then the formula $d(x, y) = |x^{-1}y|$ defines a left-invariant metric d; in turn, the value is expressible in terms of the metric, $|x| = d(e, x)$.*

Proof. If $x \neq y$, then $d(x, y) = |x^{-1}y| > 0$ because $x^{-1}y \neq e$; on the other hand, $d(x, x) = |x^{-1}x| = |e| = 0$ for all x. Symmetry: $d(y, x) = |y^{-1}x| = |(y^{-1}x)^{-1}| = |x^{-1}y| = d(x, y)$. Triangle inequality: $d(x, y) = |x^{-1}y| = |(x^{-1}z)(z^{-1}y)| \leq |x^{-1}z| + |z^{-1}y| = d(x, z) + d(z, y)$. Left-invariance: $d(ax, ay) = |(ax)^{-1}(ay)| = |x^{-1}a^{-1}ay| = |x^{-1}y| = d(x, y)$. Finally, $d(e, x) = |e^{-1}x| = |x|$. ∎

(8.4) Definition. With notation as in (8.3), we call $d(x, y) = |x^{-1}y|$ the left-invariant metric defined by the value; the topology derived from d is called the *left value topology* of the valued group $(G, |x|)$.

According to (8.3), every value gives rise to a left-invariant metric; conversely, every left-invariant metric gives rise to a value:

(8.5) Theorem. *If d is a left-invariant metric on a group G, then the formula $|x| = d(e, x)$ defines a value on G; in turn, the metric is expressible in terms of the value, $d(x, y) = |x^{-1}y|$.*

Proof. If $x \neq e$ then $|x| = d(e, x) > 0$; on the other hand, $|e| = d(e, e) = 0$. By left-invariance, $|x^{-1}| = d(e, x^{-1}) = d(xe, xx^{-1}) = d(x, e) = d(e, x) = |x|$. Also, $|xy| = d(e, xy) \leq d(e, x) + d(x, xy) = d(e, x) + d(e, y) = |x| + |y|$. Thus $x \mapsto |x|$ is a value on G. Finally, $d(x, y) = d(x^{-1}x, x^{-1}y) = d(e, x^{-1}y) = |x^{-1}y|$. ∎

(8.6) It is clear from (8.3) and (8.5) that the values $x \mapsto |x|$ and the left-invariant metrics d on a group G are in one-to-one correspondence, via the mutually inverse correspondences $d(x, y) = |x^{-1}y|$, $|x| = d(e, x)$. Since left-invariant metrics d are in one-to-one correspondence with right-invariant metrics D, via the formula $D(x, y) = d(x^{-1}, y^{-1})$, there is also a one-to-one correspondence between values $x \mapsto |x|$ and right-invariant metrics D, namely, $D(x, y) = |xy^{-1}|$, $|x| = D(e, x)$.

Suppose $x \mapsto |x|$ is a value on G, $d(x, y) = |x^{-1}y|$, $D(x, y) = |xy^{-1}|$. Regrettably, the *right value topology* (i.e., the topology derived from D) may differ from the left value topology (the topology derived from d); an example is given in (8.16), and the same example shows that a value topology need not be compatible with the group structure.

For our purposes, it will suffice to restrict attention to *left* value topologies; in any case, the most important applications are to abelian groups, where the distinction disappears:

(8.7) **Theorem.**   *If* $(G, |x|)$ *is a valued abelian group, then the left value topology coincides with the right value topology, and this topology is compatible with the group structure.*

Proof.   With notation as in (8.6), $D(x, y) = |xy^{-1}| = |(xy^{-1})^{-1}| = |yx^{-1}| = |x^{-1}y| = d(x, y)$, therefore the two value topologies coincide; the same calculation shows that $d(x^{-1}, y^{-1}) = d(x, y)$, therefore $x \mapsto x^{-1}$ is continuous. Finally, if $d(x_n, x) \to 0$ and $d(y_n, y) \to 0$, then $d(x_n y_n, xy) \to 0$ results from the calculation

$$d(x_n y_n, xy) = |(x_n y_n)^{-1}(xy)| = |y_n^{-1} x_n^{-1} xy| = |(x_n^{-1} x)(y_n^{-1} y)|$$
$$\leq |x_n^{-1} x| + |y_n^{-1} y| = d(x_n, x) + d(y_n, y);$$

thus $(x, y) \mapsto xy$ is continuous.   ∎

In particular, a valued abelian group, equipped with the value topology, is a metrizable topological group. For our purposes, the important examples of values arise from metrizable topological groups:

(8.8) **Theorem.**   *If G is a metrizable topological group, then there exists a value $x \mapsto |x|$ on G such that the left value topology and the right value topology coincide with the given topology on G.*

Proof.   Let d be a left-invariant metric compatible with the topology of G (6.3) and define $|x| = d(e, x)$. Then $x \mapsto |x|$ is a value on G (8.5), and $d(x, y) = |x^{-1} y|$ shows that the given topology coincides with the left value topology. The right value topology, which is generated by the metric $D(x, y) = |xy^{-1}| = d(x^{-1}, y^{-1})$, is the same as the topology generated by d (because $x \mapsto x^{-1}$ is a homeomorphism), namely, the given topology on G.   ∎

The topology derived from a left-invariant metric d on a group G need not be compatible with the group structure (8.16), but at least every left translation $x \mapsto ax$ is continuous (even isometric). It is useful to have a name for this limited form of compatibility:

(8.9) **Definition.**   A *left topological group* is a group G equipped with a topology such that, for each $a \in G$, the left translation $x \mapsto ax$ is continuous (and therefore a homeomorphism, with inverse mapping $x \mapsto a^{-1}x$). {Right topological groups are defined in the obvious dual way; for our purposes, the 'left' concept will suffice.}

The following proposition is immediate from the remarks preceding (8.9):

(8.10) **Theorem.**   *If d is a left-invariant metric on a group G, then G, equipped with the topology derived from d, is a left topological group.*

The next result is a slight extension of (4.4); the mixture of 'left' and 'right' is noteworthy:

(8.11) Lemma. *If G is a left topological group, H is a subgroup of G, $H\backslash G$ is the space of right cosets equipped with the quotient topology, and $\pi: G \to H\backslash G$ is the canonical mapping, then* (i) *for each $a \in G$ the neighborhoods of a are precisely the sets aV, where V is a neighborhood of the neutral element e of G,* (ii) *π is an open, continuous mapping, and* (iii) *for each $a \in G$, the neighborhoods of $\pi(a)$ are precisely the sets $\pi(aV)$, where V is a neighborhood of e.*

Proof. (i) The mapping $x \mapsto ax$ is a homeomorphism of G that maps e onto a. For use in the proof of (ii), we note the following consequence: if U is an open set in G, then aU is also open.

(ii) The canonical mapping is $\pi(x) = Hx$. By definition, the open sets of $H\backslash G$ are precisely the sets $A \subset H\backslash G$ such that $\pi^{-1}(A)$ is open in G; in particular, π is continuous. If U is open in G, then $\pi^{-1}(\pi(U)) = HU = \bigcup_{a \in H} aU$ is also open in G, therefore $\pi(U)$ is open in $H\backslash G$.

(iii) In view of (i), the proof written out in (4.4) may be used verbatim (but the notation π is to be interpreted differently). ▮

Allowing the subgroup in (8.11) to be nonnormal is, ultimately, a luxury; when the subgroup is normal, a little more can be squeezed out of the quotient group structure:

(8.12) Lemma. *If G is a left topological group, N is a normal subgroup of G, and $\pi: G \to G/N$ is the canonical mapping, then* (i) *the quotient group G/N, equipped with the quotient topology, is also a left topological group, and* (ii) *for each $a \in G$, the neighborhoods of $\pi(a)$ are precisely the sets $\pi(a)\pi(V)$, where V is a neighborhood of the neutral element e of G.*

Proof. As noted in (4.3), $G/N = N\backslash G$ and the canonical mapping is $\pi(x) = xN = Nx$.

The conclusion (ii) follows at once from part (iii) of (8.11) and the fact that π is a homomorphism.

(i) Fix $u_0, v_0 \in G/N$; the problem is to show that the mapping $v \mapsto u_0 v$ is continuous at v_0. Let A be a neighborhood of $u_0 v_0$; we seek a neighborhood B of v_0 such that $u_0 B \subset A$. Choose any $a_0 \in u_0$, $b_0 \in v_0$; thus $\pi(a_0) = u_0$, $\pi(b_0) = v_0$, and $\pi(a_0 b_0) = u_0 v_0$. Since A is a neighborhood of $u_0 v_0 = \pi(a_0 b_0)$, by (8.11) there exists a neighborhood V of e such that $\pi(a_0 b_0 V) = A$. Then $A = \pi(a_0)\pi(b_0 V) = u_0 \pi(b_0 V)$, thus $B = \pi(b_0 V)$ is a suitable neighborhood of $\pi(b_0) = v_0$. ▮

The convenience of the value concept is that there is a suggestive notation for passage to quotients:

(8.13) Theorem. *Let $(G, |x|)$ be a valued group, equip G with the left value topology (8.4), and suppose N is a closed normal subgroup of G. For $u \in G/N$ define*

(*) $|u| = \inf\{|x| : x \in u\}$.

Then (i) $u \mapsto |u|$ *is a value on* G/N, *and* (ii) *the left value topology on* G/N *coincides with the quotient topology.*

Proof. The topology on G is defined by the left-invariant metric $d(x, y) = |x^{-1}y|$; this is the topology relative to which N is assumed to be closed. Write $\pi : G \to G/N$ for the canonical mapping.

By (8.10), G is a left topological group; it follows from (8.12) that the quotient group G/N, equipped with the quotient topology, is a left topological group, and that the neighborhoods of $u \in G/N$ are the sets $u\pi(V)$, where V is a neighborhood of the neutral element e of G. Note that every left coset aN is a closed subset of G, since N is closed and $x \mapsto ax$ is a homeomorphism.

(i) Let us verify the axioms (8.1). Obviously $|u| \geq 0$ for all $u \in G/N$. The neutral element of G/N is $\pi(e) = eN = N$; since $e \in \pi(e)$, by (*) we have $0 \leq |\pi(e)| \leq |e| = 0$, thus $|\pi(e)| = 0$.

Suppose $u \in G/N$, $|u| = 0$. By (*) there exists a sequence $x_n \in u$ such that $|x_n| \to 0$. Say $u = xN$. Since $x_n \in xN$ and $d(e, x_n) = |x_n| \to 0$, it follows that $e \in xN$ (xN is closed) and therefore $xN = N$, that is, $u = \pi(e)$. Consequently $|u| > 0$ when $u \neq \pi(e)$.

Let $u \in G/N$. If $x \in u$ then $x^{-1} \in u^{-1}$, therefore $|u^{-1}| \leq |x^{-1}| = |x|$; varying x, $|u^{-1}| \leq |u|$ by (*). Replacing u by u^{-1} yields the reverse inequality, thus $|u^{-1}| = |u|$.

Let $u, v \in G/N$. If $x \in u$ and $y \in v$ then $xy \in uv$, therefore $|uv| \leq |xy| \leq |x| + |y|$; varying x and y independently, $|uv| \leq |u| + |v|$.

(ii) It remains to be shown that the quotient topology on G/N coincides with the left value topology defined by the left-invariant metric $|u^{-1}v|$. For both of these topologies, G/N is a left topological group (by (8.12) and (8.10), respectively). Therefore by part (i) of (8.11) (applied to the group G/N) it will suffice to show that the neighborhoods of $\pi(e)$ are the same for the two topologies.

A basic neighborhood of $\pi(e)$ for the left value topology is $A = \{u : |u| < \varepsilon\}$, $\varepsilon > 0$. Let $V = \{x \in G : |x| < \varepsilon\}$. Let us show that $\pi(V) = A$. If $x \in V$ then $|\pi(x)| \leq |x| < \varepsilon$, therefore $\pi(x) \in A$; conversely, if $u \in A$, that is, if $|u| < \varepsilon$, then it is clear from (*) that there exists $x \in u$ such that $|x| < \varepsilon$, thus $x \in V$ and $u = \pi(x) \in \pi(V)$. Then $A = \pi(V)$ is a neighborhood of $\pi(e)$ for the quotient topology (8.11).

On the other hand, let B be any neighborhood of $\pi(e)$ for the quotient topology. Then $\pi^{-1}(B)$ is a neighborhood of e (π is continuous). Choose $\varepsilon > 0$ so that $\{x : |x| < \varepsilon\} \subset \pi^{-1}(B)$. Writing $V = \{x : |x| < \varepsilon\}$ and $A = \{u : |u| < \varepsilon\}$, we have $V \subset \pi^{-1}(B)$ and therefore $B = \pi(\pi^{-1}(B)) \supset \pi(V) = A$; since A is a neighborhood of $\pi(e)$ for the left value topology, so is B. ∎

In the notation of left-invariant metrics, (8.13) takes the following form:

(8.14) Theorem. *Let* G *be a group, let* d *be a left-invariant metric on* G, *and equip* G *with the topology derived from* d. *Suppose* N *is a closed normal subgroup of* G. *For* $u, v \in G/N$ *define*

$$D(u, v) = \inf \{d(x, y) : x \in u, y \in v\}.$$

Then (i) *D is a left-invariant metric on G/N, and* (ii) *the topology derived from D coincides with the quotient topology. Define* $|x| = d(e, x)$ *and*

$$|u| = \inf \{|x| : x \in u\}.$$

Then (iii) $u \mapsto |u|$ *is a value on G/N, and* (iv) $D(u, v) = |u^{-1}v|$.

Proof. By (8.5), $x \mapsto |x|$ is a value on G; since $|x^{-1}y| = d(e, x^{-1}y) = d(x, y)$, the left value topology on G is precisely the topology derived from d (8.4). It follows from (8.13) that $u \mapsto |u|$ is a value on G/N, for which the left value topology coincides with the quotient topology. The left value topology on G/N is, by definition, the topology derived from the left-invariant metric $(u, v) \mapsto |u^{-1}v|$; the proof of the theorem will be completed by showing that $D(u, v) = |u^{-1}v|$.

Let $u, v \in G/N$. If $x \in u$ and $y \in v$ then $x^{-1}y \in u^{-1}v$, therefore $|u^{-1}v| \le |x^{-1}y| = d(x, y)$; varying x and y, $|u^{-1}v| \le D(u, v)$. On the other hand, suppose $z \in u^{-1}v$. Say $z = x^{-1}y$, with $x \in u$ and $y \in v$. Then $D(u, v) \le d(x, y) = |x^{-1}y| = |z|$; varying z, $D(u, v) \le |u^{-1}v|$. ∎

Exercises

(8.15) Let G be the group of *all* bijections of $[0, 1]$ (with composition product $(f \circ g)(t) = f(g(t))$) and let d be the metric on G defined by $d(f, g) = \sup |f(t) - g(t)|$ (cf. 2.22). Then (i) d is right-invariant, i.e., $d(f \circ h, g \circ h) = d(f, g)$, (ii) for each $f \in G$, the right-translation $g \mapsto g \circ f$ is d-continuous (indeed, d-isometric). Define $\|f\| = d(\iota, f)$, where ι is the identity mapping of $[0, 1]$. Then (iii) $f \mapsto \|f\|$ is a value on G, with $d(f, g) = \|f \circ g^{-1}\|$. Define $D(f, g) = \|f^{-1} \circ g\|$. Then (iv) D is a left-invariant metric, and (v) $D(f, g) = d(f^{-1}, g^{-1})$.

(8.16) Notation as in (8.15). Let f be the piecewise linear function defined by $f(t) = t$ for $0 \le t < 1/3$, $f(t) = t + 1/3$ for $1/3 \le t < 2/3$, $f(t) = t - 1/3$ for $2/3 \le t < 1$, and $f(1) = 1$. Let c_n be a sequence of real numbers such that $0 < c_n < 1/3$ and $c_n \uparrow 1/3$. For each n let g_n be the continuous, piecewise linear function whose graph is determined by the points $(0, 0)$, $(1/3, c_n)$, $(1, 1)$. Then (i) $d(g_n, \iota) \to 0$, and (ii) $(f \circ g_n)(1/3) - (f \circ \iota)(1/3) = c_n - 2/3 \to 1/3$, consequently (iii) left-translation by f is not d-continuous. But left-translation by f is D-continuous (8.10). Thus (iv) the d-topology and the D-topology are distinct, and (v) inversion in G cannot be d-continuous.

(8.17) If $(G, |x|)$ is a valued group such that the left value topology is compatible with the group structure (or if merely $x \mapsto x^{-1}$ is continuous), then the left value topology coincides with the right value topology.

(8.18) If $(G, |x|)$ is a valued group, then $|\, |x| - |y| \,| \le |x^{-1}y|$ for all $x, y \in G$ (the outer bars on the left side of the inequality stand for the ordinary absolute value). In particular, $x \mapsto |x|$ is uniformly continuous with respect to the left value metric (and similarly for the right value metric).

9. Quotient groups of metrizable complete topological groups.

To set the stage, suppose G is a complete topological group (7.3), and N is a closed normal subgroup of G. Question: Is the quotient topological group G/N complete? In general, the answer is no (9.3). If G is locally compact, then the answer is yes; indeed, G/N is itself locally compact (see (4.5)) and therefore complete (7.10). The answer is also yes if G is metrizable; the present section is devoted to a proof of this key result (9.1) and to a converse of it (9.2).

In later applications, we shall be interested in this circle of ideas as it pertains to the additive group of a topological vector space (1.6). In the vector space context, local compactness is an excessive restriction: it forces the space to be finite-dimensional (23.10). Of main interest to us is the case of metrizable groups, especially its application to metrizable complete topological vector spaces (Section 12) and in particular to Banach spaces (Section 16). Thus the present section is included primarily for its later application to Banach spaces; we indulge in a little extra generality, which costs no more and helps to see what is going on.

(9.1) Theorem. *If G is a metrizable complete topological group (7.5) and N is a closed subgroup of G, then N is a metrizable complete topological group in the relative topology. If, moreover, N is normal, then the quotient topological group G/N is complete and metrizable.*

Proof. Let d be a left-invariant compatible metric (6.3). Since (G, d) is a complete metric space (7.8) and N is a closed subset of G, it is elementary that (N, d) is a complete metric space (for brevity, we continue to write d here, in place of its restriction to $N \times N$); but d is a left-invariant metric compatible with the topology of N (namely, the relative topology (2.2)), therefore the topological group N is complete (7.8).

Assume now that N is normal and let $\pi : G \to G/N$ be the canonical mapping, where G/N denotes the quotient topological group (4.5). For $u, v \in G/N$ define

$$D(u, v) = \inf \{d(x, y) : x \in u, y \in v\};$$

by (8.14), D is a left-invariant metric on G/N, and the topology derived from D coincides with the quotient topology, that is, D is a left-invariant compatible metric on the topological group G/N. To prove that G/N is a complete topological group, it will suffice, by (7.8), to show that D is a complete metric. As in (8.14), let us write $|x| = d(e, x)$ and $|u| = \inf \{|x| : x \in u\}$; as shown there, $u \mapsto |u|$ is a value on G/N, and $|u^{-1}v| = D(u, v)$.

Let u_n be a sequence in G/N such that $D(u_m, u_n) \to 0$. We seek $u \in G/N$ such that $D(u_n, u) \to 0$; passing to a subsequence, we can suppose that $D(u_n, u_{n+1}) < 2^{-n}$ for all n, i.e.,

$$(1) \qquad\qquad |u_n^{-1}u_{n+1}| < 2^{-n}.$$

Construct a sequence x_n in G as follows. Choose $x_1 \in u_1$ arbitrarily. For $n > 1$, choose $x_n \in u_{n-1}^{-1}u_n$ so that

$$(2) \qquad\qquad |x_n| < |u_{n-1}^{-1}u_n| + 2^{-n};$$

this is possible since $|u|$ is defined as an infimum. Set $y_n = x_1 x_2 \cdots x_n$ (in additive notation, y_n would be a partial sum of a series); thus $y_1 = x_1$ and $y_{n+1} = y_n x_{n+1}$. Since $\pi(x_1) = u_1$ and $\pi(x_n) = u_{n-1}^{-1} u_n$ for $n > 1$, we have $\pi(y_1) = u_1$ and, for $n > 1$,

$$\pi(y_n) = \pi(x_2)\pi(x_2)\cdots\pi(x_n) = u_1(u_1^{-1}u_2)\cdots(u_{n-1}^{-1}u_n) = u_n$$

(a 'telescoping product'), thus

(3) $\pi(y_n) = u_n \qquad (n = 1, 2, 3, \ldots).$

The sequence y_n is Cauchy with respect to d. Indeed, citing (1), (2) and the relation $y_{n+1} = y_n x_{n+1}$, we have

$$d(y_n, y_{n+1}) = |y_n^{-1}y_{n+1}| = |x_{n+1}| < |u_n^{-1}u_{n+1}| + 2^{-(n+1)}$$
$$< 2^{-n} + 2^{-(n+1)} < 2^{-n+1};$$

then

$$d(y_n, y_{n+p}) \le \sum_{k=n}^{n+p-1} d(y_k, y_{k+1}) < \sum_{k=n}^{n+p-1} 2^{-k+1} < 2^{-n+2},$$

thus $d(y_n, y_{n+p}) \to 0$ as $n \to \infty$, uniformly in p. Since d is a complete metric, there exists $y \in G$ such that $d(y_n, y) \to 0$. Then $u = \pi(y)$ is the desired limit for the Cauchy sequence u_n: $D(u_n, u) = D(\pi(y_n), \pi(y)) \le d(y_n, y) \to 0.$ ∎
Conversely:

(9.2) Theorem.   *Suppose G is a metrizable topological group and N is a closed normal subgroup of G such that N and G/N are complete topological groups (for the relative topology and the quotient topology, respectively). Then G is a complete topological group.*

Proof.   Define d, D, $|x|$, $|u|$, π as in the proof of (9.1); as noted there, D is a left-invariant compatible metric on G/N, with $D(u, v) = |u^{-1}v|$. In view of (7.8), our hypothesis is that (N, d) and $(G/N, D)$ are complete metric spaces, and the problem is to show that (G, d) is a complete metric space.
Let x_n be a d-Cauchy sequence in G. Then $u_n = \pi(x_n)$ is a D-Cauchy sequence in G/N:

$$D(u_m, u_n) = D(\pi(x_m), \pi(x_n)) \le d(x_m, x_n) \to 0.$$

By hypothesis, there exists $u \in G/N$ such that $D(u_n, u) \to 0$. Choose any $x \in u$; thus $\pi(x) = u$. Choose $y_n \in u_n^{-1}u$ so that $|y_n| < |u_n^{-1}u| + 1/n$ (possible because $|u_n^{-1}u|$ is defined as an infimum); since $|u_n^{-1}u| = D(u_n, u) \to 0$, evidently $|y_n| \to 0$. Since $\pi(y_n) = u_n^{-1}u = \pi(x_n)^{-1}\pi(x) = \pi(x_n^{-1}x)$, one has $y_n = x_n^{-1}x z_n$ for suitable $z_n \in N$. The sequence $z_n = x^{-1}x_n y_n$ is d-Cauchy:

$$d(z_m, z_n) = d(x^{-1}x_m y_m, x^{-1}x_n y_n) = d(x_m y_m, x_n y_n) = |(x_m y_m)^{-1}(x_n y_n)|$$
$$= |y_m^{-1}x_m^{-1}x_n y_n| \le |y_m^{-1}| + |x_m^{-1}x_n| + |y_n|$$
$$= |y_m| + d(x_m, x_n) + |y_n| \to 0$$

as $m, n \to \infty$. By hypothesis, d is a complete metric on N, therefore there exists $z \in N$ such that $d(z_n, z) \to 0$. Then $x_n = xz_n y_n^{-1}$ converges to xz:

$$d(x_n, xz) = d(xz_n y_n^{-1}, xz) = d(z_n y_n^{-1}, z) = |(z_n y_n^{-1})^{-1}z|$$

$$= |y_n z_n^{-1} z| \leq |y_n| + |z_n^{-1} z| = |y_n| + d(z_n, z) \to 0$$

as $n \to \infty$. Thus the given d-Cauchy sequence x_n is convergent (to the limit xz). ∎

Exercises

*(9.3)   There exists a complete topological abelian group G with a closed subgroup N such that the quotient topological group G/N is not complete.

(9.4)   If G is a separated topological group and if N is a closed normal subgroup such that both N and G/N are complete topological groups, then G is also complete.

*10. Bi-uniform topological groups.   The coincidence of the left and right uniform structures on a topological group is bound to be interesting, and even useful (see (10.6), (10.10), (10.11)); let us give a name to this phenomenon:

(10.1) Definition.   A topological group G is called *bi-uniform* if its left and right uniform structures coincide, that is, if $\mathscr{U}_s = \mathscr{U}_r$. {See (5.6) and (5.16) for the definitions.}

An abelian topological group is trivially bi-uniform. A deeper result is that every compact topological group is bi-uniform (5.24). A locally compact topological group need not be bi-uniform ((10.8), (10.11)). The question of when a metrizable topological group is bi-uniform is answered in (10.5). Some useful reformulations of bi-uniformity:

(10.2) Theorem.   *The following conditions on a topological group G are equivalent:*

(a) *G is bi-uniform;*

(b) *the mapping $x \mapsto x^{-1}$ is uniformly continuous with respect to the left uniformity $\mathscr{U}_s$;*

(c) *the mapping $x \mapsto x^{-1}$ is uniformly continuous with respect to the right uniformity $\mathscr{U}_r$;*

(d) *if V is any neighborhood of the neutral element e, then there exists a neighborhood W of e such that $x^{-1}Wx \subset V$ for all $x \in G$;*

(e) *if V is any neighborhood of e, then $\bigcap_{x \in G} xVx^{-1}$ is also a neighborhood of e;*

(f) *there exists a fundamental system of neighborhoods of e that are invariant under every inner automorphism of G.*

Proof.   Write $Jx = x^{-1}$. To say that J is uniformly continuous with respect to $\mathscr{U}_s$ means that, given any $A \in \mathscr{U}_s$, there exists $B \in \mathscr{U}_s$ such that $(J \times J)(B) \subset A$ (i.e., $(x, y) \in B$ implies $(x^{-1}, y^{-1}) \in A$) and therefore

$(J \times J)^{-1}(A) \in \mathcal{U}_s$; equivalently, $(J \times J)^{-1}(\mathcal{U}_s) \subset \mathcal{U}_s$; since $J \times J$ is a self-inverse bijection of $G \times G$, this is in turn equivalent to $(J \times J)(\mathcal{U}_s) = \mathcal{U}_s$. Similarly, J is uniformly continuous with respect to $\mathcal{U}_r$ if and only if $(J \times J)(\mathcal{U}_r) = \mathcal{U}_r$. Since, in any case, $(J \times J)(\mathcal{U}_s) = \mathcal{U}_r$ (5.19), the equivalence of (a), (b), (c) is clear.

(a) implies (d): If V is any neighborhood of e, the relations $V_s \in \mathcal{U}_s = \mathcal{U}_r$ imply that $W_r \subset V_s$ for some neighborhood W of e (5.15). For any $x \in G$, $W_r(x) \subset V_s(x)$; since $W_r(x) = Wx$ and $V_s(x) = xV$ (see (5.9) and the proof of (5.17)), we have $Wx \subset xV$, that is, $x^{-1}Wx \subset V$ for all $x \in G$.

(d) implies (e): Let V be a neighborhood of e. By hypothesis there exists a neighborhood W of e such that $x^{-1}Wx \subset V$ for all $x \in G$. Then $W \subset xVx^{-1}$ for all $x \in G$; thus $\bigcap_{x \in G} xVx^{-1}$ includes the neighborhood W of e, hence is itself a neighborhood of e.

(e) implies (f): An arbitrary neighborhood V of e contains the set $W = \bigcap_{x \in G} xVx^{-1}$, which is, by hypothesis, a neighborhood of e; obviously $xWx^{-1} = W$ for all $x \in G$, that is, W is invariant under all inner automorphisms of G.

(f) implies (a): We show that $\mathcal{U}_s \subset \mathcal{U}_r$; the proof of the reverse inclusion is similar. Let V be a neighborhood of e. By hypothesis there exists a neighborhood W of e such that $W \subset V$ and W is invariant under all inner automorphisms of G. Then $x^{-1}Wx = W \subset V$, i.e., $Wx \subset xV$ for all $x \in G$, therefore $W_r \subset V_s$ by the calculation in the proof of '(a) implies (d).' Thus $V_s \supset W_r \in \mathcal{U}_r$ and therefore $V_s \in \mathcal{U}_r$. Since the sets V_s form a base for the filter $\mathcal{U}_s$, it follows that $\mathcal{U}_s \subset \mathcal{U}_r$. ∎

What about metrizable groups? The strongest hold we have on such a group is the Birkhoff-Kakutani theorem (6.3): each fundamental sequence of neighborhoods of e, suitably 'improved' (see the proof of (6.3)), leads to a left-invariant compatible metric d. Presumably, if one can further 'improve' the neighborhood sequence, properties of d in addition to left-invariance may be obtained. If, in particular, the group is also bi-uniform, it possesses a fundamental sequence of neighborhoods of e that are invariant under all inner automorphisms (10.2); this leads, as we show in the next theorem, to a compatible metric that is both left- and right-invariant. First let us formalize this concept:

(10.3) Definition. A metric d on a group G is called *bi-invariant* if it is both left-invariant and right-invariant: $d(ax, ay) = d(x,y) = d(xa, ya)$.

The connection with bi-uniformity is immediate:

(10.4) Lemma. *Let G be a group, d a bi-invariant metric on G, and equip G with the topology derived from d. Then G is a bi-uniform topological group.*

Proof. The continuity of $x \mapsto x^{-1}$ is shown by the calculation

$$d(x^{-1}, y^{-1}) = d(xx^{-1}, xy^{-1}) = d(xx^{-1}y, xy^{-1}y) = d(y, x) = d(x, y).$$

If $d(x_n, x) \to 0$ and $d(y_n, y) \to 0$, then

$$d(x_n y_n, xy) \le d(x_n y_n, xy_n) + d(xy_n, xy) = d(x_n, x) + d(y_n, y) \to 0,$$

therefore the mapping $(x, y) \mapsto xy$ is continuous. Thus, the topology derived from d is compatible with the group structure.

Let $\mathscr{U}_s$ and $\mathscr{U}_r$ be the left and right uniformities of the topological group G. Citing (7.6) and its obvious dualization, we have $\mathscr{U}_s = \mathscr{U}_d = \mathscr{U}_r$, thus G is bi-uniform. ∎

Lemma (10.4) and its converse are combined in the following result:

(10.5) Theorem. (V. L. Klee). *A metrizable topological group G is bi-uniform if and only if there exists a bi-invariant metric d compatible with the topology of G.*

Proof. The 'if' part is covered by the lemma. Conversely, suppose G is bi-uniform. In view of (10.2), in the proof of the Birkhoff-Kakutani metrization theorem (6.3) one can take the neighborhoods U_n—and therefore the V_n—to be invariant under every inner automorphism of G. Since $x^{-1}y \in V_n$ iff $(axa^{-1})^{-1}(aya^{-1}) = a(x^{-1}y)a^{-1} \in V_n$, it is clear that the proof of (6.3) yields a left-invariant compatible metric d such that $d(axa^{-1}, aya^{-1}) = d(x, y)$ for all a, x, y. It follows that d is also right-invariant: $d(xa, ya) = d(a(xa)a^{-1}, a(ya)a^{-1}) = d(ax, ay) = d(x, y)$. ∎

Exercises

(10.6) Every separated bi-uniform topological group G possesses a completion; that is, G may be regarded as a dense subgroup of a complete topological group (see (7.15)).

(10.7) Every compact topological group is bi-uniform (cf. 5.24).

(10.8) A complete topological group need not be bi-uniform. Indeed, there exists a topological group G that is locally compact (and therefore complete) but not bi-uniform. For example, let $\mathscr{M} = M(3, \mathbb{R})$ be the set of all 3×3 real matrices $A = (a_{ij})$, topologized by the (complete) metric

$$d(A, B) = \max \{|a_{ij} - b_{ij}| : i, j = 1, 2, 3\},$$

and let G be the multiplicative group of all $A \in \mathscr{M}$ of the form

$$A = \begin{pmatrix} 1 & a & c \\ 0 & 1 & b \\ 0 & 0 & 1 \end{pmatrix},$$

equipped with the relative topology. It is easy to see that G is a (metrizable) topological group. Since $A \mapsto (a, b, c)$ is clearly a homeomorphism of G onto $\mathbb{R}^3$, G is locally compact (even locally Euclidean), therefore G is a complete topological group (7.10). Let

$$A = \begin{pmatrix} 1 & a & c \\ 0 & 1 & b \\ 0 & 0 & 1 \end{pmatrix}, \qquad A_n = \begin{pmatrix} 1 & n & 0 \\ 0 & 1 & 0 \\ 0 & 0 & 1 \end{pmatrix}$$

$(n = 1, 2, 3, \ldots)$. The calculation

$$A_n A A_n^{-1} = \begin{pmatrix} 1 & a & c + nb \\ 0 & 1 & b \\ 0 & 0 & 1 \end{pmatrix}$$

shows that no neighborhood of the identity matrix of the form $\{B : d(B, I) < \varepsilon\}$ can contain a neighborhood of I that is invariant under all inner automorphisms of G, thus G is not bi-uniform (10.2).

(10.9) Notation as in (10.8). Then G is a metrizable complete group, d is a complete compatible metric on G, but $\mathcal{U}_d$ is distinct from both $\mathcal{U}_s$ and $\mathcal{U}_r$. For example, to see that $\mathcal{U}_d \neq \mathcal{U}_s$, consider the matrices

$$A_n = \begin{pmatrix} 1 & n & 0 \\ 0 & 1 & 0 \\ 0 & 0 & 1 \end{pmatrix}, \qquad B_n = \begin{pmatrix} 1 & n & 0 \\ 0 & 1 & 1/n \\ 0 & 0 & 1 \end{pmatrix}$$

$(n = 1, 2, 3, \ldots)$. Then

$$d(A_n, B_n) = 1/n, \qquad d(A_n^{-1}B_n, I) = 1,$$

therefore $\mathcal{U}_d \neq \mathcal{U}_s$ by (6.9).

*(10.10) (V. L. Klee)   Let G be a metrizable, bi-uniform topological group, and suppose G admits a complete compatible metric. Then G is a complete topological group (therefore G admits a bi-invariant complete compatible metric).

*(10.11) (R. Godement)   Let G be a unimodular, locally compact topological group. The following conditions on G are equivalent: (a) G is bi-uniform; (b) G is a group of finite class (i.e., for every central Radon measure μ on G of positive type, the von Neumann algebras associated with the unitary double representation derived from μ are of finite class); (c) the von Neumann algebras associated with the regular unitary double representation of G are of finite class.

(10.12) (V. L. Klee)   A metric d on a group G is bi-invariant if and only if $d(xy, ab) \leq d(x, a) + d(y, b)$ for all x, y, a, b in G.

Chapter 2

Topological Vector Spaces

Our discussion is limited to topological vector spaces over the field of real numbers $\mathbb{R}$ and the field of complex numbers $\mathbb{C}$, the fields of primary importance in analysis. When the subject is pressed to its depths, the distinction between real and complex TVS is crucial; for instance, complex TVS are susceptible to complex function-theoretic methods, whereas order-theoretic methods are prominent in the theory of real TVS. However, at the beginning, it frequently does not matter whether the base field is $\mathbb{R}$ or $\mathbb{C}$; in such situations, we use $\mathbb{K}$ to denote either $\mathbb{R}$ or $\mathbb{C}$.

11. Real and complex topological vector spaces. If E is a vector space, we shall write θ for the zero element of E, i.e., the neutral element of the additive group of E. Let us review the definition of a topological vector space (1.6), specialized to the base field $\mathbb{K}$:

(11.1) Definition. A *topological vector space* (briefly, TVS) over $\mathbb{K}$ is a vector space E over $\mathbb{K}$, equipped with a topology that is compatible with the vector space structure, i.e., such that the mappings $(x, y) \mapsto x + y$ and $(\lambda, x) \mapsto \lambda x$ are continuous. (It is assumed that $\mathbb{K}$ bears the usual topology, defined by the metric $|\lambda - \mu|$.) We refer to E as a *real* TVS or a *complex* TVS according as $\mathbb{K} = \mathbb{R}$ or $\mathbb{K} = \mathbb{C}$.

Every complex TVS E is obviously also a real TVS, since the restriction of the continuous mapping $(\lambda, x) \mapsto \lambda x$ ($\lambda \in \mathbb{C}$, $x \in E$) to $\mathbb{R} \times E$ is continuous. However, it can happen that a topology on a complex vector space is compatible with the real vector space structure but not with the complex vector space structure (11.12).

(11.2) Example. *The space $\mathbb{K}^r$.* Fix a positive integer r and let $E = \mathbb{K}^r$ be the product vector space; that is, if $x = (\lambda_k)$ and $y = (\mu_k)$ are $r$ples of elements of $\mathbb{K}$, and if $\lambda \in \mathbb{K}$, then

$$x + y = (\lambda_1 + \mu_1, \ldots, \lambda_r + \mu_r),$$
$$\lambda x = (\lambda \lambda_1, \ldots, \lambda \lambda_r).$$

Equip $\mathbb{K}^r$ with the product topology τ; thus, τ is the topology derived from any of the metrics d described in (2.14), e.g., $d(x, y) = \sum_1^r |\lambda_k - \mu_k|$. The topology τ is the 'topology of coordinatewise convergence'; that is, if $x_n = (\lambda_{nk})$ ($n = 1, 2, 3, \ldots$) and $x = (\lambda_k)$, then $x_n \to x$ if and only if $\lambda_{nk} \to \lambda_k$ for each k ($k = 1, \ldots, r$). Obviously τ is compatible with the vector space structure of $\mathbb{K}^r$. When we refer to $\mathbb{K}^r$ as a TVS, it is understood that we have in mind the above algebraic and topological structures. {However, the metric generating the topology may vary with the context; for example, the Euclidean

47

metric $d(x, y) = (\sum_1^r |\lambda_k - \mu_k|^2)^{1/2}$ is appropriate for certain geometrical considerations.}

We need not delay the exposition by working out a lengthy list of examples here; other examples of TVS will arise in abundance in later sections. Indeed, one way to build a sizeable repertory of examples of a structure is to start with a modest collection of known examples, and apply to the collection suitable constructs (e.g., substructure, quotient structure, product structure) which result in a structure of the same sort. (For topological groups, this procedure is illustrated to a certain extent in Sections 2 and 4.)

A useful reformulation of the definition of a TVS is as follows:

(11.3) Theorem. *Let E be a vector space over $\mathbb{K}$ and let τ be a topology on E. Then τ is compatible with the vector space structure of E if and only if (i) τ is compatible with the additive group structure of E, (ii) for each $x \in E$, the mapping $\lambda \mapsto \lambda x$ ($\lambda \in \mathbb{K}$) is continuous at $\lambda = 0$, (iii) for each $\lambda \in \mathbb{K}$, the mapping $x \mapsto \lambda x$ ($x \in E$) is continuous at $x = \theta$, and (iv) the mapping $(\lambda, x) \mapsto \lambda x$ ($\lambda \in \mathbb{K}$, $x \in E$) is continuous at $(0, \theta)$.*

Proof. Only if: Conditions (ii)–(iv) follow trivially from the definition of a TVS (11.1). Condition (i) results from the continuity of $(x, y) \mapsto x + y$ and $x \mapsto (-1)x = -x$ (1.2).

If: Let $\lambda_0 \in \mathbb{K}$, $x_0 \in E$; the problem is to show that $(\lambda, x) \mapsto \lambda x$ is continuous at (λ_0, x_0). Let U be a neighborhood of θ; thus $\lambda_0 x_0 + U$ is a typical neighborhood of $\lambda_0 x_0$ (by (i) and (2.9)). We seek neighborhoods A, W of 0, θ such that

$$(\lambda_0 + A)(x_0 + W) \subset \lambda_0 x_0 + U.$$

Let V be a neighborhood of θ such that $V + V + V \subset U$ (3.1). By (ii) there exists a neighborhood A_1 of θ such that $A_1 x_0 \subset V$. By (iii) there exists a neighborhood W_1 of θ such that $\lambda_0 W_1 \subset V$. By (iv) there exist a neighborhood A_2 of 0 and a neighborhood W_2 of θ such that $A_2 W_2 \subset V$. Let

$$A = A_1 \cap A_2, \qquad W = W_1 \cap W_2.$$

Evidently $Ax_0 \subset V$, $\lambda_0 W \subset V$ and $AW \subset V$, therefore, by distributivity,

$$(\lambda_0 + A)(x_0 + W) \subset \lambda_0 x_0 + Ax_0 + \lambda_0 W + AW$$

$$\subset \lambda_0 x_0 + V + V + V \subset \lambda_0 x_0 + U. \qquad \blacksquare$$

In a topological group, the translation mappings play an important role; in a TVS, the mappings that play the analogous role are the homotheties:

(11.4) Definition. *If E is a vector space over $\mathbb{K}$, a *homothetic mapping* (or *homothety*) is a mapping of the form*

$$x \mapsto \lambda x + a \qquad (x \in E),$$

where $\lambda \in \mathbb{K}$ and $a \in E$ are fixed. If $\lambda \neq 0$, such a mapping is bijective, with a homothetic inverse mapping, namely, $y \mapsto \lambda^{-1} y - \lambda^{-1} a$ ($y \in E$).

(11.5) Theorem. *In a TVS, every nonconstant homothety is a homeomorphism.*

Proof. Say the homothety is $f(x) = \lambda x + a$, where $\lambda \neq 0$. Since the inverse of f is also a homothety, it suffices to show that f is continuous, and this is evident from the factorization $x \mapsto \lambda x \mapsto \lambda x + a$ (cf. 11.1). ∎

(11.6) Theorem. *If E is a TVS over $\mathbb{K}$ and M is a linear subspace of E, then M, with the relative topology, is also a TVS over $\mathbb{K}$.*

Proof. The restrictions of the continuous mappings $(x, y) \mapsto x + y$ and $(\lambda, x) \mapsto \lambda x$, to $M \times M$ and $\mathbb{K} \times M$, respectively, are continuous for the relative topology of M. ∎

(11.7) Theorem. *In a TVS, the closure of a linear subspace is a linear subspace.*

Proof. If M is a linear subspace of a TVS E over $\mathbb{K}$, then the continuous mapping $(x, y) \mapsto x + y$, which maps $M \times M$ into M, also maps $(M \times M)^- = \overline{M} \times \overline{M}$ into $\overline{M}$, thus $x + y \in \overline{M}$ whenever $x, y \in \overline{M}$. Similarly, the continuous mapping $(\lambda, x) \mapsto \lambda x$, which maps $\mathbb{K} \times M$ into M, also maps $(\mathbb{K} \times M)^- = \mathbb{K} \times \overline{M}$ into $\overline{M}$; that is, $\lambda x \in \overline{M}$ for all $\lambda \in \mathbb{K}$ and $x \in \overline{M}$. ∎

(11.8) Theorem. *Let M be a linear subspace of a TVS E over $\mathbb{K}$, E/M the quotient vector space, $\pi : E \to E/M$ the canonical mapping, and equip E/M with the quotient topology. Then:*

(i) *E/M is a TVS over $\mathbb{K}$;*

(ii) *π is an open, continuous mapping;*

(iii) *for each $x \in E$, the neighborhoods of $\pi(x) = x + M$ are precisely the sets $\pi(x + V) = \pi(x) + \pi(V)$, where V is a neighborhood of $\theta \in E$;*

(iv) *E/M is separated if and only if M is a closed linear subspace of E.*

Proof. Since the topology of E is compatible with the additive group structure of E (11.3), the quotient topology is compatible with the additive group structure of E/M (4.5). The statements (ii)–(iv) are then immediate from (4.5).

It remains to be shown that the mapping $(\lambda, u) \mapsto \lambda u$ ($\lambda \in \mathbb{K}$, $u \in E/M$) is continuous. Fix $\lambda_0 \in \mathbb{K}$, $u_0 \in E/M$, and let A be a neighborhood of $\lambda_0 u_0$. Say $u_0 = \pi(x_0)$; then $\lambda_0 u_0 = \pi(\lambda_0 x_0)$. Let W be a neighborhood of $\lambda_0 x_0$ such that $\pi(W) = A$ (ii). Since W is a neighborhood of $\lambda_0 x_0$ and $(\lambda, x) \mapsto \lambda x$ is continuous, there exist neighborhoods U, V of λ_0, x_0, respectively, such that $UV \subset W$. Then $\pi(UV) \subset \pi(W) = A$, i.e., $U\pi(V) \subset A$; thus, we have found a neighborhood U of λ_0 and a neighborhood $\pi(V)$ of $\pi(x_0) = u_0$ (ii) such that $U\pi(V) \subset A$, which proves that $(\lambda, u) \mapsto \lambda u$ is continuous at (λ_0, u_0). ∎

(11.9) Theorem. *If $(E_\iota)_{\iota \in I}$ is a family of TVS over $\mathbb{K}$, and if $E = \prod_{\iota \in I} E_\iota$ is the product vector space, then the product topology on E is compatible with the vector space structure; that is, E, equipped with the product vector space structure and the product topology, is a TVS over $\mathbb{K}$.*

Proof. The linear operations in E are defined coordinatewise: if $x = (x_\iota)$, $y = (y_\iota)$ and $\lambda \in \mathbb{K}$, then

$$x + y = (x_\iota + y_\iota), \qquad \lambda x = (\lambda x_\iota).$$

Write $f(x, y) = x + y$, $g(\lambda, x) = \lambda x$; thus $f: E \times E \to E$ and $g: \mathbb{K} \times E \to E$, and the problem is to show that f and g are continuous (for the product topology on E). One can give a brief proof based on formal properties of product topologies (cf. the proof of (2.12)) but we prefer to give the straight-forward direct proof.

To see that f is continuous, fix $a = (a_\iota)$, $b = (b_\iota)$ in E, and let W be a neighborhood of $a + b$; we seek neighborhoods U, V of a, b such that $U + V \subset W$. We may suppose that $W = \prod_{\iota \in I} W_\iota$, where W_ι is a neighborhood of $a_\iota + b_\iota$, and $W_\iota = E_\iota$ for $\iota \notin J$, J a suitable finite subset of I (such neighborhoods are basic for the product topology). If $\iota \in J$, choose neighborhoods U_ι, V_ι of a_ι, b_ι such that $U_\iota + V_\iota \subset W_\iota$ (this is possible because addition in E_ι is continuous), and if $\iota \notin J$ define $U_\iota = E_\iota$, $V_\iota = E_\iota$. Set

$$U = \prod_{\iota \in I} U_\iota, \qquad V = \prod_{\iota \in I} V_\iota.$$

Then U, V are neighborhoods of a, b, and clearly $U + V \subset W$.

To see that g is continuous, fix λ in $\mathbb{K}$ and $a = (a_\iota)$ in E, and let W be a neighborhood of $\lambda a = (\lambda a_\iota)$; we seek neighborhoods A, V of λ, a such that $AV \subset W$. We may suppose $W = \prod_{\iota \in I} W_\iota$, where W_ι is a neighborhood of λa_ι, and $W_\iota = E_\iota$ for $\iota \notin J$, J a suitable finite subset of I. If $\iota \in J$, choose a neighborhood A_ι of λ and a neighborhood V_ι of a_ι such that $A_\iota V_\iota \subset W_\iota$, and if $\iota \notin J$ define $V_\iota = E_\iota$. Set $A = \bigcap_{\iota \in J} A_\iota$ and $V = \prod_{\iota \in I} V_\iota$. Then A, V are neighborhoods of λ, a such that $AV \subset W$. ∎

(11.10) Definition. The *uniform structure* of a TVS is defined to be the uniform structure derived from its additive topological group (11.3). Thus, if E is a TVS over $\mathbb{K}$ and V is a neighborhood of $\theta \in E$, then

$$\{(x, y) : y - x \in V\}$$

is a basic entourage for the uniform structure of E (5.16). (Since the additive group of E is abelian, the left and right uniform structures are identical.) In particular, a *complete* TVS is, by definition, a TVS whose additive topological group is complete in the sense of (7.3).

It is noteworthy that the scalar multiplication contributes nothing to the notion of uniformity in a TVS; the underlying reason for this is that the mappings $x \mapsto \lambda x$ $(x \in E)$, $\lambda \neq 0$, are already isomorphisms for the uniform structure of the additive topological group (cf. 5.12).

(11.11) Definition. A TVS is said to be *metrizable* if it is metrizable as a topological space (cf. Section 6). In view of (6.3), a separated TVS E is metrizable if and only if there exists a fundamental sequence of neighborhoods of θ, in which case there exists a metric d, compatible with the topology of E, which is *additively invariant*, i.e., $d(x + z, y + z) = d(x, y)$ for all x, y, z in E.

This by no means exhausts the topics we shall need from the general theory of topological vector spaces. Formalistically, one should take these up now; instead, purely for the sake of variety, we prefer to intersperse them with more specialized results. For example, the material developed in the preceding chapter on topological groups allows us at once to plunge rather deeply into the theory of metrizable TVS, which we proceed to do in the next five sections.

Exercise

(11.12) Regard $\mathbb{C}$ as a one-dimensional vector space over $\mathbb{C}$, and a two-dimensional vector space over $\mathbb{R}$, in the usual way. Define

$$p(\lambda) = |\operatorname{Im} \lambda| \qquad (\lambda \in \mathbb{C}).$$

For all $\mu \in \mathbb{C}$ and $\varepsilon > 0$, write

$$U_\varepsilon(\mu) = \{\lambda \in \mathbb{C} : p(\lambda - \mu) < \varepsilon\}.$$

Then (i) there exists a unique topology τ on $\mathbb{C}$ for which the sets $U_\varepsilon(\mu)$, $\varepsilon > 0$, are a fundamental system of neighborhoods of μ, (ii) τ is compatible with the real vector space structure of $\mathbb{C}$, and (iii) τ is not compatible with the complex vector space structure of $\mathbb{C}$.

12. Metrizable complete TVS.

According to the conventions established in (11.10), (11.11), a topological vector space (over $\mathbb{K} = \mathbb{R}$ or $\mathbb{C}$) is said to be *complete* if it is complete as an additive topological group, and *metrizable* if it is metrizable as a topological space. Thus a *metrizable complete TVS* is a TVS whose additive topological group is complete and metrizable.

Explicitly, let E be a vector space over $\mathbb{K}$, equipped with a topology τ. Let us write $(E, +)$ for the additive group of E. Thus, E is a TVS iff (i) $(E, +, \tau)$ is a topological group, and (ii) $(\lambda, x) \mapsto \lambda x$ is continuous. In particular, E is a metrizable complete TVS iff (i) $(E, +, \tau)$ is a metrizable complete topological group in the sense of (7.5), and (ii) $(\lambda, x) \mapsto \lambda x$ is continuous. Accordingly, with due attention paid to scalar multiples, the results in Chapter 1 on metrizable complete topological groups can be carried over to the TVS setting. To illustrate, the TVS analogues of (9.1) and (9.2) are derived in this section; these are especially important in later applications (cf. Section 16).

(12.1) Theorem. *Suppose E is a metrizable complete TVS and M is a closed linear subspace of E. Then M and E/M are metrizable complete TVS, in the relative topology and the quotient topology, respectively.*

Moreover, if d is an (additively) invariant compatible metric on E (6.3), and if one defines

$$|x| = d(x, \theta) \qquad (x \in E)$$

$$|u| = \inf\{|x| : x \in u\} \qquad (u \in E/M),$$

$$D(u, v) = \inf\{d(x, y) : x \in u, y \in v\} \qquad (u, v \in E/M),$$

then (i) $x \mapsto |x|$ *is a value on* E, (ii) $d(x, y) = |x - y|$, (iii) $u \mapsto |u|$ *is a value on* E/M, (iv) $D(u, v) = |u - v|$, (v) D *is an invariant compatible metric on* E/M, *and* (vi) (E, d), (M, d) *and* $(E/M, D)$ *are complete metric spaces.*

Proof. The heart of the matter is contained in (9.1) and its proof; the only new element is that the mappings $(\lambda, x) \mapsto \lambda x$ $(\lambda \in \mathbb{K}, x \in M)$ and $(\lambda, u) \mapsto \lambda u$ $(\lambda \in \mathbb{K}, u \in E/M)$ are continuous for the indicated topologies (11.6), (11.8). ∎

Conversely:

(12.2) Theorem. *Suppose* E *is a metrizable TVS, and* M *is a closed linear subspace of* E *such that* M *and* E/M *are complete TVS, in the relative topology and the quotient topology, respectively. Then* E *is a complete TVS.*

Proof. The TVS M and E/M are of course metrizable (cf. 6.4); the completeness of E is an immediate application of (9.2). ∎

Metrizable complete TVS appeared early in the history of functional analysis (in another guise—see the next section), and they have stayed late. From the beginning, their theory has occupied a fundamental position in functional analysis, supporting a number of the cornerstone theorems of the subject and providing some of its most powerful and appealing applications. In particular, the applications of the theory to Banach spaces and Banach algebras will be among the deepest matters we shall take up in this book.

Exercises

*(12.3) (G. Köthe)   There exists a complete TVS E with a closed linear subspace M such that the quotient TVS E/M is not complete.

(12.4) If E_n is a sequence of metrizable complete TVS, then the product TVS $\prod_{n=1}^{\infty} E_n$ is also complete and metrizable.

*13. Spaces of type (F).

In real life, a metrizable topological group generally comes overtly equipped with a natural metric. In particular, metrizable complete TVS weren't born that way; historically, they were first considered as metrized vector spaces with pleasant invariance and continuity properties hypothesized for the linear operations. Our aim in this section is to show the equivalence of the two formulations, i.e., the equivalence of the concept 'metrizable complete TVS' and the concept 'space of type (F)' defined as follows:

(13.1) Definition.   A *space of type* (F) is a pair (E, d), where E is a vector space over $\mathbb{K}$ $(= \mathbb{R}$ or $\mathbb{C})$ and d is a metric on E such that
   (i) d is a complete metric on E;
   (ii) $d(x, y) = d(x - y, \theta)$ for all $x, y \in E$;
   (iii) if $\lambda_n \in \mathbb{K}$ and $\lambda_n \to 0$, then $d(\lambda_n x, \theta) \to 0$ for each $x \in E$;
   (iv) if $x_n \in E$ and $d(x_n, \theta) \to 0$, then $d(\lambda x_n, \theta) \to 0$ for each $\lambda \in \mathbb{K}$.

We assume that a space (E, d) of type (F) is equipped with the topology derived from d. {Concerning the terminology, see the remarks at the end of the section.}

(**13.2**) Suppose E is a metrizable complete TVS over $\mathbb{K}$. Let d be an (additively) invariant compatible metric on E (6.3). Then (i) (E, d) is a complete metric space (7.8); (ii) $d(x, y) = d(x + (-y), y + (-y)) = d(x - y, \theta)$; (iii) if $\lambda_n \in \mathbb{K}$ and $\lambda_n \to 0$, then $d(\lambda_n x, \theta) \to 0$ for each $x \in E$, by the continuity of $\lambda \mapsto \lambda x$ at $\lambda = 0$; (iv) if $x_n \in E$ and $d(x_n, \theta) \to 0$, then $d(\lambda x_n, \theta) \to 0$ for each $\lambda \in \mathbb{K}$, by the continuity of $x \mapsto \lambda x$ at $x = \theta$. Thus (E, d) is a space of type (*F*), whose topology coincides with the given topology.

Briefly, every metrizable complete TVS is a space of type (*F*). The aim of this section is to prove the converse: If (E, d) is a space of type (*F*), then E, equipped with the topology derived from d, is a metrizable complete TVS. As we shall see, the only nonobvious point is that the mapping $(\lambda, x) \mapsto \lambda x$ is continuous (jointly in its variables).

(**13.3**) **Lemma.**   *Let E be a vector space over $\mathbb{K}$, with a metrizable topology such that* (i) *the topology is compatible with the additive group structure,* (ii) *for each $x \in E$, the mapping $\lambda \mapsto \lambda x$ is continuous at $\lambda = 0$, and* (iii) *for each $\lambda \in \mathbb{K}$, the mapping $x \mapsto \lambda x$ is continuous at $x = \theta$. Then the topology is compatible with the vector space structure, i.e., E is a (metrizable) TVS.*

Proof.   The point is that in the presence of metrizability, condition (iv) of (11.3) is superfluous. Thus, in view of (11.3), it will suffice to verify that $(\lambda, x) \mapsto \lambda x$ is continuous at $(0, \theta)$.

We observe first that the mapping $(\lambda, x) \mapsto \lambda x$ is separately continuous in λ and in x; this follows from (i)–(iii) and the formulas $\lambda x_0 - \lambda_0 x_0 = (\lambda - \lambda_0) x_0$, $\lambda_0 x - \lambda_0 x_0 = \lambda_0 (x - x_0)$.

Let d be any metric (not necessarily invariant) that generates the topology. Assuming $\lambda_n \to 0$ and $d(x_n, \theta) \to 0$, we wish to show that $d(\lambda_n x_n, \theta) \to 0$. Suppose to the contrary that there exists an $\varepsilon > 0$ such that $d(\lambda_n x_n, \theta) \geq \varepsilon$ for infinitely many n. Passing to a subsequence, we can suppose that $d(\lambda_n x_n, \theta) \geq \varepsilon$ for all n. Let U be a neighborhood of θ such that $d(x + y, \theta) < \varepsilon$ for all $x, y \in U$ (this is possible by (i)). We can suppose U to be closed (5.14) and symmetric $(-U = U)$. For each n, let

$$C_n = \{\lambda \in \mathbb{K} : \lambda x_i \in U \quad \text{for all} \quad i \geq n\}$$

$$= \bigcap_{i \geq n} \{\lambda : \lambda x_i \in U\};$$

since each of the sets $\{\lambda : \lambda x_i \in U\}$ is closed (it is the inverse image of the closed set U under the continuous mapping $\lambda \mapsto \lambda x_i$), it follows that the C_n are closed. Moreover,

(*) $$\bigcup_{n=1}^{\infty} C_n = \mathbb{K};$$

for, given any $\lambda \in \mathbb{K}$, one has $\lambda x_i \to \theta$ by (iii), consequently there exists an index n such that $\lambda x_i \in U$ for all $i \geq n$, that is, $\lambda \in C_n$.

According to the relation (*), $\mathbb{K}$ is the union of a sequence of closed sets C_n. It is a classical property of $\mathbb{K}$ (indeed, of any complete metric space (46.6)) that at least one of the sets C_n must have an interior point. Say C_k has an interior point λ_0; for a suitable $\delta > 0$, the open set $B = \{\lambda : |\lambda - \lambda_0| < \delta\}$

is contained in C_k. Let $A = \{\mu : |\mu| < \delta\}$; then A is a symmetric neighborhood of 0, and

$$(**) \qquad\qquad \lambda_0 + A = B \subset C_k.$$

Choose an index n so large that $\lambda_i \in A$ for all $i \geq n$ (possible because $\lambda_i \to 0$) and $\lambda_0 x_i \in U$ for all $i \geq n$ (possible by (iii)). Let $m = \max \{k, n\}$. Since $m \geq n$ we have $\lambda_m \in A$ (by the definition of n) and therefore $\lambda_0 + \lambda_m \in C_k$ by $(**)$; since also $m \geq k$, it then follows from the definition of C_k that $(\lambda_0 + \lambda_m)x_m \in U$. Also, $m \geq n$ implies $\lambda_0 x_m \in U$ (by the definition of n) and therefore $-\lambda_0 x_m \in -U = U$. Writing $x = -\lambda_0 x_m$ and $y = (\lambda_0 + \lambda_m)x_m$, we have $x, y \in U$ and $x + y = \lambda_m x_m$; then

$$d(\lambda_m x_m, \theta) = d(x + y, \theta) < \varepsilon$$

by the definition of U, whereas $d(\lambda_m x_m, \theta) \geq \varepsilon$ by the choice of ε, a contradiction. $\blacksquare$

(13.4) Theorem. *If (E, d) is a space of type (F) then E is a metrizable complete TVS in the topology derived from d.*

Proof. The metric d is additively invariant:

$$d(x + z, y + z) = d((x + z) - (y + z), \theta) = d(x - y, \theta) = d(x, y).$$

It follows that the topology is compatible with the additive group structure (10.4), therefore E is a TVS by the lemma. The completeness of E then results from (7.8) (cf. Section 12). $\blacksquare$

(13.5) Corollary. *Let E be a vector space over $\mathbb{K}$, equipped with a topology. In order that there exist a metric d on E such that (E, d) is a space of type (F) and d generates the given topology, it is necessary and sufficient that E be a metrizable complete TVS.*

Proof. This is immediate from (13.2) and (13.4). $\blacksquare$

Thus, the terms 'space of type (F)' and 'metrizable complete TVS' describe the same class of topologized vector spaces.

(13.6) The term 'espace du type (F)' is used in Banach's classical monograph (1932); such spaces were studied by Fréchet in a 1926 paper. These are the spaces we have called 'spaces of type (F).' Contemporary usage is at variance with the classical terminology. Nowadays, the term 'espace (F)' or 'espace de Fréchet'—in English, 'F-space' or 'Fréchet space'—generally refers to a subclass of the spaces of type (F), namely, the locally convex ones (locally convex TVS are defined in Section 33). In other words, 'Fréchet space' nowadays generally means 'locally convex, metrizable complete TVS.'

Some instructive classical examples of spaces of type (F) are worked out in the next two sections.

Exercises

*(13.7) (V. L. Klee)   Let E be a metrizable TVS over $\mathbb{K}$, and suppose there exists a complete compatible metric. Then there exists a compatible metric d

such that (E, d) is a space of type (F) (in other words, E is a metrizable complete TVS).

(13.8) If (E, d) is a space of type (F) then the metric d' defined by

$$d'(x, y) = \frac{d(x, y)}{1 + d(x, y)}$$

generates the same topology as d, and (E, d') is also a space of type (F).

(13.9) If (E_n, d_n) is a sequence of spaces of type (F), $E = \prod_{n=1}^{\infty} E_n$, and if, for $x = (x_n)$, $y = (y_n)$ in E one defines

$$d(x, y) = \sum_{n=1}^{\infty} \frac{2^{-n} d_n(x_n, y_n)}{1 + d_n(x_n, y_n)},$$

then (E, d) is a space of type (F), and the topology on E derived from d coincides with the product topology. {The special case that $E_n = \mathbb{K}$ for all n, with the usual metric $d_n(\lambda, \mu) = |\lambda - \mu|$, is worked out in the next section.}

(13.10) Let E and F be metrizable TVS over $\mathbb{K}$, and let $T: E \to F$ be linear. (If $\mathbb{K} = \mathbb{C}$, T can be either linear or conjugate-linear.) Let d and δ be additively invariant metrics generating the topologies of E and F. Define

$$d'(x, y) = d(x, y) + \delta(Tx, Ty) \qquad (x, y \in E).$$

Then (i) d' is an additively invariant metric on E; (ii) the d'-topology is compatible with the vector space structure of E; (iii) the d'-topology is finer than the d-topology.

(13.11) The space (s) of Section 14 is locally convex (33.1), hence is a Fréchet space.

*14. The space (s). Theorem (11.9) is a trivial source of examples of infinite-dimensional TVS; in this section we work out an instructive, classical special case.

(14.1) Definition. We write (s) for the set of all sequences $x = (\lambda_n)$ with $\lambda_n \in \mathbb{K}$ $(n = 1, 2, 3, \ldots)$. {As usual, $\mathbb{K} = \mathbb{R}$ or $\mathbb{C}$.} Sequences $x = (\lambda_n)$, $y = (\mu_n)$ are said to be equal if they are equal term-by-term; thus $x = y$ means that $\lambda_n = \mu_n$ for all n. Equipped with the term-by-term linear operations

$$(\lambda_n) + (\mu_n) = (\lambda_n + \mu_n), \qquad \lambda(\lambda_n) = (\lambda\lambda_n),$$

(s) is a vector space over $\mathbb{K}$. {Thus (s) may be regarded as the vector space of all $\mathbb{K}$-valued functions on the set of all positive integers, equipped with the pointwise linear operations.}

The main result of the section is that (s) provides a simple example of an infinite-dimensional, metrizable complete TVS:

(14.2) Theorem. *The formula*

$$d((\lambda_n), (\mu_n)) = \sum_{n=-\infty}^{\infty} 2^{-n} \frac{|\lambda_n - \mu_n|}{1 + |\lambda_n - \mu_n|}$$

defines a complete, invariant metric on the vector space (s). *Viewing the set* (s) *as the Cartesian product of* $\aleph_0$ *copies of* $\mathbb{K}$, *the product topology on* (s) *coincides with the d-topology. Equipped with the d-topology,* (s) *is a metrizable, complete TVS.*

Most of the proof is metric-space-theoretic; for clarity, we break it up into several lemmas.

(14.3) Lemma. *If* α, β, γ *are nonnegative real numbers such that* $\alpha \leq \beta + \gamma$, *then*

$$\frac{\alpha}{1 + \alpha} \leq \frac{\beta}{1 + \beta} + \frac{\gamma}{1 + \gamma}.$$

Proof. Elementary. ∎

(14.4) Lemma. *If* (X, d) *is a metric space, the formula*

$$D(x, y) = \frac{d(x, y)}{1 + d(x, y)}$$

defines an equivalent metric D on X. In particular, a sequence in X is convergent for d iff it is convergent for D. Moreover, a sequence in X is Cauchy for d iff it is Cauchy for D; thus X is complete for d iff it is complete for D.

Proof. In view of (14.3), the triangle inequality for D results from that for d. The rest is straightforward. ∎

(14.5) Lemma. *The product of a sequence of metrizable topological spaces is metrizable. Explicitly, suppose* (X_n, d_n) $(n = 1, 2, 3, \ldots)$ *is a sequence of metric spaces and* $X = \prod_{n=1}^{\infty} X_n$ *is the Cartesian product of the sets* X_n. *For* $x = (x_n)$, $y = (y_n)$ *in* X, *define*

$$d(x, y) = \sum_{n=1}^{\infty} 2^{-n} \frac{d_n(x_n, y_n)}{1 + d_n(x_n, y_n)}.$$

Then (i) *d is a metric on X;* (ii) *the topology on X derived from d coincides with the product topology;* (iii) *a sequence* $x^k \in X$ $(k = 1, 2, 3, \ldots)$ *is Cauchy [convergent to* $x \in X$] *if and only if, for each n, the nth coordinate of* x^k *is Cauchy [converges to the nth coordinate of* x].

Proof. For brevity, we write

$$D_n(x_n, y_n) = \frac{d_n(x_n, y_n)}{1 + d_n(x_n, y_n)} \qquad (x_n, y_n \in X_n);$$

thus D_n is a metric on X_n equivalent to d_n (14.4), and

$$d(x, y) = \sum_{n=1}^{\infty} 2^{-n} D_n(x_n, y_n)$$

for $x = (x_n)$, $y = (y_n)$ in X.

(i) Clear.

(ii) Write τ for the product topology on X (derived from the d_n-topologies) and τ_d for the d-topology; it is to be shown that $\tau = \tau_d$.

If $a = (a_n) \in X$ and $\varepsilon > 0$, we write

$$U_\varepsilon(a) = \{x \in X : d(x, a) < \varepsilon\}.$$

The balls $U_\varepsilon(a)$ are a neighborhood base at a for the d-topology, so the inclusion $\tau_d \subset \tau$ will be proved if we show that $U_\varepsilon(a)$ is a τ-neighborhood of a. Thus, we seek a τ-open set U with $a \in U \subset U_\varepsilon(a)$. Choose a positive integer N such that

$$\sum_{n=N+1}^{\infty} 2^{-n} < \frac{\varepsilon}{2},$$

and let

$$U = \left\{x = (x_n) \in X : \sum_{n=1}^{N} 2^{-n} D_n(x_n, a_n) < \frac{\varepsilon}{2}\right\}$$

(the points of U are unrestricted in the coordinates from $N + 1$ onward). Clearly U is a τ-open set containing a, so it will suffice to show that $U \subset U_\varepsilon(a)$. If $x = (x_n) \in U$ then

$$d(x, a) = \sum_{n=1}^{N} 2^{-n} D_n(x_n, a_n) + \sum_{n=N+1}^{\infty} 2^{-n} D_n(x_n, a_n)$$

$$< \frac{\varepsilon}{2} + \sum_{n=N+1}^{\infty} 2^{-n} < \frac{\varepsilon}{2} + \frac{\varepsilon}{2},$$

thus $x \in U_\varepsilon(a)$. This completes the proof that $\tau_d \subset \tau$.

To prove that $\tau \subset \tau_d$, suppose U is any τ-open set and $a = (a_n) \in U$; it will suffice to find an $\varepsilon > 0$ such that $U_\varepsilon(a) \subset U$. Since U is a τ-neighborhood of a, there exist an index N and open sets $U_1, \ldots, U_N$ in $X_1, \ldots, X_N$, respectively, such that

(1) $$a \in U_1 \times \cdots \times U_N \times \prod_{n=N+1}^{\infty} X_n \subset U.$$

In view of (14.4) one can suppose that, for a suitable $\varepsilon > 0$,

$$U_n = \{x_n \in X_n : 2^{-n} D_n(x_n, a_n) < \varepsilon\} \qquad (n = 1, \ldots, N).$$

Suppose $x = (x_n) \in U_\varepsilon(a)$. For all n, we have

$$2^{-n} D_n(x_n, a_n) \le d(x, a) < \varepsilon;$$

in particular, $x_n \in U_n$ for $n = 1, \ldots, N$, thus $x \in U$ by (1). This completes the proof that $\tau = \tau_d$.

(iii) Let x^k be a sequence in X, say

$$x^k = (x_n^k) \qquad (k = 1, 2, 3, \ldots),$$

and let $x = (x_n) \in X$. We are to show that $x^k \to x$ for the metric d if and only if, for each n, $x_n^k \to x_n$ for the metric d_n. {This could be derived from properties of the product topology, but the following explicit proof can be imitated for the assertion concerning Cauchy sequences.} For brevity, let us write

$$\alpha_{kn} = d_n(x_n{}^k, x_n), \qquad \alpha_k = d(x^k, x)$$

for $k, n = 1, 2, 3, \ldots$. By the definition of d,

$$(2) \qquad \alpha_k = \sum_{n=1}^{\infty} 2^{-n} \frac{\alpha_{kn}}{1 + \alpha_{kn}} \qquad (k = 1, 2, 3, \ldots);$$

the problem is to show that

$$(3) \qquad \alpha_k \to 0 \quad \text{as} \quad k \to \infty$$

if and only if

$$(4) \qquad \text{for each } n, \quad \alpha_{kn} \to 0 \quad \text{as} \quad k \to \infty.$$

Assume (3) holds, and fix n. From (2) we have

$$0 \le \frac{\alpha_{kn}}{1 + \alpha_{kn}} \le 2^n \alpha_k;$$

since $2^n \alpha_k \to 0$ as $k \to \infty$, it follows that $\alpha_{kn} \to 0$ as $k \to \infty$. This proves (4).

Conversely, assume (4) holds, and let $\varepsilon > 0$; we seek an index K such that $\alpha_k \le \varepsilon$ for all $k \ge K$. Choose an index N such that

$$(5) \qquad \sum_{n=N+1}^{\infty} 2^{-n} \le \frac{\varepsilon}{2}.$$

The assumption (4) for $n = 1, \ldots, N$ yields

$$\sum_{n=1}^{N} 2^{-n} \frac{\alpha_{kn}}{1 + \alpha_{kn}} \to 0 \quad \text{as} \quad k \to \infty,$$

thus there exists an index K such that

$$(6) \qquad \sum_{n=1}^{N} 2^{-n} \frac{\alpha_{kn}}{1 + \alpha_{kn}} \le \frac{\varepsilon}{2} \qquad \text{for all } k \ge K.$$

From (2), (5) and (6) we see that $\alpha_k \le \varepsilon$ for all $k \ge K$. This proves (3).

The assertion concerning Cauchy sequences is proved similarly. ∎

Proof of Theorem 14.2. Applying (14.5) with $X_n = \mathbb{K}$ and $d_n(\lambda, \mu) = |\lambda - \mu|$ for all n, we conclude that d is a metric on (s) that generates the product topology. Thus, as a topological-algebraic structure, (s) equipped with the d-topology is the product of a sequence of copies of the one-dimensional TVS $\mathbb{K}$, therefore (s) is also a TVS (11.9). It is obvious from the defining formula that d is (additively) invariant; since, moreover, d is a complete metric, it follows that (s) is a metrizable complete TVS (see (7.8), (11.11) and the definition of metrizable complete TVS in Section 12). ∎

Exercise

(14.6) For each positive integer n, the linear form $\varphi_n \colon (s) \to \mathbb{K}$ that sends x to its nth coordinate is continuous. Fact: Every continuous linear form on (s) is a linear combination of finitely many of the φ_n.

*15. The space (*S*). An example of a metrizable, complete TVS arising in measure theory is worked out in this section. The example is known mainly for the properties it doesn't possess, i.e., it is a useful counterexample (cf. (15.10), (15.11), (15.13), (26.7)). What makes the example especially appealing is the agreeable way in which classical measure-theoretic facts fit into place in the topological-algebraic picture. {Historically, it must have been the reverse: what made the abstract topological-algebraic structures seem worthy of formulation and study was the way in which many known facts fit neatly into place.}

For the rest of the section, $(X, \mathscr{S}, \mu)$ is a finite measure space (cf. [10], [65] for the measure-theoretic terminology); thus X is a set, $\mathscr{S}$ is a σ-ring of subsets of X, and μ is a countably additive, nonnegative real-valued function on $\mathscr{S}$. {Towards the end of the section, we assume, in addition, that $X \in \mathscr{S}$.} For simplicity, we stick to real-valued functions, the extension to complex-valued functions being obvious.

(15.1) Definition. We denote by $\mathscr{M}$ the set of all functions $f: X \to \mathbb{R}$ that are measurable with respect to $\mathscr{S}$. For $f \in \mathscr{M}$ we write

$$N(f) = \{x \in X : f(x) \neq 0\}.$$

We denote by $\mathscr{N}$ the set of all $f \in \mathscr{M}$ that vanish a.e. (with respect to μ), that is, $\mu(N(f)) = 0$. With respect to the pointwise operations, $\mathscr{M}$ is a real algebra and $\mathscr{N}$ is an algebra ideal of $\mathscr{M}$. If $f \in \mathscr{M}$ we write

$$[f] = f + \mathscr{N}$$

for the canonical image of f in the quotient algebra $\mathscr{M}/\mathscr{N}$; thus $[f] = [g]$ if and only if $f = g$ a.e. For brevity, we write $[\mathscr{M}] = \mathscr{M}/\mathscr{N}$; elements of the algebra $[\mathscr{M}]$ are denoted $u, v, w, \ldots$.

If $f \in \mathscr{M}$ then the function

$$\frac{|f|}{1 + |f|}$$

is measurable and bounded, hence integrable (μ is finite); the value of the integral is unaltered if f is replaced by any other function in $[f]$. This paves the way for the following definition:

(15.2) Definition. If $u, v \in [\mathscr{M}]$, say $u = [f]$, $v = [g]$, we define

$$d(u, v) = \int \frac{|f - g|}{1 + |f - g|} \, d\mu.$$

(15.3) Lemma. *d is an additively invariant metric on $[\mathscr{M}]$.*

Proof. If, in (15.2), $d(u, v) = 0$, then the integrand, being nonnegative, must vanish a.e., whence $f = g$ a.e., that is, $u = v$. So the main point to be established is the triangle inequality. Suppose $u = [f]$, $v = [g]$, $w = [h]$.

By (14.3) we have

$$\frac{|f(x) - h(x)|}{1 + |f(x) - h(x)|} \leq \frac{|f(x) - g(x)|}{1 + |f(x) - g(x)|} + \frac{|g(x) - h(x)|}{1 + |g(x) - h(x)|}$$

for all $x \in X$; integration yields $d(u, w) \leq d(u, v) + d(v, w)$.

Obviously $d(u, v) = d(u - v, \theta)$, so d is invariant, that is,

$$d(u + w, v + w) = d(u, v)$$

(cf. (6.1), (11.11)). ∎

The interest of (15.3) is that convergence with respect to d is equivalent to a classical notion:

(15.4) Lemma. *Suppose* $u_n \in [\mathcal{M}]$ *(*$n = 1, 2, 3, \ldots$*) and* $u \in [\mathcal{M}]$*; say* $u_n = [f_n]$ *and* $u = [f]$. *Then* (i) $u_n \to u$ *for* d *iff* $f_n \to f$ *in measure;* (ii) u_n *is Cauchy for* d *iff* f_n *is fundamental in measure.*

Proof. (i) By the invariance of d, we can suppose $u = \theta$; thus the problem is to show that $d(u_n, \theta) \to 0$ iff $f_n \to 0$ in measure.

Suppose $d(u_n, \theta) \to 0$, that is,

$$\int \frac{|f_n|}{1 + |f_n|} \, d\mu \to 0.$$

Let $\eta > 0$ and write $E_n = \{x : |f_n(x)| \geq \eta\}$; it is to be shown that $\mu(E_n) \to 0$. If $|f_n(x)| \geq \eta$ then

$$\frac{|f_n(x)|}{1 + |f_n(x)|} \geq \frac{\eta}{1 + \eta};$$

thus, writing χ_{E_n} for the characteristic function of E_n, we have

$$\chi_{E_n} \frac{|f_n|}{1 + |f_n|} \geq \frac{\eta}{1 + \eta} \chi_{E_n},$$

whence

$$d(u_n, \theta) = \int \frac{|f_n|}{1 + |f_n|} \, d\mu \geq \int_{E_n} \frac{|f_n|}{1 + |f_n|} \, d\mu \geq \frac{\eta}{1 + \eta} \mu(E_n).$$

Thus $\mu(E_n) \leq \eta^{-1}(1 + \eta)d(u_n, \theta)$ for all n, therefore $\mu(E_n) \to 0$.

Conversely, assuming $f_n \to 0$ in measure, we are to show that $d(u_n, \theta) \to 0$. Let $E = \bigcup_{n=1}^{\infty} N(f_n)$. Given any $\varepsilon > 0$, it will suffice to find an index n_0 such that

$$d(u_n, \theta) \leq \varepsilon + \varepsilon\mu(E) \quad \text{for all} \quad n \geq n_0.$$

Let $E_n = \{x : |f_n(x)| \geq \varepsilon\}$. By hypothesis, $\mu(E_n) \to 0$; choose an index n_0 such that $\mu(E_n) \leq \varepsilon$ for all $n \geq n_0$. Fix $n \geq n_0$ and write $g_n = |f_n|(1 + |f_n|)^{-1}$ for brevity. Since $N(g_n) = N(f_n) \subset E$, we have

$$d(u_n, \theta) = \int g_n \, d\mu = \int_E g_n \, d\mu$$

$$= \int_{E \cap E_n} g_n \, d\mu + \int_{E - E_n} g_n \, d\mu.$$

We estimate the two terms separately. Since $g_n \leq 1$ we have

$$\int_{E \cap E_n} g_n \, d\mu \leq \mu(E \cap E_n) \leq \mu(E_n) \leq \varepsilon.$$

For $x \in \complement E_n$ we have $g_n(x) \leq |f_n(x)| < \varepsilon$, therefore

$$\int_{E - E_n} g_n \, d\mu \leq \varepsilon \mu(E - E_n) \leq \varepsilon \mu(E).$$

Thus $d(u_n, \theta) \leq \varepsilon + \varepsilon \mu(E)$.

(ii) The proof is similar to (i). ∎

(15.5) Lemma. *d is a complete metric on $[\mathcal{M}]$.*

Proof. Let u_n be a Cauchy sequence in $[\mathcal{M}]$, say $u_n = [f_n]$. According to (15.4), f_n is fundamental in measure; by the Riesz-Weyl theorem, there exists $f \in \mathcal{M}$ with $f_n \to f$ in measure [**10**, p. 67, Th. 4]. Writing $u = [f]$, we have $u_n \to u$ by (15.4). ∎

From (15.3) and (15.5), we know that the additive group of $[\mathcal{M}]$, equipped with the d-topology, is a metrizable, complete topological group (cf. (10.4), (7.8)). To prove that $[\mathcal{M}]$ is a TVS, we need only look after the scalar multiplication:

(15.6) Theorem. *$[\mathcal{M}]$ is a metrizable, complete TVS.*

Proof. By the preceding remarks, it remains only to show that the mapping $(\alpha, u) \mapsto \alpha u$ $(\alpha \in \mathbb{R}, u \in [\mathcal{M}])$ is continuous. In view of the identity

$$\alpha u - \alpha_0 u_0 = (\alpha - \alpha_0)(u - u_0) + \alpha_0(u - u_0) + (\alpha - \alpha_0)u_0$$

and the continuity of addition, it clearly suffices to prove the following three assertions:

$(\alpha, u) \mapsto \alpha u$ is continuous at $(0, \theta)$;
for each α, $u \mapsto \alpha u$ is continuous at θ;
for each u, $\alpha \mapsto \alpha u$ is continuous at 0.

By metrizability, we may restrict attention to sequential convergence. Thus, assuming $\alpha_n \to 0$, $u_n \to \theta$, $\alpha \in \mathbb{R}$, $u \in [\mathcal{M}]$, it is to be shown that

(i) $\alpha_n u_n \to \theta$,
(ii) $\alpha u_n \to \theta$,
(iii) $\alpha_n u \to \theta$.

We can suppose α and the α_n are nonzero. Say $u = [f_n]$, $u = [f]$.

To prove (i), suppose $\varepsilon > 0$ and let

$$E_n = \{x : |\alpha_n f_n(x)| \geq \varepsilon\} = \{x : |f_n(x)| \geq \varepsilon/|\alpha_n|\};$$

it is to be shown that $\mu(E_n) \to 0$. Let

$$F_n = \{x : |f_n(x)| \geq 1\};$$

from $u_n \to \theta$ we know that $\mu(F_n) \to 0$. Since $\alpha_n \to 0$, there exists an index n_0 such that $\varepsilon/|\alpha_n| \geq 1$ for all $n \geq n_0$; then $n \geq n_0$ implies $E_n \subset F_n$ and hence $\mu(E_n) \leq \mu(F_n)$. It follows at once that $\mu(E_n) \to 0$.

To prove (ii), suppose $\varepsilon > 0$ and let

$$E_n = \{x : |\alpha f_n(x)| \geq \varepsilon\} = \{x : |f_n(x)| \geq \varepsilon/|\alpha|\};$$

then $\mu(E_n) \to 0$ results from the hypothesis $u_n \to \theta$.

To prove (iii), suppose $\varepsilon > 0$ and let

$$E_n = \{x : |\alpha_n f(x)| \geq \varepsilon\} = \{x : |f(x)| \geq \varepsilon/|\alpha_n|\}.$$

Assume to the contrary that $\mu(E_n) \nrightarrow 0$. Passing to a subsequence, we can suppose, for suitable $\eta > 0$, that

(*) $$\mu(E_n) \geq \eta \quad \text{for all } n.$$

Let

$$E = \limsup E_n = \{x : x \in E_n \quad \text{for infinitely many } n\};$$

since μ is finite, it results from (*) that

(**) $$\mu(E) \geq \eta$$

(Arzela-Young theorem [10, p. 53, Th. 2]). In particular, $E \neq \varnothing$. Let $x \in E$. This means that $|f(x)| \geq \varepsilon/|\alpha_n|$ for infinitely many n. Since $\alpha_n \to 0$, we conclude that $|f(x)| = +\infty$, a contradiction. {Even if we were admitting functions that are only finite a.e., (**) would still lead to a contradiction.} ∎

The multiplication in $[\mathscr{M}]$ is also continuous:

(15.7) Theorem. $[\mathscr{M}]$ *is a topological ring.*

Proof. In view of the identity

$$uv - u_0 v_0 = (u - u_0)(v - v_0) + (u - u_0)v_0 + u_0(v - v_0),$$

it suffices to prove the following assertions:

$$(u, v) \mapsto uv \text{ is continuous at } (\theta, \theta);$$

$$\text{for each } v, u \mapsto uv \text{ is continuous at } \theta$$

(recall that multiplication is commutative). Thus, assuming $u_n \to \theta$, $v_n \to \theta$ and $v \in [\mathscr{M}]$, it is to be shown that

(i) $u_n v_n \to \theta$,

(ii) $u_n v \to \theta$.

Say $u_n = [f_n]$, $v_n = [g_n]$, $v = [g]$.

The proof of (i) is immediate from the relation

$$\{x : |f_n(x)g_n(x)| \geq \varepsilon\} \subset \{x : |f_n(x)| \geq \varepsilon\} \cup \{x : |g_n(x)| \geq 1\}.$$

The proof of (ii) is more subtle. Assume to the contrary that $d(u_n v, \theta)$ is not a null sequence. Passing to a subsequence, we can suppose $d(u_n v, \theta) \geq \varepsilon > 0$ for all n. Since $f_n \to 0$ in measure, we have $f_{n_k} \to 0$ a.e. for a suitable subsequence (by the Riesz-Weyl theorem). Then also

$$\frac{|f_{n_k} g|}{1 + |f_{n_k} g|} \to 0 \text{ a.e.},$$

therefore

$$\int \frac{|f_{n_k} g|}{1 + |f_{n_k} g|} \, d\mu \to 0$$

by Lebesgue's bounded convergence theorem [cf. **10**, p. 102, Th. 1]. Thus $d(u_{n_k} v, \theta) \to 0$, contrary to $d(u_{n_k} v, \theta) \geq \varepsilon$. ∎

That's the good side of the story. The pathology lurking in $[\mathscr{M}]$ is brought out by special hypotheses on μ.

(15.8) Theorem. *Suppose μ assumes arbitrarily small values. If V is any neighborhood of θ in $[\mathscr{M}]$, then there exists a nonzero vector u such that $\{\alpha u : \alpha \in \mathbb{R}\} \subset V$.*

Proof. Choose $\varepsilon > 0$ so that $U_\varepsilon(\theta) \subset V$, where $U_\varepsilon(\theta) = \{u : d(u, \theta) < \varepsilon\}$. By the hypothesis on μ, there exists a measurable set E such that $0 < \mu(E) < \varepsilon$. Let $f = \chi_E$, $u = [f]$. For any $\alpha \in \mathbb{R}$, evidently

$$\frac{|\alpha f|}{1 + |\alpha f|} \leq f;$$

integration yields $d(\alpha u, \theta) \leq \int f \, d\mu = \mu(E) < \varepsilon$, thus $\alpha u \in U_\varepsilon(\theta) \subset V$. ∎

An application of (15.8): If μ assumes arbitrarily small values, then the TVS $[\mathscr{M}]$ is not 'normable' (26.7).

(15.9) Lemma. *Suppose μ has the property that, for some $\varepsilon > 0$, each measurable set is the union of finitely many sets of measure $< \varepsilon$. Then $[\mathscr{M}]$ is the closed linear span of the set $\{[f] : f = \chi_E, E \in \mathscr{S}, \mu(E) < \varepsilon\}$.*

Proof. Let $u \in [\mathscr{M}]$, say $u = [f]$. Choose a sequence of simple functions f_n such that $f_n \to f$ pointwise; since μ is finite, it follows that $f_n \to f$ in measure [**10**, p. 60, Th. 1], thus $d([f_n], [f]) \to 0$ by (15.4). By the hypothesis on μ we can suppose, moreover, that each f_n is a linear combination of characteristic functions of sets (disjoint, if we like) of measure $< \varepsilon$. Writing $u_n = [f_n]$, we have $d(u_n, u) \to 0$ with u_n in the linear span of the indicated set. ∎

(15.10) Theorem. *Assume μ has the property that, for every $\varepsilon > 0$, each measurable set is the union of finitely many measurable sets of measure $< \varepsilon$. Then the only continuous linear form $\varphi : [\mathscr{M}] \to \mathbb{R}$ is $\varphi = 0$.*

Proof. Assume to the contrary that $[\mathscr{M}]$ admits a nonzero continuous linear form φ. It follows from the lemma that for each $\varepsilon > 0$ there exists a measurable set E such that $\mu(E) < \varepsilon$ and $\varphi([\chi_E]) \neq 0$. For $n = 1, 2, 3, \ldots,$

choose a measurable set E_n such that $\mu(E_n) < 1/n$ and $\varphi([\chi_{E_n}]) \neq 0$. Write $\alpha_n = \varphi([\chi_{E_n}])$ and set $f_n = (1/\alpha_n)\chi_{E_n}$. If $\eta > 0$ then

$$\{x : |f_n(x)| \geq \eta\} \subset E_n,$$

thus it is clear that $f_n \to 0$ in measure; writing $u_n = [f_n]$, we have $d(u_n, \theta) \to 0$ by (15.4). Since φ is continuous, $\varphi(u_n) \to \varphi(\theta) = 0$. But $\varphi(u_n) = (1/\alpha_n)\varphi([\chi_{E_n}]) = 1$ for all n, a contradiction. ∎

Suppose now that $(X, \mathscr{S}, \mu)$ is a *totally* finite measure space, that is, $X \in \mathscr{S}$ and $\mu(X) < \infty$. Then the constant function 1 is measurable and [1] is a unity element for the ring $[\mathscr{M}]$. We write briefly $[1] = 1$. It is routine to check that an element $u = [f]$ is invertible (i.e., $uv = 1$ for some $v \in [\mathscr{M}]$) if and only if $\mu(\{x : f(x) = 0\}) = 0$ (cf. [10, p. 43, Exer. 2]). The invertible elements of $[\mathscr{M}]$ obviously form an abelian group under multiplication. The following result shows that the topological ring $[\mathscr{M}]$ is generally very different from the topological rings that occur in the theory of 'Banach algebras' discussed in Chapter 6 (cf. 50.5):

(15.11) Theorem. *If μ is totally finite and assumes arbitrarily small values, then the group of invertible elements in $[\mathscr{M}]$ is not an open set in $[\mathscr{M}]$.*

Proof. Let E_n be a sequence of measurable sets such that $\mu(E_n) > 0$, $\mu(E_n) \to 0$. Write $F_n = X - E_n, f_n = \chi_{F_n}, u_n = [f_n]$. If $\varepsilon > 0$ then

$$\{x : |f_n(x) - 1| \geq \varepsilon\} \subset \{x : f_n(x) = 0\} = E_n,$$

thus it is clear that $f_n \to 1$ in measure, that is, $u_n \to 1$. But

$$\mu(\{x : f_n(x) = 0\}) = \mu(E_n) > 0$$

shows that u_n is not invertible. Thus, no neighborhood of 1 can consist entirely of invertible elements. ∎

An example to which all results in this section are applicable: Lebesgue measure on the unit interval [0, 1]. For this example, there is a traditional notation:

(15.12) Definition. If $X = [0, 1]$, $\mathscr{S}$ is the class of Lebesgue-measurable subsets of [0, 1], and μ is Lebesgue measure on $\mathscr{S}$, then the space $[\mathscr{M}]$ of (15.6) is denoted (S).

Exercise

(15.13) With μ as in (15.11), $[\mathscr{M}]$ is not a Q-ring; it follows that the topology of the valued abelian group $[\mathscr{M}]$ (cf. 8.5) cannot be derived from a value $u \mapsto |u|$ satisfying $|uv| \leq |u| \, |v|$ for all u, v. {An element u in a ring is said to be *quasiregular* if there exists an element v such that $u + v - uv = v + u - vu = 0$. A topological ring is called a Q-*ring* if the set of quasiregular elements is open.}

16. Normed spaces, Banach spaces.
Normed spaces are metrizable TVS, with additional geometric structure. To motivate the definition, suppose E is a metrizable TVS, and let d be an additively invariant compatible

metric (6.3). The invariance property $d(x + z, y + z) = d(x, y)$ means that for each vector z, the translation mapping $x \mapsto x + z$ is an isometry. Let us switch to the suggestive value notation (8.5): $|x| = d(x, \theta)$. The notation $|x|$, and its properties $|-x| = |x|$ and $|x + y| \le |x| + |y|$, invite us to think of $|x|$ as the 'length' of the vector x. However, all is not well with scalar multiples; our intuition expects $|\lambda x| = |\lambda| |x|$, which is generally not the case (16.13). When this property does hold (a suggestive name for it is 'absolute homogeneity') the value is called a *norm*:

(16.1) Definition.   Let E be a vector space over $\mathbb{K}$. A *norm* on E is a non-negative, real-valued function $x \mapsto \|x\|$ $(x \in E)$ such that

(i) $\|\theta\| = 0$,
(ii) $\|x\| > 0$ when $x \ne \theta$,
(iii) $\|\lambda x\| = |\lambda| \|x\|$ for all $\lambda \in \mathbb{K}$, $x \in E$,
(iv) $\|x + y\| \le \|x\| + \|y\|$ for all $x, y \in E$.

In words, a norm is *strictly positive, absolutely homogeneous* and *subadditive*.

(16.2) Definition.   A *normed space* (over $\mathbb{K}$) is a pair $(E, \|x\|)$, where E is a vector space over $\mathbb{K}$ and $x \mapsto \|x\|$ is a norm on E. Ordinarily, we say simply that 'E is a normed space'; unless otherwise specified, we tacitly assume that the norm function is notated $\|x\|$. The terms 'real normed space' and 'complex normed space' are used according as $\mathbb{K} = \mathbb{R}$ or $\mathbb{K} = \mathbb{C}$. (Synonyms for 'normed space': normed linear space, normed vector space.)

(16.3) Definition.   If E is a normed space over $\mathbb{K}$, the norm function $x \mapsto \|x\|$ is evidently a value on the additive group of E (8.1), with the extra feature of absolute homogeneity. In particular, $d(x, y) = \|x - y\|$ defines a metric on E that is additively invariant: $d(x + z, y + z) = d(x, y)$ (8.3). The topology derived from d (called the value topology in Section 8) is called the *norm topology* on E. Thus '$x_n \to x$ for the norm topology' means that $\|x_n - x\| \to 0$ as $n \to \infty$; 'x_n is Cauchy for the norm metric' means that $\|x_m - x_n\| \to 0$ as $m, n \to \infty$.

Prior to showing that the norm topology on a normed space is compatible with the vector space structure, we note that the norm function itself has pleasant continuity properties:

(16.4) Proposition.   *If E is a normed space then*

$$| \|x\| - \|y\| | \le \|x - y\|$$

for all $x, y \in E$. Consequently, (i) $x \mapsto \|x\|$ is uniformly continuous with respect to the norm metric, (ii) if $x_n \to x$ for the norm topology, then $\|x_n\| \to \|x\|$, and (iii) if x_n is Cauchy for the norm metric, then $\|x_n\|$ is a Cauchy sequence of real numbers.

Proof.   Since $x = (x - y) + y$, we have $\|x\| \le \|x - y\| + \|y\|$, thus $\|x\| - \|y\| \le \|x - y\|$; interchange of x and y yields $\|y\| - \|x\| \le \|y - x\|$

$= \|x - y\|$, thus $-\|x - y\| \le \|x\| - \|y\| \le \|x - y\|$, which is the desired inequality. The properties (i)–(iii) then follow at once. $\blacksquare$

(16.5) Theorem. *If E is a normed space, then the norm topology on E is compatible with the vector space structure; thus E, equipped with the norm topology, is a metrizable TVS.*

Proof. The norm topology is compatible with the additive group structure by (8.7). Assuming $\lambda_n \to \lambda$ and $x_n \to x$, it is to be shown that $\lambda_n x_n \to \lambda x$. One has

$$\lambda_n x_n - \lambda x = \lambda_n(x_n - x) + (\lambda_n - \lambda)x,$$

therefore

$$\|\lambda_n x_n - \lambda x\| \le |\lambda_n|\, \|x_n - x\| + |\lambda_n - \lambda|\, \|x\|;$$

since $|\lambda_n - \lambda| \to 0$ and $\|x_n - x\| \to 0$, and since $|\lambda_n|$ is bounded, clearly $\|\lambda_n x_n - \lambda x\| \to 0$. $\blacksquare$

The question, of when the topology of a metrizable TVS can be given by a norm, is answered by a theorem of Kolmogorov (see Section 26).

(16.6) If E is a normed space and M is a linear subspace of E, then M, equipped with the relative norm (i.e., the restriction to M of the norm function on E), is also a normed space, and the norm topology on M coincides with the relative topology. Quotients work out well, too:

(16.7) Theorem. *Let E be a normed space and let M be a closed linear subspace of E. For $u \in E/M$ define*

$$\|u\| = \inf\{\|x\| : x \in u\}.$$

Then (i) $u \mapsto \|u\|$ *is a norm on E/M, and* (ii) *the norm topology on E/M coincides with the quotient topology.*

Proof. In view of (8.13), it remains only to show that $\|\lambda u\| = |\lambda|\, \|u\|$. This is obvious if $\lambda = 0$. If $\lambda \ne 0$ then $\lambda u = \{\lambda x : x \in u\}$, therefore $\|\lambda u\| = \inf\{\|\lambda x\| : x \in u\} = \inf\{|\lambda|\, \|x\| : x \in u\} = |\lambda| \inf\{\|x\| : x \in u\} = |\lambda|\, \|u\|$. $\blacksquare$

(16.8) Definition. With notation as in (16.7), we call $\|u\|$ the *quotient norm* on E/M.

(16.9) Definition. A *Banach space* (over $\mathbb{K}$) is a normed space E for which the norm metric $\|x - y\|$ is complete. (The terms 'real Banach space' and 'complex Banach space' are used according as $\mathbb{K} = \mathbb{R}$ or $\mathbb{K} = \mathbb{C}$.)

(16.10) Theorem. *A normed space E is a Banach space if and only if E is a complete TVS for the norm topology.*

Proof. By definition, E is a complete TVS iff its additive topological group is complete (11.10), thus the theorem is immediate from (7.8) and the additive invariance of the metric $\|x - y\|$. $\blacksquare$

The next two theorems specialize the results of Section 12 to the normed space setting.

(16.11) Theorem. *If E is a Banach space and M is a closed linear subspace of E, then M and E/M are also Banach spaces (for the relative norm and the quotient norm, respectively). Moreover, the norm topologies on M and E/M coincide with the relative topology and the quotient topology, respectively.*

Proof. In view of (16.6), (16.7) and (16.10), the theorem is immediate from (12.1). However, this reasoning makes use of the concept of complete TVS and alludes to the Birkhoff–Kakutani theorem; it is preferable to indicate a direct argument, as follows.

The assertion about M is straightforward: a closed subset of a complete metric space is complete.

The quotient norm on E/M yields the quotient topology by (16.7); the completeness of E/M for the metric $\|u - v\|$ can be shown by repeating (in additive notation) the argument in the proof of (9.1). ∎

Conversely:

(16.12) Theorem. *If E is a normed space and if M is a closed linear subspace such that M and E/M are Banach spaces (for the relative norm and the quotient norm, respectively), then E is a Banach space.*

Proof. In view of (16.7) and (16.10), the theorem is immediate from (12.2). For a more direct proof, repeat (in additive notation) the argument in the proof of (9.2). ∎

A substantial number of examples of normed spaces and Banach spaces are worked out in the text; accordingly, to save repetition, the list of examples given in the following exercises is kept to a minimum.

Exercises

(16.13) If $(E, \|x\|)$ is a normed space, then the formula $|x| = \|x\|/(1 + \|x\|)$ defines a value on the additive group of E such that the value topology coincides with the norm topology. However, the value $x \mapsto |x|$ is not absolutely homogeneous.

(16.14) If $E = \mathbb{K}^r$ and $x = (\lambda_k) \in E$, then the formula

$$\|x\|_\infty = \max\{|\lambda_k| : 1 \le k \le r\}$$

defines a Banach space norm on E. Another norm is

$$\|x\|_1 = \sum_1^r |\lambda_k|.$$

The relation

$$\|x\|_\infty \le \|x\|_1 \le r\|x\|_\infty$$

implies that the two norm topologies coincide (with the product topology). If p is a fixed real number, $p > 1$, then the formula

$$\|x\|_p = \left(\sum_1^r |\lambda_k|^p\right)^{1/p}$$

defines yet another norm which generates the same topology (cf. 2.14). {The subadditivity property $\|x + y\|_p \le \|x\|_p + \|y\|_p$ is known as *Minkowski's inequality*; for $p = 2$, it is *Cauchy's inequality*.}

(16.15)   Let T be a nonempty set. A function $x: T \to \mathbb{K}$ is said to be *bounded* if there exists a constant M such that $|x(t)| \leq M$ for all $t \in T$. Denote by $\mathscr{B}_{\mathbb{K}}(T)$ the set of all bounded functions x. With the pointwise linear operations

$$(x + y)(t) = x(t) + y(t), \qquad (\lambda x)(t) = \lambda x(t),$$

$\mathscr{B}_{\mathbb{K}}(T)$ is a vector space over $\mathbb{K}$. Define

$$\|x\|_{\infty} = \sup \{|x(t)| : t \in T\}$$

for $x \in \mathscr{B}_{\mathbb{K}}(T)$. Then $(\mathscr{B}_{\mathbb{K}}(T), \|x\|_{\infty})$ is a Banach space.

(16.16)   Let T be a topological space, $\mathscr{B}_{\mathbb{K}}(T)$ the Banach space described in (16.15). Let $\mathscr{C}_{\mathbb{K}}^{\infty}(T)$ be the set of all $x \in \mathscr{B}_{\mathbb{K}}(T)$ that are continuous. Then $\mathscr{C}_{\mathbb{K}}^{\infty}(T)$ is a closed linear subspace of $\mathscr{B}_{\mathbb{K}}(T)$ and is therefore a Banach space for the norm $\|x\|_{\infty}$.

(16.17)   Let $I = [a, b]$ be a nondegenerate closed interval of $\mathbb{R}$, and let $\mathscr{C}_{\mathbb{K}}(I)$ be the linear space of all continuous functions $x: I \to \mathbb{K}$ (cf. (16.16); such functions are bounded, by the compactness of I, so the superscript ∞ is superfluous). For $x \in \mathscr{C}_{\mathbb{K}}(I)$ define

$$\|x\|_1 = \int_a^b |x(t)| \, dt$$

(Riemann integral); then $(\mathscr{C}_{\mathbb{K}}(I), \|x\|_1)$ is a normed space, but not a Banach space.

(16.18)   Any normed space E may be regarded as a dense linear subspace of a suitable Banach space F; indeed, if F is the metric space completion of E, then the linear operations of E may be extended to F.

(16.19)   In a normed space, $\{x : \|x - x_0\| \leq \varepsilon\}$ is the closure of $\{x : \|x - x_0\| < \varepsilon\}$ (here x_0 is any fixed vector and $\varepsilon > 0$).

17. Neighborhoods of θ.

This section is the vector space analogue of Section 3: we wish to characterize, in vector-space-theoretic terms, the class of neighborhoods of θ in a TVS. The characterization involves two geometrical concepts (balanced set, absorbent set) of general utility in vector spaces.

(17.1) Definition.   Let E be a vector space over $\mathbb{K}$. A subset A of E is said to be *balanced* if $\lambda x \in A$ whenever $x \in A$, $\lambda \in \mathbb{K}$ and $|\lambda| \leq 1$ (briefly, $\lambda A \subset A$ whenever $|\lambda| \leq 1$).

(17.2)   Suppose A is balanced. If $|\lambda| \leq |\mu|$ then $\lambda A \subset \mu A$. {Proof: This is trivial if $\mu = 0$, and if $\mu \neq 0$ then $|\lambda \mu^{-1}| \leq 1$, $\lambda \mu^{-1} A \subset A$.} If $|\gamma| = 1$ then $\gamma A = A$. {Proof: Apply the preceding remark to $\lambda = \gamma$, $\mu = 1$ and to $\lambda = 1$, $\mu = \gamma$.} In particular, A is symmetric, i.e., $A = -A$.

(17.3) Definition.   Let E be a vector space over $\mathbb{K}$, S any subset of E. Obviously (i) there exist balanced sets (for example E) containing S, and (ii) the intersection of any family of balanced sets is balanced. It follows that there exists a smallest balanced set A containing S (namely, A is the intersection of all balanced subsets of E that contain S); it is called the *balanced hull* of S, denoted $A = \text{bal } S$. Thus, A is characterized by the following properties: (1) A is balanced, (2) $S \subset A$, and (3) if B is balanced and $S \subset B$ then $A \subset B$. It is easy to describe A explicitly:

(17.4) Theorem. *If S is a subset of a vector space E over $\mathbb{K}$, then the balanced hull of S is the set $\{\lambda x : x \in S, \lambda \in \mathbb{K}, |\lambda| \leq 1\}$.*

Proof. Let A be the balanced hull of S. Writing $D = \{\lambda \in \mathbb{K} : |\lambda| \leq 1\}$ (D is a closed interval or a closed disc, according as $\mathbb{K} = \mathbb{R}$ or $\mathbb{K} = \mathbb{C}$), the problem is to show that $A = DS$. Since $S \subset A$ one has $DS \subset DA \subset A$ by the definition of balanced set; on the other hand, it is clear that DS is a balanced set containing S, therefore $A \subset DS$ by the definition of A. ∎

(17.5) Theorem. *In a TVS, the closure of a balanced set is balanced.*

Proof. Suppose A is a balanced subset of the TVS E over $\mathbb{K}$. Let $\lambda \in \mathbb{K}$, $|\lambda| \leq 1$. Then $\lambda A \subset A$ and the problem is to show that $\lambda \overline{A} \subset \overline{A}$; this is immediate from the continuity of the mapping $x \mapsto \lambda x$ $(x \in E)$. ∎

(17.6) Definition. Let E be a TVS over $\mathbb{K}$, S any subset of E. Obviously (i) there exist closed balanced sets (for example, E) containing S, and (ii) the intersection of any family of closed balanced sets is closed and balanced. It follows that there exists a smallest closed balanced set containing S; it is called the *closed balanced hull* of S.

(17.7) Corollary. *If S is a subset of a TVS E, and A is the balanced hull of S, then the closed balanced hull of S is $\overline{A}$.*

Proof. Let B be the closed balanced hull of S. In particular, B is a balanced superset of S, so $A \subset B$; then $\overline{A} \subset \overline{B} = B$. On the other hand, $\overline{A}$ is a closed and balanced (17.5) superset of S, therefore $B \subset \overline{A}$. ∎

In any TVS, the closed balanced neighborhoods of θ are basic:

(17.8) Lemma. *Let E be a TVS over $\mathbb{K}$, $\mathscr{B}$ any fundamental system of neighborhoods of θ. Let $\mathscr{B}'$ denote the class of all closed balanced hulls of the sets in $\mathscr{B}$. Then $\mathscr{B}'$ is also a fundamental system of neighborhoods of θ.*

Proof. Given any neighborhood V of θ, we seek a neighborhood $S \in \mathscr{B}$ whose closed balanced hull is contained in V.

Let W be a closed neighborhood of θ such that $W \subset V$ (5.14). By the continuity of $(\lambda, x) \mapsto \lambda x$ at $(0, \theta)$, there exist $\varepsilon > 0$ and $U \in \mathscr{B}$ such that $\lambda x \in W$ whenever $|\lambda| \leq \varepsilon$ and $x \in U$. Thus, writing $D_\varepsilon = \{\lambda : |\lambda| \leq \varepsilon\}$, we have $D_\varepsilon U \subset W$. Since $D_\varepsilon U$ is a neighborhood of θ (e.g., it contains εU) there exists $S \in \mathscr{B}$ with $S \subset D_\varepsilon U$. Thus,

$$S \subset D_\varepsilon U \subset W \subset V.$$

Let A be the balanced hull of S. Since $D_\varepsilon U$ is obviously balanced, we have $A \subset D_\varepsilon U \subset W \subset V$ and therefore

$$\overline{A} \subset \overline{W} = W \subset V.$$

By (17.7), $\overline{A}$ is the closed balanced hull of S; thus $V \supset \overline{A} \in \mathscr{B}'$. ∎

(17.9) Theorem. *Let E be any TVS over $\mathbb{K}$ and let $\mathscr{B}$ be the class of all neighborhoods of θ that are closed and balanced. Then* (i) *$\mathscr{B}$ is a fundamental system of neighborhoods of θ, and* (ii) *$\lambda V \in \mathscr{B}$ for all $V \in \mathscr{B}$, $\lambda \in \mathbb{K}$, $\lambda \neq 0$.*

Proof. $\mathscr{B}$ is a neighborhood base at θ by (17.8). If $\lambda \in \mathbb{K}$, $\lambda \neq 0$, the mapping $x \mapsto \lambda x$ obviously preserves closed sets, balanced sets, and neighborhoods of θ; in particular, $V \in \mathscr{B}$ implies $\lambda V \in \mathscr{B}$. ∎

(17.10) Definition. Let E be a vector space over $\mathbb{K}$. A subset A of E is said to be *absorbent* if, for each $x \in E$, there exists an $\varepsilon > 0$ such that $\lambda x \in A$ whenever $|\lambda| \leq \varepsilon$. It follows (from either $\lambda \theta = \theta$ or $0x = \theta$) that $\theta \in A$.

(17.11) Lemma. *In a TVS, every neighborhood of θ is absorbent.*

Proof. Let V be a neighborhood of θ and let $x \in E$. Since $\lambda \mapsto \lambda x$ is continuous at $\lambda = 0$, and since $0x = \theta$, there exists an $\varepsilon > 0$ such that $\lambda x \in V$ whenever $|\lambda| \leq \varepsilon$. ∎

If E is any TVS over $\mathbb{K}$ and $\mathscr{B}$ is the class of all closed balanced neighborhoods of θ, then, since $\mathscr{B}$ is a neighborhood base at θ (17.9), $V \in \mathscr{B}$ implies that there exists $W \in \mathscr{B}$ with $W + W \subset V$ (3.1). Moreover, every $V \in \mathscr{B}$ is absorbent (17.11). Conversely:

(17.12) Theorem. *Let E be a vector space over $\mathbb{K}$, and suppose $\mathscr{B}$ is a base of a filter of subsets of E, such that* (i) *every $V \in \mathscr{B}$ is balanced and absorbent,* (ii) *$\lambda V \in \mathscr{B}$ whenever $V \in \mathscr{B}$, $\lambda \in \mathbb{K}$, $\lambda \neq 0$, and* (iii) *$V \in \mathscr{B}$ implies there exists $W \in \mathscr{B}$ with $W + W \subset V$. Then there exists a unique topology τ on E such that* (1) *τ is compatible with the vector space structure of E, and* (2) *$\mathscr{B}$ is a fundamental system of τ-neighborhoods of θ.*

Proof. If $V \in \mathscr{B}$ then $\theta \in V$ (17.10) and $V = -V$ (17.2); citing (iii), it follows from (3.3) that there exists a unique topology τ on E such that (1′) τ is compatible with the additive group structure of E, and (2) $\mathscr{B}$ is a fundamental system of τ-neighborhoods of θ. It remains only to show that τ is compatible with the vector space structure of E. In view of (11.3) it suffices to show that (a) $(\lambda, x) \mapsto \lambda x$ is continuous at $(0, \theta)$, and that, for fixed $\lambda_0 \in \mathbb{K}$ and $x_0 \in E$, (b) $\lambda \mapsto \lambda x_0$ is continuous at $\lambda = 0$, and (c) $x \mapsto \lambda_0 x$ is continuous at $x = \theta$.

(a) Let $V \in \mathscr{B}$. Thus V is a basic neighborhood of $\theta = 0\theta$. If $D = \{\lambda : |\lambda| \leq 1\}$ then $DV \subset V$ because V is balanced; since D is a neighborhood of 0, this proves (a).

(b) Let $V \in \mathscr{B}$. Thus V is a basic neighborhood of $\theta = 0x_0$. Since V is absorbent, there exists $\varepsilon > 0$ such that $\lambda x_0 \in V$ whenever $|\lambda| \leq \varepsilon$. Since $D_\varepsilon = \{\lambda : |\lambda| \leq \varepsilon\}$ is a neighborhood of 0, this proves (b).

(c) The assertion is trivial if $\lambda_0 = 0$. Assume $\lambda_0 \neq 0$ and let $V \in \mathscr{B}$. Thus V is a basic neighborhood of $\theta = \lambda_0 \theta$. Setting $W = \lambda_0^{-1}V$, we have $\lambda_0 W = V$; since W is a neighborhood of θ by (ii), this proves (c). ∎

A useful application (which is also easy to prove directly):

(17.13) Corollary. *If $f: E \to F$ is a linear mapping, where E is a vector space over $\mathbb{K}$ and F is a TVS over $\mathbb{K}$, then the initial topology for f is compatible with the vector space structure of E.*

Proof. Equip E with the initial topology for f, i.e., the coarsest topology on E for which f is continuous; the open sets of E are the sets $f^{-1}(U)$, where U is open in F.

Let $\mathscr{C}$ be the class of all balanced neighborhoods of θ in F. By (17.8), $\mathscr{C}$ is a neighborhood base at θ; therefore, for each $y \in F$, the class $\{y + V : V \in \mathscr{C}\}$ is a neighborhood base at y. In particular, for each $x \in E$, the class $\{f(x) + V : V \in \mathscr{C}\}$ is a neighborhood base at $f(x)$; it follows from the definition of the initial topology that the class

$$\mathscr{B}_x = \{f^{-1}(f(x) + V) : V \in \mathscr{C}\}$$

is a neighborhood base at x. Since $f^{-1}(f(x) + V) = x + f^{-1}(V)$ by the linearity of f, we have

(*) $$\mathscr{B}_x = \{x + f^{-1}(V) : V \in \mathscr{C}\}.$$

In particular, the class

$$\mathscr{B}_\theta = \{f^{-1}(V) : V \in \mathscr{C}\}$$

is a fundamental system of neighborhoods of θ in E. It is easy to see from the properties of $\mathscr{C}$, and the linearity of f, that $\mathscr{B}_\theta$ satisfies the conditions (i)–(iii) of (17.12), therefore there is a unique topology τ on E such that $\mathscr{B}_\theta$ is a fundamental system of τ-neighborhoods of θ, and τ is compatible with the vector space structure; in view of (*), the initial topology for f coincides with τ (cf. 2.9) and is therefore compatible with the vector space structure. ∎

In the same vein:

(17.14) Theorem. *If $f: E \to F$ is a surjective linear mapping, where E is a TVS over $\mathbb{K}$ and F is a vector space over $\mathbb{K}$, then the final topology for f is compatible with the vector space structure of F.*

Proof. The topology τ in question is the finest topology on F for which f is continuous: a subset U of F is τ-open if and only if $f^{-1}(U)$ is open in E.

Let $N = \{x \in E : f(x) = \theta\}$ be the null space of f, let $\pi: E \to E/N$ be the canonical mapping, and let $g: E/N \to F$ be the unique bijective linear mapping such that $f = g \circ \pi$. Equip E/N with the quotient topology, i.e., the final topology for the mapping π; thus, a subset A of E/N is open if and only if $\pi^{-1}(A)$ is open in E.

Let $U \subset F$. Since $U = g(g^{-1}(U)) = (g^{-1})^{-1}(g^{-1}(U))$, U is open for the initial topology of g^{-1} iff $g^{-1}(U)$ is open in E/M iff $\pi^{-1}(g^{-1}(U))$ is open in E, i.e., $f^{-1}(U)$ is open in E, i.e., U is τ-open. Thus τ coincides with the initial topology for g^{-1}; since E/N is a TVS (11.8), it follows from (17.13) that τ is compatible with the vector space structure of F. ∎

Exercises

(17.15) The word "surjective" in (17.14) cannot be omitted.

(17.16) If E is a separated TVS over $\mathbb{K}$, and S is a compact subset of E, then the balanced hull of S is compact and coincides with the closed balanced hull of S.

(17.17) If A is an absorbent subset of the vector space E over $\mathbb{K}$, then A is a system of generators for E (i.e., the linear subspace spanned by A is E).

(17.18) If A is a balanced subset of a vector space E over $\mathbb{K}$, the following conditions are equivalent: (a) A is absorbent; (b) for each $x \in E$ there exists $\lambda \in \mathbb{K}$, $\lambda \neq 0$, such that $\lambda x \in A$.

*18. Continuity of multilinear mappings.   If E and F are topological vector spaces over $\mathbb{K}$ and if $f: E \to F$ is a linear mapping, a simple criterion for the continuity of f is given by (5.1): f is continuous iff it is continuous at zero. For vector functions of more than one variable, the situation is more complicated; indeed, the appropriate notion of linearity is more complicated:

(18.1) Definition.   A mapping $f: E_1 \times \cdots \times E_n \to F$, where $E_1, \ldots, E_n$, F are vector spaces over $\mathbb{K}$, is said to be *multilinear* if it is linear in each of its variables separately; this means, precisely, that for each index k and for each selection of $n - 1$ vectors $a_j \in E_j$ $(j \neq k)$, the mapping $E_k \to F$ defined by

$$x_k \mapsto f(a_1, \ldots, a_{k-1}, x_k, a_{k+1}, \ldots, a_n) \qquad (x_k \in E_k)$$

is linear.

Our aim in this section is to show that the simple criterion for continuity mentioned above extends to multilinear mappings:

(18.2) Theorem.   *Let* $f: E_1 \times \cdots \times E_n \to F$ *be a multilinear mapping, where* $E_1, \ldots, E_n$, F *are TVS over* $\mathbb{K}$ *and* $E_1 \times \cdots \times E_n$ *bears the product topology. Then* f *is a continuous mapping if and only if it is continuous at* $(\theta, \ldots, \theta)$.

The fact that $E_1 \times \cdots \times E_n$ has also a vector space structure is immaterial here; what counts is the product topological structure alone. It is useful to separate out a combinatorial lemma that depends only on the additive structure:

(18.3) Lemma.   *Let* $f: E_1 \times \cdots \times E_n \to F$ *be a multilinear mapping, where* $E_1, \ldots, E_n$, F *are vector spaces over* $\mathbb{K}$. *For each* $k = 1, \ldots, n$, *fix a pair of vectors* a_k, z_k *in* E_k. *For each subset* H *of the index set* $I = \{1, \ldots, n\}$, *define a vector* u_H *in* F *by*

$$u_H = f(y_1, \ldots, y_n),$$

where $y_k = a_k$ *for* $k \in H$ *and* $y_k = z_k$ *for* $k \notin H$. *Then*

$$f(a_1 + z_1, \ldots, a_n + z_n) = \sum_H u_H,$$

where H varies over all subsets of I.

Proof. In particular, $u_I = f(a_1, \ldots, a_n)$ and $u_\varnothing = f(z_1, \ldots, z_n)$. If $n = 1$ then $f: E_1 \to F$ is a linear mapping and

$$\sum_H u_H = u_I + u_\varnothing = f(a_1) + f(z_1) = f(a_1 + z_1).$$

Assume inductively that the desired formula holds for $n - 1$. Invoking additivity in the first coordinate, we have

(i) $f(a_1 + z_1, \ldots, a_n + z_n) =$

$$f(a_1, a_2 + z_2, \ldots, a_n + z_n) + f(z_1, a_2 + z_2, \ldots, a_n + z_n).$$

Since the mapping

$$(x_2, \ldots, x_n) \mapsto f(a_1, x_2, \ldots, x_n)$$

is multilinear from $E_2 \times \cdots \times E_n$ to F, the inductive hypothesis yields

(ii) $$f(a_1, a_2 + z_2, \ldots, a_n + z_n) = \sum_J u_J,$$

where J ranges over all subsets of I with $1 \in J$. Similarly,

(iii) $$f(z_1, a_2 + z_2, \ldots, a_n + z_n) = \sum_K u_K,$$

where K ranges over all subsets of I with $1 \notin K$. Substitution of (ii) and (iii) into (i) yields the desired formula. ∎

Proof of Theorem (18.2). Fix $(a_1, \ldots, a_n) \in E_1 \times \cdots \times E_n$; assuming that f is continuous at $(\theta, \ldots, \theta)$, let us show that f is continuous at $(a_1, \ldots, a_n)$. Given a neighborhood W of θ in F, we seek neighborhoods $V_1, \ldots, V_n$ of θ in $E_1, \ldots, E_n$ such that

(*) $$f((a_1 + V_1) \times \cdots \times (a_n + V_n)) \subset f(a_1, \ldots, a_n) + W.$$

Let U be a neighborhood of θ in F such that

(1) $$U + \cdots + U \subset W,$$

with $2^n - 1$ terms on the left (3.1); we can suppose that U is balanced, i.e., $\gamma U \subset U$ whenever $|\gamma| \leq 1$ (17.9). Since f is continuous at $(\theta, \ldots, \theta)$, there exist neighborhoods $U_1, \ldots, U_n$ of θ in $E_1, \ldots, E_n$ such that

(2) $$f(U_1 \times \cdots \times U_n) \subset U.$$

For each $k = 1, \ldots, n$, choose a nonzero scalar λ_k such that $\lambda_k a_k \in U_k$ (17.11); let λ be a scalar such that $|\lambda| > 1$ and $|\lambda| > |\lambda_k|^{-1}$ for all k. Define

(3) $$V_k = \lambda^{-n} U_k \qquad (k = 1, \ldots, n);$$

it will be shown that the neighborhoods $V_1, \ldots, V_n$ satisfy (*).

Choose $z_k \in V_k$ $(k = 1, \ldots, n)$; the problem is to show that

$$f(a_1 + z_1, \ldots, a_n + z_n) - f(a_1, \ldots, a_n) \in W.$$

Adopting the notations of the lemma, we have

$$f(a_1 + z_1, \ldots, a_n + z_n) - f(a_1, \ldots, a_n) = \sum_{H \neq I} u_H;$$

in view of (1), it will suffice to show that $u_H \in U$ for all $H \neq I$.

Fix $H \subset I$, $H \neq I$, and let p be the number of indices omitted by H; thus $p = \operatorname{card}(I - H) \geq 1$. According to the notations of the lemma, we have

$$u_H = f(y_1, \ldots, y_n),$$

where $y_k = a_k$ for $k \in H$ and $y_k = z_k$ for $k \notin H$. For each $k = 1, \ldots, n$, define $x_k \in U_k$ as follows. If $k \in H$ then $y_k = a_k = \lambda_k^{-1}(\lambda_k a_k)$; setting $x_k = \lambda_k a_k$, we have $x_k \in U_k$ by the definition of λ_k. If $k \notin H$ then $y_k = z_k \in V_k = \lambda^{-n} U_k$, therefore $y_k = \lambda^{-n} x_k$ for suitable $x_k \in U_k$. Then $y_k = \lambda_k^{-1} x_k$ for $k \in H$ and $y_k = \lambda^{-n} x_k$ for $k \notin H$; since f is homogeneous,

$$u_H = f(y_1, \ldots, y_n) = \left(\prod_{k \in H} \lambda_k^{-1}\right)(\lambda^{-n})^p f(x_1, \ldots, x_n).$$

Thus, setting $\mu = \left(\prod_{k \in H} \lambda_k^{-1}\right)(\lambda^{-n})^p$, we have

(4) $u_H = \mu f(x_1, \ldots, x_n),$

where $f(x_1, \ldots, x_n) \in U$ by (2); to show that $u_H \in U$ it will suffice, in view of (4) and the fact that U is balanced, to show that $|\mu| < 1$. We have

$$|\mu| = |\lambda|^{-np} \prod_{k \in H} |\lambda_k|^{-1} \leq |\lambda|^{-np} |\lambda|^{n-p} = |\lambda|^{-np+n-p}$$

by the definition of λ; $|\mu| < 1$ results from the fact that $|\lambda| > 1$ and the exponent $-np + n - p = n - p(n + 1)$ is < 0. ∎

Exercises

(18.4) If A is an algebra over $\mathbb{K}$, with a topology τ such that (i) τ is compatible with the vector space structure of A, and (ii) the multiplication mapping $(x, y) \mapsto xy$ is continuous at (θ, θ), then τ is compatible with the algebra structure of A, that is, (A, τ) is a topological algebra in the sense of (1.7).

(18.5) Let r be a positive integer and let $E = \mathbb{K}^r$. The determinant function on $r \times r$ matrices may be viewed as a multilinear mapping

$$\det: E \times \cdots \times E \to \mathbb{K}$$

(r factors in the product), where $\det(u_1, \ldots, u_r)$ is the determinant of the matrix whose column vectors are $u_1, \ldots, u_r$. {The determinant of a matrix is a polynomial function of n^2 variables, so one need not resort to (18.2) to establish its continuity.}

19. TVS defined by a family of linear mappings.

We are concerned here with the notion of initial topology, in the vector space setting.

(19.1) First, let us review the notion for general topological spaces. One is given a set E, a family $(E_\iota)_{\iota \in I}$ of topological spaces, and a family of mappings $f_\iota: E \to E_\iota$ ($\iota \in I$). The *initial topology* on E for the family of mappings $(f_\iota)_{\iota \in I}$ is the coarsest topology on E for which all of the mappings f_ι are continuous; a base for the initial topology is given by the sets

$$\bigcap_{\iota \in J} f_\iota^{-1}(U_\iota),$$

where J is a finite subset of I and, for each $\iota \in J$, U_ι is an open set in E_ι. An alternative description of this topology is as follows. Let $F = \prod_{\iota \in I} E_\iota$ be the product topological space and let

$$f: E \to F$$

be the mapping defined by $f(x) = (f_\iota(x))_{\iota \in I}$. A base for the topology of F is given by the sets

$$U = \prod_{\iota \in I} U_\iota,$$

where, for a suitable finite subset J of I, $U_\iota = E_\iota$ for $\iota \notin J$, and U_ι is an open subset of E_ι for $\iota \in J$; evidently

$$f^{-1}(U) = \bigcap_{\iota \in J} f_\iota^{-1}(U_\iota),$$

thus the initial topology for the mapping f coincides with the initial topology for the family of mappings $(f_\iota)_{\iota \in I}$. Summarizing:

(19.2) Lemma. *If E is a set, $(E_\iota)_{\iota \in I}$ is a family of topological spaces, and $f_\iota: E \to E_\iota$ ($\iota \in I$) is a family of mappings, then the initial topology for the family $(f_\iota)_{\iota \in I}$ coincides with the initial topology for the mapping*

$$f: E \to \prod_{\iota \in I} E_\iota$$

defined by $f(x) = (f_\iota(x))_{\iota \in I}$.

Continuing with the above notations, suppose in addition that E is a vector space over $\mathbb{K}$, the E_ι are TVS over $\mathbb{K}$, and the f_ι are linear mappings. Then F, with the product vector space structure and the product topology, is a TVS (11.9), and the mapping $f: E \to F$ is obviously linear. The initial topology for f is compatible with the vector space structure of E (17.13); in view of (19.2), we have proved the following:

(19.3) Theorem. *If E is a vector space over $\mathbb{K}$, $(E_\iota)_{\iota \in I}$ is a family of TVS over $\mathbb{K}$, and $f_\iota: E \to E_\iota$ ($\iota \in I$) is a family of linear mappings, then the initial topology for the family $(f_\iota)_{\iota \in I}$ is compatible with the vector space structure of E.*

(19.4) Corollary. *The supremum of a family of compatible topologies is compatible. More precisely, if E is a vector space over $\mathbb{K}$, and $(\tau_\iota)_{\iota \in I}$ is a family of topologies on E that are compatible with the vector space structure of E, then the supremum τ of the topologies τ_ι is also compatible with the vector space structure of E.*

Proof. For each ι, view τ_ι as the class of open sets for that topology; thus, τ is the coarsest topology on E that contains every τ_ι. Let E_ι denote the

TVS (E, τ_ι), and let $f_\iota : E \to E_\iota$ be the identity mapping $f_\iota(x) = x$. Then $f_\iota^{-1}(\tau_\iota) = \tau_\iota$ shows that τ is the coarsest topology for which all the f_ι are continuous, i.e., τ is the initial topology for the family $(f_\iota)_{\iota \in I}$. Quote (19.3). ∎

(19.5) An especially important application of (19.3) is the case of the topology on a vector space E defined by a family of linear mappings $f_\iota : E \to \mathbb{K}$ ($\iota \in I$), with $\mathbb{K}$ regarded as a one-dimensional TVS over itself; such topologies, called 'weak topologies,' will be studied in detail later on. For the present, we note the simplification that occurs when the E_ι coincide. In the general case, a basic neighborhood of θ in E is given by

$$\bigcap_{\iota \in J} f_\iota^{-1}(V_\iota),$$

where J is a finite subset of I and, for each $\iota \in J$, V_ι is a neighborhood of θ in E_ι. Suppose, in addition, that the E_ι coincide, i.e., suppose that there exists a TVS G such that $E_\iota = G$ for all $\iota \in I$; then

$$V = \bigcap_{\iota \in J} V_\iota$$

is also a neighborhood of θ in G, and

$$\bigcap_{\iota \in J} f_\iota^{-1}(V) \subset \bigcap_{\iota \in J} f_\iota^{-1}(V_\iota),$$

therefore the sets

$$\bigcap_{\iota \in J} f_\iota^{-1}(V)$$

are a fundamental system of neighborhoods of θ in E. Summarizing:

(19.6) Theorem. *If E is a vector space over $\mathbb{K}$, G is a TVS over $\mathbb{K}$, and $f_\iota : E \to G$ ($\iota \in I$) is a family of linear mappings, then the initial topology for the family $(f_\iota)_{\iota \in I}$ is compatible with the vector space structure of E, and a basic neighborhood of θ in E is given by*

$$\{x \in E : f_\iota(x) \in V \quad \text{for all} \quad \iota \in J\},$$

where V is a basic neighborhood of θ in G, and J is a finite subset of I. In particular, if $G = \mathbb{K}$ then a basic neighborhood of θ in E is given by

$$\{x \in E : |f_\iota(x)| \leq \varepsilon \quad \text{for all} \quad \iota \in J\},$$

where $\varepsilon > 0$ and J is a finite subset of I.

It is important to know when the initial topology described in (19.3) is separated:

(19.7) Theorem. *Let E be a vector space over $\mathbb{K}$, $(E_\iota)_{\iota \in I}$ a family of separated TVS over $\mathbb{K}$, and $f : E \to E_\iota$ ($\iota \in I$) a family of linear mappings. Equip E with the initial topology for the family $(f_\iota)_{\iota \in I}$. In order that E be separated, it is necessary and sufficient that the intersection of the null spaces of the f_ι be $\{\theta\}$, that is, if $x \in E$, $x \neq \theta$, then there exist an index ι such that $f_\iota(x) \neq \theta$.*

Proof. Write $f: E \to F = \prod_{\iota \in I} E_\iota$, $f(x) = (f_\iota(x))_{\iota \in I}$, as in the proof of (19.3). Since the E_ι are separated, so is F; it follows easily that the initial topology for f is separated if and only if f is injective. ∎

Exercises

(19.8) An 'open-set-free' proof of (19.2) can be given making use of the coordinate projections $\pi_\iota: F \to E_\iota$, and the fact that the product topology on F is the initial topology for the family of mappings $(\pi_\iota)_{\iota \in I}$.

(19.9) The hypothesis that the E_ι are separated cannot be omitted in (19.7).

(19.10) There is an obvious variant of the last conclusion of (19.6) with $\mathbb{K}$ replaced by any normed space (or any metrizable TVS).

20. Topologically supplementary subspaces.

The notion of topological supplement is fundamental for the structure theory of topological algebraic structures; the notion in pure algebra to which it corresponds is the internal direct product (or direct sum). We restrict the discussion here to topological vector spaces, but the principles involved are clearly more general.

(20.1) Let us begin by reviewing the purely algebraic situation. Suppose E is a vector space over $\mathbb{K}$, and M is a linear subspace of E. A linear subspace N is said to be *supplementary* to M if $E = M + N$ and $M \cap N = \{\theta\}$. Such subspaces N exist (choose a basis of M, expand it to a basis of E, and let N be the linear span of the basis vectors not in M); indeed, it is clear that if $M \neq E$ and $M \neq \{\theta\}$, then there exist infinitely many such subspaces N. Fix such a subspace N, equip $M \times N$ with the product vector space structure, and consider the mapping

$$T: M \times N \to E$$

defined by $T(y, z) = y + z$. Evidently T is linear, surjective (because $M + N = E$) and injective (because $M \cap N = \{\theta\}$), i.e., T is a vector space isomorphism. Thus E is structurally decomposed, with M as one of the factors.

Suppose, in addition, that E is a TVS. Equip M and N with the relative topologies, and $M \times N$ with the product topology. Then T is clearly continuous. We may ask: is it bicontinuous, i.e., is T an isomorphism for the TVS structures? In general the answer is negative. Indeed, as we shall see below, it can happen that for a particular subspace M, no choice of the supplementary subspace N renders the mapping T bicontinuous. Thus, there is room for the following definition:

(20.2) Definition. Let E be a TVS, M and N supplementary linear subspaces of E, i.e., $E = M + N$ and $M \cap N = \{\theta\}$. If the mapping $T: M \times N \to E$ defined by $T(y, z) = y + z$ is bicontinuous, then E is said to be the *topological direct sum* of M and N; when this is the case, N is called a *topological supplement* of M.

Our aim in this section is to give several useful reformulations of this concept.

(20.3) Theorem. *Let E be a TVS and let M, N be supplementary linear subspaces of E; thus, every vector x in E has a unique representation $x = y + z$ with $y \in M$ and $z \in N$. The following conditions are equivalent:*
(a) *E is the topological direct sum of M and N;*
(b) *the mapping $y + z \mapsto y$ ($y \in M$, $z \in N$) is continuous;*
(c) *the mapping $y + z \mapsto z$ ($y \in M$, $z \in N$) is continuous.*

Proof. Define mappings $u, v : E \to E$ as follows: if $x \in E$, write $x = y + z$ with $y \in M$, $z \in N$ and define $u(x) = y$, $v(x) = z$; by the uniqueness of such representations $x = y + z$, u and v are well-defined and linear. Since $u(x) + v(x) = x$ for all x, clearly u is continuous iff v is continuous, thus (b) and (c) are equivalent. If $T : M \times N \to E$ is the mapping $T(y, z) = y + z$, then

$$T^{-1}(y + z) = (y, z) = (u(y + z), v(y + z))$$

for all $y \in M$, $z \in N$, that is $T^{-1}(x) = (u(x), v(x))$ for all $x \in E$. Thus, u and v are the compositions of T^{-1} with the coordinate projections of $M \times N$, therefore T^{-1} is continuous if and only if u and v are continuous; but the continuity of T^{-1} is, by definition, precisely the condition (a). ∎

(20.4) Suppose E is a vector space over $\mathbb{K}$, and M, N are supplementary linear subspaces of E. Let u and v be the linear mappings defined in the proof of (20.3). If $y \in M$ obviously $u(y) = y$; therefore $u(u(x)) = u(x)$ for all $x \in E$, that is, $u \circ u = u$. Similarly $v \circ v = v$. Such mappings have a name:

(20.5) Definition. Let E be a vector space. A linear mapping $p : E \to E$ is called a *projector* if $p \circ p = p$.

(20.6) Theorem. *Let E be a TVS and let M be a linear subspace of E. In order that M possess a topological supplement, it is necessary and sufficient that there exist a continuous projector with range M.*

Proof. If M has a topological supplement N, then, with notations as in (20.4), u is a projector with $u(E) = M$, and u is continuous by (20.3).
Conversely, suppose $p : E \to E$ is a continuous projector such that $p(E) = M$. Since $p \circ p = p$, M is the set of vectors fixed under p:

$$M = \{x : p(x) = x\}.$$

Let N be the null space of p:

$$N = \{x : p(x) = \theta\}.$$

Obviously $M \cap N = \{\theta\}$; moreover, for any vector $x \in E$, one has

(*) $x = p(x) + (x - p(x))$,

where $p(x) \in M$ (because M is the range of p) and $x - p(x) \in N$ (because $p(x - p(x)) = p(x) - p(p(x)) = \theta$), thus M and N are supplementary subspaces. Since, for the decomposition (*), the mapping $x \mapsto p(x)$ is continuous by assumption, it follows from (20.3) that E is the topological direct sum of M and N. ∎

(20.7) **Corollary.** *If E is a separated TVS and M is a linear subspace of E that possesses a topological supplement, then M is a closed subspace of E.*

Proof. If p is a continuous projector with range M (20.6), then M is the inverse image, under the continuous mapping $x \mapsto x - p(x)$, of the closed set $\{\theta\}$. ∎

If M is a nonclosed linear subspace of a separated TVS E, then M cannot have a topological supplement in E (20.7). Worse, in a TVS (even in a Banach space) there may exist closed linear subspaces without topological supplements (20.10).

(20.8) Suppose E is a vector space and M, N are supplementary linear subspaces of E. Let $\pi : E \to E/M$ be the canonical mapping, and let π_0 be the restriction of π to N. Then $\pi_0 : N \to E/M$ is surjective (because $M + N = E$) and injective (because $M \cap N = \{\theta\}$), thus π_0 is a vector space isomorphism. If, in addition, E is a TVS, then π_0 is obviously continuous (for the relative topology on N and the quotient topology on E/M); for π_0 to be bicontinuous, it is necessary and sufficient that the direct sum be topological:

(20.9) **Theorem.** *Let E be a TVS and let M, N be supplementary linear subspaces of E. In order that E be the topological direct sum of M and N, it is necessary and sufficient that the canonical vector space isomorphism of N onto E/M be bicontinuous.*

Proof. Adopt the notations v, π, π_0 of (20.4) and (20.8). The point at issue is the continuity of π_0^{-1}. In any case, if $x \in E$, say $x = y + z$ ($y \in M$, $z \in N$), then

$$\pi_0^{-1}(\pi(x)) = z = v(x),$$

thus $v = \pi_0^{-1} \circ \pi$. Since E/M bears the final topology for the mapping π (4.4), it follows that π_0^{-1} is continuous if and only if $\pi_0^{-1} \circ \pi = v$ is continuous; but v is continuous if and only if the direct sum is topological (20.3). ∎

The preceding theorem yields an alternative proof of (20.7): if M and N are supplementary subspaces of a separated TVS E, then N is also separated, therefore E/M is separated by (20.9), therefore M is closed (4.5).

Exercises

(20.10) There exists a Banach space E with a closed linear subspace M such that M has no topological supplement.

(20.11) There exists a Banach space E with closed linear subspaces M and N such that $M \cap N = \{\theta\}$, $M + N$ is dense in E, and $M + N \neq E$.

21. Hyperplanes and linear forms. The present section is algebraic background for the next; here we consider abstract vector spaces over $\mathbb{K}$ ($= \mathbb{R}$ or $\mathbb{C}$), but we are ultimately interested only in the case of topological vector spaces.

To save repetition, let E denote a vector space over $\mathbb{K}$ throughout the section; it is understood that $\mathbb{K}$ may be either $\mathbb{R}$ or $\mathbb{C}$, but in the last part of the section (21.10–21.16) the field $\mathbb{C}$ is specified. The underlying strategy in this chapter is to keep the real and complex cases moving forward together. {An alternative strategy is to work out the real theory first, then go over the material again, making the necessary adaptations for the complex case.}

(21.1) Definition. A linear subspace M of E is said to be *maximal* if (i) $M \neq E$, and (ii) if N is a linear subspace of E such that $M \subset N$, then either $N = M$ or $N = E$.

(21.2) Definition. A *linear form* (or 'linear functional') is a scalar-valued function f on E such that $f(x + y) = f(x) + f(y)$ and $f(\lambda x) = \lambda f(x)$ for all x, y in E and all scalars λ. In other words, a linear form on E is a linear mapping $f: E \to \mathbb{K}$, where $\mathbb{K}$ is regarded as a one-dimensional vector space over itself.

(21.3) Theorem. *If M is a linear subspace of E, the following conditions on M are equivalent:*
 (a) *M is a maximal linear subspace of E;*
 (b) *M is the null space of some nonzero linear form f;*
 (c) *E/M is one-dimensional;*
 (d) *there exists a vector $z \notin M$ such that $E = M + \mathbb{K}z$;*
 (e) *$M \neq E$ and, for each vector $x \notin M$, $M + \mathbb{K}x = E$.*

If f, g are nonzero linear forms, then f and g have the same null space if and only if they are proportional, i.e., $g = \rho f$ for some nonzero scalar ρ; thus the linear form f in (b) is unique up to proportionality.

Proof. Let $\pi: E \to E/M$ be the canonical mapping. We prove the circle of implications (a) $\Rightarrow$ (e) $\Rightarrow$ (d) $\Rightarrow$ (c) $\Rightarrow$ (b) $\Rightarrow$ (a).

(a) implies (e): Suppose M is maximal. In particular, $M \neq E$. If $x \notin M$ then $M + \mathbb{K}x$ is a linear subspace containing M properly, therefore $M + \mathbb{K}x = E$ by maximality.

(e) implies (d) trivially.

(d) implies (c): Suppose $z \notin M$ and $M + \mathbb{K}z = E$. Let $u = \pi(z)$; since $z \notin M$, u is nonzero. Every $x \in E$ can be written in the form $x = y + \lambda z$ with $y \in M$ and $\lambda \in \mathbb{K}$, and one has $\pi(x) = \theta + \lambda \pi(z) = \lambda u$. Thus $E/M = \{\lambda u : \lambda \in \mathbb{K}\}$ is one-dimensional.

(c) implies (b): Suppose E/M is one-dimensional. Let u be a nonzero vector in E/M; thus, every vector of E/M is uniquely expressible as a scalar multiple of u. For each x in E, let $f(x)$ be the unique scalar such that $\pi(x) = f(x)u$. Clearly f is a linear form with null space M.

(b) implies (a): Let $f: E \to \mathbb{K}$ be a nonzero linear form with null space M. Since $\mathbb{K}$ has no linear subspaces other than $\{0\}$ and $\mathbb{K}$, and since the range of f is a nonzero linear subspace of $\mathbb{K}$, necessarily $f(E) = \mathbb{K}$. Thus, the quotient vector space E/M is isomorphic to $\mathbb{K}$ by the 'first isomorphism theorem', and the absence of subspaces of $\mathbb{K}$ between $\{0\}$ and $\mathbb{K}$ precludes the existence of subspaces of E between M and E.

Finally, suppose f and g are nonzero linear forms with $M = \{x : f(x) = 0\} = \{x : g(x) = 0\}$. Choose any $z \notin M$. Then $f(z) \neq 0$ and we may define $\rho = g(z)/f(z)$. Given any $x \in E$, we assert that $g(x) = \rho f(x)$. By (e) we may write $x = y + \lambda z$ with $y \in M$ and $\lambda \in \mathbb{K}$; then $f(y) = g(y) = 0$, therefore $g(x) = \lambda g(z) = [g(z)/f(z)][\lambda f(z)] = \rho f(x)$. ∎

(21.4) Definition. A *linear variety* in E is a set V of the form $V = x + N$, where $x \in E$ and N is a linear subspace of E. Thus, a linear variety is a translate of some linear subspace, in other words, a coset of E relative to some linear subspace. Clearly $N = \{y - z : y, z \in V\}$; in particular, the linear subspace N is uniquely determined by the variety V. Obviously $\theta \in V$ if and only if V is a linear subspace; a linear subspace is called a *homogeneous variety* (a variety V is homogeneous if and only if $\lambda V \subset V$ for all scalars λ).

An inclusion relation between linear varieties implies an inclusion relation between the corresponding linear subspaces:

(21.5) Theorem. *If* $x_1 + N_1 \subset x_2 + N_2$, *where* N_1 *and* N_2 *are linear subspaces of* E, *then* $N_1 \subset N_2$.

Proof. Writing $V_1 = x_1 + N_1$ and $V_2 = x_2 + N_2$, the inclusion $V_1 \subset V_2$ implies that

$$\{y - z : y, z \in V_1\} \subset \{y - z : y, z \in V_2\},$$

that is, $N_1 \subset N_2$. ∎

(21.6) Definition. A *hyperplane* in E is a linear variety H that is maximal in the sense that (i) $H \neq E$, and (ii) if V is a linear variety such that $V \supset H$, then either $V = H$ or $V = E$.

(21.7) Theorem. *Let H be a linear variety in E. The following conditions on H are equivalent:*
(a) *H is a hyperplane;*
(b) *H is a translate of some maximal linear subspace;*
(c) *there exist a nonzero linear form f and a scalar λ such that $H = \{x : f(x) = \lambda\}$.*

If H is a hyperplane, then the pair (f, λ) representing H as in (c) is unique up to proportionality; that is, if

$$H = \{x : f(x) = \lambda\} = \{x : g(x) = \mu\},$$

where f, g are nonzero linear forms and λ, μ are scalars, then there exists a nonzero scalar ρ such that $g = \rho f$ and $\mu = \rho \lambda$.

Proof. (a) implies (b): Say $H = x + M$, where M is a linear subspace. Since H is a proper subset of E, so is M. Suppose N is a linear subspace such that $M \subset N$; it is to be shown that $N = M$ or $N = E$. Since $x + N$ is a linear variety and $N + x \supset M + x = H$, it follows from the maximality of H that $N + x = E$ or $N + x = M + x$, that is, $N = E$ or $N = M$.

(b) implies (c): Assuming $H = z + M$, where M is a maximal linear subspace, let f be a linear form with null space M (21.3) and set $\lambda = f(z)$. Then

$$\{x : f(x) = \lambda\} = \{x : f(x) = f(z)\} = \{x : f(x - z) = 0\}$$
$$= \{x : x - z \in M\} = M + z = H.$$

(c) implies (a): Suppose $H = \{x : f(x) = \lambda\}$, where f is a nonzero linear form and $\lambda \in \mathbb{K}$. Let M be the null space of f and choose any $z \in H$. Then $H = z + M$ by an argument similar to that in the preceding paragraph; since M is a maximal linear subspace (21.3) it follows that H is a hyperplane, by an argument similar to that in the first paragraph of the proof.

Finally, suppose $H = \{x : f(x) = \lambda\} = \{x : g(x) = \mu\}$, where $\lambda, \mu \in \mathbb{K}$ and f, g are nonzero linear forms. Choose $u \in H$ and let M be the null space of f; as noted above, $H = u + M$. Similarly, if N is the null space of g, then $H = u + N$. Then $M = N$ by (21.5), therefore $g = \rho f$ for a suitable scalar ρ (21.3), whence $\mu = g(u) = \rho f(u) = \rho \lambda$. ∎

In view of (21.4), (21.6) and (21.7), 'homogeneous hyperplane' means the same as 'maximal linear subspace.'

Every linear variety is the intersection of the hyperplanes containing it; the proof is based on the homogeneous case:

(21.8) Lemma. *If N is any linear subspace of E, then*

$$N = \bigcap M,$$

where M varies over the family of all maximal linear subspaces of E that contain N.

Proof. Obviously N is contained in the intersection. Assuming $x \notin N$, it remains to show that there exists a maximal linear subspace M such that $N \subset M$ and $x \notin M$. Let S be a basis of N. Since $x \notin N$, the set $S \cup \{x\}$ is independent; therefore $S \cup \{x\}$ can be enlarged to a basis B of E. Write $B = S \cup \{x\} \cup T$ with T disjoint from $S \cup \{x\}$. Let f be the unique linear form on E such that $f(x) = 1$ and $f = 0$ on $S \cup T$, and let M be the null space of f. Then $x \notin M$ (because $f(x) = 1$) and $N \subset M$ (because $f = 0$ on S). ∎

(21.9) Theorem. *If V is any linear variety in E, then*

$$V = \bigcap H,$$

where H varies over the family of all hyperplanes that contain V.

Proof. Say $V = x + N$, where N is a linear subspace. If M is a maximal linear subspace with $M \supset N$, then $H = x + M$ is a hyperplane (21.7) with

$H \supset x + N = V$. Conversely, suppose H is a hyperplane with $H \supset V$. Say $H = z + M$, where M is a maximal linear subspace. Since $z + M = H \supset V = x + N$, one has $M \supset N$ by (21.5); moreover, since $x \in z + M$ one has $x + M = z + M = H$.

Summarizing: as M varies over all maximal linear subspaces containing N, $H = x + M$ varies over all hyperplanes containing $V = x + N$; since $\bigcap M = N$ by (21.8), we have

$$\bigcap H = \bigcap (x + M) = x + \bigcap M = x + N = V.$$

{If $V = E$ then no such H exist, but we are rescued by the convention that the intersection of an empty family of subsets of E is equal to E.} ∎

For the rest of the section (except the very last result), E denotes a *complex vector space*. One can also regard E as a real vector space, by restricting scalar multiplication. We are interested in the connection between the linear forms for the two vector space structures.

(21.10) Definition. Let E be a vector space over $\mathbb{C}$. We view E also as a vector space over $\mathbb{R}$, by restricting the scalar multiplication.

The terms $\mathbb{C}$-*linear subspace* and $\mathbb{R}$-*linear subspace* are self-explanatory, as are the terms $\mathbb{C}$-*variety*, $\mathbb{R}$-*variety*, $\mathbb{C}$-*hyperplane*, $\mathbb{R}$-*hyperplane*.

A linear form for the complex vector space structure will be called a $\mathbb{C}$-*form*; thus, a $\mathbb{C}$-form is a mapping $f: E \to \mathbb{C}$ such that $f(x + y) = f(x) + f(y)$ and $f(\lambda x) = \lambda f(x)$ for all $x, y \in E$ and $\lambda \in \mathbb{C}$.

A linear form for the real vector space structure will be called an $\mathbb{R}$-*form*; thus, an $\mathbb{R}$-form is a mapping $g: E \to \mathbb{R}$ such that $g(x + y) = g(x) + g(y)$ and $g(\alpha x) = \alpha g(x)$ for all $x, y \in E$ and $\alpha \in \mathbb{R}$.

Thus, a function on E is a $\mathbb{C}$-form iff it is complex-valued, additive, and complex-homogeneous; it is an $\mathbb{R}$-form iff it is real-valued, additive, and real-homogeneous. {Another concept, not pertinent here, is the notion of a complex-valued, additive, real-homogeneous function—for example, a function $x \mapsto ig(x)$, where g is an $\mathbb{R}$-form.}

(21.11) Theorem. *Let E be a vector space over $\mathbb{C}$ and let f be a $\mathbb{C}$-form on E.*

(i) *There exist unique $\mathbb{R}$-forms g and h on E such that $f(x) = g(x) + ih(x)$ for all $x \in E$.*

(ii) *Necessarily $h(x) = -g(ix)$, thus $f(x) = g(x) - ig(ix)$ for all $x \in E$.*

(iii) *If, moreover, E is a TVS over $\mathbb{C}$, then f is continuous iff g is continuous iff h is continuous.*

Proof. Regard $\mathbb{C}$ as a two-dimensional vector space over $\mathbb{R}$. If $\lambda = \alpha + i\beta$ ($\alpha, \beta \in \mathbb{R}$), the formulas

$$\text{Re } \lambda = \alpha, \qquad \text{Im } \lambda = \beta$$

define $\mathbb{R}$-forms Re and Im on $\mathbb{C}$.

(i) Define $g(x) = \text{Re } f(x)$, $h(x) = \text{Im } f(x)$; clearly g, h are $\mathbb{R}$-forms on E, and $f = g + ih$. The uniqueness of such a representation is obvious.

(ii) Since $f(ix) = if(x)$, we have

$$g(ix) + ih(ix) = i[g(x) + ih(x)] = -h(x) + ig(x);$$

comparing real and imaginary parts, $g(ix) = -h(x)$ and $h(ix) = g(x)$.

(iii) Immediate from (i) and (ii). ∎

(21.12) Definition. With notations as in (21.11), we write $g = \operatorname{Re} f$ and $h = \operatorname{Im} f$.

Conversely:

(21.13) Theorem. *Let E be a vector space over $\mathbb{C}$. If g is an $\mathbb{R}$-form on E, then the formula*

$$f(x) = g(x) - ig(ix) \qquad (x \in E)$$

defines a $\mathbb{C}$-form f such that $\operatorname{Re} f = g$.

Proof. Clearly f is additive and real-homogeneous, and $\operatorname{Re} f(x) = g(x)$. It remains to show that f is complex-homogeneous; it is sufficient, by additivity and real-homogeneity, to show that $f(ix) = if(x)$. Indeed,

$$f(ix) = g(ix) - ig(i^2 x) = i[-ig(ix) - g(-x)]$$
$$= i[g(x) - ig(ix)] = if(x). \qquad ∎$$

The correspondences $f \mapsto g$ and $g \mapsto f$ described in (21.11) and (21.13) are mutually inverse; they define a bijection between the set of all $\mathbb{C}$-forms f and the set of all $\mathbb{R}$-forms g. There is induced a correspondence between $\mathbb{C}$-hyperplanes and $\mathbb{R}$-hyperplanes, as we see in the next theorem.

(21.14) Lemma. *Let E be a vector space over $\mathbb{C}$. A subset M of E is a maximal $\mathbb{C}$-linear subspace if and only if $M = M_0 \cap (iM_0)$ with M_0 a maximal $\mathbb{R}$-linear subspace.*

Proof. Suppose M is a maximal $\mathbb{C}$-linear subspace. Let f be a nonzero $\mathbb{C}$-form with null space M (21.3), and let $g = \operatorname{Re} f$ (21.12). Since

(*) $$f(x) = g(x) - ig(ix)$$

for all $x \in E$ (21.11), it follows that the $\mathbb{R}$-form g is nonzero, therefore $M_0 = \{x : g(x) = 0\}$ is a maximal $\mathbb{R}$-linear subspace (21.3). In view of (*) we have $f(x) = 0$ iff $g(x) = 0$ and $g(ix) = 0$, iff $x \in M_0$ and $ix \in M_0$, iff

$$x \in M_0 \cap ((1/i)M_0) = M_0 \cap (-iM_0) = M_0 \cap (iM_0),$$

thus $M = M_0 \cap (iM_0)$.

Conversely, assume $M = M_0 \cap (iM_0)$ with M_0 a maximal $\mathbb{R}$-linear subspace. Let g be a nonzero $\mathbb{R}$-form with null space M_0, and let f be the $\mathbb{C}$-form with $\operatorname{Re} f = g$ (21.13). The above calculation shows that the null space of f is $M_0 \cap (iM_0) = M$, therefore M is a maximal $\mathbb{C}$-linear subspace by (21.3). ∎

(21.15) Theorem. *Let E be a vector space over $\mathbb{C}$, and let H be a subset of E. The following conditions on H are equivalent:*

(a) *H is a $\mathbb{C}$-hyperplane;*

(b) *there exist a maximal* $\mathbb{R}$-*linear subspace* M_0 *and vectors* $x_0, y_0 \in E$, *such that* $H = (x_0 + M_0) \cap (y_0 + iM_0)$;

(c) *there exist a maximal* $\mathbb{R}$-*linear subspace* M_0 *and a vector* $z_0 \in E$, *such that* $H = z_0 + M_0 \cap (iM_0)$.

Proof. The sets $M_0 \cap (iM_0)$ described in (c) are precisely the maximal $\mathbb{C}$-linear subspaces (21.14), therefore their translates are the $\mathbb{C}$-hyperplanes (21.7); thus (c) and (a) are equivalent.

(c) implies (b): Since translation by z_0 is a bijection,

$$z_0 + M_0 \cap (iM_0) = (z_0 + M_0) \cap (z_0 + iM_0).$$

(b) implies (c): Suppose H has the form described in (b). Let g be an $\mathbb{R}$-form with null space M_0 (21.3) and let f be the $\mathbb{C}$-form with $\operatorname{Re} f = g$ (21.13); thus $f(x) = g(x) - ig(ix)$ for all $x \in E$. Set $\lambda = g(x_0) - ig(iy_0)$. Then $f(x) = \lambda$ iff $g(x) = g(x_0)$ and $g(ix) = g(iy_0)$; this is equivalent, by an elementary calculation, to $x \in (x_0 + M_0) \cap (y_0 + iM_0)$. Thus

$$H = (x_0 + M_0) \cap (y_0 + iM_0) = \{x : f(x) = \lambda\},$$

which is a $\mathbb{C}$-hyperplane by (21.7). ∎

(21.16) Corollary. *If E is a vector space over $\mathbb{C}$, and H is a $\mathbb{C}$-hyperplane in E, then there exist $\mathbb{R}$-hyperplanes H_1, H_2 such that $H = H_1 \cap H_2$.*

Proof. With notation as in (b) of (21.15), let $H_1 = x_0 + M_0$, $H_2 = y_0 + iM_0$. Then H_1 is an $\mathbb{R}$-hyperplane by (21.7). Moreover, $H_2 = i(-iy_0 + M_0)$; since $-iy_0 + M_0$ is an $\mathbb{R}$-hyperplane, and since $x \mapsto ix$ is an isomorphism for the real vector space structure of E, it follows that H_2 is also an $\mathbb{R}$-hyperplane. ∎

It is not asserted, nor is it true, that the intersection of any two distinct $\mathbb{R}$-hyperplanes is a $\mathbb{C}$-hyperplane.

The final result of the section is a piece of elementary linear algebra, parked here for want of a better place; it is not needed until the proof of (38.7):

(21.17) Theorem. *Let $f_1, \ldots, f_n$ be linear forms on a vector space E. In order that a linear form f on E be a linear combination of $f_1, \ldots, f_n$, it is necessary and sufficient that f vanish on the intersection of the null spaces of the f_k.*

Proof. The necessity of the condition is obvious.

Sufficiency: Let M_k be the null space of f_k; assuming f is a linear form on E such that $f = 0$ on $M_1 \cap \cdots \cap M_n$, it is to be shown that there exist scalars $\lambda_1, \ldots, \lambda_n$ such that $f = \sum_1^n \lambda_k f_k$. We can suppose without loss of generality that the f_k are linearly independent.

The proof is by induction on n. The case that $n = 1$ is noted in (21.3): f is a multiple of f_1. Assume inductively that all is well with $n - 1$. For each k, f_k is not a linear combination of the f_j with $j \neq k$; by the induction hypothesis, there exists a vector x_k such that $f_k(x_k) = 1$ and $f_j(x_k) = 0$ for $j \neq k$. Let

$\lambda_k = f(x_k)$ $(k = 1, \ldots, n)$; we show that $f = \sum_1^n \lambda_k f_k$. For each $x \in E$, the vector

$$x - \sum_1^n f_k(x)x_k$$

is annihilated by every f_j, therefore it is annihilated by f:

$$0 = f(x) - \sum_1^n f_k(x)f(x_k) = f(x) - \sum_1^n \lambda_k f_k(x),$$

i.e., $f(x) = (\sum_1^n \lambda_k f_k)(x)$. ∎

Exercise

(21.18) Let E be a complex vector space and let M be a maximal $\mathbb{C}$-linear subspace of E. Suppose M_0, M_0' are maximal $\mathbb{R}$-linear subspaces such that

$$M = M_0 \cap (iM_0) = M_0' \cap (iM_0')$$

(cf. 21.14). Let g_0, g_0' be $\mathbb{R}$-forms with null spaces M_0, M_0', and define $h_0(x) = g_0(ix)$, $h_0'(x) = g_0'(ix)$ (cf. 21.13). Then g_0, h_0 are real linear combinations of $\tilde{g}_0'$, h_0', and vice versa. This is the measure of uniqueness of the representation in (21.14); namely, the two-dimensional real vector space spanned by g_0 and h_0 is uniquely determined by M.

22. Closed hyperplanes and continuous linear forms in a TVS.

In the preceding section we considered the relation between linear forms and hyperplanes, in the setting of an abstract vector space over $\mathbb{K}$. Here we are interested in the TVS setting. In a TVS, the linear forms of maximal interest are the continuous ones; there may not be any (cf. 15.10), but, in any case, they coexist with, and correspond to, the closed hyperplanes. More precisely, we shall show that a linear form on a TVS over $\mathbb{K}$ is continuous if and only if its null space is closed. The proof rests on the fact that, from a topological-algebraic point of view, the only one-dimensional separated TVS over $\mathbb{K}$ is $\mathbb{K}$ itself:

(22.1) Theorem. *If E is a one-dimensional, separated TVS over $\mathbb{K}$, and if a is any nonzero vector in E, then the mapping $\lambda \mapsto \lambda a$ is a bicontinuous vector space isomorphism of $\mathbb{K}$ onto E.*

Proof. Write $f(\lambda) = \lambda a$ $(\lambda \in \mathbb{K})$; regarding $\mathbb{K}$ as a one-dimensional TVS over itself, f is obviously a continuous vector space isomorphism. The problem is to show that f is open; to this end, it will suffice to show that if A is a neighborhood of 0 in $\mathbb{K}$, then $f(A)$ is a neighborhood of θ in E (5.1). We can suppose that $A = \{\lambda : |\lambda| \leq \varepsilon\}$, $\varepsilon > 0$. Choose any $\mu \in \mathbb{K}$ with $0 < |\mu| < \varepsilon$. Since $\mu a \neq \theta$ and E is separated, there exists a neighborhood V of θ such that $\mu a \notin V$; we can suppose, moreover, that V is balanced (17.9). The proof is concluded by showing that $f(A) \supset V$, i.e., $V \subset Aa$. Let $x \in V$, write $x = \lambda a$, and assume to the contrary that $\lambda \notin A$, i.e., $|\lambda| > \varepsilon$. Then $|\mu| < \varepsilon < |\lambda|$, $|\mu\lambda^{-1}| < 1$; since $x \in V$ and V is balanced, $(\mu\lambda^{-1})x \in (\mu\lambda^{-1})V \subset V$, thus V contains $\mu(\lambda^{-1}x) = \mu a$, contrary to the choice of V. ∎

(22.2) Corollary. *Let E be a TVS over* $\mathbb{K}$, *let f be a linear form on E, and let M be the null space of f. Then f is continuous if and only if M is closed.*

Proof. Since $M = f^{-1}(\{0\})$ and $\{0\}$ is closed, the continuity of f implies that M is also closed.

Suppose, conversely, that M is closed. If f is identically zero, there is nothing to prove. Assuming f nonzero, E/M is one-dimensional (21.3). Let $\pi: E \to E/M$ be the canonical mapping and equip E/M with the quotient topology; then E/M is a TVS and, since M is closed, E/M is separated (11.8). Let $g: E/M \to \mathbb{K}$ be the unique vector space isomorphism such that $g(\pi(x)) = f(x)$ for all $x \in E$. Then g is continuous by the theorem, therefore $f = g \circ \pi$ is also continuous. ∎

The following corollary is for use in the next section:

(22.3) Corollary. *Let E be a TVS over* $\mathbb{K}$, *let M be a closed, maximal linear subspace of E, and let N be any algebraic supplement of M. Then E is the topological direct sum of M and N, that is, N is a topological supplement of M.*

Proof. Since M is closed, the quotient TVS E/M is separated (11.8). Also, $M \cap N$ is closed in N (for the relative topology on N); since $M \cap N = \{\theta\}$, this means that N is separated (3.4). Let $\pi: E \to E/M$ be the canonical mapping, and let π_0 be the restriction of π to N. As noted in (20.8), $\pi_0: N \to E/M$ is a vector space isomorphism; since N and E/M are one-dimensional and separated, it is clear from (22.1) that π_0 is bicontinuous, therefore E is the topological direct sum of M and N (20.9). ∎

In view of (22.2), most of the results in Section 21 readily imply their TVS analogues, as we now show (but see the remarks at the end of the section).

(22.4) Theorem. *Let E be a TVS over* $\mathbb{K}$ *and let M be a linear subspace of E. The following conditions on M are equivalent:*
(a) *M is a closed, maximal linear subspace of E;*
(b) *M is the null space of some nonzero continuous linear form f;*
(c) *M is closed and E/M is one-dimensional.*

Proof. This is immediate from (21.3) and (22.2). ∎

(22.5) Theorem. *Let E be a TVS over* $\mathbb{K}$ *and let* $H \subset E$. *The following conditions on H are equivalent:*
(a) *H is a closed hyperplane;*
(b) *H is a translate of some closed, maximal linear subspace;*
(c) *there exist a nonzero continuous linear form f and a scalar* λ, *such that* $H = \{x : f(x) = \lambda\}$.

Proof. This is immediate from (21.7), (22.2), and the fact that a translation is a homeomorphism. ∎

(22.6) Theorem. *If E is a TVS over* $\mathbb{C}$, *then the formulas*

$$g(x) = \operatorname{Re} f(x) \qquad (x \in E)$$

and

$$f(x) = g(x) - ig(ix) \qquad (x \in E)$$

define mutually inverse bijective mappings $f \mapsto g$ and $g \mapsto f$ between the set of all continuous $\mathbb{C}$-forms f and the set of all continuous $\mathbb{R}$-forms g.

Proof. This is immediate from the remark following (21.13) and the fact that f is continuous if and only if g is continuous. ∎

(22.7) Lemma. *Let E be a TVS over $\mathbb{C}$. A subset M of E is a closed, maximal $\mathbb{C}$-linear subspace if and only if $M = M_0 \cap (iM_0)$ with M_0 a closed, maximal $\mathbb{R}$-linear subspace.*

Proof. In view of (22.2), the linear forms appearing in the proof of (21.14) are continuous. ∎

(22.8) Theorem. *Let E be a TVS over $\mathbb{C}$ and let $H \subset E$. The following conditions on H are equivalent:*

(a) *H is a closed $\mathbb{C}$-hyperplane;*

(b) *there exist a closed, maximal $\mathbb{R}$-linear subspace M_0 and vectors $x_0, y_0 \in E$, such that $H = (x_0 + M_0) \cap (y_0 + iM_0)$;*

(c) *there exist a closed, maximal $\mathbb{R}$-linear subspace M_0 and a vector $z_0 \in E$, such that $H = z_0 + M_0 \cap (iM_0)$.*

Proof. This follows from the lemma in the same way that (21.15) follows from (21.14), on noting that translations are homeomorphisms. ∎

(22.9) Corollary. *If E is a TVS over $\mathbb{C}$ and if H is a closed $\mathbb{C}$-hyperplane in E, then there exist closed $\mathbb{R}$-hyperplanes H_1, H_2 such that $H = H_1 \cap H_2$.*

Proof. This follows from (22.8) in the same way that (21.16) follows from (21.15). ∎

The TVS analogues of (21.8) and (21.9) are generally false. The trouble is that an arbitrary TVS need not have any nonzero continuous linear forms (cf. 15.10), let alone 'sufficiently many' of them. This situation is repaired if the TVS is separated and locally convex (33.13).

Exercises

(22.10) In a TVS, the closure of a linear variety is a linear variety.

(22.11) A hyperplane in a TVS is either closed or dense.

23. Finite-dimensional TVS, Riesz's theorem. From the viewpoint of topological algebra, there is, up to isomorphism, only one separated, one-dimensional TVS over $\mathbb{K}$, namely, $\mathbb{K}$ itself (22.1). The central result of the present section is that the analogous result holds for finite-dimensional separated TVS; up to (bicontinuous, linear) isomorphism, the only r-dimensional separated TVS over $\mathbb{K}$ is $\mathbb{K}^r$:

(23.1) Theorem. (Tihonov) *If E is a separated TVS over $\mathbb{K}$, of finite dimension r, then every vector space isomorphism $u: \mathbb{K}^r \to E$ is bicontinuous.*

Proof. The proof is by induction on r, the case $r = 1$ being covered by (22.1). Assume inductively that the theorem holds for $r - 1$.

Let $e_1, \ldots, e_r$ be the canonical basis of $\mathbb{K}^r$, that is, $e_k = (\delta_{jk})_{1 \le j \le r}$. Define $x_k = u(e_k)$ $(k = 1, \ldots, r)$; then $x_1, \ldots, x_r$ is a basis of E, and, for any (λ_k) in $\mathbb{K}^r$,

$$u((\lambda_k)) = u\left(\sum_1^r \lambda_k e_k\right) = \sum_1^r \lambda_k u(e_k) = \sum_1^r \lambda_k x_k.$$

Let M be the linear subspace of E spanned by $x_1, \ldots, x_{r-1}$; with the relative topology, M is also a separated TVS. By induction, the vector space isomorphism $\mathbb{K}^{r-1} \to M$ defined by

$$(\lambda_1, \ldots, \lambda_{r-1}) \mapsto \sum_1^{r-1} \lambda_k x_k$$

is bicontinuous. Since $\mathbb{K}^{r-1}$ is a complete TVS (7.8), its isomorph M is also complete (5.12); since E is separated, it follows that M is closed in E. Thus, M is a closed, maximal linear subspace of E. Let $N = \mathbb{K}x_r$ be the one-dimensional subspace spanned by x_r; then E is the topological direct sum of M and N (22.3), that is, the canonical vector space isomorphism of $M \times N$ onto E given by

(*) $$\left(\sum_1^{r-1} \lambda_k x_k, \lambda_r x_r\right) \mapsto \sum_1^r \lambda_k x_k$$

is bicontinuous. The vector space isomorphisms $\mathbb{K}^{r-1} \to M$ and $\mathbb{K} \to N$, defined by

$$(\lambda_1, \ldots, \lambda_{r-1}) \mapsto \sum_1^{r-1} \lambda_k x_k \quad \text{and} \quad \lambda_r \mapsto \lambda_r x_r,$$

are bicontinuous (by the inductive hypothesis and the case $r = 1$), therefore the vector space isomorphism $\mathbb{K}^{r-1} \times \mathbb{K} \to M \times N$ defined by

$$((\lambda_1, \ldots, \lambda_{r-1}), \lambda_r) \mapsto \left(\sum_1^{r-1} \lambda_k x_k, \lambda_r x_r\right)$$

is bicontinuous; composing with (*), we see that

(**) $$((\lambda_1, \ldots, \lambda_{r-1}), \lambda_r) \mapsto \sum_1^r \lambda_k x_k$$

is a bicontinuous vector space isomorphism $\mathbb{K}^{r-1} \times \mathbb{K} \to E$. Also, the canonical vector space isomorphism $\mathbb{K}^r \to \mathbb{K}^{r-1} \times \mathbb{K}$, namely,

$$(\lambda_1, \ldots, \lambda_r) \mapsto ((\lambda_1, \ldots, \lambda_{r-1}), \lambda_r),$$

is bicontinuous; composing with (**), we see that the mapping

$$(\lambda_1, \ldots, \lambda_r) \mapsto \sum_1^r \lambda_k x_k,$$

i.e., the mapping u, is a bicontinuous vector space isomorphism. ∎

Tihonov's theorem is, in a sense, an algebraic characterization of $\mathbb{K}^r$ as a separated TVS; the algebraic invariant of dimension is completely determining. Riesz's theorem, to be proved below, is, in effect, a topological character-ization of the class of spaces $\mathbb{K}^r$ ($r = 1, 2, 3, \ldots$); it asserts that a separated TVS is finite-dimensional (and therefore isomorphic to $\mathbb{K}^r$ for some r) if and only if it is locally compact. Before proving Riesz's theorem, we reap the harvest of corollaries to Tihonov's theorem.

(23.2) Corollary. *If E and F are finite-dimensional separated TVS over $\mathbb{K}$, and if $\dim_{\mathbb{K}} E = \dim_{\mathbb{K}} F$, then every vector space isomorphism $u : E \to F$ is bicontinuous.*

(23.3) Corollary. *If E is a finite-dimensional vector space over $\mathbb{K}$, and if τ_1, τ_2 are separated topologies on E that are compatible with the vector space structure, then $\tau_1 = \tau_2$.*

Proof. Apply (23.2) to the identity mapping $(E, \tau_1) \to (E, \tau_2)$. ∎

(23.4) Corollary. *If (E, τ) is a finite-dimensional separated TVS over $\mathbb{K}$, then* (i) *(E, τ) is normable; indeed, every norm on E yields the given topology τ;* (ii) *all norms on E are complete, thus E is normable to be a Banach space;* (iii) *(E, τ) is a metrizable complete TVS.*

Proof. (i) Let $x \mapsto \|x\|$ be any norm on E. (For example, choose a basis $x_1, \ldots, x_r$ of E and define, for any $x = \sum_1^r \lambda_k x_k$ in E, $\|x\| = \sum_1^r |\lambda_k|$.) Then the topology derived from the norm is separated and compatible with the vector space structure (16.5), therefore it coincides with τ by (23.3).

(ii), (iii) With the preceding notations, the mapping $(\lambda_k) \mapsto \sum_1^r \lambda_k x_k$ is an isometric vector space isomorphism of $\mathbb{K}^r$ onto $(E, \|x\|)$, where $\mathbb{K}^r$ is equipped with the norm $(\lambda_k) \mapsto \sum_1^r |\lambda_k|$; since the latter is clearly a Banach space norm on $\mathbb{K}^r$ (cf. 16.14), it follows trivially that $(E, \|x\|)$ is a Banach space, therefore (E, τ) is a complete TVS (16.10). ∎

(23.5) Corollary. *If E is a separated TVS over $\mathbb{K}$, and F is a finite-dimensional linear subspace of E, then F is closed in E.*

Proof. In the relative topology, F is a complete, separated TVS (23.4); since E is separated, it follows that F is closed in E. ∎

The sum of two closed linear subspaces may not be closed (20.11), but it is when one of the summands is finite-dimensional:

(23.6) Corollary. *Let E be a TVS over $\mathbb{K}$. If N, F are linear subspaces of E such that N is closed and F is finite-dimensional, then the linear subspace $N + F$ is also closed.*

Proof. {Note that F need not be closed (it is not assumed that E is separated).} Let $\pi : E \to E/N$ be the canonical mapping; since N is closed, E/N is separated (11.8). Since π is linear, $\pi(F)$ is a finite-dimensional linear subspace of E/N, therefore $\pi(F)$ is closed in E/N by (23.5); then $N + F = \pi^{-1}(\pi(F))$ is closed in E by the continuity of π. ∎

If E is a TVS over $\mathbb{K}$, M is a closed linear subspace of E of codimension 1, and N is any algebraic supplement of M, then E is the topological direct sum of M and N (22.3). This generalizes to arbitrary finite codimension:

(23.7) Corollary. *If E is a TVS over $\mathbb{K}$, M is a closed linear subspace of finite codimension, and N is any algebraic supplement of M, then E is the topological direct sum of M and N.*

Proof. The finite codimensionality of M means, by definition, that E/M is finite-dimensional. Let $\pi: E \to E/M$ be the canonical mapping and let π_0 be the restriction of π to N; thus, $\pi_0: N \to E/M$ is a continuous vector space isomorphism, and the problem is to show that π_0 is bicontinuous (20.9). Since M is closed in E, $M \cap N = \{\theta\}$ is closed in N, therefore N is separated (3.4); since, moreover, E/M is separated (11.8), it follows from (23.2) that π_0 is bicontinuous. ∎

(23.8) Corollary. *If $u: E \to F$ is a surjective continuous linear mapping, where E and F are TVS over $\mathbb{K}$, and F is separated and finite-dimensional, then u is open. If N is the null space of u, then the vector space isomorphism $E/N \to F$ induced by u is bicontinuous.*

Proof. Since F is separated, $\{\theta\}$ is closed in F, therefore $N = u^{-1}(\{\theta\})$ is closed in E by the continuity of u; consequently E/N is separated (11.8). Let $\pi: E \to E/N$ be the canonical mapping and let $v: E/N \to F$ be the unique vector space isomorphism such that $v \circ \pi = u$. By (23.2), v is bicontinuous; since, moreover, π is open (11.8), it follows that $u = v \circ \pi$ is also open. ∎

The next corollary is in a sense dual to (23.8):

(23.9) Corollary. *If $u: F \to E$ is a linear mapping, where E and F are separated TVS over $\mathbb{K}$, and F is finite-dimensional, then u is continuous.*

Proof. Replacing E by $u(F)$ in the relative topology, one can suppose that u is surjective. Then $u(F) = E$ is a finite-dimensional, separated TVS. Let M be the null space of u, let $\pi: F \to F/M$ be the canonical mapping, and let $v: F/M \to E$ be the unique vector space isomorphism such that $v \circ \pi = u$.

By (23.5), M is closed in F, therefore F/M is separated (11.8). The vector space isomorphism $v: F/M \to E$ is therefore (bi)continuous by (23.2); since π is continuous, it follows that $u = v \circ \pi$ is also continuous. ∎

The next theorem extends a classical result of F. Riesz (1918) for normed spaces:

(23.10) Theorem. *Let E be a separated TVS over $\mathbb{K}$. The following conditions on E are equivalent:*
(a) *E is finite-dimensional;*
(b) *E is locally compact;*
(c) *E admits a compact neighborhood of θ;*
(d) *E admits a totally bounded neighborhood of θ.*

Proof. (a) implies (b): If E has finite dimension r, then E is homeomorphic with $\mathbb{K}^r$ (23.1).

(b) implies (c) trivially.

(c) implies (d): Recall that a subset A of a uniform space is totally bounded iff, for every entourage V, there exists a finite covering $A \subset \bigcup_1^n A_k$ such that the A_k are small of order V (i.e., $A_k \times A_k \subset V$). It will suffice to show that any compact subset A of a uniform space is totally bounded. Given any entourage V, let W be a symmetric entourage such that $W \circ W \subset V$. For each $x \in A$, $W(x) = \{y : (x, y) \in W\}$ is a neighborhood of x; by compactness, $A \subset \bigcup_1^n W(x_k)$ for a suitable finite set $x_1, \ldots, x_n$ in A. The sets $A_k = W(x_k)$ are evidently small of order V.

(d) implies (a): Suppose there exists a totally bounded neighborhood U of θ in E. Let V be the entourage determined by the neighborhood $\frac{1}{2}U$ of θ (cf. 5.3):

$$V = \{(x, y) : -x + y \in \tfrac{1}{2}U\}.$$

Then, for any $x \in E$, $V(x) = x + \frac{1}{2}U$ (cf. 5.9). Since U is totally bounded, there exists a finite set $x_1, \ldots, x_n$ in U such that

(*) $$U \subset \bigcup_{k=1}^n V(x_k) = \bigcup_{k=1}^n (x_k + \tfrac{1}{2}U).$$

Let M be the linear subspace of E spanned by $x_1, \ldots, x_n$; it will suffice to show that $M = E$. In any case, M is closed in E (23.5), therefore E/M is a separated TVS (11.8). Let $\pi: E \to E/M$ be the canonical mapping, and let $W = \pi(U)$; since π is open (11.8), W is a neighborhood of θ in E/M. Moreover, since π is uniformly continuous (5.12) and U is totally bounded, it is easy to see that W is also totally bounded. From (*), it is clear that $U \subset M + \frac{1}{2}U$; applying π and noting that $\pi = 0$ on M, we have $W \subset \frac{1}{2}W$, i.e., $2W \subset W$. By induction, $2^j W \subset W$ for $j = 1, 2, 3, \ldots$. It follows that $W = E/M$; for, if $u \in E/M$ then for sufficiently large j one has $2^{-j}u \in W$ and so $u \in 2^j W \subset W$. In particular, E/M is totally bounded; it will suffice to show that this implies $E/M = \{\theta\}$. If, on the contrary, E/M contained a nonzero vector u, then the one-dimensional subspace $\mathbb{K}u$ spanned by u would also be totally bounded, contrary to the fact that it is linearly homeomorphic with $\mathbb{K}$ (cf. (22.1), (5.12)). ∎

24. Completion of a separated TVS.

Since the uniform structure of a TVS is derived from its additive topological group (11.10), a better understanding of the TVS notion of 'completion' is gained by approaching it through several layers of generality: uniform structures, 'completable' topological groups, and abelian topological groups. Omitted proofs are referenced at the end of the book. We stick to separated uniform structures; one can discuss completion for arbitrary structures (at a considerable cost in technical fussing), but all of our applications are to separated TVS.

The basic construct is the general-topological abstraction of Cantor's construction of the reals:

(24.1) Proposition. *If $(X, \mathcal{U})$ is any separated uniform structure, there exist a complete uniform structure $(Y, \mathcal{V})$ and a mapping $f: X \to Y$ such that:*
  (i) *f is injective;*
  (ii) *$f(X)$ is dense in Y;*
  (iii) *the mapping $f: X \to f(X)$ and its inverse are uniformly continuous, where $f(X)$ is equipped with the relative uniform structure induced by $\mathcal{V}$.*

With notation as in (24.1), let us clarify several points of (iii). The uniformity that defines the relative uniform structure on $f(X)$ is

$$\mathcal{V} \cap [f(X) \times f(X)] = \{V \cap [f(X) \times f(X)] : V \in \mathcal{V}\}.$$

Thus $f: X \to f(X)$ is a *unimorphism* (i.e., an isomorphism for the indicated uniform structures), and the entourages $U \in \mathcal{U}$ are precisely the sets

$$(f \times f)^{-1}(V \cap [f(X) \times f(X)]) = \{(x, x') : (f(x), f(x')) \in V\},$$

where V varies over $\mathcal{V}$.

Proposition (24.1) assigns a triple $(Y, \mathcal{V}, f)$ to the given uniform structure $(X, \mathcal{U})$; the essential uniqueness of such a triple hinges on the following extension theorem:

(24.2) Proposition. *Let $(X, \mathcal{U})$ be a uniform structure, $(Y, \mathcal{V})$ a complete, separated uniform structure, A a dense subset of X, and $f: A \to Y$ a uniformly continuous mapping (for the relative uniformity on A). Then f may be extended to a unique continuous mapping $\bar{f}: X \to Y$, and $\bar{f}$ is also uniformly continuous.*

The triple $(Y, \mathcal{V}, f)$ constructed in (24.1) is called a *completion* of $(X, \mathcal{U})$; its essential uniqueness is an easy consequence of (24.2):

(24.3) Proposition. *If $(X, \mathcal{U})$ is a separated uniform structure, and if each of the triples $(Y, \mathcal{V}, f)$, $(Y', \mathcal{V}', f')$ satisfies conditions (i), (ii), (iii) of (24.1), then there exists a unimorphism $\varphi: Y \to Y'$ such that $f' = \varphi \circ f$.*

In the sense permitted by (24.3), we may speak of *the* completion of a separated uniform structure. Identifying X with $f(X)$, in the notation of (24.1), one may regard X as a dense subset of Y, and $\mathcal{U}$ as the relative uniformity induced on X by $\mathcal{V}$. It is convenient to codify these maneuvers in a fixed notation:

(24.4) Definition. The *completion* of a separated uniform structure $(X, \mathcal{U})$ is a complete, separated uniform structure $(\hat{X}, \hat{\mathcal{U}})$ such that (i) $X \subset \hat{X}$, (ii) X is dense in $\hat{X}$, and (iii) $\mathcal{U} = \hat{\mathcal{U}} \cap [X \times X]$.

The existence and essential uniqueness of the completion is settled by (24.1), (24.3). A typical use of the notation in (24.4) is the following application of (24.2):

(24.5) Proposition. *Suppose $f: X \to Y$, where $(X, \mathcal{U})$ is a separated uniform structure and $(Y, \mathcal{V})$ is a complete, separated uniform structure. If f is uniformly continuous, then f may be extended to a unique continuous mapping $\hat{f}: \hat{X} \to Y$, and $\hat{f}$ is also uniformly continuous.*

We turn now to topological groups. Suppose G is a separated topological group and $\mathscr{U}_s(G)$ is the left uniformity of G. Consider the uniform structure completion $(\hat{G}, (\mathscr{U}_s(G))^{\wedge})$. The problem is that it may not be possible to extend the group operations of G to suitable operations on $\hat{G}$; here, "suitable" means that $\hat{G}$ should become a topological group, and $(\mathscr{U}_s(G))^{\wedge}$ should coincide with the left uniform structure of $\hat{G}$ derived from its topological group structure, i.e., $(\mathscr{U}_s(G))^{\wedge} = \mathscr{U}_s(\hat{G})$. When this miracle can be performed, $\hat{G}$ becomes a complete, separated topological group, and G is identified as a dense subgroup of $\hat{G}$. The following term is useful in this circle of ideas:

(24.6) Definition. A separated topological group G is said to be *completable* if there exists a complete, separated topological group H such that (i) G is a dense subgroup of H, and (ii) the original topology of G coincides with the relative topology on G induced by the topology of H. (It follows at once that the original left uniformity of G coincides with the relative uniformity induced on G by the left uniformity of H.)

When the topological group H of (24.6) exists, it is unique:

(24.7) Proposition. *If G is a completable, separated topological group, and H, H' are complete, separated topological groups each of which satisfies conditions* (i), (ii) *of* (24.6), *then the identity mapping of G may be extended to a bicontinuous group isomorphism of H onto H'.*

Proof. By (24.3), there exists a unimorphism $\varphi \colon (H, \mathscr{U}_s(H)) \to (H', \mathscr{U}_s(H'))$ such that $\varphi(a) = a$ for all $a \in G$. It remains to show that $\varphi(xy) = \varphi(x)\varphi(y)$ for all $x, y \in H$. Indeed, the mapping $f \colon H \times H \to H'$ defined by $f(x, y) = \varphi(y)^{-1}\varphi(x)^{-1}\varphi(xy)$ is a continuous mapping such that, for all $a, b \in G$, $f(a, b) = b^{-1}a^{-1}ab = e$ (the common neutral element), thus $G \times G \subset f^{-1}(\{e\})$; since H' is separated and f is continuous, it follows that $f^{-1}(\{e\})$ is closed and therefore contains $(G \times G)^{-} = \bar{G} \times \bar{G} = H \times H$, that is, φ is a group homomorphism. $\blacksquare$

Completability can be formulated as follows:

(24.8) Theorem. *Let G be a separated topological group and let $(\hat{G}, (\mathscr{U}_s(G))^{\wedge})$ be the completion of the uniform structure $(G, \mathscr{U}_s(G))$. In order that G be completable, it is necessary and sufficient that there exist on $\hat{G}$ a group structure such that* (1) *G is a subgroup of $\hat{G}$,* (2) *the topology on $\hat{G}$ derived from the uniformity $(\mathscr{U}_s(G))^{\wedge}$ is compatible with the group structure on $\hat{G}$, and* (3) *$\mathscr{U}_s(\hat{G}) = (\mathscr{U}_s(G))^{\wedge}$. When such a group structure exists, it is unique.*

Proof. *Sufficiency:* By hypothesis, $\hat{G}$ is a topological group under the topology derived from $(\mathscr{U}_s(G))^{\wedge}$, and the left uniform structure $(\hat{G}, \mathscr{U}_s(\hat{G})) = (\hat{G}, (\mathscr{U}_s(G))^{\wedge})$ is complete, therefore $\hat{G}$ is a complete topological group (7.3), in which G is identified as a dense subgroup.

Necessity: Assuming G is completable, choose H as in (24.6). In view of (24.3), there exists a unimorphism $\varphi \colon (\hat{G}, (\mathscr{U}_s(G))^{\wedge}) \to (H, \mathscr{U}_s(H))$ such that $\varphi(a) = a$ for all $a \in G$. In particular, $(\varphi \times \varphi)^{-1}(\mathscr{U}_s(H)) = (\mathscr{U}_s(G))^{\wedge}$. Defining $xy = \varphi^{-1}(\varphi(x)\varphi(y))$ $(x, y \in \hat{G})$, $\hat{G}$ becomes a group containing G as a sub-

group, and φ is a bicontinuous group isomorphism; citing (5.12), we have
$\mathscr{U}_s(\hat{G}) = (\varphi \times \varphi)^{-1}(\mathscr{U}_s(H)) = (\mathscr{U}_s(G))^\wedge$.

Uniqueness: Quote (24.7). ∎

(24.9) Suppose G is a completable, separated topological group, and let
$\mathscr{G}$ be a filter on G that is Cauchy with respect to $\mathscr{U}_s(G)$. Choose H as in (24.6).
Then $\mathscr{G}$ may be viewed as the base of a filter $\mathscr{H}$ on H, and it is obvious that
$\mathscr{H}$ is Cauchy with respect to $\mathscr{U}_s(H)$. Since H is complete, $\mathscr{H}$ is convergent.
By continuity of inversion, the filter $\mathscr{H}^{-1} = \{A^{-1} : A \in \mathscr{H}\}$ is also convergent, and is therefore Cauchy with respect to $\mathscr{U}_s(H)$. It follows at once
that $\mathscr{G}^{-1}$ is Cauchy with respect to $\mathscr{U}_s(G)$. Summarizing, a completable,
separated topological group G has the following property: (*) If a filter $\mathscr{G}$ on
G is Cauchy with respect to $\mathscr{U}_s(G)$, then so is its image $\mathscr{G}^{-1}$ under the inversion
mapping.

It turns out that the condition (*) is also sufficient for completability
(24.17), but the following simpler result is adequate for our purposes:

(24.10) Lemma. *If G is an abelian topological group, then the mappings*
$x \mapsto x^{-1}$ *and* $(x, y) \mapsto xy$ *are uniformly continuous.*

Proof. It is understood that G is equipped with the uniformity $\mathscr{U} =$
$\mathscr{U}_s(G) = \mathscr{U}_r(G)$, and $G \times G$ with the product uniformity.

The uniform continuity of $x \mapsto x^{-1}$ is immediate from either (5.19) or
(5.12). Thus $x \mapsto x^{-1}$ is a unimorphism of $(G, \mathscr{U})$.

Write $f(x, x') = xx'$. Let V be a neighborhood of e; thus $V_s =$
$\{(x, x') : x^{-1}x' \in V\}$ is a basic entourage of $\mathscr{U}$. Let W be a neighborhood of e
such that $WW \subset V$, and let A be the following subset of $(G \times G) \times (G \times G)$:

$$A = \{((x, y), (x', y')) : (x, x') \in W_s \,\&\, (y, y') \in W_s\};$$

then A is an entourage for the product uniformity on $G \times G$, and
$((x, y), (x', y')) \in A$ iff $x^{-1}x' \in W$ and $y^{-1}y' \in W$. The uniform continuity of
f is settled by showing that $(f \times f)(A) \subset V_s$; indeed, if $((x, y), (x', y')) \in A$
then, citing commutativity, we have

$$(f(x, y))^{-1}f(x', y') = (xy)^{-1}(x'y') = y^{-1}x^{-1}x'y'$$

$$= (x^{-1}x')(y^{-1}y') \in WW \subset V,$$

thus $(f(x, y), f(x', y')) \in V_s$. ∎

(24.11) Theorem. *Every abelian, separated topological group is completable.*

Proof. Let G be an abelian, separated topological group, $\mathscr{U}$ the canonical
uniformity of G, $(G, \hat{\mathscr{U}})$ the completion of the uniform structure $(G, \mathscr{U})$.
Write $f(x, y) = xy$ and $g(x) = x^{-1}$. By the lemma, f and g are uniformly
continuous; therefore, by (24.5), f and g may be extended to unique con-
tinuous mappings $\hat{f}$ and $\hat{g}$, on $(G \times G)^\wedge = \hat{G} \times \hat{G}$ and $\hat{G}$, respectively, and

it is clear that $\hat{g}$ is a unimorphism of $(\hat{G}, \hat{\mathcal{U}})$. Define $xy = \hat{f}(x, y)$ and $x^{-1} = \hat{g}(x)$ for x, y in $\hat{G}$; it is routine to verify that the extended operations also satisfy the axioms for a group. $\blacksquare$

(24.12) Definition. If G is an abelian, separated topological group, the *completion* of G is the topological group $\hat{G}$ constructed in (24.11). Since G is a dense subgroup of $\hat{G}$, obviously $\hat{G}$ is also abelian. Thus: Every abelian, separated topological group G may be regarded as a dense subgroup of a complete, abelian, separated topological group, and the latter is essentially unique (24.7).

Consider now a separated TVS E. The uniform structure of E is derived from its additive topological group structure. In view of (24.11) and (24.8), we may regard $\hat{E}$, in a unique way, as a complete, separated topological group containing E as a dense subgroup and inducing on E its original topology. We write the group structure of $\hat{E}$ in additive notation too, noting that it is abelian. Naturally, we wish $\hat{E}$ to be a TVS including E as a linear subspace; the problem is to extend scalar multiplication to $\hat{E}$, and this is accomplished through the following lemma:

(24.13) Lemma. *Let A, B, C be complete, abelian, separated topological groups (notated additively), let A_0 and B_0 be dense subgroups of A and B, and suppose $f: A_0 \times B_0 \to C$ is bi-additive in the sense that*

$$f(a + a', b) = f(a, b) + f(a', b)$$
$$f(a, b + b') = f(a, b) + f(a, b')$$

for all a, $a' \in A_0$ and b, $b' \in B_0$. If f is continuous, then f may be extended to a unique continuous mapping $\bar{f}: A \times B \to C$, and $\bar{f}$ is also bi-additive.

Suppose now E is a separated TVS over $\mathbb{K}$, and $\hat{E}$ is the completion of its additive topological group. Write $f(\lambda, x) = \lambda x$ $(\lambda \in \mathbb{K}, x \in E)$, and apply the lemma with $A_0 = A = \mathbb{K}$, $B_0 = E$, $B = \hat{E}$, $C = \hat{E}$; it is routine to verify that the definition $\lambda x = \bar{f}(\lambda, x)$ $(\lambda \in \mathbb{K}, x \in \hat{E})$ makes $\hat{E}$ into a TVS, with E as a linear subspace. We summarize in the form of a definition:

(24.14) Definition. If E is a separated TVS over $\mathbb{K}$, the TVS $\hat{E}$ described above is called the *completion* of E. Thus, E may be identified (algebraically and topologically) as a dense linear subspace of a complete, separated TVS $\hat{E}$.

The next results are for application in Section 35.

(24.15) Proposition. *Let $(X, \mathcal{U})$ be a separated uniform structure, $(\hat{X}, \hat{\mathcal{U}})$ its completion; let $A \subset X$ and let $\bar{A}$ be the closure of A in $\hat{X}$. Then the identity mapping $i: A \to A$ may be extended to a unimorphism of $\hat{A}$ with $\bar{A}$, where $\hat{A}$ and $\bar{A}$ are equipped with their natural uniformities.*

Proof. Viewing A as a uniform structure with the relative uniformity $\mathcal{V} = \mathcal{U} \cap [A \times A]$, $(\hat{A}, \hat{\mathcal{V}})$ denotes the completion of $(A, \mathcal{V})$; on the other hand, $\bar{A}$ is viewed as a uniform structure with the relative uniformity $\mathcal{W} = \hat{\mathcal{U}} \cap [\bar{A} \times \bar{A}]$. Thus the problem is to show that there exists a unimorphism

$\varphi: (\hat{A}, \hat{\mathscr{V}}) \rightarrow (\overline{A}, \mathscr{W})$ such that $\varphi(a) = a$ for all $a \in A$. This follows at once from (24.3) and the fact that $(\overline{A}, \mathscr{W})$ is also complete. ∎

Recall that a separated uniform structure is said to be *precompact* if its completion is compact.

(24.16) Corollary. *With notation as in (24.15), A is precompact if and only if $\overline{A}$ is compact.*

Proof. With notation as in the proof of (24.15), we have $(A, \mathscr{V})$ precompact iff $(\hat{A}, \hat{\mathscr{V}})$ is compact (i.e., $\hat{A}$ is compact for the topology derived from $\hat{\mathscr{V}}$) iff the homeomorphic image $\overline{A}$ of $\hat{A}$ is compact. ∎

Exercises

(24.17) A separated topological group is completable if and only if it satisfies the condition (*) of (24.9). In particular, every bi-uniform, separated topological group is completable; however, a complete topological group need not be bi-uniform (10.8).

(24.18) (J. Dieudonné) There exist separated (even metrizable) topological groups that are not completable.

(24.19) In (24.13), the hypothesis that A and B are complete can be dropped.

Chapter 3

Convexity

25. Convex sets. Every normed space is a metrizable TVS (16.5). The converse is false; there exist metrizable TVS whose topology cannot be derived from a norm (cf. (26.7), (26.8)). Thus, in the class of metrizable TVS, normed spaces are special. One of the principal reasons is the role, in normed spaces, of the notion of convexity. This role is made precise in the next section. In the present section, we derive some of the basic propositions on convexity; the main applications will occur in the theory of locally convex TVS. *To save repetition, we assume for the rest of the section that E denotes a vector space over* $\mathbb{K}$.

(25.1) Definition. If $x, y \in E$, the *segment* from x to y is the set of all vectors of the form

$$\alpha x + (1 - \alpha)y,$$

where $0 \leq \alpha \leq 1$; such a vector is called an *internal point* of the segment if $0 < \alpha < 1$. A subset A of E is said to be *convex* if, for any $x, y \in A$, it contains the segment from x to y.

The empty subset is convex by default. In Euclidean space, the notion of segment—and therefore of convexity—agrees with geometric intuition (25.24); in general vector spaces, it is simply a useful algebraic concept.

(25.2) Example. If E is a normed space and $\varepsilon > 0$, then each of the sets

$$\{x : \|x\| < \varepsilon\}, \qquad \{x : \|x\| \leq \varepsilon\}$$

is convex; the proof is immediate from the fact that

$$\|\alpha x + (1 - \alpha)y\| \leq \alpha\|x\| + (1 - \alpha)\|y\|$$

when $0 \leq \alpha \leq 1$. When $\varepsilon = 1$ these sets are called the *open unit ball* and *closed unit ball*, respectively.

(25.3) Every singleton $\{x\}$, $x \in E$, is convex. Every linear subspace of E is convex. The intersection of any family of convex sets is convex.

(25.4) Definition. Let S be any subset of E. There exist convex supersets of S (for example, E); the intersection of the family of all convex supersets of S is a convex superset of S, called the *convex hull* of S and denoted

$$\operatorname{conv} S.$$

The convex hull of S is characterized by the properties (i) $S \subset \operatorname{conv} S$, (ii) $\operatorname{conv} S$ is convex, and (iii) if $S \subset A$ and A is convex, then $\operatorname{conv} S \subset A$. Evidently a subset A of E is convex if and only if $\operatorname{conv} A = A$; in particular, $\operatorname{conv} \varnothing = \varnothing$.

(25.5) Lemma. *If $A \subset E$ and $a \in E$, then A is convex iff $A + a$ is convex. It follows that*

$$\text{conv } (S + a) = (\text{conv } S) + a$$

for any subset $S \subset E$.

Proof. Write $f(x) = x + a$ $(x \in E)$; f is bijective, with inverse mapping $g(x) = x - a$. Since

$$\alpha f(x) + (1 - \alpha)f(y) = \alpha(x + a) + (1 - \alpha)(y + a)$$
$$= \alpha x + (1 - \alpha)y + a = f(\alpha x + (1 - \alpha)y),$$

and similarly for g, clearly A is convex if and only if $f(A) = A + a$ is convex.

Let S be any subset of E. By the assertion just proved, $(\text{conv } S) + a$ is a convex superset of $S + a$, therefore

$$(\text{conv } S) + a \supset \text{conv } (S + a).$$

Similarly, $[\text{conv } (S + a)] + (-a)$ is a convex superset of $(S + a) + (-a) = S$, therefore

$$[\text{conv } (S + a)] + (-a) \supset \text{conv } S,$$

thus $\text{conv } (S + a) \supset (\text{conv } S) + a$. ∎

(25.6) Theorem. *Let E and F be vector spaces over $\mathbb{K}$ and let $u: E \to F$ be a linear mapping.*

(i) *If A is a convex subset of E, then $u(A)$ is a convex subset of F.*
(ii) *If B is a convex subset of F, then $u^{-1}(B)$ is a convex subset of E.*
(iii) *If S is any subset of E, then $u(\text{conv } S) = \text{conv } u(S)$.*
(iv) *If T is any subset of F, then $u^{-1}(\text{conv } T) \supset \text{conv } u^{-1}(T)$. If, moreover, $T \subset u(E)$, then $u^{-1}(\text{conv } T) = \text{conv } u^{-1}(T)$.*

Proof. (i) and (ii) follow easily from the relation

$$u(\alpha x + (1 - \alpha)y) = \alpha u(x) + (1 - \alpha)u(y).$$

(iii) Citing (i), $u(\text{conv } S)$ is a convex superset of $u(S)$, therefore

$$u(\text{conv } S) \supset \text{conv } u(S).$$

On the other hand, since $u(S) \subset \text{conv } u(S)$, that is,

$$S \subset u^{-1}(\text{conv } u(S)),$$

and since the right side of the inclusion is convex by (ii), it follows that

$$\text{conv } S \subset u^{-1}(\text{conv } u(S)),$$

that is, $u(\text{conv } S) \subset \text{conv } u(S)$.

(iv) By (ii), $u^{-1}(\text{conv } T)$ is a convex superset of $u^{-1}(T)$, therefore

$$u^{-1}(\text{conv } T) \supset \text{conv } u^{-1}(T).$$

Suppose, in addition, that $T \subset u(E)$, and let $S = u^{-1}(T)$; then $u(S) = T$, and, by (iii),

$$\text{conv } T = \text{conv } u(S) = u(\text{conv } S) = u(\text{conv } u^{-1}(T)),$$

therefore

$$u^{-1}(\text{conv } T) = u^{-1}u(\text{conv } u^{-1}(T)).$$

The proof will be concluded by showing that

$$u^{-1}u(\text{conv } u^{-1}(T)) = \text{conv } u^{-1}(T);$$

in other words, if $N = \{x \in E : u(x) = \theta\}$ is the null space of u, it is to be shown that

$$N + \text{conv } u^{-1}(T) = \text{conv } u^{-1}(T).$$

If $a \in N$ then $a + u^{-1}(T) = u^{-1}(T)$ (because $u^{-1}(T)$ is a union of cosets modulo N), therefore, by the lemma,

$$a + \text{conv } u^{-1}(T) = \text{conv } (a + u^{-1}(T)) = \text{conv } u^{-1}(T). \quad \blacksquare$$

(25.7)   With notation as in (25.6), let b be a fixed vector in F and consider the mapping $v \colon E \to F$ defined by

$$v(x) = u(x) + b \qquad (x \in E).$$

(Such mappings are called *affine*.) Thus v is a linear mapping followed by a translation mapping. Evidently

$$v(\alpha x + (1 - \alpha)y) = \alpha v(x) + (1 - \alpha)v(y);$$

replacing u by v in (25.6), it is clear that (i), (ii), (iii) and the inclusion in (iv) hold also for v. If, moreover, $T \subset v(E)$, that is, if $T - b \subset u(E)$, then

(*) $$u^{-1}(\text{conv } (T - b)) = \text{conv } u^{-1}(T - b);$$

since $u^{-1}(T - b) = v^{-1}(T)$ and

$$u^{-1}(\text{conv } (T - b)) = u^{-1}((\text{conv } T) - b) = v^{-1}(\text{conv } T),$$

the relation (*) may be written

$$v^{-1}(\text{conv } T) = \text{conv } v^{-1}(T).$$

Briefly: *Theorem (25.6) holds with u replaced by v.*

(25.8) Theorem.   *The cartesian product of a family of nonempty sets is convex if and only if each factor is convex. More precisely, let $(E_\iota)_{\iota \in I}$ be a family of vector spaces over $\mathbb{K}$ and let $E = \prod_{\iota \in I} E_\iota$ be the product vector space. For each $\iota \in I$ let A_ι be a nonempty subset of E_ι, and let $A = \prod_{\iota \in I} A_\iota$. Then A is convex if and only if every A_ι is convex.*

Proof.   For each $\iota \in I$, let $\pi_\iota \colon E \to E_\iota$ be the canonical projection; then $\pi_\iota(A) = A_\iota$ (the axiom of choice is at work here). If A is convex then, for every $\iota \in I$, $A_\iota = \pi_\iota(A)$ is convex by (25.6). Conversely, if every A_ι is convex then, by (25.6), $\pi_\iota^{-1}(A_\iota)$ is convex for all $\iota \in I$, therefore

$$A = \bigcap_{\iota \in I} \pi_\iota^{-1}(A_\iota)$$

is also convex (25.3). ∎

(25.9) Corollary. *If A and B are convex subsets of E and if λ, $\mu \in \mathbb{K}$, then the set*

$$\lambda A + \mu B = \{\lambda x + \mu y : x \in A, y \in B\}$$

is also convex.

Proof. Since $A \times B$ is a convex subset of the product vector space $E \times E$ (25.8) and the mapping $u: E \times E \to E$ defined by $u(x, y) = \lambda x + \mu y$ is linear, it follows from (25.6) that $u(A \times B) = \lambda A + \mu B$ is convex. ∎

(25.10) Theorem. *If A is a convex subset of E, and if $\alpha \geq 0$, $\beta \geq 0$, then $(\alpha + \beta)A = \alpha A + \beta A$, that is,*

$$\{(\alpha + \beta)z : z \in A\} = \{\alpha x + \beta y : x, y \in A\}.$$

Proof. If $\alpha = 0$ or $\beta = 0$ the conclusion holds trivially; assume $\alpha > 0$ $\beta > 0$. The left side is obviously contained in the right side. On the other hand, if $x, y \in A$ then, setting $\alpha_1 = \alpha/(\alpha + \beta)$ and $\beta_1 = \beta/(\alpha + \beta)$, we have

$$\alpha x + \beta y = (\alpha + \beta)(\alpha_1 x + \beta_1 y);$$

since $\alpha_1 > 0, \beta_1 > 0$ and $\alpha_1 + \beta_1 = 1$, it follows from the convexity of A that $z = \alpha_1 x + \beta_1 y$ is in A, thus $\alpha x + \beta y = (\alpha + \beta)z \in (\alpha + \beta)A$. ∎

As noted in the foregoing proof, the convexity of A means that $\lambda_1 x_1 + \lambda_2 x_2 \in A$ whenever $x_1, x_2 \in A$, $\lambda_1 \geq 0$, $\lambda_2 \geq 0$, and $\lambda_1 + \lambda_2 = 1$. One calls $\lambda_1 x_1 + \lambda_2 x_2$ a 'convex combination' of x_1 and x_2. More generally:

(25.11) Definition. A vector of the form

$$\sum_1^n \lambda_k x_k,$$

where $\sum_1^n \lambda_k = 1$ and $\lambda_k \geq 0$ for all k, is called a *convex combination* of $x_1, \ldots, x_n$.

(25.12) Theorem. *If A is a convex subset of E and if $x_1, \ldots, x_n \in A$, then A contains every convex combination of $x_1, \ldots, x_n$.*

Proof. For $n = 1$ the only convex combination is $1x_1 = x_1$. For $n = 2$, cite the definition of convexity (25.1). Assuming inductively that the assertion of the theorem is true for $n - 1$ elements, let $x = \sum_1^n \lambda_k x_k$, where $x_1, \ldots, x_n \in A$, $\sum_1^n \lambda_k = 1$ and $\lambda_k \geq 0$ for all k; the problem is to show that $x \in A$. If $\lambda_k = 0$ for some k, clearly $x \in A$ by the inductive hypothesis. Suppose $\lambda_k > 0$ for all k. In particular, setting $\mu = \sum_1^{n-1} \lambda_k$, we have $\mu + \lambda_n = 1$, where $\mu > 0$ and $\lambda_n > 0$. Then

$$x = \sum_1^n \lambda_k x_k = \sum_1^{n-1} \lambda_k x_k + \lambda_n x_n$$

$$= \mu \sum_1^{n-1} (\lambda_k/\mu)x_k + (1 - \mu)x_n.$$

Setting $y = \sum_1^{n-1} (\lambda_k/\mu)x_k$, we have

(*) $x = \mu y + (1 - \mu)x_n.$

The inductive hypothesis yields $y \in A$, because

$$\sum_1^{n-1} \lambda_k/\mu = (1/\mu) \sum_1^{n-1} \lambda_k = (1/\mu)\mu = 1$$

and $\lambda_k/\mu > 0$ for $k = 1, \ldots, n - 1$. It then follows from (*), and the convexity of A, that $x \in A$. ∎

It follows that if S is a subset of E and if $A = \operatorname{conv} S$, then every convex combination $\sum_1^n \lambda_k x_k$, where $x_1, \ldots, x_n \in S$, belongs to A. In fact, the set of all such convex combinations exhausts A; this is a corollary of the following theorem:

(25.13) Theorem.   *Let $(A_\iota)_{\iota \in I}$ be a family of convex subsets of E and let*

$$S = \bigcup_{\iota \in I} A_\iota.$$

Then conv S *coincides with the set of all convex combinations*

(*) $x = \sum_{\iota \in I} \lambda_\iota x_\iota,$

where $\lambda_\iota = 0$ for all but finitely many ι, $\lambda_\iota \geq 0$ for all ι, $\sum_{\iota \in I} \lambda_\iota = 1$, and $x_\iota \in A_\iota$ whenever $\lambda_\iota > 0$.

Proof.   Let A be the set of all such vectors x; clearly the choice of x_ι is immaterial when $\lambda_\iota = 0$ (e.g., it may as well be taken to be θ). If $S = \varnothing$ then no such vectors can be formed (the conditions on the λ_ι and x_ι are incompatible, since a sum over an empty family of indices is equal to zero by convention), therefore $A = \varnothing = \operatorname{conv} S$.

Assume $S \neq \varnothing$. If $x \in A$ then x is a convex combination of finitely many elements of S; since $S \subseteq \operatorname{conv} S$, it follows from (25.12) that $x \in \operatorname{conv} S$. Thus $A \subseteq \operatorname{conv} S$. Clearly $S \subseteq A$; to prove the decisive inclusion conv $S \subseteq A$, it will suffice to show that A is convex. Suppose $x, y \in A$ and $0 < \alpha < 1$. Say

$$x = \sum_{\iota \in I} \lambda_\iota x_\iota, \qquad y = \sum_{\iota \in I} \mu_\iota y_\iota,$$

subject to the regulations governing (*). Writing $z = \alpha x + (1 - \alpha)y$, we have

$$z = \sum_{\iota \in I} [\alpha \lambda_\iota x_\iota + (1 - \alpha)\mu_\iota y_\iota];$$

the problem is to show that $z \in A$. Define

$$\gamma_\iota = \alpha \lambda_\iota + (1 - \alpha)\mu_\iota.$$

Clearly $\gamma_\iota \geq 0$ for all ι, $\gamma_\iota = 0$ for all but finitely many ι, and

$$\sum_{\iota \in I} \gamma_\iota = \alpha \sum_{\iota \in I} \lambda_\iota + (1 - \alpha) \sum_{\iota \in I} \mu_\iota$$

$$= \alpha 1 + (1 - \alpha)1 = 1.$$

Let $J = \{\iota \in I : \gamma_\iota > 0\}$. If $\iota \notin J$, that is, if $\gamma_\iota = 0$, then it is clear from the defining formula that $\lambda_\iota = \mu_\iota = 0$, therefore $\alpha\lambda_\iota x_\iota + (1 - \alpha)\mu_\iota y_\iota = \theta$; thus,

$$z = \sum_{\iota \in J} [\alpha\lambda_\iota x_\iota + (1 - \alpha)\mu_\iota y_\iota]$$

$$= \sum_{\iota \in J} \gamma_\iota [\gamma_\iota^{-1}\alpha\lambda_\iota x_\iota + \gamma_\iota^{-1}(1 - \alpha)\mu_\iota y_\iota],$$

where $\sum_{\iota \in J} \gamma_\iota = 1$. For $\iota \notin J$ choose $z_\iota \in E$ arbitrarily (e.g., let $z_\iota = \theta$); thus, $\gamma_\iota z_\iota = 0 z_\iota = \theta$ for all $\iota \notin J$. For $\iota \in J$, set

$$z_\iota = \gamma_\iota^{-1}\alpha\lambda_\iota x_\iota + \gamma_\iota^{-1}(1 - \alpha)\mu_\iota y_\iota;$$

then $z_\iota \in A_\iota$, because $x_\iota, y_\iota \in A_\iota$ and the coefficients of x_ι, y_ι are positive with sum 1. Thus

$$z = \sum_{\iota \in J} \gamma_\iota z_\iota = \sum_{\iota \in I} \gamma_\iota z_\iota,$$

where the families $(\gamma_\iota)_{\iota \in I}$ and $(z_\iota)_{\iota \in I}$ satisfy the conditions governing (*). Thus $z \in A$. ∎

(25.14) Corollary. *If S is any nonempty subset of E, then conv S coincides with the set of all convex combinations*

$$x = \sum_1^n \lambda_k x_k,$$

where $\lambda_k \geq 0$ for $k = 1, \ldots, n$, $\sum_1^n \lambda_k = 1$, and $\{x_1, \ldots, x_n\}$ is a nonempty finite subset of S.

Proof. Let $(x_\iota)_{\iota \in I}$ be any indexing of S (e.g., the identity indexing), and let $A_\iota = \{x_\iota\}$ for all $\iota \in I$. Then the A_ι are convex, with $\bigcup_{\iota \in I} A_\iota = S$, and the assertion of the corollary is immediate from the theorem. ∎

(25.15) Corollary. *If S is a balanced subset of E, then conv S is also balanced.*

Proof. If $|\lambda| \leq 1$ and $x \in$ conv S, say $x = \sum_1^n \lambda_k x_k$ as in (25.14), then

$$\lambda x = \sum_1^n \lambda_k(\lambda x_k),$$

where $\lambda x_k \in S$ by hypothesis, thus $\lambda x \in$ conv S. ∎

The rest of the section is concerned with convex sets in a TVS (a theme that will be developed in greater depth later on).

(25.16) Definition. Let E be a TVS over $\mathbb{K}$ and let S be any subset of E. There exist closed, convex supersets of S (for example, E); the intersection of the family of all closed, convex supersets of S is itself a closed, convex superset of S, called the *closed convex hull* of S. This coincides with the closure of conv S, as is shown below (25.18).

(25.17) Theorem. *If E is a TVS over $\mathbb{K}$ and if A is a convex subset of E, then $\bar{A}$ is also convex.*

Proof. Fix α, $0 \leq \alpha \leq 1$. The mapping $u: E \times E \to E$ defined by $u(x, y) = \alpha x + (1 - \alpha)y$ is continuous, and maps $A \times A$ into A, therefore it maps $\bar{A} \times \bar{A} = (A \times A)^-$ into $\bar{A}$. Since α is arbitrary, it follows that $\bar{A}$ is convex. ∎

(25.18) Corollary. *If E is a TVS over $\mathbb{K}$ and if S is any subset of E, then the closed convex hull of S coincides with* $(\text{conv } S)^-$.

Proof. Let $A = \text{conv } S$ and let B be the closed convex hull of S. Since B is a convex superset of S, we have $A \subset B$ and therefore $\bar{A} \subset \bar{B} = B$. On the other hand, $\bar{A}$ is a closed convex superset of S by the theorem, therefore $B \subset \bar{A}$. ∎

According to (25.17), the closure of a convex set A in a TVS is convex. We show in the next theorem that the interior of A is also convex; this is trivial if the interior is empty, so we look at convex sets with nonempty interior:

(25.19) Lemma. *Let E be a TVS over $\mathbb{K}$ and let A be a convex subset of E with nonempty interior. Suppose $a \in \text{int } A$ and $b \in \bar{A}$. Then every internal point of the segment from a to b is an interior point of A.*

Proof. Let c be an internal point of the segment from a to b, say

$$c = \alpha a + (1 - \alpha)b,$$

where $0 < \alpha < 1$. The problem is to show that c is interior to A, i.e., that A is a neighborhood of c.

Let $\lambda = \alpha/(\alpha - 1)$ and let $f: E \to E$ be the homothety (11.4) with center c and weight λ, that is,

$$f(x) = \lambda(x - c) + c \qquad (x \in E);$$

thus f is a homeomorphism of E, and $f(c) = c$. By straightforward elementary calculations, $f(a) = b$ and

(*) $\alpha x + (1 - \alpha)f(x) = c$ for all $x \in E$.

Let V be an open neighborhood of a such that $V \subset A$. Then $f(V)$ is a neighborhood of $f(a) = b$; since $b \in \bar{A}$, $f(V)$ must intersect A, say $v \in V$ with $f(v) \in A$.

Let g be the homothety with center $f(v)$ and weight α, that is,

$$g(x) = \alpha(x - f(v)) + f(v) = \alpha x + (1 - \alpha)f(v) \qquad (x \in E).$$

In view of (*), $g(v) = c$; since g is a homeomorphism of E, and V is an open neighborhood of v, it follows that $g(V)$ is an open neighborhood of $g(v) = c$. The proof is concluded by showing that $g(V) \subset A$: if $x \in V$ then

$$g(x) = \alpha x + (1 - \alpha)f(v)$$

is a convex combination of the vectors $x, f(v)$ of A. ∎

(25.20) Theorem. *Let E be a TVS over $\mathbb{K}$ and suppose that A is a convex subset of E with nonempty interior. Then* (i) int A *is convex,* (ii) $\overline{A} = ($int $A)^-$, *and* (iii) int $\overline{A} = $ int A.

Proof. (i) If $a, b \in$ int A, then every internal point of the segment from a to b is also in int A by the lemma. {For a simpler proof, see (25.29).}

(ii) Obviously $\overline{A} \supset ($int $A)^-$. Conversely, if $b \in \overline{A}$ and V is any neighborhood of b, it is to be shown that V intersects int A. Choose any $a \in$ int A and define $c_\alpha = \alpha a + (1 - \alpha)b$ for $0 < \alpha < 1$. By the lemma, $c_\alpha \in$ int A for all α; it will suffice to find an α such that $c_\alpha \in V$. Informally, as $\alpha \to 0$ one has $c_\alpha \to 0a + 1b = b$, therefore $c_\alpha \in V$ for α sufficiently near 0. More precisely, $c_\alpha = \alpha(a - b) + b$, and $V - b$ is a neighborhood of θ; for α sufficiently small, $c_\alpha - b = \alpha(a - b) \in V - b$ by the continuity of scalar multiplication, i.e., $c_\alpha \in V$.

(iii) Obviously int $\overline{A} \supset$ int A. Conversely, assuming $x \in$ int $\overline{A}$, it is to be shown that $x \in$ int A. Translating by $-x$, we can suppose $x = \theta$ (a translation is a convexity-preserving (25.5) homeomorphism). Let V be a neighborhood of θ such that $V \subset \overline{A}$ and such that V is symmetric, i.e., $-V = V$ (3.1). Since $\theta \in \overline{A}$, and $\overline{A} = ($int $A)^-$ by (ii), every neighborhood of θ must intersect int A; in particular, $V \cap$ int A is nonempty, say $a \in V \cap$ int A. If $a = \theta$ then $\theta \in$ int A as asserted. Otherwise, let $b = -a$; then $b \in -V = V \subset \overline{A}$. By the lemma, int A contains every internal point of the segment from a to b; in particular, $\theta = \frac{1}{2}a + \frac{1}{2}b \in$ int A. ∎

(25.21) Definition. Let E be a TVS over $\mathbb{K}$. A *convex body* in E is a closed, convex subset of E with nonempty interior.

(25.22) Corollary. *If A is a convex body in a TVS, then $A = ($int $A)^-$.*

Proof. Citing (ii) of (25.20), we have $($int $A)^- = \overline{A} = A$. ∎

Exercises

(25.23) A convex subset of a TVS is arcwise connected.

(25.24) Let $\mathbb{R}^r$ be r-dimensional Euclidean space, with distance function

$$\|x - y\| = \left(\sum_1^r (x_k - y_k)^2 \right)^{1/2}.$$

If $x, y, z \in \mathbb{R}^r$, then z belongs to the segment from x to y if and only if

$$\|x - z\| + \|z - y\| = \|x - y\|.$$

(25.25) If E is a metrizable TVS over $\mathbb{K}$, there exists an additively invariant compatible metric d such that $d(x, y) < 1$ for all $x, y \in E$ (13.8). Writing $|x| = d(x, \theta)$, one has

$$\{x : |x| < 1\} = E.$$

Compare with the case of a norm (25.2); in a normed space, the relation $\|nx\| = n\|x\|$ $(n = 1, 2, 3, \ldots)$ precludes such behavior.

(25.26) If $\alpha\beta < 0$ the conclusion of (25.10) may fail.

(25.27)   Let E be a vector space over $\mathbb{K}$. A subset A of E is said to be *circled* if $\lambda A \subset A$ whenever $|\lambda| = 1$. Prove: A convex subset of E is circled iff it is balanced.

(25.28)   Let E be a vector space over $\mathbb{K}$. A subset A of E is said to be *absolutely convex* if $\lambda A + \mu A \subset A$ whenever $|\lambda| + |\mu| \leq 1$. The following conditions on a set A are equivalent: (a) A is absolutely convex; (b) A is balanced and convex; (c) A is circled (25.27) and convex. If S is any subset of E, the *absolutely convex hull* of S is defined in the obvious way (cf. the format of (25.4)). Prove: If S is any subset of E, then the absolutely convex hull of S coincides with the convex hull of the balanced hull of S.

(25.29)   An immediate consequence of (25.14) is that the convex hull of an open set is open. This yields a simple proof of the fact (25.20) that the interior of a convex set is convex.

(25.30)   If E is a real vector space of finite dimension n, and if S is a non-empty subset of E, then conv S is the set of all convex combinations

$$\sum_{k=0}^{n} \alpha_k x_k,$$

where $x_k \in S$, $\alpha_k \geq 0$, and $\sum_0^n \alpha_k = 1$ (the point is that sums of length $n + 1$ are sufficient).

*26. Kolmogorov's normability criterion.   Every normed space is a metrizable TVS (16.5). Which metrizable TVS are normed spaces? Since absolute homogeneity is highly perishable under remetrization (16.13), the question needs to be reformulated:

(26.1) Definition.   A topological vector space E over $\mathbb{K}$ is said to be *normable* if there exists a norm $x \mapsto \|x\|$ on E such that the norm topology (16.3) coincides with the given topology.

The proper question is: Which metrizable TVS are normable? A clue to the answer is to be found in the fact that a metrizable TVS E can always be metrized with an invariant metric d such that $d(x, y) < 1$ for all x, y (11.11), (13.8); for such a metric,

$$\{x : d(x, \theta) < 1\} = E,$$

which suggests that (i) the notion of 'bounded set' will have to be revised to be of any use in a general metrizable TVS, and (ii) nontrivial convex open sets are harder to construct in a general TVS than in a normed space (cf. 25.2). The decisive notion of boundedness is as follows:

(26.2) Definition.   A subset A of a TVS is said to be *bounded* if, for every neighborhood V of θ, there exists a scalar λ such that $A \subset \lambda V$.

In a normed space, the above notion of boundedness coincides with the classical one:

(26.3) Lemma.   *A subset A of a normed space is bounded in the sense of (26.2) if and only if there exists a constant $M > 0$ such that $\|x\| \leq M$ for all x in A.*

Proof. Let $U = \{x : \|x\| \leq 1\}$. If A is bounded in the sense of (26.2), let λ be a nonzero scalar such that $A \subset \lambda U$, and set $M = |\lambda|$; for all $x \in A$ one has $x \in \lambda U$, $\lambda^{-1}x \in U$, $\|\lambda^{-1}x\| \leq 1$, $\|x\| \leq |\lambda| = M$.

Conversely, suppose there exists a positive constant M such that $\|x\| \leq M$ for all $x \in A$, i.e., $A \subset MU$, and let V be any neighborhood of θ. Choose $\varepsilon > 0$ so that $\{x : \|x\| \leq \varepsilon\} \subset V$, i.e., $\varepsilon U \subset V$. Then $A \subset MU \subset (M/\varepsilon)V$, thus $\lambda = M/\varepsilon$ meets the requirement of (26.2). ∎

The set U in the foregoing proof is a bounded, convex neighborhood of θ. The existence of such a neighborhood is characteristic of normable spaces:

(26.4) Theorem. (A. Kolmogorov)   *A topological vector space E over $\mathbb{K}$ is normable if and only if it is separated and possesses a bounded convex neighborhood of zero.*

Proof. If E is normable, and if $x \mapsto \|x\|$ is a norm on E that generates the topology, then $U = \{x : \|x\| \leq 1\}$ is a bounded (26.3) and convex (25.2) neighborhood of θ. Obviously E is separated (as is any metrizable space).

Conversely, suppose E is a separated TVS possessing a bounded convex neighborhood U of θ. Let V be a balanced neighborhood of θ such that $V \subset U$ (17.9). Then conv $V \subset$ conv $U = U$, where conv V is a balanced (25.15) convex neighborhood of θ. Clearly, any subset of a bounded set is bounded; thus, replacing U by conv V, we may suppose that U is a *bounded, convex, balanced neighborhood of θ.*

For every nonzero vector x, define

$$A(x) = \{\lambda \in \mathbb{K} : x \notin \lambda U\};$$

obviously $0 \in A(x)$. Define $A(\theta) = \{0\}$.

If x is any nonzero vector, we assert that $A(x)$ contains nonzero scalars. Indeed, if V is a neighborhood of θ such that $x \notin V$ (E is separated) and α is a nonzero scalar such that $U \subset \alpha V$ (U is bounded), i.e., $\alpha^{-1}U \subset V$, then $x \notin \alpha^{-1}U$ and so $\alpha^{-1} \in A(x)$.

Define a function $x \mapsto \|x\|$ by the formula

$$\|x\| = \sup\{|\lambda| : \lambda \in A(x)\} \qquad (x \in E).$$

Clearly $\|\theta\| = 0$ and $\|x\| > 0$ when $x \neq \theta$. It is elementary that

$$A(\mu x) = \mu A(x) \qquad (\mu \in \mathbb{K}, x \in E),$$

therefore $\|\mu x\| = |\mu|\,\|x\|$. To show that $x \mapsto \|x\|$ is a norm, it remains to verify that $\|x\| < \infty$ and that $\|x + y\| \leq \|x\| + \|y\|$. To this end, we first show that

(*) $$\{\lambda : |\lambda| < \|x\|\} \subset A(x).$$

If $x = \theta$ then the left side of (*) is empty. Assume $x \neq \theta$, and suppose $|\lambda| < \|x\|$. Since $0 \in A(x)$ we can suppose $0 < |\lambda| < \|x\|$. Then, by the definition of $\|x\|$, there exists $\mu \in A(x)$ such that $|\lambda| < |\mu| \leq \|x\|$. Thus, $x \notin \mu U$; since U is balanced, $\lambda U \subset \mu U$ (17.2), therefore $x \notin \lambda U$, i.e., $\lambda \in A(x)$.

Let $x \in E$. We assert that $\|x\| < \infty$. Since U is absorbent (17.11), there exists a nonzero scalar λ such that $\lambda x \in U$, i.e., $1 \notin A(\lambda x)$; then $\|\lambda x\| \leq 1$ by (*), thus $|\lambda| \|x\| \leq 1$, which shows that $\|x\|$ is finite.

Let $x, y \in E$. We assert that $\|x + y\| \leq \|x\| + \|y\|$. We can suppose x, y and $x + y$ to be nonzero. Given any $\varepsilon > 0$, it will suffice to show that

$$\|x + y\| \leq \|x\| + \|y\| + 2\varepsilon.$$

Let $\alpha = \|x\| + \varepsilon, \beta = \|y\| + \varepsilon$. Since $\alpha > \|x\|$ it follows from the definition of $\|x\|$ that $\alpha \notin A(x)$, i.e., $x \in \alpha U$. Similarly, $y \in \beta U$. Since U is convex, it follows (25.10) that

$$x + y \in \alpha U + \beta U = (\alpha + \beta)U,$$

thus $\alpha + \beta \notin A(x + y)$; in view of (*), this implies

$$\alpha + \beta \geq \|x + y\|,$$

and the assertion is proved.

Summarizing, $x \mapsto \|x\|$ is a norm on E. It remains to show that the norm topology coincides with the given topology. Since both topologies are compatible with the additive group structure, it is sufficient to verify that their neighborhood systems at θ coincide.

Suppose V is any neighborhood of θ for the given topology. Choose a nonzero scalar λ such that $U \subset \lambda V$. If $\|x\| < |\lambda|^{-1}$ then $\lambda^{-1} \notin A(x)$ (by the definition of $\|x\|$), i.e., $x \in \lambda^{-1}U \subset V$; thus

$$\{x : \|x\| < |\lambda|^{-1}\} \subset V,$$

which shows that V is a neighborhood of θ for the norm topology.

Conversely, let V be any neighborhood of θ for the norm topology. Choose $\varepsilon > 0$ so that $\{x : \|x\| \leq \varepsilon\} \subset V$. If $x \in \varepsilon U$ then $\varepsilon \notin A(x)$, therefore $\|x\| \leq \varepsilon$ by (*); thus $\varepsilon U \subset V$. Since εU is a neighborhood of θ for the given topology, so is V. ∎

Exercises

(26.5)   A subset A of a TVS is bounded iff for every sequence $x_n \in A$, and for every sequence of scalars λ_n such that $\lambda_n \to 0$, one has $\lambda_n x_n \to \theta$.

(26.6)   Let E and F be TVS over $\mathbb{K}$ and let $T: E \to F$ be a continuous linear mapping. If A is a bounded subset of E then $T(A)$ is a bounded subset of F.

(26.7)   The space (S) of Section 15 is not normable.

(26.8)   The space (s) of Section 14 is not normable; indeed, no neighborhood of θ is bounded in the sense of (26.2).

27. Convex cones.
Throughout this section E denotes a vector space over $\mathbb{R}$.

(27.1) Definition.   A subset A of E is called a *cone*, with vertex θ, if $\alpha A \subset A$ whenever $\alpha > 0$.

(27.2) More generally, a 'cone with vertex x' may be defined as a set of the form $x + A$, where A is a cone with vertex θ. We shall have no occasion to use the more general term; for our purposes, 'cone' will mean 'cone with vertex θ.'

(27.3) Definition. A cone A in E is said to be *pointed* if $\theta \in A$, *unpointed* if $\theta \notin A$.

(27.4) Every linear subspace of E is a cone. The empty set is a cone. If A is a cone then $A \cup \{\theta\}$ is a pointed cone, and $A - \{\theta\}$ is an unpointed cone.

Of special importance are the convex cones, for which one has the following characterization:

(27.5) Theorem. *A subset A of E is a convex cone if and only if* (i) $\alpha A \subset A$ *for all $\alpha > 0$, and* (ii) $A + A \subset A$.

Proof. Condition (i) is the definition of a cone. If, moreover, A is convex, then condition (ii) is a consequence of condition (i); indeed, citing (25.10), we have $A + A = 1A + 1A = (1 + 1)A = 2A \subset A$.

Conversely, assuming A satisfies (i) and (ii), it is to be shown that A is convex. Indeed, if $0 < \alpha < 1$ then $\alpha A + (1 - \alpha)A \subset A + A \subset A$. ∎

(27.6) Definition. Let S be any subset of E. There exist convex cones containing S (for example, E); the intersection of the family of all convex cones containing S is a convex cone containing S, called the convex cone *generated* by S. Explicitly:

(27.7) Corollary. *If S is any subset of E, then the convex cone generated by S is the set*

$$\bigcup_{\alpha > 0} \alpha(\text{conv } S).$$

Proof. Let $A = \text{conv } S$, let

$$B = \bigcup_{\alpha > 0} \alpha A,$$

and let C be the convex cone generated by S; the problem is to show that $B = C$. Obviously $C \supset B$ and $S \subset B$; to prove the decisive inclusion $C \subset B$ it will suffice to show that B is a convex cone. If $\beta > 0$ then

$$\beta B = \bigcup_{\alpha > 0} \beta(\alpha A) = \bigcup_{\alpha > 0} (\beta\alpha)A = \bigcup_{\gamma > 0} \gamma A = B,$$

thus condition (i) of (27.5) is verified. If $\alpha > 0$ and $\beta > 0$ then, citing (25.10), we have

$$\alpha A + \beta A = (\alpha + \beta)A \subset B;$$

it follows that $B + B \subset B$, thus condition (ii) of (27.5) is verified. ∎

In particular, if A is a convex subset of E then the convex cone generated by A is $\bigcup_{\alpha > 0} \alpha A$.

Another consequence of (27.5) is a simple description of the linear subspace spanned by a convex cone:

(27.8) Corollary. *If A is a nonempty convex cone in E, then the linear subspace of E generated by A is the set $A - A = \{x - y : x \in A, y \in A\}$.*

Proof. (The use of the minus sign in the present context is exceptional; ordinarily we reserve it for set-theoretic differences.) Let N be the linear subspace generated by A. Obviously $N \supset A - A$. On the other hand, $A \subset A - A$ (if $x \in A$ then $x = 2x - x \in A - A$); to establish the decisive inclusion $N \subset A - A$ it will suffice to show that $A - A$ is a linear subspace of E. Since A is nonempty, we have $\theta \in A - A$. Obviously $-(A - A) = A - A$, i.e., $A - A$ contains negatives. If $\alpha > 0$ then $\alpha(A - A) = \alpha A - \alpha A \subset A - A$; combined with the preceding remarks, this shows that $\lambda(A - A) \subset A - A$ for all real λ. Finally, citing (27.5), we have

$$(A - A) + (A - A) = (A + A) - (A + A) \subset A - A,$$

thus $A - A$ is closed under addition. ∎

For a pointed convex cone, there is another naturally related linear subspace:

(27.9) Corollary. *If A is a pointed convex cone in E, then $A \cap (-A)$ is a linear subspace of E; it is the largest linear subspace contained in A.*

Proof. Let $N = A \cap (-A)$. Since A is pointed, $\theta \in N$. Obviously $-N = N$, i.e., N contains negatives. If $\alpha > 0$ then $\alpha N = (\alpha A) \cap (-\alpha A) \subset A \cap (-A) = N$; combined with the preceding remarks, this shows that $\lambda N \subset N$ for all real λ. Moreover, citing (27.5), we have

$$N + N \subset (A + A) \cap (-A - A) \subset A \cap (-A) = N,$$

thus N is a linear subspace.

If M is any linear subspace with $M \subset A$, then also $M = -M \subset -A$, thus $M \subset A \cap (-A) = N$; this proves the maximality of N. ∎

If A is an unpointed convex cone, then $A \cup \{\theta\}$ is clearly a (pointed) convex cone. However, if A is a pointed convex cone, then $A - \{\theta\}$ is a cone but it may fail to be convex. {Example: A a nonzero linear subspace of E.} Precisely, the situation is as follows:

(27.10) Theorem. *If A is a pointed convex cone in E, the following conditions on A are equivalent:*
(a) $A - \{\theta\}$ *is a convex cone;*
(b) $A \cap (-A) = \{\theta\}$;
(c) *A contains no nonzero linear subspace.*

Proof. The equivalence of (b) and (c) is clear from (27.9).

(a) implies (b): If $A \cap (-A)$ contains a nonzero vector x, then $x \in A - \{\theta\}$, $-x \in A - \{\theta\}$, and $\frac{1}{2}x + \frac{1}{2}(-x) = \theta$, therefore $A - \{\theta\}$ is not convex.

(b) implies (a): Assuming $A \cap (-A) = \{\theta\}$, suppose $x, y \in A - \{\theta\}$ and

$0 < \alpha < 1$; it is to be shown that $\alpha x + (1 - \alpha)y \in A - \{\theta\}$. Since A is convex, $\alpha x + (1 - \alpha)y \in A$; $\alpha x + (1 - \alpha)y = \theta$ would imply

$$\alpha x = -(1 - \alpha)y \in A \cap (-A),$$

a contradiction. ∎

(27.11) Definition. A pointed convex cone A is said to be *salient* if $A \cap (-A) = \{\theta\}$.

Exercises

(27.12) Let E and F be vector spaces over $\mathbb{R}$, let A, B be convex cones in E, F, respectively, and let $f: E \to F$ be a linear mapping. Prove:

(i) $f(A)$ is a convex cone in F.

(ii) $f^{-1}(B)$ is a convex cone in E.

(iii) If A is pointed then so is $f(A)$; if, moreover, A is salient and f is injective, then $f(A)$ is also salient.

(iv) If B is pointed, then so is $f^{-1}(B)$; if, moreover, B is salient, then $f^{-1}(B)$ is salient if and only if f is injective.

(27.13) Let $(A_\iota)_{\iota \in I}$ be a family of convex cones in E and let

$$S = \bigcup_{\iota \in I} A_\iota.$$

Then the convex generated by S consists of the set of all sums $x = \sum_{\iota \in J} x_\iota$, where J is a nonempty finite subset of E, and $x_\iota \in A_\iota$ for all $\iota \in J$.

(27.14) If S is any subset of E, then the convex cone generated by S consists of the set of all sums $x = \sum_{\iota \in J} \alpha_\iota y_\iota$, where J is a nonempty finite subset of I, $y_\iota \in S$ ($\iota \in J$), and $\alpha_\iota > 0$ ($\iota \in J$).

(27.15) Let E be a TVS over $\mathbb{R}$ and let A be a convex cone in E. (i) If $A \neq \varnothing$ then $\bar{A}$ is a pointed convex cone. (ii) If $A \neq E$ then int A is an unpointed convex cone.

28. Hahn-Banach theorem.

The Hahn-Banach theorem is so fundamental for functional analysis that it is worthwhile to present several formulations and approaches to its proof.

In its classical version, the Hahn-Banach theorem concerns the extension of linear forms: one is given a linear form g defined on a linear subspace M of a vector space E, and the problem is to extend g to a linear form f on all of E. The problem is compounded by imposing a constraint on g, and requiring that the extension f satisfy the same constraint. The classical Hahn-Banach theorem asserts that this can be done, for a suitable type of constraint. This is the so-called 'analytic form' of the Hahn-Banach theorem, to be proved later in this section (28.6), (28.8).

Another approach to the Hahn-Banach theorem exploits the connection between linear forms and hyperplanes (Sections 21, 22). Here the setting is a topological vector space E. One is given a continuous linear form g defined on a linear subspace M of E, and one seeks an extension of g to a continuous linear form f on all of E. In general, this is not possible (28.10), but it is if E is 'locally convex' in the sense of (33.1). This result will be proved in a later section (34.8); the topological-algebraic core of the argument is the following result, which is known as the 'geometric form' of the Hahn-Banach theorem:

(28.1) Theorem. *Let E be a TVS over $\mathbb{K}$, let A be a nonempty open convex subset of E, and let M be a linear variety in E such that $M \cap A = \varnothing$. Then there exists a closed hyperplane H in E such that $H \supset M$ and $H \cap A = \varnothing$.*

It is convenient to separate out two elementary lemmas.

(28.2) Lemma. *If E is a TVS over $\mathbb{R}$ and if $\dim_{\mathbb{R}} E \geq 2$, then $E - \{\theta\}$ is connected.*

Proof. Given $x, y \in E - \{\theta\}$, it will suffice to show that there exists a connected subset C of $E - \{\theta\}$ containing both x and y.

Suppose first that x and y are linearly independent. Then the mapping $f : [0, 1] \to E$ defined by $f(\alpha) = (1 - \alpha)x + \alpha y$ does not assume the value θ, that is, the range $C = f([0, 1])$ of f is contained in $E - \{\theta\}$. Since f is continuous and $[0, 1]$ is connected, C is also connected (for the relative topology). Moreover, C contains $x = f(0)$ and $y = f(1)$.

Now suppose that x and y are linearly dependent. Since x and y are nonzero, and $\dim_{\mathbb{R}} \geq 2$, there exists a vector z such that x, z are linearly independent and y, z are linearly independent. By the first part of the proof, there exist connected subsets C_1, C_2 of $E - \{\theta\}$ with $x, z \in C_1$ and $y, z \in C_2$. Then $x, y \in C_1 \cup C_2$ and, since $C_1 \cap C_2$ is nonempty (it contains z), $C_1 \cup C_2$ is connected. {The proof shows that $E - \{\theta\}$ is in fact arcwise connected.} ∎

(28.3) Lemma. *If X is a connected topological space and if S is a nonempty proper subset of X, then S has at least one boundary point.*

Proof. The assumption is that $S \neq \varnothing$, $S \neq X$; it is to be shown that $\bar{S} \cap (X - S)^- \neq \varnothing$. Indeed, if this intersection were empty, then $X = \bar{S} \cup (X - S)^-$ would be a disconnection of X. ∎

Proof of Theorem (28.1). One can suppose, after translation, that M is a linear subspace of E.

We consider first the case that $\mathbb{K} = \mathbb{R}$. Let $\mathscr{P}$ be the class of all linear subspaces N of E such that $N \supset M$ and $N \cap A = \varnothing$. Observe that if $N \in \mathscr{P}$ then $\bar{N} \in \mathscr{P}$; for, $\bar{N}$ is a linear subspace (11.7) and, since A is open, the relation $N \subset E - A$ implies $\bar{N} \subset (E - A)^- = E - A$, therefore $\bar{N} \cap A = \varnothing$. Order $\mathscr{P}$ by inclusion; by an obvious application of Zorn's lemma, $\mathscr{P}$ contains a maximal element H. Since $H \subset \bar{H} \in \mathscr{P}$, necessarily $H = \bar{H}$ by maximality, thus H is closed; it is to be shown that H is a (homogeneous) hyperplane. Let $F = E/H$ be the quotient topological vector space, and let $\pi : E \to F$ be the canonical mapping (11.8). Since $H \cap A = \varnothing$ and A is nonempty, we have $H \neq E$, thus $F \neq \{\theta\}$; the problem is to show that F is one-dimensional (21.3). Let $B = \pi(A)$; since π is linear and open (11.8), B is convex (25.6) and open in F. The condition $H \cap A = \varnothing$ implies that $\theta \notin B$. If N is a linear subspace of E such that $N \supset H$ properly, then $N \cap A \neq \varnothing$ by maximality, and therefore $\pi(N) \cap B \neq \varnothing$; passing to quotients, this means that if F_0 is any nonzero linear subspace of F, then $F_0 \cap B \neq \varnothing$.

Summarizing: F is a real TVS of dimension ≥ 1, and B is a convex open set in F such that $\theta \notin B$ and such that every nonzero linear subspace of F intersects B. It will suffice to show that these conditions imply that F is one-dimensional.

Assume to the contrary that $\dim_{\mathbb{R}} F \geq 2$, and let $G = F - \{\theta\}$. Let C be the convex cone generated by the convex set B; by (27.7),

$$C = \bigcup\nolimits_{\alpha > 0} \alpha B.$$

The condition $\theta \notin B$ implies $\theta \notin C$, thus C is unpointed; since C is convex, it follows that C cannot be symmetric. But G is obviously symmetric, therefore $C \neq G$. Thus, C is a nonempty proper subset of G; since G is connected (28.2) it follows from (28.3) that the boundary of C in the space G contains at least one point, say b. The closure of C in G is $\bar{C} \cap G$, and the closure of $G - C$ in G is $(G - C)^{-} \cap G$, thus

(*) $$b \in \bar{C} \cap (G - C)^{-} \cap G;$$

since C is obviously open in F, it is also open in G, therefore $G - C$ is closed in G, i.e., $(G - C)^{-} \cap G = G - C$, so that (*) simplifies to

(**) $$b \in \bar{C} \cap (G - C).$$

In particular, $b \in G = F - \{\theta\}$, thus $b \neq \theta$. Let $F_0 = \mathbb{R}b$; since F_0 is a nonzero linear subspace of F, necessarily $F_0 \cap B \neq \varnothing$ (see the preceding paragraph). All the more, $F_0 \cap C \neq \varnothing$; say $\alpha b \in C$. If $\alpha = 0$ we have the contradiction $\theta = 0b \in C$. If $\alpha > 0$ then $b = \alpha^{-1}(\alpha b) \in \alpha^{-1}C \subset C$, thus $b \in C$, contrary to (**). We are left with the possibility that $\alpha < 0$; then $(-\alpha^{-1})(\alpha b) \in (-\alpha^{-1})C \subset C$, that is, $-b \in C$. Thus $-b \in C = \text{int } C$, and $b \in \bar{C}$ (by (**)); it follows from (25.19) that C contains every internal point of the segment from $-b$ to b, in particular, $\theta \in C$, a contradiction. This concludes the proof in the case that $\mathbb{K} = \mathbb{R}$.

Now suppose $\mathbb{K} = \mathbb{C}$. Since E is also a real TVS and M is also $\mathbb{R}$-linear, there exists, by the case already considered, a closed $\mathbb{R}$-hyperplane H_0 such that $H_0 \supset M$ and $H_0 \cap A = \varnothing$. Then also $iH_0 \supset iM = M$; thus, setting $H = H_0 \cap (iH_0)$, we have $H \supset M$ and $H \cap A = \varnothing$, where H is a closed $\mathbb{C}$-hyperplane by (22.8). $\blacksquare$

(28.4) **Corollary.** *Let E be a TVS over $\mathbb{R}$, let A be a nonempty open convex set in E, and let M be a linear subspace of E such that $M \cap A = \varnothing$. Then there exists a continuous linear form g on E such that $g = 0$ on M and $g > 0$ on A.*

Proof. By (28.1) there exists a closed maximal linear subspace H of E such that $H \supset M$ and $H \cap A = \varnothing$. Let g be a continuous linear form with null space H (22.4). Then $g(A)$ is a convex set in $\mathbb{R}$ (25.6), i.e., an interval.

Since $H \cap A = \varnothing$, $0 \notin g(A)$; therefore $g(A) \subset (0, \infty)$ or $g(A) \subset (-\infty, 0)$
Thus either $g > 0$ on A, or $g < 0$ on A (in which case $-g$ meets the require-
ments of the corollary). ∎

The complex variant of (28.4) is as follows:

(28.5) Corollary. *Let E be a TVS over $\mathbb{C}$, let A be a nonempty open
convex set in E, and let M be a linear subspace of E such that $M \cap A = \varnothing$.
Then there exists a continuous linear form f on E such that $f = 0$ on M and
$\mathrm{Re}\, f > 0$ on A.*

Proof. Viewing E as a real TVS, there exists, by (28.4), a continuous
$\mathbb{R}$-form g on E such that $g = 0$ on M and $g > 0$ on A. Let f be the $\mathbb{C}$-form
such that $\mathrm{Re}\, f = g$ (22.6), that is,

$$f(x) = g(x) - ig(ix) \qquad (x \in E).$$

Since $iM = M$, clearly $f = 0$ on M. ∎

We now take up the Hahn-Banach theorem in its classical form:

(28.6) Theorem. (Hahn-Banach) *Let E be a vector space over $\mathbb{R}$ and
suppose p is a real-valued function on E such that (i) $p(x + y) \leq p(x) + p(y)$
for all $x, y \in E$, and (ii) $p(\alpha x) = \alpha p(x)$ for all $x \in E$ and $\alpha > 0$. If g is a linear
form defined on a linear subspace M of E, such that*

$$g(y) \leq p(y) \qquad \text{for all } y \in M,$$

then g may be extended to a linear form f on E such that

(*) $$f(x) \leq p(x)$$

for all $x \in E$.

Proof. Conditions (i) and (ii) may be verbalized as follows: p is subadditive and positively homogeneous. Negative values are permitted, and p
may fail to be absolutely homogeneous. Let us note some properties of p
that derive from (i) and (ii). Putting $x = \theta$ and $\alpha = 2$ in (ii), we have $p(\theta) =
2p(\theta)$, therefore $p(\theta) = 0$. Thus $p(\alpha x) = \alpha p(x)$ for $\alpha \geq 0$. Putting $y = -x$
in (i) yields $-p(-x) \leq p(x)$.

If $M = E$ there is nothing to prove. Assuming $a \notin M$, let us show how to
extend g to a linear form f on the linear subspace $T = M + \mathbb{R}a$ generated
by M and a, in such a way that f satisfies (*) for all x in T. The elements of
T are the vectors $x = y + \alpha a$, with $y \in M$ and $\alpha \in \mathbb{R}$ uniquely determined
by x; the essential problem is to define $f(a)$, since the value of $f(x)$ will then
be determined by linearity and the condition $f|M = g$. To motivate the
definition, suppose such an extension to T were possible; let $\beta = f(a)$. Then,
writing $x = y + \alpha a$ as above, we have $f(x) = f(y) + \alpha f(a) = g(y) + \alpha \beta$.
Citing (*),

(1) $$g(y) + \alpha \beta \leq p(y + \alpha a) \qquad (y \in M, \alpha \in \mathbb{R}).$$

In particular, for $\alpha = 1$,

(2) $$\beta \leq -g(y) + p(y + a) \qquad (y \in M).$$

In (1), replace y by $-y$, and take $\alpha = -1$; then $-g(y) - \beta \leq p(-y - a)$, thus

(3) $\qquad\qquad -g(y) - p(-y - a) \leq \beta \qquad (y \in M)$.

Combining (2) and (3),

(4) $\qquad -g(z) - p(-z - a) \leq \beta \leq -g(y) + p(y + a) \qquad (y, z \in M)$.

This shows that $\beta = f(a)$ cannot be specified arbitrarily; it must be chosen so as to satisfy (4). To show that such a choice is possible, it is necessary to show that

(5) $\qquad -g(z) - p(-z - a) \leq -g(y) + p(y + a) \qquad (y, z \in M)$;

the verification of (5) consists in the calculation $g(y) - g(z) = g(y - z) \leq p(y - z) = p[(y + a) + (-z - a)] \leq p(y + a) + p(-z - a)$.

We are ready to define an extension f on T. In view of (5), there exists a scalar β satisfying (4). (For example, one can take $\beta = \inf\{-g(y) + p(y + a) : y \in M\}$.) Define

$$f(y + \alpha a) = g(y) + \alpha\beta \qquad (y \in M, \alpha \in \mathbb{R}).$$

Clearly f is a linear form on T extending g. It remains to verify that (*) holds for all $x \in T$, that is,

(6) $\qquad\qquad g(y) + \alpha\beta \leq p(y + \alpha a) \qquad (y \in M, \alpha \in \mathbb{R})$.

If $\alpha = 0$ this is a property of g. Assuming $\alpha \neq 0$, replace y and z in (4) by $\alpha^{-1}y$; then

$$-g(\alpha^{-1}y) - p(-\alpha^{-1}y - a) \leq \beta \leq -g(\alpha^{-1}y) + p(\alpha^{-1}y + a),$$

thus

(6a) $\qquad\qquad g(\alpha^{-1}y) + \beta \leq p(\alpha^{-1}y + a)$,

(6b) $\qquad\qquad -g(\alpha^{-1}y) - \beta \leq p(-\alpha^{-1}y - a)$.

If $\alpha > 0$, multiplication by α in (6a) yields (6); if $\alpha < 0$, multiplication by $-\alpha$ in (6b) yields (6).

We have thus shown that an extension of the desired sort can be performed one dimension at a time; to get all the way up to E, we employ a transfinite argument in the standard Zorn format. Let $\mathscr{P}$ be the family of all ordered pairs (N, h), where N is a linear subspace of E containing M, h is a linear form on N such that $h|M = g$, and $h(x) \leq p(x)$ for all $x \in N$. For instance, (M, g) itself is such a pair. Order $\mathscr{P}$ by extension, that is, define $(N, h) \leq (N', h')$ in case $N \subset N'$ and $h'|N = h$. Every simply ordered subset $\mathscr{C}$ of $\mathscr{P}$ has an upper bound in $\mathscr{P}$: namely, if $\mathscr{C} = ((N_\iota, h_\iota))_{\iota \in I}$ is any indexing of $\mathscr{C}$, then $N = \bigcup_{\iota \in I} N_\iota$ is a linear subspace of E, and the function h on N, defined by $h(x) = h_\iota(x)$ if $x \in N_\iota$, is well defined and linear; clearly $(N, h) \in \mathscr{P}$ and $(N_\iota, h_\iota) \leq (N, h)$ for all $\iota \in I$. By Zorn's lemma, $\mathscr{P}$ has at least one maximal element (N, f). Necessarily $N = E$, since otherwise the construction of the preceding paragraph could be used to contradict maximality. ∎

The complex variant of (28.6) requires a revision of the conditions imposed on the function p. The appropriate conditions are of interest in both the real and complex case; they are a weakening of the conditions satisfied by a norm (16.1):

(28.7) Definition. Let E be a vector space over $\mathbb{K}$ ($=\mathbb{R}$ or $\mathbb{C}$). A *seminorm* on E is a real-valued function p on E such that (1) $p(x) \geq 0$ for all $x \in E$, (2) $p(x + y) \leq p(x) + p(y)$ for all $x, y \in E$, and (3) $p(\lambda x) = |\lambda| p(x)$ for all $\lambda \in \mathbb{K}$, $x \in E$. In words, p is *positive*, *subadditive*, and *absolutely homogeneous*.

A seminorm obviously satisfies conditions (i), (ii) of (28.6). A seminorm p is a norm if and only if it is strictly positive, i.e., $p(x) > 0$ whenever $x \neq \theta$.

(28.8) Theorem. (Hahn-Banach-Bohnenblust-Sobczyk) *Let E be a vector space over $\mathbb{K}$, let p be a seminorm on E, and let M be a linear subspace of E. If g is a linear form on M such that*

$$|g(y)| \leq p(y) \qquad \text{for all } y \in M,$$

then g may be extended to a linear form f on E such that

$$|f(x)| \leq p(x) \qquad \text{for all } x \in E.$$

Proof. Suppose first that $\mathbb{K} = \mathbb{R}$. By the Hahn-Banach theorem (28.6), g may be extended to a linear form f on E such that $f(x) \leq p(x)$ for all $x \in E$. Then also $-f(x) = f(-x) \leq p(-x) = p(x)$, thus $|f(x)| \leq p(x)$.

Consider now the case that $\mathbb{K} = \mathbb{C}$. Define $g'(y) = \operatorname{Re} g(y)$ $(y \in M)$; thus g' is an $\mathbb{R}$-form on E, and

$$(*) \qquad\qquad g(y) = g'(y) - ig'(iy) \qquad \text{for all } y \in M$$

by (21.11). Since $|g'(y)| \leq |g(y)| \leq p(y)$ for all $y \in M$, it follows from the real case that g' may be extended to an $\mathbb{R}$-form f' on E such that

$$|f'(x)| \leq p(x) \qquad \text{for all } x \in E.$$

Define $f(x) = f'(x) - if'(ix)$ $(x \in E)$; then f is a $\mathbb{C}$-form on E (21.13), and it is clear from (*) that f extends g. It remains only to show that $|f(x)| \leq p(x)$ for all $x \in E$. Given any $x \in E$, write $f(x) = \alpha\mu$ with $\alpha \geq 0$ and $|\mu| = 1$. Then $|f(x)| = \alpha = \mu^* f(x) = f(\mu^* x) = f'(\mu^* x) - if'(i\mu^* x)$; since α is real, necessarily $f'(i\mu^* x) = 0$, thus $|f(x)| = f'(\mu^* x) \leq p(\mu^* x) = |\mu^*| p(x) = p(x)$. ∎

(28.9) Corollary. *Let E be a vector space over $\mathbb{K}$ and let p be a seminorm on E. Given any $a \in E$, there exists a linear form f on E such that $f(a) = p(a)$ and $|f(x)| \leq p(x)$ for all $x \in E$.*

Proof. If $a = \theta$, then $p(a) = 0$ and we may take $f = 0$. Otherwise, let $M = \mathbb{K}a$ and define g on M by

$$g(\lambda a) = \lambda p(a) \qquad (\lambda \in \mathbb{K}).$$

Since $|g(\lambda a)| = |\lambda| p(a) = p(\lambda a)$ $(\lambda \in \mathbb{K})$, the conditions of (28.8) are fulfilled. ∎

The techniques of the proof of the 'geometric form' of the Hahn-Banach theorem (28.1) are totally different from those used in the proof of the classical Hahn-Banach theorem (28.6). The connection between the two results is far from obvious; its clarification is a part of the theory of locally convex TVS, and is deferred until Section 37 (see (37.22)). The following two remarks may at least make the existence of a connection plausible. (1) If p is a continuous seminorm on a TVS, then $A = \{x : p(x) < 1\}$ is a nonempty open convex set. (2) In the case of a normed space, (28.9) makes a direct connection with the topology: If E is a normed space and $a \in E$, then there exists a linear form f on E such that $f(a) = \|a\|$ and $|f(x)| \le \|x\|$ for all $x \in E$, and the continuity of f is shown by the calculation $|f(x) - f(y)| = |f(x - y)| \le \|x - y\|$.

The Hahn-Banach theorem is one of the cornerstones of functional analysis; the application in the next section is one of many illustrations of its remarkable power.

Exercises

(28.10) Let E be a TVS over $\mathbb{K}$ that has no nonzero continuous linear forms (cf. Section 15) and let x be a nonzero vector in E. Then $\lambda x \mapsto \lambda$ is a continuous linear form g on the one-dimensional subspace $M = \mathbb{K}x$, but g cannot be extended to a continuous linear form f on E.

(28.11) Let E be a vector space over $\mathbb{R}$, and suppose that p is a real-valued function on E satisfying conditions (i), (ii) of (28.6). Given any $a \in E$, there exists a linear form f on E such that $f(a) = p(a)$ and $f(x) \le p(x)$ for all $x \in E$.

(28.12) Let E be the real vector space of all bounded real sequences $x = (\alpha_n)$, with equality and linear operations defined term-by-term, and define

$$p(x) = \limsup \alpha_n.$$

Prove: (1) p satisfies conditions (i), (ii) of (28.6). (2) There exists a linear form f on E such that $f(x) = \lim \alpha_n$ whenever $x = (\alpha_n)$ is a convergent sequence. (Cf. Section 29.)

(28.13) Let E be a normed space, let M be a linear subspace of E, and let g be a continuous linear form on M. Prove: (i) There exists a positive constant k such that $|g(y)| \le k\|y\|$ for all $y \in M$. (ii) g may be extended to a continuous linear form on E.

*29. Invariant means, generalized limits. The subject of invariant means has its origins in the theory of almost-periodic functions, ergodic theory, and the implausible wish to assign a limit to a sequence that doesn't have one. The unifying theme is the notion of semigroup. To save space we move quickly to a substantive general result, leaving the motivating examples as applications.

(29.1) Definition. A *semigroup* is a nonempty set T with an associative binary operation $T \times T \to T$; thus, denoting the operation additively,

$(s, t) \mapsto s + t$, the sole axiom is $r + (s + t) = (r + s) + t$. The semigroup T is *abelian* if $s + t = t + s$ for all s, t in T.

(29.2) Definition. Let T be a semigroup and let $\mathscr{B} = \mathscr{B}_{\mathbb{R}}(T)$ be the Banach space of all bounded, real-valued functions x on T, with the pointwise linear operations and the sup norm $\|x\|_\infty$ (16.15). (Actually, only the vector space structure is essential for this section, but it would be negligent not to mention the Banach space overtones.) For $x \in \mathscr{B}$ we write

$$\inf x = \inf \{x(t) : t \in T\}, \qquad \sup x = \sup \{x(t) : t \in T\}.$$

If $x \in \mathscr{B}$ and $s \in T$, the *right translate* of x by s is the function x_s defined by

$$x_s(t) = x(t + s) \qquad (t \in T);$$

evidently $x_s \in \mathscr{B}$ and $\|x_s\|_\infty \leq \|x\|_\infty$ (with equality holding when T is a group). Moreover,

$$(x + y)_s = x_s + y_s, \qquad (\alpha x)_s = \alpha x_s,$$

and, by the associativity of the semigroup operation,

$$x_{s+t} = (x_t)_s.$$

Also, $x \geq 0$ implies $x_s \geq 0$.

(29.3) Definition. A *right-invariant mean* on a semigroup T is a linear form f on $\mathscr{B} = \mathscr{B}_{\mathbb{R}}(T)$ such that

(i) $f(x_s) = f(x)$,

(ii) $\inf x \leq f(x) \leq \sup x$,

for all $x \in \mathscr{B}$ and $s \in T$. A semigroup is called *right-amenable* if it possesses a right-invariant mean. The terms *left-invariant mean* and *left-amenable* have the obvious dual meanings, based on the consideration of the left translate of a function x by an element s (namely, the function $t \mapsto x(s + t)$). An *invariant mean* on T is a linear form on $\mathscr{B}$ that is both a left-invariant mean and a right-invariant mean; a semigroup possessing an invariant mean is said to be *amenable*. For abelian semigroups, the left/right distinctions may be dropped; this is the only case in which we are interested in the text. (The left/right terminology is mentioned for the sake of the exercises, and consistency with the literature.)

Three facts should be reported at once: (1) there exist semigroups that are neither right-amenable nor left-amenable; (2) even if a semigroup is right-amenable, the right-invariant mean is in general highly nonunique; (3) every abelian semigroup is amenable. The main result of this section is a proof of (3); the proof itself will exhibit the nonuniqueness that one must expect in general, even in the abelian case.

The following key lemma constructs a functional p suitable for application of the Hahn-Banach theorem; the style of proof has its roots in the theory of almost-periodic functions:

(29.4) Lemma. *If T is an abelian semigroup, there exists a real-valued function p on $\mathcal{B} = \mathcal{B}_{\mathbb{R}}(T)$ such that*

(1) $p(x + y) \le p(x) + p(y)$ *for all $x, y \in \mathcal{B}$,*

(2) $p(\alpha x) = \alpha p(x)$ *for all $x \in \mathcal{B}$, $\alpha > 0$,*

(3) $\inf x \le p(x) \le \sup x$ *for all $x \in \mathcal{B}$,*

(4) $p(x - x_s) = p(x_s - x) = 0$ *for all $x \in \mathcal{B}$, $s \in T$.*

Moreover,

(5) $p(\alpha 1) = \alpha$ *for all $\alpha \in \mathbb{R}$,*

(6) $x \ge 0$ *implies $p(x) \ge 0$, and $x \le 0$ implies $p(x) \le 0$.*

Proof. Denote by $\mathcal{F}$ the set of all $n$ples $F = (t_1, \ldots, t_n)$ of elements of T, with n arbitrary and with repetitions allowed; that is, $\mathcal{F} = \bigcup_{n=1}^{\infty} T^n$, where T^n is the cartesian product of n copies of the set T. The first step of the proof is to construct, for each $F \in \mathcal{F}$, a certain functional $x \mapsto M_F(x)$ on $\mathcal{B}$. If $x \in \mathcal{B}$ and $s_1, \ldots, s_n \in T$, then

$$\inf x \le (1/n) \sum_{i=1}^{n} x(s_i) \le \sup x.$$

Therefore, if $F = (t_1, \ldots, t_n)$ and $x \in \mathcal{B}$, we may define

$$M_F(x) = \sup \left\{ (1/n) \sum_{1}^{n} x(t + t_i) : t \in T \right\} = \sup \left[(1/n) \sum_{1}^{n} x_{t_i} \right],$$

and we have

(i) $$\inf x \le M_F(x) \le \sup x.$$

Note that $M_F(x)$ is unchanged by a permutation of the elements of F.

If $F = (t_1, \ldots, t_n)$ and $G = (s_1, \ldots, s_m)$, we write $F + G$ for the mnple whose elements are the $t_i + s_j$ in some order (by the preceding remark, the particular order is immaterial). The next step of the proof is to verify that for any $F, G \in \mathcal{F}$ and $x, y \in \mathcal{B}$,

(ii) $$M_{F+G}(x + y) \le M_F(x) + M_G(y).$$

Indeed (commutativity of T is used in the first step)

$$(1/mn) \sum_{i,j} (x + y)_{t_i + s_j}$$

$$= (1/mn) \sum_{i,j} x_{s_j + t_i} + (1/mn) \sum_{i,j} y_{t_i + s_j}$$

$$= (1/m) \sum_{j} \left[(1/n) \sum_{i} (x_{t_i})_{s_j} \right] + (1/n) \sum_{i} \left[(1/m) \sum_{j} (y_{s_j})_{t_i} \right]$$

$$= (1/m) \sum_{j} \left[(1/n) \sum_{i} x_{t_i} \right]_{s_j} + (1/n) \sum_{i} \left[(1/m) \sum_{j} y_{s_j} \right]_{t_i}$$

$$\le (1/m) \sum_{j} [M_F(x) \cdot 1]_{s_j} + (1/n) \sum_{i} [M_G(y) \cdot 1]_{t_i}$$

$$= (1/m) \sum_{j} [M_F(x) \cdot 1] + (1/n) \sum_{i} [M_G(y) \cdot 1]$$

$$= M_F(x) \cdot 1 + M_G(y) \cdot 1,$$

therefore

$$M_{F+G}(x + y) = \sup\left[(1/mn)\sum_{i,j}(x + y)_{t_i+s_j}\right]$$

$$\leq M_F(x) + M_G(y).$$

Define the function $p: \mathscr{B} \to \mathbb{R}$ by the formula

$$p(x) = \inf\{M_F(x) : F \in \mathscr{F}\}.$$

In view of (i), we have $\inf x \leq p(x) \leq \sup x$, thus condition (3) is verified. Conditions (5) and (6) are obvious from the definition of p (they also follow at once from (3)).

To verify (1), suppose $x, y \in \mathscr{B}$ and let $\varepsilon > 0$; it will suffice to show that

$$p(x + y) \leq p(x) + p(y) + 2\varepsilon.$$

By the definition of p, there exist $F, G \in \mathscr{F}$ such that

$$M_F(x) \leq p(x) + \varepsilon, \qquad M_G(y) \leq p(y) + \varepsilon;$$

citing (ii), we have

$$p(x + y) \leq M_{F+G}(x + y) \leq M_F(x) + M_G(y) \leq p(x) + \varepsilon + p(y) + \varepsilon.$$

Condition (2) follows at once from the observation that $M_F(\alpha x) = \alpha M_F(x)$ for all $x \in \mathscr{B}$ and $\alpha > 0$.

To verify (4), suppose $x \in \mathscr{B}$, $s \in T$. We assert that

(iii) $p(x - x_s) \leq 0.$

Given any positive integer n, it will suffice to show that

$$p(x - x_s) \leq (1/n)(\sup x - \inf x).$$

Set $t_1 = s, t_2 = 2s, \ldots, t_n = ns$, where, by definition,

$$is = s + \cdots + s \qquad (i \text{ terms}).$$

Let $F = (t_1, \ldots, t_n)$. For any $t \in T$,

$$(1/n)\sum_{i=1}^n (x - x_s)(t + t_i) = (1/n)\sum x(t + t_i) - (1/n)\sum x_s(t + t_i)$$

$$= (1/n)\sum x(t + is) - (1/n)\sum x_s(t + is)$$

$$= (1/n)\sum x(t + is) - (1/n)\sum x(t + (i + 1)s)$$

$$= (1/n)x(t + s) - (1/n)x(t + (n + 1)s)$$

$$\leq (1/n)\sup x - (1/n)\inf x;$$

since t is arbitrary,

$$(1/n)\sum_1^n (x - x_s)_{t_i} \leq (1/n)(\sup x - \inf x),$$

therefore

$$p(x - x_s) \leq M_F(x - x_s) = \sup (1/n)\sum_1^n (x - x_s)_{t_i} \leq (1/n)(\sup x - \inf x),$$

thus (iii) is verified. An almost identical calculation yields

(iv) $p(x_s - x) \leq 0.$

Since $-p(-y) \leq p(y)$ for all $y \in \mathscr{B}$ (see the first paragraph of the proof of (28.6)), we have, citing (iv) and (iii),

$$0 \leq -p(x_s - x) \leq p(x - x_s) \leq 0,$$

thus $p(x_s - x) = p(x - x_s) = 0.$ ∎

(29.5) **Theorem.** *Every abelian semigroup is amenable.*

Proof. Let T be an abelian semigroup and let p be the functional on $\mathscr{B} = \mathscr{B}_{\mathbb{R}}(T)$ constructed in the lemma. By the Hahn-Banach theorem (28.6), applied to the functional p and the zero subspace of $\mathscr{B}$, there exists a linear form f on $\mathscr{B}$ such that $f(x) \leq p(x)$ for all $x \in \mathscr{B}$. Thus $f(x) \leq p(x) \leq \sup x$ $(x \in B)$; replacing x by $-x$, we have $-f(x) \leq p(-x) \leq \sup(-x)$, therefore

$$\inf x = -\sup(-x) \leq -p(-x) \leq f(x) \leq p(x) \leq \sup x.$$

In particular, condition (ii) in the definition of an invariant mean (29.3) is verified. If $x \in \mathscr{B}$ and $s \in T$, then

$$0 = -p(x_s - x) = -p(-(x - x_s)) \leq f(x - x_s) \leq p(x - x_s) = 0,$$

therefore $f(x - x_s) = 0$; thus $f(x_s) = f(x)$, which is condition (i) of (29.3). ∎

For the application to sequences, we specialize to the additive semigroup of all positive integers, $\mathbb{P} = \{1, 2, 3, \ldots\}$. Then $\mathscr{B}_{\mathbb{R}}(\mathbb{P})$ is the Banach space of all bounded sequences $x = (\alpha_k)$; the traditional symbol for this Banach space is (m). If $x = (\alpha_k) \in (m)$ and $j \in \mathbb{P}$, the translate x_j is simply the 'truncated' sequence (α_{j+k}), that is, $\alpha_{j+1}, \alpha_{j+2}, \ldots$.

(29.6) **Corollary.** *Let (m) be the Banach space of all bounded sequences $x = (\alpha_k)$ of real numbers. There exists a linear form L on (m) such that*
   (i) $L(x_j) = L(x)$ *for all $x \in (m)$ and $j = 1, 2, 3, \ldots$,*
   (ii) $\inf \alpha_k \leq L(x) \leq \sup \alpha_k$ *for all $x = (\alpha_k) \in (m)$.*
It follows that
   (iii) $\lim \inf \alpha_k \leq L(x) \leq \lim \sup \alpha_k$
for all $x = (\alpha_k) \in (m)$, therefore
   (iv) $L(x) = \lim \alpha_k$ *whenever $x = (\alpha_k)$ is convergent.*

Proof. Since $\mathbb{P} = \{1, 2, 3, \ldots\}$ is an abelian semigroup with respect to addition, by (29.5) there exists a linear form L on (m) satisfying (i) and (ii).

To verify (iii), suppose $x = (\alpha_k) \in (m)$. The notations $\sup x$, $\lim \sup x$, etc., have the usual meanings. For every positive integer j,

$$L(x) = L(x_j) \leq \sup x_j,$$

therefore

$$L(x) \leq \inf\{\sup x_j : j = 1, 2, 3, \ldots\} = \lim \sup x;$$

applying this to $-x$ yields

$$\lim \inf x = -\lim \sup(-x) \leq -L(-x) = L(x). \qquad ∎$$

(29.7) Definition. A *Banach generalized limit* is a linear form L on (m) satisfying the conditions of (29.6).

It is easily verified that if L is a Banach generalized limit, then $|L(x)| \leq \|x\|_\infty = \sup |\alpha_k|$ for all $x = (\alpha_k)$ in (m). In particular,

$$|L(x) - L(y)| = |L(x - y)| \leq \|x - y\|_\infty,$$

thus L is a continuous linear form on the Banach space (m). Moreover, if $x = (\alpha_k)$ and if $\alpha_k \geq 0$ for all k (or from some index onward), then $L(x) \geq 0$.

Exercises

(29.8) The extension to complex scalars is straightforward.

(29.9) A semigroup is left-amenable if and only if the opposite semigroup is right-amenable.

*__(29.10)__ A right-amenable group is obviously left-amenable; in fact, it is amenable. More generally, if a semigroup is both left- and right-amenable, then it is amenable.

*__(29.11)__ There exist groups that are not amenable.

*__(29.12)__ It is elementary that every finite group is amenable. Better yet (but not elementary), every compact topological group is amenable.

(29.13) (i) Every subgroup of an amenable group is amenable. (ii) Every quotient group of an amenable group is amenable. (iii) If N is a normal subgroup of a group G such that both N and G/N are amenable, then G is amenable. (iv) Every solvable group is amenable.

(29.14) The functional p on (m), defined by $p(x) = \lim \sup x$, is subadditive, positively homogeneous, and truncation-invariant, but it is not suitable for proving (29.6).

*__(29.15)__ Every uniformly bounded representation of an amenable group, as invertible operators on a Hilbert space, is similar to a unitary representation.

(29.16) A Banach generalized limit L cannot satisfy $L(xy) = L(x)L(y)$ for all x, y in (m).

30. Half-spaces, separation of convex sets in a real TVS.

The concepts in this section impinge on the notion of hyperplane, and therefore of linear form (cf. Section 21). We consider both real and complex vector spaces, but the heart of the matter lies in the real case.

(30.1) Definition. A subset S of a real vector space E is called a *half-space* if there exist a nonzero linear form g on E and a real number α such that S may be represented in one of the following forms:

$$S_1 = \{x : g(x) \leq \alpha\}$$
$$S_2 = \{x : g(x) \geq \alpha\}$$
$$S_3 = \{x : g(x) < \alpha\}$$
$$S_4 = \{x : g(x) > \alpha\}.$$

Let H be the hyperplane determined by g and α (21.7), that is,

$$H = \{x : g(x) = \alpha\}.$$

S_1 and S_2 are called the *closed half-spaces*, S_3 and S_4 the *open half-spaces*, determined by H.

(30.2) With notation as in (30.1), $H = S_1 \cap S_2$. Every half-space is convex; for example, S_2 is the inverse image, under the linear mapping g, of the interval $[\alpha, \infty)$, and is therefore convex (25.6). If E is a real TVS, and if g is continuous—equivalently, if H is closed (22.5)—then S_1, S_2 are closed subsets of E, and S_3, S_4 are open subsets of E. (Cf. (30.13).)

(30.3) **Definition.** With notation as in (30.1), let A be a subset of E. We say that A *lies to one side of* H if either $A \subset S_1$ or $A \subset S_2$; A *lies strictly to one side of* H if either $A \subset S_3$ or $A \subset S_4$. Let A, B be subsets of H; we say that H *separates* A and B if either $A \subset S_1$ and $B \subset S_2$, or $A \subset S_2$ and $B \subset S_1$; we say that H *separates* A and B *strictly* if either $A \subset S_3$ and $B \subset S_4$, or $A \subset S_4$ and $B \subset S_3$.

In particular, the separation of convex sets is of major importance; for example, as we shall see below (30.6) and in Section 34, the Hahn-Banach theorem belongs to this circle of ideas.

(30.4) **Theorem.** *Let H be a hyperplane in the real vector space E and let A be a convex subset of E. Then A lies strictly to one side of H if and only if $A \cap H = \varnothing$.*

Proof. Say $H = \{x : g(x) = \alpha\}$, where g is a suitable nonzero linear form (21.7), and adopt the notation of (30.1). If $A \subset S_3$ or $A \subset S_4$, obviously $A \cap H = \varnothing$.

Conversely, suppose $A \cap H = \varnothing$, that is, $\alpha \notin g(A)$. Since $g(A)$ is an interval (25.6) that excludes α, necessarily $g(A) \subset (-\infty, \alpha)$ or $g(A) \subset (\alpha, \infty)$, that is, $A \subset S_3$ or $A \subset S_4$. {Cf. the proof of (28.4).} ∎

(30.5) **Theorem.** *Let E be a real TVS, let S be a subset of E with non-empty interior, and suppose H is a hyperplane in E such that S lies to one side of H. Then (1) H is closed, (2) $\bar{S}$ lies to one side of H, and (3) int S lies strictly to one side of H.*

Proof. Say $H = \{x : g(x) = \alpha\}$, g a suitable nonzero linear form, and suppose $S \subset \{x : g(x) \geq \alpha\}$.

(1) Choose any $z \in E$ with $g(z) = 1 - \alpha$. Since the half-space $\{x : g(x) > \alpha - 1\}$ contains S, it has nonempty interior, therefore the half-space

$$\{y : g(y) > 0\} = \{y : g(y - z) > \alpha - 1\} = \{x : g(x) > \alpha - 1\} + z$$

also has nonempty interior. Since $H \subset \bar{H}$ and $\bar{H}$ is a linear variety (cf. 11.7), and since $\bar{H} \neq E$ (because the complement of H has nonempty interior), it follows that $H = \bar{H}$ by the maximality of H (21.6).

(2) Since H is closed, g is continuous (22.5); thus $\{x : g(x) \geq \alpha\}$ is a closed set containing S, and therefore $\bar{S}$.

(3) Since g is an open mapping (cf. (23.8) or the proof of (22.2)), $g(\text{int } S)$ is open in $\mathbb{R}$; since $g(\text{int } S) \subset g(S) \subset [\alpha, \infty)$, this implies $g(\text{int } S) \subset (\alpha, \infty)$, i.e., int $S \subset \{x : g(x) > \alpha\}$. ∎

(30.6) Theorem. *Let E be a real TVS, let A and B be nonempty convex subsets of E such that $A \cap B = \varnothing$, and suppose that A is open. Then (1) there exists a closed hyperplane H separating A and B; (2) A lies strictly to one side of H; (3) if B is also open, then H separates A and B strictly.*

Proof. Let D be the set of all differences $a - b$ $(a \in A, b \in B)$. Since A is open, so is $D = \bigcup_{b \in B} A + (-b)$; moreover, since A and $-B$ are convex, so is $D = A + (-B)$ (25.10). The condition $A \cap B = \varnothing$ means that $\theta \notin D$. Summarizing: D is a nonempty, open convex set that is disjoint from the linear subspace $M = \{\theta\}$. By Corollary (28.4) of the geometric form of the Hahn-Banach theorem, there exists a continuous linear form g such that $g > 0$ on D, that is, $g(a) > g(b)$ for all $a \in A, b \in B$. Setting $\alpha = \inf g(A)$, we have

(*) $$g(b) \leq \alpha \leq g(a) \qquad (a \in A, b \in B).$$

Consider the closed hyperplane $H = \{x : g(x) = \alpha\}$ (22.5). In the notation of (30.1), we have $B \subset S_1$ and $A \subset S_2$ by (*); citing (30.5), the last inclusion may be strengthened to $S_4 \supset \text{int } A = A$. This proves (1) and (2). If, moreover, B is open, then also $S_3 \supset \text{int } B = B$, in which case H separates A and B strictly. ∎

(30.7) Corollary. *Let E be a complex TVS, let A and B be nonempty convex subsets of E such that $A \cap B = \varnothing$, and suppose that A is open. Then there exist a continuous linear form f and a real number α such that $\text{Re} f > \alpha$ on A and $\text{Re} f \leq \alpha$ on B; if B is also open, then $\text{Re} f < \alpha$ on B.*

Proof. Regarding E as a real TVS, adopt the notations of the proof of (30.6) and let f be the $\mathbb{C}$-form with $\text{Re} f = g$ (22.6). ∎

(30.8) Definition. Let E be a real vector space and let A be a subset of E. A hyperplane H in E is said to be *tangent* to A if (i) A lies to one side of H, and (ii) $A \cap H \neq \varnothing$. In view of the definitions (cf. (21.7), (30.3)), this means that there exist a nonzero linear form g and a real number α such that (i) $g \leq \alpha$ on A, and (ii) $g(a) = \alpha$ for at least one $a \in A$.

(30.9) Definition. Let E be a real vector space and let H_1, H_2 be hyperplanes in E. One says that H_1 and H_2 are *parallel* if they are translates of each other, that is, $H_2 = H_1 + x$ for some $x \in E$; equivalently, H_1 and H_2 are cosets of the same maximal linear subspace M (21.7): $H_1 = x_1 + M$, $H_2 = x_2 + M$. Still another formulation (21.7): there exist a nonzero linear form g and real numbers α_1, α_2 such that $H_k = \{x : g(x) = \alpha_k\}$ $(k = 1, 2)$.

A criterion for the existence of a tangent hyperplane in a specified 'direction':

(30.10) Lemma.   *Let E be a real vector space, A a nonempty subset of E, g a nonzero linear form, and $M = \{x : g(x) = 0\}$. In order that A admit a tangent hyperplane parallel to M, it is necessary and sufficient that $g(A)$ contain a largest element or a smallest element.*

Proof.   Suppose there exists a hyperplane H that is tangent to A and parallel to M. Say $H = \{x : g(x) = \alpha\}$. If, for instance, $g \geq \alpha$ on A, then $g(A)$ has a smallest element (namely, α).

Conversely, suppose $g(A)$ has, say, a smallest element α. Choose any $a \in A$ with $g(a) = \alpha$. Then $g \geq \alpha$ on A, thus the hyperplane $H = \{x : g(x) = \alpha\}$ is tangent to A.   ∎

To rephrase the criterion in (30.10), what is needed is that g assume on A a minimum value or a maximum value; there is a topological situation in which this happens automatically:

(30.11) Theorem.   *Let E be a real TVS and let A be a nonempty compact subset of E. If H is any closed hyperplane in E, then there exists a hyperplane parallel to H and tangent to A.*

Proof.   If $H = \{x : g(x) = \alpha\}$, g a continuous linear form (22.5), then $g(A)$ is a nonempty compact subset of $\mathbb{R}$, therefore it contains a smallest (and a largest) element. Quote the lemma. {Incidentally, A need only be assumed quasicompact, i.e., it need not be assumed separated.}   ∎

The complex variant of (30.11) would read as follows: If A is a nonempty compact subset of a complex TVS E, and if f is any continuous linear form on E, then there exists a real number α such that $\operatorname{Re} f \geq \alpha$ on A and $(\operatorname{Re} f)(a) = \alpha$ for at least one point a of A. However, one need not use (30.11) in the proof, since the result holds trivially for any continuous complex-valued function f on E (the continuous real-valued function $\operatorname{Re} f$ assumes a minimum on the compact set A).

So far, (30.6) is the only substantial result of the section; we close the section with another:

(30.12) Theorem.   *Let E be a real TVS and let A be a convex body in E. Then* (1) *every hyperplane tangent to A is closed;* (2) *if b is any boundary point of A, there exists a hyperplane H tangent to A and passing through b;* (3) *A is the intersection of the family of all the closed half-spaces that contain A and are determined by hyperplanes tangent to A.*

Proof.   The hypothesis is that A is a closed convex set with nonempty interior (25.21).

(1) is immediate from (30.5).

(2) int A is a nonempty open convex set (25.20) and $b \notin$ int A, that is, int A is disjoint from the linear variety $\{b\}$, therefore by the geometric form of the Hahn-Banach theorem (28.1) there exists a closed hyperplane H such that $b \in H$ and (int A) $\cap H = \varnothing$. Then int A lies to one side of H (30.4), therefore so does (int $A)^-$ (30.5); but (int $A)^- = A$ (25.22), thus A lies to one side of H. Since, moreover, $b \in A \cap H$, H is tangent to A.

(3) Assuming $z \notin A$, we seek a hyperplane H tangent to A, such that z belongs to the open half-space opposite to A. The first problem is to locate a suitable point of tangency b. Let $a \in \text{int } A$. Informally, the closed segment from a to z intersects A in a closed segment, from a to some point b; so to speak, b is the point of A where the segment from a to z exits A, and b is the desired point of tangency. More precisely, consider the mapping $\varphi : [0, 1] \to E$ defined by $\varphi(\alpha) = (1 - \alpha)a + \alpha z$, and let $\beta = \sup\{\alpha : \varphi(\alpha) \in A\} = \sup \varphi^{-1}(A)$. Since φ is continuous, $\varphi^{-1}(A)$ is closed, therefore $\beta \in \varphi^{-1}(A)$, $\varphi(\beta) \in A$. Let $b = \varphi(\beta) = (1 - \beta)a + \beta z$. Since $z \notin A$, necessarily $\beta < 1$. If $\beta < \alpha < 1$, then $\varphi(\alpha) \notin A$ by the definition of β; since $\varphi(\alpha) \to \varphi(\beta) = b$ as $\alpha \downarrow \beta$, it follows that b is a boundary point of A. In particular, $b \notin \text{int } A$, therefore $b \neq a$; thus $\beta > 0$, so that b is an internal point of the segment from a to z.

By (2) there exists a hyperplane H tangent to A and passing through b, and H is closed by (1). Say $H = \{x : g(x) = \gamma\}$, g a suitable continuous linear form (22.5). We may suppose $A \subset S = \{x : g(x) \leq \gamma\}$; the proof will be concluded by showing that $z \notin S$. Assume to the contrary that $z \in S$; since $a \in \text{int } A \subset \text{int } S$ and S is convex, it follows from (25.19) that int S contains every internal point of the segment from a to z; in particular, $b \in \text{int } S$. Since S lies to one side of H, int S lies strictly to one side of H (30.5); in particular, $b \notin H$, contrary to the choice of H. ∎

Exercises

(30.13) Let E be a real TVS and let $S = \{x : g(x) \leq \alpha\}$, where g is a nonzero linear form and α is a real number. If S is a closed subset of E, then g is continuous.

(30.14) Regard $\mathbb{C}$ as a two-dimensional real vector space and let A be a convex set in $\mathbb{C}$ (not necessarily closed). Suppose that λ is a boundary point of A and that $\lambda \in A$. Then there exists a line through λ tangent to A.

*31. Ordered real vector spaces.

An *ordering* of a set is a binary relation $\leq$ on the set that is reflexive ($x \leq x$), transitive ($x \leq y$ and $y \leq z$ imply $x \leq z$) and antisymmetric ($x \leq y$ and $y \leq x$ imply $x = y$); an *ordered set* is a set equipped with an ordering. (Other commonly used terms: partial ordering, partially ordered set.) If the condition of antisymmetry is dropped, we speak of a *preordering*, and of a *preordered set*. Another notation for $x \leq y$ is $y \geq x$.

Throughout the section, E denotes a vector space over $\mathbb{R}$.

(31.1) Definition. A preordering [ordering] of E is said to be *compatible* with the vector space structure if (i) $x \leq y$ implies $x + z \leq y + z$ for all z, and (ii) $x \leq y$ and $\alpha \geq 0$ imply $\alpha x \leq \alpha y$; so to speak, it is translation-invariant and positively homogeneous. (Equivalently, any relation $x \leq y$ is preserved under every homothety $z \mapsto \alpha z + a$ with nonnegative weight α.) A *preordered [ordered] real vector space* is a pair $(E, \leq)$, where E is a real vector space and $\leq$ is a preordering [ordering] of E that is compatible with the vector space structure.

(31.2) In a preordered vector space, (1) $x \geq \theta$ and $y \geq \theta$ imply $x + y \geq \theta$; (2) $x \geq \theta$ and $\alpha \geq 0$ imply $\alpha x \geq \theta$; (3) $x \leq y$ iff $y - x \geq \theta$; (4) $x \leq y$ iff $-y \leq -x$; (5) $x \leq y$ and $\alpha \leq 0$ imply $\alpha x \geq \alpha y$. {Proofs: (1) $x + y \geq \theta + y = y \geq \theta$; (2) $\alpha x \geq \alpha\theta = \theta$; (3) consider $z = -x$ in (i) of (31.1); (4) $-x - (-y) = y - x$; (5) $\alpha x - \alpha y = (-\alpha)(y - x)$.}

(31.3) A preordering $\leq$ on E is compatible with the vector space structure iff (i) $x \leq y$ implies $x + z \leq y + z$ for all z, and (ii') $x \geq \theta$ and $\alpha \geq 0$ imply $\alpha x \geq \theta$. {Proof: Suppose (i) and (ii') hold. If $x \leq y$ and $\alpha \geq 0$ then $y - x = y + (-x) \geq x + (-x) = \theta$, therefore $\alpha(y - x) \geq \theta$; then $\alpha x = \theta + \alpha x \leq \alpha(y - x) + \alpha x = \alpha y$. Thus the conditions of (31.1) are verified. The converse is noted in (31.2).}

The next two theorems establish a bijective correspondence between the set of all pointed convex cones in E and the set of all compatible preorderings of E; under this correspondence, the salient pointed convex cones correspond to the compatible orderings.

(31.4) **Theorem.**   Let $(E, \leq)$ be a preordered real vector space, and let $P = \{x : x \geq \theta\}$. Then P is a pointed convex cone, and $x \leq y$ iff $y - x \in P$. For $\leq$ to be an ordering, it is necessary and sufficient that P be salient.

Proof.   If $x, y \in P$ and $\alpha \geq 0$, then $x + y \in P$ and $\alpha x \in P$ by (1) and (2) of (31.2), thus P is a pointed convex cone (cf. (27.5), (27.3)). Also, $x \leq y$ iff $y - x \in P$ by (3) of (31.2).

Suppose $\leq$ is an ordering. If $x \in P \cap (-P)$ then $x \geq \theta$ and $-x \geq \theta$, thus $x \geq \theta$ and $x \leq \theta$ (31.2), therefore $x = \theta$; that is, P is salient (27.11).

Conversely, suppose P is salient. If $x \leq y$ and $y \leq x$, then $y - x \in P$ and $-(y - x) = x - y \in P$, thus $y - x \in P \cap (-P) = \{\theta\}$, $x = y$; that is, $\leq$ is an ordering.   ∎

(31.5) **Theorem.**   Let P be a pointed convex cone in a real vector space E, and define $x \leq y$ iff $y - x \in P$. Then $\leq$ is a preordering of E, compatible with the vector space structure, and $P = \{x : x \geq \theta\}$.

Proof.   If $x \leq y$ and $y \leq z$ then $z - x = (z - y) + (y - x) \in P + P \subset P$, thus $x \leq z$. Since $x - x = \theta \in P$, $x \leq x$. Thus $\leq$ is a preordering.

If $x \leq y$ then $(y + z) - (x + z) = y - x \in P$, that is, $x + z \leq y + z$, for all z. If $x \leq y$ and $\alpha \geq 0$, then $\alpha y - \alpha x = \alpha(y - x) \in \alpha P \subset P$, whence $\alpha x \leq \alpha y$. Thus $\leq$ is compatible with the vector space structure.

Finally, $x \geq \theta$ iff $x - \theta \in P$.   ∎

The following definition is for use in the next section:

(31.6) **Definition.**   A preordered vector space $(E, \leq)$ is said to be *Archimedean* if there exists a vector u in E with the following property: if $x \in E$ then there exists a positive integer n (depending on x) such that $-nu \leq x \leq nu$. Such a vector u is called an *order unit* of E.

Exercises

(31.7) Let $(E, \le)$ be a preordered real vector space and let

$$N = \{x : x \ge \theta \ \& \ x \le \theta\}.$$

Then (i) N is a linear subspace of E, and (ii) E/N, with the natural ordering, is an ordered vector space.

(31.8) If $(E, \le)$ is an Archimedean preordered vector space, there exists an order unit v such that $v \ge \theta$.

*32. Extension of positive linear forms.

***32. Extension of positive linear forms.** Throughout this section, E denotes a real vector space.

Given a linear subspace M of E and a linear form g on M, consider the problem of extending g to a linear form on all of E. For general vector spaces, the problem is always solvable, by an easy basis argument. However, if E is a TVS and g is continuous, we are interested in extensions that preserve continuity. If E admits a functional p satisfying (i), (ii) of (28.6), and if g is dominated by p in the sense of (28.6), we are interested in extensions that are also dominated by p. If E is a preordered vector space and g is a positive linear form (in the sense defined below), we are interested in extensions that preserve positivity. In general, this can't be done (32.5), but it can if suitable additional conditions are imposed. The present section is devoted to proving two theorems of this type; in the first, the extra conditions are order-theoretic, in the second, topological.

(32.1) Definition. A linear form f on a preordered real vector space $(E, \le)$ is said to be *positive* if $x \ge \theta$ implies $f(x) \ge 0$ (equivalently, $x \le y$ implies $f(x) \le f(y)$). (Briefly, f is a PLF.)

(32.2) Theorem. *Let $(E, \le)$ be an Archimedean preordered real vector space, let u be an order unit of E, and let M be a linear subspace of E such that $u \in M$. Then every positive linear form g on M may be extended to a positive linear form f on E.*

Proof. The proof is patterned after the classical Hahn-Banach theorem (28.6). As in that proof, the essential point is to show that one can perform the extension to the linear subspace $T = M + \mathbb{R}a$ generated by M and a vector a not in M. Define

$$A = \{y \in M : y \le a\}, \qquad B = \{z \in M : z \ge a\}.$$

A and B are nonempty; indeed, if n is any positive integer such that $-nu \le a \le nu$, then, since $u \in M$, we have $-nu \in A$ and $nu \in B$.

If $y \in A$ and $z \in B$, then $y \le a \le z$, therefore $g(y) \le g(z)$; thus,

$$\sup g(A) \le \inf g(B).$$

Let β be any real number such that

$$\sup g(A) \le \beta \le \inf g(B);$$

thus,

(*) $g(y) \leq \beta \leq g(z)$ $(y \in A, z \in B).$

We are ready to extend g to a linear form f on T. If $x \in T$, define $f(x)$ as follows: write $x = y + \alpha a$ for suitable (unique) $y \in M$ and $\alpha \in \mathbb{R}$, and set

$$f(x) = g(y) + \alpha \beta.$$

Clearly f is a linear form on T and $f|M = g$. It remains to show that f is positive. Suppose $x \in T$, say $x = y + \alpha a$, $y \in M$; assuming $x \geq \theta$, it is to be shown that $g(y) + \alpha \beta \geq 0$. If $\alpha = 0$ this is a property of g. We consider separately the cases $\alpha > 0$ and $\alpha < 0$.

If $\alpha > 0$ then the assumption $y + \alpha a \geq \theta$ may be written $\alpha^{-1}y + a \geq \theta$, i.e., $a \geq -\alpha^{-1}y$; then $-\alpha^{-1}y \in A$, therefore by (*) we have $g(-\alpha^{-1}y) \leq \beta$, $-g(y) \leq \alpha\beta$, $g(y) + \alpha\beta \geq 0$.

If $\alpha < 0$ the assumption $y + \alpha a \geq \theta$ may be written $\alpha^{-1}y + a \leq \theta$, $a \leq -\alpha^{-1}y$; then $-\alpha^{-1}y \in B$, therefore by (*) we have $g(-\alpha^{-1}y) \geq \beta$, $-\alpha^{-1}g(y) \geq \beta$, $g(y) \geq -\alpha\beta$, $g(y) + \alpha\beta \geq 0$.

The proof proceeds as in (28.6). ∎

The second of the promised extension theorems depends on the following lemma to the effect that, under certain conditions, continuity is an automatic consequence of positivity:

(32.3) Lemma.   *Let $(E, \leq)$ be a preordered TVS over $\mathbb{R}$, let $P = \{x : x \geq \theta\}$, and suppose int $P \neq \varnothing$. Then every positive linear form f on E is continuous; if, moreover, f is nonzero, then $f > 0$ on int P.*

Proof.   In view of (31.4) and (31.5), the matter can be put as follows. We are given a real TVS E and a pointed convex cone P in E such that int $P \neq \varnothing$. Assuming f is a nonzero linear form on E such that $f \geq 0$ on P, we are to show that f is continuous. Setting $H = \{x : f(x) = 0\}$, it is equivalent to show that the hyperplane H is closed (22.2). By hypothesis, P lies to one side of H (30.3); since P has nonempty interior, it follows from (30.5) that H is closed and int P lies strictly to one side of H, thus $f > 0$ on int P. ∎

(32.4) Theorem.   *Let $(E, \leq)$ be a preordered TVS over $\mathbb{R}$, let $P = \{x : x \geq \theta\}$, and suppose int $P \neq \varnothing$. Let M be a linear subspace of E such that $M \cap$ int $P \neq \varnothing$, and suppose g is a positive linear form on M. Then g may be extended to a positive linear form f on E, and f is necessarily continuous.*

Proof.   The assumption is that $g \geq 0$ on $M \cap P$; we seek a linear form f on E such that $f|M = g$ and $f \geq 0$ on P. We can suppose $g \neq 0$.

Since $M \cap$ int P is nonempty, the interior of $M \cap P$ relative to M is also nonempty, namely,

$$\text{int}_M (M \cap P) \supset M \cap \text{int } P \neq \varnothing.$$

It follows from the lemma that g is continuous, and $g > 0$ on $M \cap$ int P (in fact, on $\text{int}_M (M \cap P)$). Thus $N = \{x \in M : g(x) = 0\}$ is a closed hyperplane in M, and $N \cap$ int $P = \varnothing$.

Summarizing, $N \cap \text{int } P = \varnothing$, where $\text{int } P$ is a nonempty open convex set in E, and N is a linear subspace of E. By the geometric form of the Hahn-Banach theorem (28.1) there exists a closed maximal linear subspace H of E such that $H \supset N$ and $H \cap \text{int } P = \varnothing$.

Let h be a continuous linear form such that $H = \{x : h(x) = 0\}$ (22.4). It will be shown that a suitable scalar multiple of h is the desired extension f of g.

Note that H does not contain M; indeed, $H \cap \text{int } P = \varnothing$ whereas $M \cap \text{int } P \neq \varnothing$. Then $N = N \cap H \subset M \cap H \neq M$; since N is a maximal linear subspace of M, it follows that $N = M \cap H$. Thus if $g' = h|M$, then the null space of g', namely, $M \cap H$, coincides with the null space N of g; it follows by elementary algebra that g and g' are scalar multiples of each other. Say $g = \alpha g'$, $\alpha \neq 0$. Set $f = \alpha h$. Then f is a continuous linear form on E such that $f|M = g$. Now,

$$\{x : f(x) = 0\} = \{x : h(x) = 0\} = H;$$

since $H \cap \text{int } P = \varnothing$, $\text{int } P$ lies strictly to one side of H (30.4), and since $f = g > 0$ on $M \cap \text{int } P$ it follows that $f > 0$ on $\text{int } P$. Citing (25.20) and the continuity of f, we have $f \geq 0$ on $(\text{int } P)^- = \bar{P}$; in particular, $f \geq 0$ on P, that is, f is a positive linear form on E. Finally, the continuity of f is no accident (32.3). ∎

Exercise

(32.5)   Let (m) be the Banach space of all bounded sequences $x = (\alpha_k)$ of real numbers, with the linear operations defined coordinatewise, and $\|x\| = \sup |\alpha_k|$ (cf. (16.15), (29.6)). Order (m) by defining $x \leq y$ iff $\alpha_k \leq \beta_k$ for all k, where $x = (\alpha_k)$ and $y = (\beta_k)$. Every positive linear form on (m) is continuous. Let (c_{00}) be the set of all finitely nonzero sequences x, i.e., such that $\alpha_k = 0$ for all but finitely many k. Then (c_{00}) is a linear subspace of (m), and the formula $g(x) = \sum \alpha_k$ defines a positive linear form on (c_{00}) that cannot be extended to a positive linear form on all of (m).

33. Locally convex TVS.   We consider both real and complex spaces. As usual, $\mathbb{K}$ denotes either $\mathbb{R}$ or $\mathbb{C}$.

(33.1) **Definition.**   A TVS E over $\mathbb{K}$ is said to be *locally convex* if the convex neighborhoods of θ are basic.

(33.2)   Since translations preserve convexity (25.5), in a locally convex TVS every point has a fundamental system of convex neighborhoods.

(33.3) **Example.**   Every normed space is locally convex; specifically, the open balls (or the closed balls) centered at θ are basic convex neighborhoods of θ (25.2).

The construction of further examples is expedited by some general theory.

(33.4) **Theorem.**   *Every subspace and every quotient space of a locally convex TVS is locally convex.*

Proof. Let E be a locally convex TVS over $\mathbb{K}$ and let M be a linear subspace of E. If V is a convex neighborhood of θ in E, then $V \cap M$ is a convex neighborhood of θ in M; since the V's are basic in E, the $V \cap M$'s are basic in M. Thus M is locally convex. Let E/M be the quotient TVS, $\pi: E \to E/M$ the canonical mapping. If V is a convex neighborhood of θ in E, then $\pi(V)$ is convex (25.6); since the V's are basic in E, the $\pi(V)$'s are basic neighborhoods of $\pi(\theta)$ in E/M by the definition of the quotient topology (11.8). ∎

(33.5) Theorem. *The product of any family of locally convex TVS is locally convex.*

Proof. Let $(E_\iota)_{\iota \in I}$ be a family of locally convex TVS over $\mathbb{K}$ and let $E = \prod_{\iota \in I} E_\iota$ be the product TVS (11.9). A basic neighborhood of θ in E has the form $V = \prod_{\iota \in I} V_\iota$, where, for a suitable finite subset J of I, $V_\iota = E_\iota$ for $\iota \notin J$, and V_ι is a convex neighborhood of θ in E_ι for $\iota \in J$. Such a set V is convex (25.8). ∎

(33.6) Lemma. *If $f: E \to F$ is a linear mapping, where E is a vector space over $\mathbb{K}$ and F is a locally convex TVS over $\mathbb{K}$, then the initial topology for f is locally convex.*

Proof. The topology in question is compatible with the vector space structure of E (17.13). A basic neighborhood of θ in E has the form $f^{-1}(V)$, where V is a convex neighborhood of θ in F; such sets are convex (25.6). ∎

(33.7) Theorem. *If E is a vector space over $\mathbb{K}$, $(E_\iota)_{\iota \in I}$ is a family of locally convex TVS over $\mathbb{K}$, and if, for each $\iota \in I$, $f_\iota: E \to E_\iota$ is a linear mapping, then E, equipped with the initial topology for the family $(f_\iota)_{\iota \in I}$, is a locally convex TVS.*

Proof. Let $F = \prod_{\iota \in I} E_\iota$ be the product TVS and let $f: E \to F$ be the mapping $f(x) = (f_\iota(x))_{\iota \in I}$. The initial topology τ for the family $(f_\iota)_{\iota \in I}$ coincides with the initial topology for f (19.2) and is compatible with the vector space structure of E (19.3); since, moreover, the topology on F is locally convex (33.5), it follows from the lemma that τ is also locally convex. ∎

(33.8) Corollary. *The supremum of a family of locally convex compatible topologies is a locally convex compatible topology.*

Proof. This is clear from (33.7) and the proof of (19.4). ∎

A key application for the theory of duality (see, e.g., Section 38):

(33.9) Corollary. *Let E be a vector space over $\mathbb{K}$ and let $f_\iota: E \to \mathbb{K}$ ($\iota \in I$) be any family of linear forms on E. Then:*

(i) E, equipped with the initial topology τ for the family $(f_\iota)_{\iota \in I}$, is a locally convex TVS;

(ii) a basic neighborhood of θ in E for the topology τ is given by

$$\{x \in E : |f_\iota(x)| \le \varepsilon \quad \text{for all } \iota \in J\},$$

where $\varepsilon > 0$ and J is any finite subset of I;

(iii) τ *is the coarsest topology on E for which all the f_ι are continuous;*

(iv) τ *is separated iff, for each nonzero vector x in E, there exists an index ι such that $f_\iota(x) \neq 0$.*

Proof. (iii) simply recalls the definition of initial topology (19.1). In view of (19.6) and (19.7), it remains only to observe that τ is locally convex. This follows either from (33.7) and the fact that $\mathbb{K}$ is locally convex, or directly from the fact that the sets described in (ii) are obviously convex. ∎

In a locally convex TVS, the filter of neighborhoods of θ has a pleasant base:

(33.10) Theorem. *In a locally convex TVS, the convex, balanced, closed neighborhoods of θ are basic.*

Proof. Let U be any neighborhood of θ; we seek a neighborhood V of θ such that $V \subset U$ and V has the listed properties.

Let U_1 be a closed neighborhood of θ with $U_1 \subset U$ (5.14), let U_2 be a convex neighborhood of θ with $U_2 \subset U_1$, and let U_3 be a balanced neighborhood of θ with $U_3 \subset U_2$. Set $W = \operatorname{conv} U_3$. Then W is a convex, balanced (25.15) neighborhood of θ such that $W = \operatorname{conv} U_3 \subset \operatorname{conv} U_2 = U_2$; therefore $V = \overline{W}$ is a closed, convex (25.17), balanced (17.5) neighborhood of θ such that $V = \overline{W} \subset \overline{U}_2 \subset \overline{U}_1 = U_1 \subset U$. ∎

In the reverse direction, there is a simple set of neighborhood axioms for a locally convex TVS:

(33.11) Theorem. *Let E be a vector space over $\mathbb{K}$ and suppose $\mathscr{B}$ is a base of a filter on E such that $\theta \in V$ for every $V \in \mathscr{B}$, and such that every $V \in \mathscr{B}$ is convex, balanced and absorbent. Then (i) the class of sets*

$$\mathscr{B}' = \{\alpha V : \alpha > 0,\ V \in \mathscr{B}\}$$

is a base of a filter on E; (ii) there exists a unique topology τ on E such that τ is compatible with the vector space structure of E and $\mathscr{B}'$ is a fundamental system of neighborhoods of θ; (iii) the topology τ is locally convex.

Proof. (i) Assuming $\alpha_1, \alpha_2 > 0$ and $V_1, V_2 \in \mathscr{B}$, it is to be shown that $\alpha V \subset (\alpha_1 V_1) \cap (\alpha_2 V_2)$ for suitable $\alpha > 0$ and $V \in \mathscr{B}$. Choose $V \in \mathscr{B}$ with $V \subset V_1 \cap V_2$ and let $\alpha = \min \{\alpha_1, \alpha_2\}$. Citing (17.2) we have $\alpha V \subset \alpha V_j \subset \alpha_j V_j$ $(j = 1, 2)$.

(ii) The sets in $\mathscr{B}'$ are evidently balanced and absorbent. We verify the remaining hypotheses of (17.12). If $\alpha > 0$, $V \in \mathscr{B}$ and $\lambda \in \mathbb{K}$, $\lambda \neq 0$, then, citing (17.2), we have

$$\lambda(\alpha V) = (\lambda\alpha)V = |\lambda\alpha|V \in \mathscr{B}'.$$

If $\alpha_j > 0$ and $V_j \in \mathscr{B}$ $(j = 1, 2)$, then, choosing $V \in \mathscr{B}$ with $V \subset V_1 \cap V_2$, it follows from (25.10) and the convexity of V that

$$\alpha_1 V_1 + \alpha_2 V_2 \supset \alpha_1 V + \alpha_2 V = (\alpha_1 + \alpha_2)V \in \mathscr{B}'.$$

The proof of (ii) is then immediate from (17.12).

(iii) The sets in $\mathscr{B}'$ are convex. ∎

In a vector space, every linear variety is the intersection of the hyperplanes that contain it (21.9). As promised at the end of Section 22, the topological analogue holds in a locally convex TVS:

(33.12) Theorem. *If E is a locally convex TVS over $\mathbb{K}$ and if M is a closed linear variety in E, then M is the intersection of all the closed hyperplanes that contain it.*

Proof. We can suppose, by translation, that $\theta \in M$, i.e., that M is a closed linear subspace of E. In view of (22.4), the theorem may be reformulated thus: assuming $a \notin M$, we seek a continuous linear form f on E such that $f = 0$ on M and $f(a) \neq 0$.

Since M is closed and $a \notin M$, there exists a neighborhood A of a with $A \cap M = \varnothing$. We can suppose A is convex. Then the neighborhood int A of a is also convex (25.20), so we can suppose A is open. By the geometric form of the Hahn-Banach theorem ((28.4) or (28.5), according as $\mathbb{K} = \mathbb{R}$ or $\mathbb{K} = \mathbb{C}$), there exists a continuous linear form f on E such that $f = 0$ on M and f is never 0 on A—in particular, $f(a) \neq 0$. ∎

It follows that a separated locally convex TVS has 'sufficiently many' continuous linear forms:

(33.13) Corollary. *If E is a separated, locally convex TVS, then, given any nonzero vector a in E, there exists a continuous linear form f on E such that $f(a) \neq 0$.*

Proof. Since $\{\theta\}$ is a closed linear subspace (3.4), the theorem is applicable with $M = \{\theta\}$. ∎

Exercises

(33.14) In a locally convex TVS, the convex, balanced, open neighborhoods of θ are basic.

(33.15) A locally convex, metrizable complete TVS is called a Fréchet space. In view of (33.5), the space (s) of Section 14 is a Fréchet space, but it is not normable (26.8).

(33.16) The space (S) of Section 15 is separated (even metrizable) but has no nonzero continuous linear forms (15.10), hence it cannot be locally convex (33.13). Thus (S) is an example of a space of type (F) (see Section 13) that is not a Fréchet space.

(33.17) Let $(E_\iota)_{\iota \in I}$ be a family of locally convex TVS over $\mathbb{K}$, let E be a vector space over $\mathbb{K}$, and let $f_\iota : E_\iota \to E$ ($\iota \in I$) be a family of linear mappings.

(i) Let $\mathscr{B}$ be the set of all convex, balanced, absorbent subsets V of E such that $f_\iota^{-1}(V)$ is a neighborhood of θ in E_ι for every $\iota \in I$. Then $\mathscr{B}$ is a neighborhood base at θ for a compatible topology τ on E.

(ii) τ is the finest locally convex topology on E for which every f_ι is continuous.

(iii) Let F be a locally convex TVS over $\mathbb{K}$. In order that a linear mapping $f: E \to F$ be continuous for τ, it is necessary and sufficient that $f \circ f_\iota: E_\iota \to F$ be continuous for all $\iota \in I$.

(33.18) Let E be a vector space over $\mathbb{K}$ and let $(E_\iota)_{\iota \in I}$ be an increasingly directed family of linear subspaces of E (i.e., the index set I is directed to the right, and $\iota \leq \varkappa$ implies $E_\iota \subset E_\varkappa$). Assume that, for each $\iota \in I$, τ_ι is a locally convex, compatible topology on E_ι; assume, moreover, that if $\iota \leq \varkappa$ then the relative topology on E_ι induced by $\tau_\varkappa$ is coarser than τ_ι (i.e., the identity injection $E_\iota \to E_\varkappa$ is continuous). Write $f_\iota: E_\iota \to E$ for the identity injection. The finest locally convex compatible topology τ on E that renders every f_ι continuous (33.17) is called the *inductive limit* of the τ_ι.

34. Separation of convex sets in a locally convex TVS.

We pursue in locally convex spaces (both real and complex) a theme initiated in Section 30 for general TVS.

To set the stage, suppose E is a real TVS and A, B are nonempty convex sets in E such that $A \cap B = \varnothing$. The problem is to find a closed hyperplane H that separates A and B. In general, no such H exists (34.11), but we have already recorded some positive results based on the Hahn-Banach theorem. Namely, assuming A, B as above, a closed separating hyperplane exists in the following situations. Theorem (30.6): A or B open. Theorem (30.12): A a convex body, $B = \{z\}$. The Hahn-Banach theorem itself may be viewed as a separation theorem (28.4): A open, B a linear subspace of E. At first glance, Theorem (32.4) is not a theorem of this genre, since the critical hypothesis is that the two convex sets in question—namely, int P and M—have nonempty intersection; but a second glance at the proof shows that this is really a separation theorem in disguise. Still another separation theorem (33.12): E locally convex, A a closed linear variety, $B = \{z\}$. In the present section we exploit local convexity to prove further results of this type; a major application occurs in Section 36 (the Kreĭn-Mil'man theorem).

(34.1) Theorem. *Let E be a locally convex TVS over $\mathbb{R}$, let A and B be nonempty convex sets such that $A \cap B = \varnothing$, and suppose that A is closed and B is compact. Then there exists a closed hyperplane H that separates A and B strictly.*

Proof. Equip E with the uniform structure derived from its additive topological group structure (11.10). It is a theorem in general topology that $W(A) \cap W(B) = \varnothing$ for a suitable entourage W. {Recall that $W(A)$ is the set of all y such that $(a, y) \in W$ for some $a \in A$.} We can suppose $W = \{(x, y) : y - x \in V\}$, where V is a convex open neighborhood of θ (by hypothesis, such entourages are basic (5.7)). Since $W(A) = A + V$ and $W(B) = B + V$ (5.9), we have

$$(A + V) \cap (B + V) = \varnothing,$$

where $A + V$ and $B + V$ are convex (25.9) open sets; by (30.6) there exists a closed hyperplane H that strictly separates $A + V$ and $B + V$, and therefore also their subsets A and B. ∎

The complex variant of (34.1):

(34.2) Corollary. *Let E be a locally convex TVS over $\mathbb{C}$, let A and B be nonempty convex sets with $A \cap B = \varnothing$, and suppose that A is closed and B is compact. Then there exist a continuous linear form f on E and a real number α, such that $\mathrm{Re}\, f < \alpha$ on A and $\mathrm{Re}\, f > \alpha$ on B.*

Proof. Viewing E as a TVS over $\mathbb{R}$, there exist, by the theorem, a continuous $\mathbb{R}$-form g on E and a real number α, such that $g < \alpha$ on A and $g > \alpha$ on B. Let f be the $\mathbb{C}$-form with $g = \mathrm{Re}\, f$ (22.6). ∎

Suppose E is any TVS over $\mathbb{R}$. The intersection of any family of topologically closed half-spaces is obviously a closed convex set A. In the reverse direction, if A is a convex body in E, then A is the intersection of the family of all topologically closed half-spaces that contain it (30.12). In a locally convex space, *every* closed convex set can be obtained in this way:

(34.3) Theorem. *If E is a locally convex TVS over $\mathbb{R}$ and if A is a nonempty closed convex set in E, then A is the intersection of the family of all topologically closed half-spaces that contain it.*

Proof. In view of Definition (30.1), given any $b \notin A$ we seek a continuous linear form g and a real number β such that $g \leq \beta$ on A and $g(b) > \beta$. Since $B = \{b\}$ is a compact convex set disjoint from A, our assertion is immediate from (34.1), which even yields $g < \beta$ on A. ∎

The complex variant of (34.3):

(34.4) Corollary. *Suppose E is a locally convex TVS over $\mathbb{C}$, and A is a nonempty closed convex set in E. If $b \notin A$ then there exist a continuous linear form f on E and a real number β, such that $\mathrm{Re}\, f < \beta$ on A and $\mathrm{Re}\, f(b) > \beta$.*

Proof. With notations as in the proof of (34.3), let f be the $\mathbb{C}$-form with $\mathrm{Re}\, f = g$ (22.6). ∎

Theorem (34.3) can be improved significantly in the presence of compactness:

(34.5) Theorem. *Suppose E is a locally convex, separated TVS over $\mathbb{R}$, and A is a nonempty, compact convex subset of E. Then A is the intersection of the topologically closed half-spaces that contain it and are determined by closed hyperplanes tangent to A.*

Proof. Let $b \notin A$. Since A is a closed subset of E, by (34.3) there exist a continuous linear form g on E and a real number β, such that $g < \beta$ on A and $g(b) > \beta$. Let $\alpha = \sup g(A)$. Since $g(A)$ is a compact subset of $\mathbb{R}$, one has $\alpha \in g(A)$; then $\alpha = g(a)$ for suitable $a \in A$, and in particular $\alpha < \beta$. Consider the closed hyperplane $H = \{x : g(x) = \alpha\}$. Since $g \leq \alpha$ on A, A lies to one side of H; moreover, $a \in H \cap A$, thus H is tangent to A. Finally, $g(b) > \beta > \alpha$, thus b is excluded by the half-space $g \leq \alpha$. ∎

The complex variant of (34.5):

(34.6) Corollary. *Suppose E is a locally convex, separated TVS over* $\mathbb{C}$, *and A is a nonempty, compact convex subset of E. If* $b \notin A$ *then there exist a continuous linear form f on E and a real number* α, *such that* $\operatorname{Re} f \leq \alpha$ *on A,* $\operatorname{Re} f(a) = \alpha$ *for at least one* $a \in A$, *and* $\operatorname{Re} f(b) > \alpha$.

Proof. With notation as in the proof of (34.5), let f be the $\mathbb{C}$-form with $\operatorname{Re} f = g$. ∎

For use in the proof of the Kreĭn-Mil'man theorem (36.9), we note the following consequence of (34.3):

(34.7) Theorem. *Let E be a locally convex TVS over* $\mathbb{R}$, *A a nonempty closed convex set in E, and S a subset of E. In order that* $S \subset A$ *it is necessary and sufficient that* $g(S) \subset g(A)$ *for every continuous linear form g on E.*

Proof. The necessity of the condition is trivial. Conversely, suppose the condition holds. If $b \notin A$ then by (34.3) there exist a continuous linear form g and a real number β, such that $g \leq \beta$ on A and $g(b) > \beta$, and therefore $g(b) \notin g(A)$; since $g(S) \subset g(A)$ by hypothesis, it follows that $b \notin S$. Thus $\complement A \subset \complement S$, i.e., $S \subset A$. ∎

The complex variant of (34.7) is noted in an exercise (34.12).

The Hahn-Banach theorem for locally convex spaces belongs to this circle of ideas; it takes the following appealing form:

(34.8) Theorem. *If E is a locally convex TVS over* $\mathbb{K}$ *and if M is a linear subspace of E, then every continuous linear form on M may be extended to a continuous linear form on E.*

Proof. Suppose first that $\mathbb{K} = \mathbb{R}$ and let $g: M \to \mathbb{R}$ be a continuous linear form. Since $\mathbb{R}$ is a separated, complete uniform space (7.8) and g is uniformly continuous for the relative uniform structure (5.12), it follows that g may be extended (uniquely) to a continuous mapping $\bar{g}: \overline{M} \to \mathbb{R}$ (see (24.2)), and, by an elementary argument, $\bar{g}$ is also linear. In other words, we can suppose that M is closed. We can also suppose $g \neq 0$ (if $g = 0$ the extension problem is solved trivially). The linear subspace

$$N = \{y \in M : g(y) = 0\}$$

is closed in M, and therefore in E (because M is closed in E). Choose any $a \in M$ with $g(a) = 1$; thus $M = N + \mathbb{R}a$. Since $a \notin N$ it follows from (33.12) that there exists a closed hyperplane H in E with $H \supset N$ and $a \notin H$. Say

$$H = \{x \in E : h(x) = 0\},$$

h a continuous linear form on E. Since $a \notin H$, $h(a) \neq 0$; replacing h by $(h(a))^{-1}h$, we can suppose that $h(a) = 1$. Then $h = g$ on $N + \mathbb{R}a = M$, that is, $h|M = g$.

Suppose now that $\mathbb{K} = \mathbb{C}$ and let $f: M \to \mathbb{C}$ be a continuous linear form. We may also view E as a TVS over $\mathbb{R}$, and M is an $\mathbb{R}$-linear subspace of E. Then $g = \operatorname{Re} f$ is a continuous $\mathbb{R}$-form on M, therefore by the first part of the proof there exists a continuous $\mathbb{R}$-form g' on E that extends g. Let f' be the (necessarily continuous) $\mathbb{C}$-form on E with $\mathrm{Re}\,f' = g'$ (22.6). Then

$$(\mathrm{Re}\,f')|M = g'|M = g = \mathrm{Re}\,f;$$

thus $f'|M$ and f are $\mathbb{C}$-forms on M having the same real part, therefore $f'|M = f$ by (21.11). ∎

The question of replacing $\mathbb{K}$ in such extension theorems by more complicated spaces (possibly with E replaced by some other type of TVS) is a large subject, to which the following corollary belongs:

(34.9) Corollary. *Suppose E is a locally convex TVS over $\mathbb{K}$, I is a nonempty set of indices, and $\mathbb{K}^I$ is the product TVS $\prod_{\iota \in I} E_\iota$, where $E_\iota = \mathbb{K}$ for all $\iota \in I$. If M is a linear subspace of E and if $u: M \to \mathbb{K}^I$ is a continuous linear mapping, then u may be extended to a continuous linear mapping $\bar{u}: E \to \mathbb{K}^I$.*

Proof. For each $\iota \in I$, let $\pi_\iota: \mathbb{K}^I \to \mathbb{K}$ be the ιth coordinate projection, and set $f_\iota = \pi_\iota \circ u$. Thus, $u = (f_\iota)_{\iota \in I}$, that is,

$$u(y) = (f_\iota(y))_{\iota \in I} \qquad (y \in M).$$

Since π_ι is a continuous linear form on $\mathbb{K}^I$, f_ι is a continuous linear form on M; citing (34.8), there exists a continuous linear form $\bar{f}_\iota$ on E such that $\bar{f}_\iota|M = f_\iota$. Define $\bar{u}: E \to \mathbb{K}^I$ by $\bar{u} = (\bar{f}_\iota)_{\iota \in I}$, that is,

$$\bar{u}(x) = (\bar{f}_\iota(x))_{\iota \in I} \qquad (x \in E).$$

Then $\bar{u}$ is linear, and $\bar{u}|M = u$. Moreover, since each of the compositions $\pi_\iota \circ \bar{u} = \bar{f}_\iota$ is continuous, it follows that $\bar{u}$ is continuous. {Recall that the product topology is the initial topology for the family of coordinate projections (cf. (11.9), (19.1)).} ∎

An application:

(34.10) Corollary. *If E is a locally convex, separated TVS over $\mathbb{K}$, then every finite-dimensional linear subspace of E has a topological supplement.*

Proof. Let M be a finite-dimensional linear subspace of E; we seek a linear subspace N of E such that E is the topological direct sum of M and N (20.2). In view of (20.6), it suffices to show that E admits a continuous projector p with range M.

Say $\dim_{\mathbb{K}} M = n$. Let $u: M \to \mathbb{K}^n$ be any bijective linear mapping. Since M is separated, u is bicontinuous (23.1). Citing (34.9), u may be extended to a (surjective) continuous linear mapping $\bar{u}: E \to \mathbb{K}^n$. Define $p: E \to E$ by $p = u^{-1} \circ \bar{u}$. Clearly p is a continuous linear mapping, $p(E) = u^{-1}(\bar{u}(E)) = u^{-1}(\mathbb{K}^n) = M$, and $y \in M$ implies $p(y) = u^{-1}(u(y)) = y$; it follows that $p \circ p = p$, thus p is a continuous projector with range M. ∎

Merging (34.10) with (23.7), we have the following satisfyingly symmetric proposition: *In a separated, locally convex TVS, every closed linear subspace of finite dimension, or of finite codimension, has a topological supplement.*

Exercises

(34.11) Let g be a discontinuous linear form on a real TVS E, and let $A = \{x : g(x) \leq 0\}$, $B = \{x : g(x) > 0\}$. Then A and B are disjoint convex sets that cannot be separated by a closed hyperplane.

(34.12) Let E be a locally convex TVS over $\mathbb{C}$, A a closed convex set in E, and $S \subset E$. The following conditions are equivalent: (a) $S \subset A$; (b) $f(S) \subset f(A)$ for every continuous linear form f on E; (c) $\operatorname{Re} f(S) \subset \operatorname{Re} f(A)$ for every continuous linear form f on E.

(34.13) In Theorem (34.3) the assumption that $A \neq \varnothing$ cannot in general be omitted; if $A = \varnothing$, the assertion of the theorem holds if and only if E admits a nonzero continuous linear form.

35. Compact convex sets in a TVS.

Among the principal themes in functional analysis are the notions of linearity, completeness, compactness and convexity; these are merged in the powerful notion of compact convex set, to which this section and the next are devoted. At center stage in this circle of ideas is the Kreĭn-Mil'man theorem, proved in the next section; among its many applications, we mention that it is one of the keys to modern harmonic analysis.

In general, the compactness of a set in a TVS does not imply compactness of its convex hull, nor even of its closed convex hull. There are, however, some partial results of this character, to which the present section is devoted.

(35.1) Theorem. *Let E be a separated TVS over $\mathbb{K}$, let A_k $(k = 1, \ldots, n)$ be a finite number of compact convex subsets of E, and let $S = \bigcup_1^n A_k$. Then $\operatorname{conv} S$ is also compact.*

Proof. Let B be the set of all $(\lambda_1, \ldots, \lambda_n)$ in $\mathbb{K}^n$ such that $\lambda_k \geq 0$ and $\sum_1^n \lambda_k = 1$; clearly B is a compact subset of $\mathbb{K}^n$. Define $f \colon B \times \prod_1^n A_k \to E$ by

$$f((\lambda_1, \ldots, \lambda_n), (a_1, \ldots, a_n)) = \sum_1^n \lambda_k a_k.$$

Then f is a continuous mapping of a compact space into the separated space E, hence its range is compact. But the range of f is precisely $\operatorname{conv} S$ (25.13). ∎

(35.2) Corollary. *In a separated TVS, the convex hull of a finite set is compact.*

Proof. If $S = \{s_1, \ldots, s_n\}$, apply the theorem with $A_k = \{s_k\}$. ∎

(35.3) Theorem. *Let E be a locally convex, separated TVS over $\mathbb{K}$. If S is any precompact subset of E, then the convex hull of S is also precompact.*

Proof. Recall that a separated uniform structure $(X, \mathcal{U})$ is called precompact iff its completion is compact (cf. Section 24); equivalently, $(X, \mathcal{U})$ is totally bounded, in the sense that if U is any basic entourage, X is expressible as the union of finitely many sets that are small of order U.

Let W be any neighborhood of θ, and let $W_r = \{(x, y) : y - x \in W\}$; we seek a representation conv $S = T_1 \cup \cdots \cup T_m$ such that each T_j is small of order W_r, i.e., $T_j \times T_j \subset W_r$.

Choose a convex, symmetric neighborhood V of θ such that $V + V + V + V \subset W$. Since S is precompact, there exists a finite system $s_1, \ldots, s_n \in S$ such that

$$S \subset \bigcup_1^n V_r(s_k) = \bigcup_1^n (s_k + V).$$

Writing $T = \bigcup_1^n (s_k + V)$, we have $S \subset T$. Let $A = \text{conv } \{s_1, \ldots, s_n\}$. Since A and V are convex, so is $A + V$ (25.9); obviously $T \subset A + V$, therefore

(*) $\qquad\qquad\qquad$ conv $S \subset$ conv $T \subset A + V$.

Since A is compact (35.2), there exists a finite system $a_1, \ldots, a_m \in A$ with $A \subset \bigcup_1^m (a_j + V)$; then, citing (*), we have

$$\text{conv } S \subset \left[\bigcup_1^m (a_j + V)\right] + V = \bigcup_1^m [(a_j + V) + V],$$

where the sets $T_j = (a_j + V) + V = a_j + (V + V)$ are small of order W_r (because V is symmetric and $(V + V) + (V + V) \subset W$). ∎

(35.4) Corollary. *Let E be a complete, locally convex, separated TVS over $\mathbb{K}$. If S is a subset of E with compact closure, then the closed convex hull of S is compact.*

Proof. The condition on S is that it is precompact (24.16), therefore conv S is also precompact by the theorem; since E is complete, it follows from (24.16) that (conv $S)^-$ is compact. ∎

(35.5) Lemma. *Let E be a separated TVS over $\mathbb{K}$. If S is a precompact subset of E then the balanced hull of S is also precompact.*

Proof. Regard E as a dense linear subspace of a complete, separated TVS $\hat{E}$ as in (24.14); we write $\bar{S}$ for the closure of S in $\hat{E}$, thus $\bar{S}$ is compact (24.16). Denote the balanced hull of S by bal (S). Thus, if $D = \{\lambda \in \mathbb{K} : |\lambda| \leq 1\}$ and $h: D \times \hat{E} \to \hat{E}$ is defined by $h(\lambda, x) = \lambda x$, we know that bal $(S) = h(D \times S)$ (17.4). Since $\bar{S}$ is compact, so is bal $(\bar{S}) = h(D \times \bar{S})$; then bal $(S) \subset$ bal $(\bar{S})$ shows that the closure of bal (S) in $\hat{E}$ is compact, therefore bal (S) is precompact. ∎

(35.6) Theorem. *Let E be a locally convex, separated TVS over $\mathbb{K}$. If S is a precompact subset of E, then the balanced convex hull of S is also precompact.*

Proof. Let A be the balanced convex hull of S, that is, the smallest superset of S that is balanced and convex. Let $B = $ bal (S) be the balanced hull of S; since conv B is also balanced (25.15), clearly conv $B = A$. Since S is precompact, so is $B = $ bal (S) by the lemma, therefore $A = $ conv B is also precompact (35.3). ∎

(35.7) Corollary. *Let E be a complete, locally convex, separated TVS over* $\mathbb{K}$. *If S is a subset of E with compact closure, then the closed, balanced convex hull of S is compact.*

Proof. Let A be the balanced convex hull of S; since $\bar{A}$ is also balanced (17.5) and convex (25.17), it is the closed, balanced convex hull of S. By the theorem, A is precompact, therefore $\bar{A}$ is compact (24.16). {It follows at once that the closed convex hull of S, namely, $(\text{conv } S)^{-}$, is also compact; indeed, every closed subset of $\bar{A}$ is compact.} ∎

Exercise

(35.8) If E is a finite-dimensional, separated TVS over $\mathbb{K}$ and if S is a compact subset of E, then conv S is also compact.

36. Extremal points of a convex set; Kreĭn-Mil′man theorem.

The concept of extremal point is quite intuitive: for example, the extremal points of a convex polygon in the plane are its vertices, the characteristic feature of the vertices being that no vertex can belong to an open interval whose endpoints are in the polygon. To formalize and generalize the concept is the first task of this section. What is decidedly not intuitive is the spectacular role played by the extremal points of convex sets arising in specific topological-algebraic contexts. When the context is sufficiently rich (notably, when compactness is juxtaposed with convexity, in a sufficiently structured TVS), extremal points exist in abundance and have extraordinary algebraic properties; in such contexts, the representability of general points of the convex set in terms of its extremal points leads to structure theorems of great beauty and power. The principal theorem of this section is an abstract result of this type (the Kreĭn-Mil′man theorem), on which, for example, are anchored some of the deepest considerations in abstract harmonic analysis and operator theory (cf. (67.27), (69.26)).

It is useful to refine the concept of segment in a vector space (25.1):

(36.1) Definition. Let E be a vector space over $\mathbb{K}$ and let $u, v \in E$. The *closed segment* with endpoints u, v is defined to be the set

$$[u, v] = \{\alpha u + (1 - \alpha)v : 0 \le \alpha \le 1\}.$$

The *open segment* with endpoints u, v is defined to be the set

$$\langle u, v \rangle = \{\alpha u + (1 - \alpha)v : 0 < \alpha < 1\}.$$

A vector of the form $\alpha u + (1 - \alpha)v$, where $0 < \alpha < 1$, is called an *internal point* of the segments $[u, v]$ and $\langle u, v \rangle$; thus $\langle u, v \rangle$ consists entirely of internal points, and is the set of internal points of $[u, v]$. Observe that

$$\langle u, u \rangle = [u, u] = \{u\};$$

such a segment is called *degenerate*. If $u \ne v$ then $\langle u, v \rangle$ and $[u, v]$ are called *nondegenerate* segments.

It is easy to see that $[u, v]$ and $\langle u, v \rangle$ are convex, and that $[u, v]$ is the convex hull of the set $\{u, v\}$. If E is a separated TVS over $\mathbb{K}$ then every closed segment $[u, v]$ is a compact, and therefore closed, subset of E (it is the image of the compact interval $[0, 1]$ under the continuous mapping $\alpha \mapsto \alpha u + (1 - \alpha)v$).

Note that $u \in \langle u, v \rangle$ iff $u = v$. {Proof: If $u \in \langle u, v \rangle$, say $u = \alpha u + (1 - \alpha)v$ with $0 < \alpha < 1$, then $(1 - \alpha)u = (1 - \alpha)v$ and so $u = v$ results from $\alpha \neq 1$; conversely, if $u = v$ then $\langle u, v \rangle = \{u\}$.}

(36.2) Definition. Let E be a vector space over $\mathbb{K}$ and let A be a convex subset of E. A point $a \in A$ is called an *extremal point* of A if there exists no nondegenerate open segment $\langle u, v \rangle$ such that $a \in \langle u, v \rangle \subset A$. In other words, $a \in A$ is an extremal point of A iff the relations $a \in \langle u, v \rangle \subset A$ imply $u = v$.

In the definition of extremal point, it is no restriction to suppose $u, v \in A$. {Proof: Suppose $a \in \langle u, v \rangle \subset A$, say $a = \alpha u + (1 - \alpha)v$, $0 < \alpha < 1$. Let $u' = \frac{1}{2}u + \frac{1}{2}a$, $v' = \frac{1}{2}a + \frac{1}{2}v$. Then

$$\alpha u' + (1 - \alpha)v' = \alpha(\tfrac{1}{2}u + \tfrac{1}{2}a) + (1 - \alpha)(\tfrac{1}{2}a + \tfrac{1}{2}v)$$
$$= \tfrac{1}{2}[\alpha u + (1 - \alpha)v] + \tfrac{1}{2}a$$
$$= \tfrac{1}{2}a + \tfrac{1}{2}a = a,$$

thus $a \in \langle u', v' \rangle$. Also,

$$u' = \tfrac{1}{2}u + \tfrac{1}{2}[\alpha u + (1 - \alpha)v] = \tfrac{1}{2}(1 + \alpha)u + \tfrac{1}{2}(1 - \alpha)v$$

shows that $u' \in \langle u, v \rangle$, and similarly $v' \in \langle u, v \rangle$, hence $u', v' \in A$.}

Intuitively, one can sometimes arrive at extremal points by a process of dimensional retreat: for example, in the case of a convex (solid) polyhedron in Euclidean 3-space, one arrives at an extremal point by retreating to one of its faces, then to an edge of the face, then to an endpoint of the edge. In nontrivial cases, the dimensional retreat must be performed transfinitely, with only compactness to prevent one from retreating into a vacuum (see the proof of (36.8)). The concept appropriate to describe uniformly the intermediate stages of retreat is as follows:

(36.3) Definition. Let E be a vector space over $\mathbb{K}$ and let A be a nonempty convex subset of E. A *face* of A is a nonempty convex subset F of A with the following property: any open segment in E, that intersects F and is contained in A, is contained in F; that is, the relations $u \in E, v \in E, \langle u, v \rangle \subset A$, $\langle u, v \rangle \cap F \neq \varnothing$ imply $\langle u, v \rangle \subset F$. This is a weakening of the concept of extremal point:

(36.4) Proposition. *Let E be a vector space over $\mathbb{K}$, A a convex subset of E, and suppose $F = \{a\}$, where $a \in A$. Then F is a face of A iff a is an extremal point of A.*

Proof. Suppose first that $\{a\}$ is a face of A. If $a \in \langle u, v \rangle \subset A$ then in particular $\langle u, v \rangle \cap \{a\} \neq \varnothing$, therefore $\langle u, v \rangle \subset \{a\}$ and so $u = v = a$. This shows that a is an extremal point of A.

Conversely, suppose a is an extremal point of A. If $\langle u, v \rangle \subset A$ and $\langle u, v \rangle \cap \{a\} \neq \varnothing$, i.e., if $a \in \langle u, v \rangle \subset A$, then $u = v = a$ by hypothesis, and so $\langle u, v \rangle \subset \{a\}$. This shows that $\{a\}$ is a face of A. ∎

(36.5) The following remarks pertaining to faces are elementary (notation as in (36.3)). (1) A is a face of itself. (2) If $(F_\iota)_{\iota \in I}$ is a family of faces of A with nonempty intersection, then $\bigcap_{\iota \in I} F_\iota$ is also a face of A. (3) If F is a face of A, and if G is a face of F, then G is also a face of A. An important example of a face:

(36.6) Proposition. *Let E be a vector space over $\mathbb{K}$, let A be a nonempty convex subset of E, and suppose H is an $\mathbb{R}$-hyperplane of E that is tangent to A. Then $H \cap A$ is a face of A.*

Proof. Let $F = H \cap A$; by the assumptions, F is a nonempty, convex subset of A. Choose an $\mathbb{R}$-form g on E and a real number α such that $H = \{x : g(x) = \alpha\}$ and $g \leq \alpha$ on A (30.8). Suppose $\langle u, v \rangle \subset A$ and $\langle u, v \rangle \cap F \neq \varnothing$; it is to be shown that $\langle u, v \rangle \subset F$. Choose $z \in \langle u, v \rangle \cap F$; say $z = (1 - \nu)u + \nu v$, $0 < \nu < 1$. Given any $x \in \langle u, v \rangle$, let us show that $x \in F = H \cap A$; since $x \in \langle u, v \rangle \subset A$, it will suffice to show that $x \in H$, i.e., that $g(x) = \alpha$.

Say $x = (1 - \lambda)u + \lambda v$, $0 < \lambda < 1$. If $\lambda = \nu$ then $x = z \in F$, as desired. Suppose $\lambda \neq \nu$, say $\lambda < \nu$. Choose any μ with $\nu < \mu < 1$, and set $y = (1 - \mu)u + \mu v$. Thus $0 < \lambda < \nu < \mu < 1$, hence $z \in \langle x, y \rangle$, say $z = (1 - \rho)x + \rho y$, $0 < \rho < 1$. Since $x, y \in \langle u, v \rangle \subset A$ we have $g(x) \leq \alpha$, $g(y) \leq \alpha$; also, $z \in F \subset H$, thus $g(z) = \alpha$; then

$$\alpha = g(z) = (1 - \rho)g(x) + \rho g(y) \leq (1 - \rho)\alpha + \rho\alpha = \alpha,$$

therefore $g(x) = g(y) = \alpha$, i.e., $x, y \in H$. ∎

The following link between faces and extremal points turns out to be decisive in the existence proof for extremal points to be given below:

(36.7) Lemma. *Let E be a locally convex, separated TVS over $\mathbb{K}$, let A be a nonempty, compact convex set in E, and let F be a minimal closed face of A (i.e., a closed face of A that contains no other closed face of A). Then $F = \{a\}$, where a is an extremal point of A.*

Proof. We can suppose $\mathbb{K} = \mathbb{R}$; the complex case follows from the real case by restriction of scalars.

Note that F is also compact and convex. In view of (36.4), it will suffice to show that F is a singleton; since E has sufficiently many continuous linear forms to distinguish between distinct points (33.13), it will suffice to show that every continuous linear form is constant on F.

Let g be any nonzero continuous linear form on E; it is to be shown that $g(F)$ is a singleton. In any case, $g(F)$ is compact and convex, by the continuity and linearity of g, therefore $g(F) = [\alpha, \beta]$ for suitable $\alpha, \beta \in \mathbb{R}$. Let

$$H = \{x \in E : g(x) = \alpha\};$$

since H is a closed hyperplane (22.5) tangent to F (cf. 30.10), it follows from (36.6) that $H \cap F$ is a closed face of F, and therefore of A (36.5). In view of the minimality of F, it follows that $H \cap F = F$; thus $F \subset H$, i.e., $g = \alpha$ on F. ∎

A fundamental existence theorem for extremal points:

(36.8) Theorem. *Let E be a locally convex, separated TVS over $\mathbb{K}$, and let A be any nonempty, compact convex subset of E. If H is any closed $\mathbb{R}$-hyperplane tangent to A, then H contains at least one extremal point of A.*

Proof. {This is an existence theorem because such hyperplanes H exist (33.13), (30.11).} The proof is a minimality argument of Zorn type. Let $\mathscr{F}$ be the family of all closed faces F of A such that $F \subset H$; e.g., $H \cap A \in \mathscr{F}$ by (36.6). In view of the lemma, the objective is to find an element of $\mathscr{F}$ that is minimal with respect to inclusion; to fit the usual maximality language of Zorn's lemma, we reverse the order relation: define $F_1 \leq F_2$ iff $F_1 \supset F_2$. The 'induction step' rests on the following assertion: if $(F_\iota)_{\iota \in I}$ is a family in $\mathscr{F}$ that is simply ordered (by $\leq$ or by inclusion—it comes to the same thing) then the intersection $F = \bigcap_{\iota \in I} F_\iota$ also belongs to $\mathscr{F}$; indeed, F is nonempty by the compactness of A, thus F is a closed face of A (36.5) such that $F_\iota \leq F$ for all $\iota \in I$. By Zorn's lemma, $\mathscr{F}$ has an element F_0 that is maximal for $\leq$, i.e., minimal for $\subset$; citing the lemma, we have $F_0 = \{a\}$, a an extremal point of A. ∎

(36.9) Theorem. (Kreĭn-Mil'man) *Let E be a locally convex, separated TVS over $\mathbb{K}$, let A be a nonempty, compact convex set in E, and let A_{ep} be the set of all extremal points of A. Then A is the closed convex hull of A_{ep}, that is, $A = (\mathrm{conv}\, A_{\mathrm{ep}})^-$.*

Proof. We can suppose $\mathbb{K} = \mathbb{R}$. Let B be the closed convex hull of A_{ep}; thus, $B = (\mathrm{conv}\, A_{\mathrm{ep}})^-$ (25.18), and the problem is to show that $B = A$. Evidently $B \subset A$. To show that $A \subset B$ we invoke (34.7): given any nonzero continuous linear form g on E, it will suffice to show that $g(A) \subset g(B)$. Since g is continuous and linear, $g(A)$ is also compact and convex, say $g(A) = [\alpha, \beta]$. Then $H = \{x : g(x) = \alpha\}$ is a closed hyperplane tangent to A; by (36.8), H contains a point $a \in A_{\mathrm{ep}}$, therefore $\alpha = g(a) \in g(A_{\mathrm{ep}}) \subset g(B)$. Similarly $\beta \in g(B)$, thus $g(A) = [\alpha, \beta] \subset g(B)$ by the convexity of $g(B)$. ∎

The gist of the next result is that any compact 'system of generators' of a compact convex set must already contain the extremal points:

(36.10) Theorem. *Let E be a locally convex, separated TVS over $\mathbb{K}$, and let S be a compact subset of E whose closed convex hull A is also compact. Then S contains every extremal point of A.*

Proof. We can suppose $\mathbb{K} = \mathbb{R}$ and A nonempty. Let a be an extremal point of A; the problem is to show that $a \in S$. Since S is closed,

$$S = \bigcap_V (S + V),$$

where V runs over any neighborhood base at θ (5.13). In the present instance, the convex, closed, symmetric neighborhoods of θ are basic (33.10); thus, given any such neighborhood V, it will suffice to show that $a \in S + V$.

Since S is compact, we may choose a finite system $s_1, \ldots, s_n$ in S such that

$$S \subset \bigcup_1^n (s_k + V).$$

Let A_k be the closed convex hull of $S \cap (s_k + V)$; evidently $A_k \subset A$, therefore A_k is also compact. Set

$$T = \bigcup_1^n A_k.$$

Since $S = \bigcup_1^n [S \cap (s_k + V)]$, clearly

$$(\operatorname{conv} S)^- = \left(\operatorname{conv}\left(\bigcup_1^n A_k\right)\right)^-,$$

that is, $A = (\operatorname{conv} T)^-$; moreover, $\operatorname{conv} T$ is closed (35.1), thus

$$A = \operatorname{conv} T.$$

In particular, $a \in \operatorname{conv} T$; citing (25.13), there exists a representation

$$(*) \qquad\qquad a = \sum_1^n \alpha_k a_k,$$

where $a_k \in A_k$, $\alpha_k \geq 0$, and $\sum_1^n \alpha_k = 1$. Since a is an extremal point of A, an easy induction on the definition shows that the representation $(*)$ must be degenerate: $\alpha_j = 1$ for some j and $\alpha_k = 0$ for all $k \neq j$, thus $a = a_j \in A_j$. But $s_j + V$ is a closed convex set containing $S \cap (s_j + V)$, therefore

$$s_j + V \supset (\operatorname{conv} [S \cap (s_j + V)])^- = A_j,$$

thus $a \in A_j \subset s_j + V \subset S + V$. ∎

If the space E in (36.10) is complete, then the hypothesis that A is compact is redundant:

(36.11) Corollary. *Let E be a complete, locally convex, separated TVS over $\mathbb{K}$, let S be a compact subset of E, and let A be the closed convex hull of S, i.e., $A = (\operatorname{conv} S)^-$. Then:*

(i) *A is a compact convex set;*
(ii) *A is the closed convex hull of its extremal points;*
(iii) *S contains every extremal point of A.*

Proof. A is compact by (35.4), therefore A is the closed convex hull of A_{ep} by the Kreĭn-Mil'man theorem (36.9) and $A_{\mathrm{ep}} \subset S$ by the theorem. ∎

Exercises

(36.12) There exists a compact convex set in $\mathbb{R}^3$ whose set of extremal points is not closed.

(36.13) The set of extremal points of a compact convex set in $\mathbb{R}^2$ is closed.

(36.14) In a uniformly convex normed space, the set of extremal points of a closed ball $\{x : \|x - x_0\| \leq r\}$ coincides with its surface $\{x : \|x - x_0\| = r\}$.

(A normed space is said to be *uniformly convex* if, for every $\varepsilon > 0$, there exists a $\delta > 0$ such that the relations $\|x\| = \|y\| = 1$ and $\|\frac{1}{2}x + \frac{1}{2}y\| \geq 1 - \delta$ imply $\|x - y\| \leq \varepsilon$.)

(36.15)   In a uniformly convex normed space (36.14), every bare point of a convex set is an extremal point. (Let A be a convex subset of a normed space E. A point $b \in A$ is called a *bare point* of A if there exist $x_0 \in E$ and $r > 0$ such that $\|b - x_0\| = r$ and $\|a - x_0\| \leq r$ for all $a \in A$, i.e., b lies on the surface of a closed ball that contains A.)

37. Seminorms and local convexity.

Seminorms play a role in the theory of TVS analogous to the role of pseudometrics in general topology. More precisely, the uniformizable topological spaces are ordinarily defined in terms of entourages and may be characterized as those topological spaces whose topology is definable by a family of pseudometrics; analogously, having defined locally convex TVS in terms of neighborhoods of zero (33.1), we show in this section that these spaces may be characterized as the TVS whose topology is definable by a family of seminorms. For the definition of seminorm, see (28.7).

(37.1)   **Theorem.**   *If E is a vector space over $\mathbb{K}$ and if p is a seminorm on E, then* (i) *the sets*

$$B = \{x : p(x) \leq 1\}, \qquad U = \{x : p(x) < 1\}$$

are balanced and convex; (ii) *one has*

$$|p(x) - p(y)| \leq p(x - y)$$

for all x, y in E. If, moreover, E is a TVS and p is continuous at θ, then (iii) *p is uniformly continuous on E,* (iv) *B is closed, indeed, $B = \overline{U}$, and* (v) *U is open, indeed, $U = \text{int } B$.*

Proof.   If $0 \leq \alpha \leq 1$ then

$$p(\alpha x + (1 - \alpha)y) \leq p(\alpha x) + p((1 - \alpha)y) = \alpha p(x) + (1 - \alpha)p(y),$$

thus the convexity of B and U is clear; if $|\lambda| \leq 1$ then $p(\lambda x) = |\lambda| p(x) \leq p(x)$ shows that B and U are balanced.

(ii) Cf. the proof of (16.4).

(iii) Follows at once from (ii).

(iv), (v) The fact that U is open and B is closed follows from (iii) and the formulas

$$U = p^{-1}((-1, 1)), \qquad B = p^{-1}([-1, 1]).$$

Since $U \subset B$, one has $\overline{U} \subset \overline{B} = B$. On the other hand, if $x \in B$ then for every α with $0 < \alpha < 1$ one has $p(\alpha x) = \alpha p(x) \leq \alpha < 1$, thus $\alpha x \in U$; since every neighborhood of x contains such vectors αx for α sufficiently near 1 (by the continuity of scalar multiplication), it follows that $x \in \overline{U}$. Thus $B = \overline{U}$. Since U is convex and int $U = U \neq \varnothing$, it follows from (25.20) that int $B = \text{int } \overline{U} = \text{int } U = U$.   ∎

Theorem (37.1) says that if p is a continuous seminorm on a TVS, then the 'closed unit ball' $\{x : p(x) \leq 1\}$ is a balanced convex body. In the next theorem we show, conversely, that every balanced convex body in a TVS is the closed unit ball of a suitable seminorm.

(37.2) Lemma. *If E is a vector space over $\mathbb{K}$ and if p, q are seminorms on E such that*

$$\{x : p(x) \leq 1\} = \{x : q(x) \leq 1\},$$

then $p = q$.

Proof. Let $x \in E$. If $\varepsilon > 0$ and $y = (p(x) + \varepsilon)^{-1}x$, then $p(y) = (p(x) + \varepsilon)^{-1}p(x) < 1$, therefore $q(y) \leq 1$, i.e., $q(x) \leq p(x) + \varepsilon$. Since ε is arbitrary, $q(x) \leq p(x)$. Similarly $p(x) \leq q(x)$. ∎

(37.3) Lemma. *If E is a vector space over $\mathbb{K}$, p is a seminorm on E, and if, for each $\varepsilon > 0$,*

$$B_\varepsilon = \{x : p(x) \leq \varepsilon\},$$

then $B_\varepsilon = \varepsilon B_1$.

Proof. The following conditions imply one another: $x \in B_\varepsilon$, $p(x) \leq \varepsilon$, $p(\varepsilon^{-1}x) \leq 1$, $\varepsilon^{-1}x \in B_1$, $x \in \varepsilon B_1$. ∎

(37.4) Theorem. (Minkowski functional) *Let E be a TVS over $\mathbb{K}$ and let B be a balanced, convex body in E (i.e., a balanced, closed convex set with nonempty interior). Then there exists a unique seminorm p on E such that*

$$B = \{x : p(x) \leq 1\}.$$

Moreover, p is uniformly continuous on E, and $\text{int } B = \{x : p(x) < 1\}$.

Proof. Uniqueness is covered by (37.2). We begin the proof of existence (cf. the proof of Kolmogorov's normability criterion (26.4)) by observing that B is a neighborhood of θ, i.e., that θ is an interior point of B; indeed, if b is an interior point of B, then $-b = (-1)b$ is an interior point of $(-1)B = B$, and, since int B is convex (25.20), $\theta = \frac{1}{2}b + \frac{1}{2}(-b) \in \text{int } B$.
    For each $x \in E$, define

$$B(x) = \{\alpha > 0 : x \in \alpha B\} = \{\alpha > 0 : \alpha^{-1}x \in B\};$$

since B is absorbent (17.11), $B(x) \neq \varnothing$, thus we may define

(*) $p(x) = \inf B(x).$

We now derive the various properties of the correspondence $x \mapsto B(x)$ on which the desired properties of p depend.
    (i) $B(\theta) = (0, \infty) = \{\alpha : \alpha > 0\}$. This is immediate from $\theta \in B$.
    (ii) If $\mu \in \mathbb{K}$, $|\mu| = 1$, then $B(\mu x) = B(x)$. This follows from the fact that B is circled (25.27); indeed, if $\alpha > 0$ then the following relations imply one another: $\alpha \in B(\mu x)$, $\mu x \in \alpha B$, $x \in \alpha \mu^{-1}B = \alpha B$, $\alpha \in B(x)$.

(iii) If $\rho > 0$ then $B(\rho x) = \rho B(x)$. Indeed, if $\alpha > 0$ then the following relations imply one another: $\alpha \in B(\rho x)$, $\rho x \in \alpha B$, $x \in \rho^{-1} \alpha B$, $\rho^{-1} \alpha \in B(x)$, $\alpha \in \rho B(x)$.

(iv) If $\lambda \in \mathbb{K}$, $\lambda \neq 0$, then $B(\lambda x) = |\lambda| B(x)$. Indeed, setting $\rho = |\lambda|$ and $\mu = \rho^{-1}\lambda$, we have $|\mu| = 1$ and so $B(\lambda x) = B(\mu \rho x) = B(\rho x) = \rho B(x)$ by (ii) and (iii).

(v) $B(x) + B(y) \subset B(x + y)$. That is, if $\alpha \in B(x)$ and $\beta \in B(y)$, then $\alpha + \beta \in B(x + y)$; indeed, citing (25.10), we have $x + y \in \alpha B + \beta B = (\alpha + \beta)B$.

(vi) If $\alpha \in B(x)$ and $\beta > \alpha$, then $\beta \in B(x)$. Indeed, $x \in \alpha B \subset \beta B$ because B is balanced (17.2).

We now verify that p is a seminorm. Obviously $p(x) \geq 0$ by the defining formula (*), and $p(\theta) = 0$ by (i). The relation $p(\lambda x) = |\lambda| p(x)$ is obvious when $\lambda = 0$, and follows from (iv) when $\lambda \neq 0$. If $\alpha \in B(x)$ and $\beta \in B(y)$ then, citing (v), we have $\alpha + \beta \in B(x + y)$, therefore $p(x + y) \leq \alpha + \beta$; varying α and β, it results that $p(x + y) \leq p(x) + p(y)$. Thus, p is a seminorm.

We assert that $B = \{x : p(x) \leq 1\}$. Indeed, if $x \in B = 1B$ then $1 \in B(x)$ and so $p(x) \leq 1$. Conversely, suppose $p(x) \leq 1$. If $\beta > 1$, i.e., if

$$\beta > 1 \geq p(x) = \inf B(x),$$

then there exists $\alpha \in B(x)$ with $p(x) \leq \alpha < \beta$; by (vi) we have $\beta \in B(x)$, i.e., $\beta^{-1}x \in B$. Thus $\beta^{-1}x \in B$ for all $\beta > 1$; since $\beta^{-1}x \to x$ as $\beta \to 1$, it follows that $x \in \bar{B} = B$.

It follows from (37.3) that for every $\varepsilon > 0$, $\{x : p(x) \leq \varepsilon\} = \varepsilon B$, which is a neighborhood of θ; consequently p is continuous at θ, and the last assertion of the theorem follows from (37.1). $\blacksquare$

Every seminorm defines a locally convex compatible topology:

(37.5) Theorem. *Let E be a vector space over $\mathbb{K}$ and let p be a seminorm on E.*

(1) *There exists a unique topology τ_p on E, compatible with the vector space structure of E, for which the sets*

$$B_\varepsilon = \{x : p(x) \leq \varepsilon\} \qquad (\varepsilon > 0)$$

are a fundamental system of neighborhoods of θ.

(2) *(E, τ_p) is a locally convex TVS.*

(3) *p is a (uniformly) continuous function on (E, τ_p), and*

$$\operatorname{int} B_\varepsilon = \{x : p(x) < \varepsilon\}.$$

(4) *τ_p is the coarsest topology on E that makes p continuous at θ and is compatible with the additive group structure of E.*

Proof. (1) We verify that the sets B_ε satisfy the criteria of (17.12). Obviously $\theta \in B_\varepsilon$ for all $\varepsilon > 0$. If $0 < \varepsilon \leq \eta$ then $B_\varepsilon \subset B_\eta$; it follows that the family $(B_\varepsilon)_{\varepsilon > 0}$ is the base of a filter of sets containing θ.

Write $B = B_1 = \{x : p(x) \leq 1\}$. Since B is balanced and convex (37.1) and since $B_\varepsilon = \varepsilon B$ (37.3), it follows that the B_ε are balanced and convex.

Moreover, each B_ε is absorbent: if $x \in E$ and if $\alpha > 0$ is chosen so small that $\alpha p(x) \leq \varepsilon$, then $p(\alpha x) \leq \varepsilon$, $\alpha x \in B_\varepsilon$. If $\lambda \in \mathbb{K}$, $\lambda \neq 0$, then, writing $\lambda = \rho\mu$ with $\rho > 0$ and $|\mu| = 1$, we have $\lambda B_\varepsilon = \lambda \varepsilon B = \rho\varepsilon(\mu B) = \rho\varepsilon B = B_{\rho\varepsilon}$, thus the family (B_ε) satisfies condition (ii) of (17.12). Finally, if $\varepsilon > 0$ then, setting $\eta = \frac{1}{2}\varepsilon$, we have $B_\eta + B_\eta = \eta B + \eta B = (\eta + \eta)B = \varepsilon B$ by (25.10). The assertion (1) now follows at once from (17.12).

(2) Since, moreover, the B_ε are convex, the topology τ_p is locally convex (33.1).

(3) Since each B_ε is a neighborhood of θ for τ_p, it is immediate that p is continuous at θ; it follows from (37.1) that p is uniformly continuous on E and that int $B_\varepsilon = $ int $(\varepsilon B) = \varepsilon$ int $B = \varepsilon\{x : p(x) < 1\} = \{x : p(x) < \varepsilon\}$.

(4) Suppose τ is any topology on E that is compatible with the additive group structure of E and such that p is continuous at θ for τ. It follows that the sets $B_\varepsilon = \{x : p(x) \leq \varepsilon\}$ are τ-neighborhoods of θ; moreover, for each $a \in E$, $a + B_\varepsilon$ is a τ-neighborhood of a by additive compatibility. Since the sets $a + B_\varepsilon$ ($\varepsilon > 0$) are a neighborhood base at a for τ_p, it follows that $\tau_p \subset \tau$. ∎

An important application:

(37.6) Corollary. *Let E be a vector space over $\mathbb{K}$, f a linear form on E, and p the seminorm on E defined by $p(x) = |f(x)|$. Then τ_p coincides with the initial topology for the mapping f.*

Proof. Let τ be the initial topology for the mapping $f: E \to \mathbb{K}$; since f is linear, τ is compatible with the vector space structure of E (17.13). Since, in particular, f is continuous for τ, clearly so is p, therefore $\tau_p \subset \tau$ by (37.5). On the other hand, for every $\varepsilon > 0$, $\{x : |f(x)| \leq \varepsilon\} = \{x : p(x) \leq \varepsilon\}$ is a τ_p-neighborhood of θ, thus f is τ_p-continuous at θ; since τ_p is compatible with the vector space structure, it follows that f is τ_p-continuous (5.1), therefore $\tau \subset \tau_p$. ∎

In a 'seminormed TVS' the continuous linear forms are characterized by a boundedness condition:

(37.7) Theorem. *Let E be a vector space over $\mathbb{K}$, p a seminorm on E, and f a linear form on E. The following conditions on f are equivalent:*

(a) *f is continuous for τ_p;*

(b) *there exists a constant $M > 0$ such that $|f(x)| \leq Mp(x)$ for all x in E;*

(c) *the set $\{f(x) : p(x) \leq 1\}$ is bounded, i.e., f is bounded on the closed unit ball for p.*

Proof. (a) implies (b): Since f is τ_p-continuous at θ, the set $\{x : |f(x)| \leq 1\}$ is a τ_p-neighborhood of θ; in view of the definition of τ_p (37.5) there exists an $\varepsilon > 0$ such that

$$\{x : p(x) \leq \varepsilon\} \subset \{x : |f(x)| \leq 1\},$$

that is, $p(x) \leq \varepsilon$ implies $|f(x)| \leq 1$. Set $M = 1/\varepsilon$. If $x \in E$ and $\alpha > 0$ then the vector $y = \varepsilon(p(x) + \alpha)^{-1}x$ satisfies $p(y) = \varepsilon(p(x) + \alpha)^{-1}p(x) < \varepsilon$, therefore

$|f(y)| \leq 1$, i.e., $|f(x)| \leq M(p(x) + \alpha)$; since $\alpha > 0$ is arbitrary, $|f(x)| \leq Mp(x)$.

(b) implies (c): Obvious.

(c) implies (a): Say $M > 0$ is a constant such that $p(x) \leq 1$ implies $|f(x)| \leq M$. Let $\varepsilon > 0$. If $p(x) \leq \varepsilon/M$, i.e., if $p((M/\varepsilon)x) \leq 1$, then $|f((M/\varepsilon)x| \leq M$ i.e., $|f(x)| \leq \varepsilon$. Thus

$$\{x : |f(x)| \leq \varepsilon\} \supset \{x : p(x) \leq \varepsilon/M\},$$

therefore $\{x : |f(x)| \leq \varepsilon\}$ is a τ_p-neighborhood of θ (37.5). This shows that f is continuous at θ and is therefore a continuous mapping (5.1). ∎

In a TVS, a linear form is continuous if and only if it is dominated by some continuous seminorm:

(37.8) Corollary. *If E is a TVS over $\mathbb{K}$ and f is a linear form on E, the following conditions are equivalent:* (a) *f is continuous;* (b) *there exists a continuous seminorm p on E such that $|f(x)| \leq p(x)$ for all x in E.*

Proof. (a) implies (b): Let $p(x) = |f(x)|$.

(b) implies (a): Let τ be the given topology on E. Since f is continuous for τ_p by (37.7) (with $M = 1$) and since $\tau_p \subset \tau$ by (37.5), it follows that f is continuous for τ. ∎

So far, we have considered the topology τ_p generated by a single seminorm p. We now turn to the topology generated by a set of seminorms; this is defined as follows:

(37.9) Definition. If E is a vector space over $\mathbb{K}$ and if $\mathscr{F}$ is a set of seminorms on E, the topology on E *generated* by $\mathscr{F}$ is the topology τ generated by the topologies τ_p $(p \in \mathscr{F})$, i.e., τ is the coarsest topology on E such that $\tau_p \subset \tau$ for all $p \in \mathscr{F}$ (in other words, τ is the supremum of the topologies τ_p). Notation: $\tau = \tau(\mathscr{F})$. In view of (37.5), a basic neighborhood of θ for $\tau(\mathscr{F})$ is given by

$$\{x \in E : p_i(x) \leq \varepsilon_i \text{ for } i = 1, \ldots, n\},$$

where $p_i \in \mathscr{F}$ and $\varepsilon_i > 0$ $(i = 1, \ldots, n)$.

(37.10) Lemma. *The topology generated by a set of seminorms is compatible with the vector space structure and is locally convex.*

Proof. With notation as in (37.9), each of the topologies τ_p $(p \in \mathscr{F})$ is compatible with the vector space structure of E and is locally convex (37.5), therefore so is their supremum (33.8). ∎

Relevant to the supremum indicated in (37.9) is the fact that the correspondence $p \mapsto \tau_p$ is monotone:

(37.11) Lemma. *If E is a vector space over $\mathbb{K}$ and if p, q are seminorms on E such that $p \leq q$, then $\tau_p \subset \tau_q$.*

Proof. For every $\varepsilon > 0$, $\{x : q(x) \leq \varepsilon\} \subset \{x : p(x) \leq \varepsilon\}$; it follows from (37.5) that every τ_p-neighborhood of θ is a τ_q-neighborhood of θ, therefore $\tau_p \subset \tau_q$ by translation. ∎

The set of all seminorms is closed under the formation of finite suprema:

(37.12) Lemma. *If E is a vector space over $\mathbb{K}$ and $p_1, \ldots, p_n$ are seminorms on E, and if*

$$p(x) = \max \{p_1(x), \ldots, p_n(x)\} \qquad (x \in E),$$

then p is also a seminorm on E. Moreover, p is continuous with respect to the topology generated by the set $\{p_1, \ldots, p_n\}$.

Proof. It is clear that p is a seminorm. For each $\varepsilon > 0$,

$$\{x : p(x) \le \varepsilon\} = \{x : p_k(x) \le \varepsilon \quad \text{for } k = 1, \ldots, n\}$$

$$= \bigcap_1^n \{x : p_k(x) \le \varepsilon\},$$

therefore $\{x : p(x) \le \varepsilon\}$ is a neighborhood of θ for the topology τ generated by the set $\{p_1, \ldots, p_n\}$; it follows that p is continuous at θ for the topology τ. Since τ is compatible with the vector space structure (37.10) it follows from (37.5) that $\tau_p \subset \tau$ and that p is τ-continuous on E. ∎

Generalizing (37.7), we have:

(37.13) Lemma. *If E is a vector space over $\mathbb{K}$ and if p, q are seminorms on E, then the following conditions are equivalent:*
(a) *p is continuous for τ_q;*
(b) *$\tau_p \subset \tau_q$;*
(c) *there exists a constant $M > 0$ such that $p(x) \le Mq(x)$ for all x in E.*

Proof. Since τ_q is a compatible topology, (a) and (b) are equivalent by part (4) of (37.5).

(c) implies (a): For each $\varepsilon > 0$, clearly

$$\{x : p(x) \le \varepsilon\} \supset \{x : q(x) \le \varepsilon/M\},$$

therefore $\{x : p(x) \le \varepsilon\}$ is a τ_q-neighborhood of θ; thus p is continuous at θ for τ_q, hence $\tau_p \subset \tau_q$ by (37.5).

(a) implies (c): By hypothesis, $\{x : p(x) \le 1\}$ is a τ_q-neighborhood of θ, therefore, for suitable $\varepsilon > 0$,

$$\{x : p(x) \le 1\} \supset \{x : q(x) \le \varepsilon\}.$$

Setting $M = 1/\varepsilon$, one has $p(x) \le Mq(x)$ for all $x \in E$, as in the proof of (37.7). ∎

(37.14) Definition. Let E be a vector space over $\mathbb{K}$. A set $\mathscr{F}$ of seminorms on E is said to be *saturated* if it satisfies the following condition: if $p_1, \ldots, p_n \in \mathscr{F}$ and p is a seminorm on E such that $p \le p_1 + \cdots + p_n$, then also $p \in \mathscr{F}$. The set of all seminorms on E is trivially saturated, and the intersection of any family of saturated sets is saturated; therefore, given any set $\mathscr{F}$ of seminorms, there exists a smallest saturated set $\mathscr{F}^*$ containing $\mathscr{F}$; we call $\mathscr{F}^*$ the *saturation* of $\mathscr{F}$.

The saturation of a set of seminorms can be described quite explicitly:

(37.15) Lemma. *If E is a vector space over $\mathbb{K}$ and $\mathscr{F}$ is any set of seminorms on E, then a seminorm p belongs to $\mathscr{F}^*$ iff there exist finitely many seminorms $p_1, \ldots, p_n \in \mathscr{F}$ such that $p \leq p_1 + \cdots + p_n$.*

Proof. Let $\mathscr{G}$ be the set of all seminorms p such that $p \leq p_1 + \cdots + p_n$ for suitable $p_1, \ldots, p_n \in \mathscr{F}$. Since $\mathscr{F} \subset \mathscr{F}^*$ and $\mathscr{F}^*$ is saturated, clearly $\mathscr{G} \subset \mathscr{F}^*$. On the other hand, it is elementary that $\mathscr{G}$ is saturated, and $\mathscr{G} \supset \mathscr{F}$, therefore $\mathscr{G} \supset \mathscr{F}^*$. ∎

A set of seminorms and its saturation generate the same topology:

(37.16) Lemma. *If E is a vector space over $\mathbb{K}$ and if $\mathscr{F}$ is any set of seminorms on E, then $\tau(\mathscr{F}^*) = \tau(\mathscr{F})$.*

Proof. Since $\mathscr{F} \subset \mathscr{F}^*$, obviously $\tau(\mathscr{F}) \subset \tau(\mathscr{F}^*)$. Conversely, assuming $p \in \mathscr{F}^*$, it is to be shown that $\tau_p \subset \tau(\mathscr{F})$. Choose $p_1, \ldots, p_M \in \mathscr{F}$ with $p \leq p_1 + \cdots + p_M$, and write $q = p_1 + \cdots + p_M$. In view of (37.11), it will suffice to show that $\tau_q \subset \tau(\mathscr{F})$. Set $r(x) = \max\{p_1(x), \ldots, p_M(x)\}$; by (37.12), the seminorm r is continuous with respect to $\tau(\{p_1, \ldots, p_M\}) \subset \tau(\mathscr{F})$, therefore $\tau_r \subset \tau(\mathscr{F})$ by (37.5). Evidently $q \leq Mr$, therefore citing (37.13) we have $\tau_q \subset \tau_r \subset \tau(\mathscr{F})$. ∎

The saturation of a set of seminorms can also be described in topological terms:

(37.17) Theorem. *Let E be a vector space over $\mathbb{K}$, let $\mathscr{F}$ be any set of seminorms on E, and let $\tau(\mathscr{F})$ be the topology on E generated by $\mathscr{F}$ (37.9). Then:*

(i) *$(E, \tau(\mathscr{F}))$ is a locally convex TVS, i.e., $\tau(\mathscr{F})$ is compatible and locally convex:*

(ii) *$\tau(\mathscr{F}^*) = \tau(\mathscr{F})$;*

(iii) *a seminorm p on E is continuous for $\tau(\mathscr{F})$ iff $p \in \mathscr{F}^*$.*

Proof. (i) and (ii) are (37.10) and (37.16). If $p \in \mathscr{F}^*$ then $\tau_p \subset \tau(\mathscr{F}^*) = \tau(\mathscr{F})$, therefore p is continuous for $\tau(\mathscr{F})$ (37.5).

Conversely, assuming p is continuous for $\tau(\mathscr{F})$, let us show that $p \in \mathscr{F}^*$. Since $\{x : p(x) \leq 1\}$ is a neighborhood of θ for $\tau(\mathscr{F})$, it follows from the definition of $\tau(\mathscr{F})$ that there exist $p_1, \ldots, p_n \in \mathscr{F}$ and $\varepsilon_1 > 0, \ldots, \varepsilon_n > 0$ such that

$$\{x : p(x) \leq 1\} \supset \bigcap_1^n \{x : p_k(x) \leq \varepsilon_k\}.$$

Setting $\varepsilon = \min\{\varepsilon_1, \ldots, \varepsilon_n\}$, we have

$$\{x : p(x) \leq 1\} \supset \bigcap_1^n \{x : p_k(x) \leq \varepsilon\},$$

that is, $p(x) \leq 1$ whenever $p_k(x) \leq \varepsilon$ for all $k = 1, \ldots, n$. Set

$$q(x) = \max\{p_1(x), \ldots, p_n(x)\} \qquad (x \in E);$$

clearly $\{x : p(x) \leq 1\} \supset \{x : q(x) \leq \varepsilon\}$, therefore $p \leq (1/\varepsilon)q$ as in the proof of (37.13). Let m be a positive integer such that $1/\varepsilon \leq m$. Then $p \leq (1/\varepsilon)q \leq mq$; obviously $q \leq p_1 + \cdots + p_n$, thus

$$p \leq m(p_1 + \cdots + p_n) = (p_1 + \cdots + p_n) + \cdots + (p_1 + \cdots + p_n),$$

which shows that p is $\leq$ the sum of finitely many (namely, mn) elements of $\mathscr{F}$, i.e., $p \in \mathscr{F}^*$. ∎

(37.18) Corollary. *If E is a vector space over $\mathbb{K}$ and if $\mathscr{F}$, $\mathscr{G}$ are saturated sets of seminorms on E, then $\tau(\mathscr{F}) = \tau(\mathscr{G})$ iff $\mathscr{F} = \mathscr{G}$.*

Proof. If $\tau(\mathscr{F}) = \tau(\mathscr{G})$, then a seminorm on E is continuous for $\tau(\mathscr{F})$ iff it is continuous for $\tau(\mathscr{G})$; in view of (37.17), this means that $\mathscr{F}^* = \mathscr{G}^*$. But $\mathscr{F} = \mathscr{F}^*$ and $\mathscr{G} = \mathscr{G}^*$ by hypothesis, thus $\mathscr{F} = \mathscr{G}$. ∎

Every locally convex topology may be generated by a suitable set of seminorms:

(37.19) Theorem. *If (E, τ) is a locally convex TVS over $\mathbb{K}$, and $\mathscr{F}$ is the set of all continuous seminorms on E, then $\tau = \tau(\mathscr{F})$. In particular, every locally convex, compatible topology on E may be generated by a suitable set of seminorms.*

Proof. By (37.5), $\tau_p \subset \tau$ for every $p \in \mathscr{F}$, thus $\tau(\mathscr{F}) \subset \tau$.

If, for the topology τ, V is any closed, balanced, convex neighborhood of θ, by (37.4) there exists $p \in \mathscr{F}$ such that $V = \{x : p(x) \leq 1\}$ and the τ-interior of V is $\{x : p(x) < 1\} \in \tau_p \subset \tau(\mathscr{F})$, thus V is a neighborhood of θ for $\tau(\mathscr{F})$; since such V's are basic (33.10), it follows that $\tau \subset \tau(\mathscr{F})$. ∎

Combining (37.17)–(37.19), we have:

(37.20) Theorem. *Let E be a vector space over $\mathbb{K}$ and let $\mathscr{F}$, $\mathscr{G}$, ... denote sets of seminorms on E.*

(i) *The correspondence $\mathscr{F} \mapsto \tau(\mathscr{F})$ maps the class of all such $\mathscr{F}$ onto the class of all locally convex, compatible topologies on E;*

(ii) *for each $\mathscr{F}$ there is one and only one saturated set $\mathscr{G}$ such that $\tau(\mathscr{G}) = \tau(\mathscr{F})$, namely, $\mathscr{G} = \mathscr{F}^*$;*

(iii) *the correspondence $\mathscr{F} \mapsto \tau(\mathscr{F})$ maps the class of all saturated sets of seminorms bijectively onto the class of all locally convex, compatible topologies on E.*

A useful criterion for $\tau(\mathscr{F})$ to be separated:

(37.21) Theorem. *Let E be a vector space over $\mathbb{K}$ and let $\mathscr{F}$ be a set of seminorms on E. The topology $\tau(\mathscr{F})$ is separated iff for each nonzero vector x there exists $p \in \mathscr{F}$ with $p(x) > 0$.*

Proof. Assuming $\mathscr{F}$ satisfies the latter condition, let us show that $\tau(\mathscr{F})$ is separated. Given any nonzero vector x, choose $p \in \mathscr{F}$ so that $p(x) > 0$; then $V = \{y : p(y) < p(x)\}$ is a $\tau(\mathscr{F})$-neighborhood of θ that excludes x, thus $\tau(\mathscr{F})$ is separated (3.4).

Conversely, suppose $\tau(\mathscr{F})$ is separated and x is a nonzero vector. Since $\tau(\mathscr{F})$ is locally convex, there exists a closed, balanced, convex neighborhood V of θ with $x \notin V$. Let p be a continuous seminorm on E such that $V = \{y : p(y) \leq 1\}$ (37.4); since $x \notin V$, $p(x) > 1$. By (37.17), $p \in \mathscr{F}^*$; choose $p_1, \ldots, p_n \in \mathscr{F}$ with $p \leq p_1 + \cdots + p_n$. Since $p(x) > 1$, necessarily $p_k(x) > 0$ for some k. ∎

The gist of (37.20) is that the concept of locally convex TVS can be defined either in terms of neighborhoods of θ (as in Section 33) or in terms of seminorms. Sections 33–36 illustrate that one can operate quite efficiently without mentioning seminorms; but the seminorm formulation is useful too, because in many applications the topology is most naturally introduced by means of an explicitly given family of seminorms.

(37.22) As an application of the results of this section, we give here an alternative proof of the 'analytical form' of the Hahn-Banach theorem (28.8). Suppose E is a vector space over $\mathbb{K}$, p is a seminorm on E, M is a linear subspace of E, and g is a linear form on M such that $|g(y)| \leq p(y)$ for all y in M; we seek a linear form f on E such that $f|M = g$ and $|f(x)| \leq p(x)$ for all x in E.

We can suppose $g \neq 0$. As in the proof of (28.8), it is sufficient to consider the case that $\mathbb{K} = \mathbb{R}$. Equip E with the locally convex topology τ_p and let $A = \{x : p(x) < 1\}$; since p is continuous for τ_p (37.5), A is an open, non-empty convex set. If $y \in M$ and $g(y) = 1$, then $p(y) \geq |g(y)| = 1$, therefore $y \notin A$; thus, setting $N = \{y \in M : g(y) = 1\}$, N is a linear variety (in M, therefore in E) such that $N \cap A = \varnothing$. By the 'geometric form' of the Hahn-Banach theorem (28.1) there exists a closed hyperplane H in E such that $H \supset N$ and $H \cap A = \varnothing$. Then $H = \{x \in E : f(x) = \alpha\}$ for a suitable τ_p-continuous linear form f and nonzero real number α (22.5). Replacing f by $\alpha^{-1}f$, we can suppose $H = \{x \in E : f(x) = 1\}$; then $f = g = 1$ on N, and it follows that $f|M = g$. It remains to show that $|f(x)| \leq p(x)$ on E. Since $H \cap A = \varnothing$ and A is convex, A lies strictly to one side of H (30.4); more precisely, since $\theta \in A$ and $f(\theta) = 0 < 1$, necessarily $A \subset \{x \in E : f(x) < 1\}$. Then $f(x) = 1$ implies that $x \notin A$ and therefore $p(x) \geq 1$. Given any $x \in E$, let us show that $|f(x)| \leq p(x)$. If $f(x) = 0$ there is nothing to prove. Assuming $f(x) \neq 0$, set $x' = (f(x))^{-1}x$; then $f(x') = 1$, therefore $p(x') \geq 1$, i.e., $p(x) \geq |f(x)|$.

Exercises

(37.23) A set $\mathscr{F}$ of seminorms on a vector space E is saturated if and only if it satisfies the following three conditions: (i) if p and q are seminorms such that $q \in \mathscr{F}$ and $p \leq q$, then $p \in \mathscr{F}$; (ii) if $p, q \in \mathscr{F}$ and r is the seminorm defined by $r(x) = \max\{p(x), q(x)\}$, then $r \in \mathscr{F}$; (iii) if $p \in \mathscr{F}$ then $2p \in \mathscr{F}$. It follows that if $\mathscr{F}$ is any set of seminorms on E, then a seminorm p belongs to $\mathscr{F}^*$ iff there exist $p_1, \ldots, p_n \in \mathscr{F}$ and a constant $M > 0$ such that $p(x) \leq M \max\{p_1(x), \ldots, p_n(x)\}$ for all $x \in E$.

(37.24) If E and F are locally convex TVS over $\mathbb{K}$ and if $T: E \rightarrow F$ is a linear mapping, then the following conditions are equivalent: (a) T is continuous; (b) for each continuous seminorm q on F, the seminorm $p(x) = q(Tx)$ $(x \in E)$ is continuous on E.

(37.25) Let E be a vector space over $\mathbb{K}$, let p be a seminorm on E, equip E with the topology τ_p of (37.5), and let $N_p = \{x \in E : p(x) = 0\}$. Then (i) N_p

is a linear subspace of E, closed for the topology τ_p; (ii) the formula $\|x + N_p\| = p(x)$ defines a norm on E/N_p, such that (iii) the norm topology on E/N_p coincides with the quotient topology (of τ_p by N_p, as in (11.8)).

(37.26)   A separated TVS E is locally convex iff it is linearly homeomorphic with a subspace of the product of a family of Banach spaces, i.e., iff there exists a family of Banach spaces $(F_\iota)_{\iota \in I}$ and an injective linear mapping $T: E \to \prod_{\iota \in I} F_\iota$ such that T maps E homeomorphically onto $T(E)$ (equipped with the relative topology).

(37.27)   Let E be a vector space over $\mathbb{K}$. If p, q are seminorms on E, define $p \prec q$ iff $p \leq Mq$ for a suitable constant $M > 0$; define $p \sim q$ iff $p \prec q$ and $q \prec p$. Then $p \sim q$ iff $\tau_p = \tau_q$.

(37.28)   Let E be a TVS over $\mathbb{K}$ and write E' for the vector space of all continuous linear forms f on E (the linear operations on E' are defined pointwise).
   (i) If $A \subset E$ is bounded in the sense of (26.2) then, for each $f \in E'$, the set $f(A)$ is bounded in $\mathbb{K}$; the formula

$$p_A(f) = \sup \{|f(x)| : x \in A\} \qquad (f \in E')$$

defines a seminorm p_A on E'.
   (ii) If $\mathfrak{S}$ is any set of bounded subsets of E, the $\mathfrak{S}$-topology on E' is defined to be the topology generated by the family of seminorms p_A ($A \in \mathfrak{S}$). The $\mathfrak{S}$-topology is always locally convex; it is separated iff the relations $f \in E'$, $f = 0$ on A for all $A \in \mathfrak{S}$, imply $f = 0$.
   (iii) When $\mathfrak{S}$ is the set of all bounded subsets A of E, the $\mathfrak{S}$-topology is called the *strong topology* on E'; it is separated and locally convex.

(37.29)   Let E be a vector space over $\mathbb{K}$, let $\mathscr{F}$ be a family of seminorms on E, and let $\tau = \tau(\mathscr{F})$ be the locally convex topology on E generated by $\mathscr{F}$. If (x_j) is a net in E and if $x \in E$, then the following conditions are equivalent: (a) $x_j \to x$ for the topology τ; (b) $p(x_j - x) \to 0$ for each $p \in \mathscr{F}$; (c) $p(x_j - x) \to 0$ for each continuous seminorm p.

(37.30)   (i) Let U be a nonempty open subset of $\mathbb{C}$, and write $\mathscr{O}(U)$ for the set of all functions $f: U \to \mathbb{C}$ that are holomorphic on U (i.e., differentiable at every point of U); equipped with the pointwise linear operations, $\mathscr{O}(U)$ is a complex vector space. Write $\mathscr{S}$ for the class of all nonempty compact subsets S of U. For each $S \in \mathscr{S}$, the formula

$$p_S(f) = \sup \{|f(\lambda)| : \lambda \in S\}$$

defines a seminorm p_S on $\mathscr{O}(U)$. The locally convex topology τ_U on $\mathscr{O}(U)$ generated by the family of seminorms $(p_S)_{S \in \mathscr{S}}$ is called the *topology of compact convergence*. A seminorm p on $\mathscr{O}(U)$ is continuous for τ_U iff there exist $S \in \mathscr{S}$ and a constant $M > 0$ such that $p \leq Mp_S$. A net $f_j \in \mathscr{O}(U)$ converges to $f \in \mathscr{O}(U)$ for the topology τ_U iff $f_j(\lambda) \to f(\lambda)$ uniformly on each compact subset of U.
   (ii) Let K be a nonempty compact subset of $\mathbb{C}$. Let $\mathscr{U}$ be the class of all open supersets U of K, $K \subset U \subset \mathbb{C}$. With $U \leq V$ defined by $U \supset V$, $\mathscr{U}$ is directed to the right.
   If $U \leq V$ then the restriction mapping $f \mapsto f|V$ defines a linear mapping $\mathscr{O}(U) \to \mathscr{O}(V)$; this mapping is obviously continuous for the topologies of compact convergence defined in (i).

Let $\mathscr{H}(K)$ be the direct union of the family $(\mathcal{O}(U))_{U \in \mathscr{U}}$,

$$\mathscr{H}(K) = \bigcup_{U \in \mathscr{U}} \mathcal{O}(U)$$

(functions with different domains are regarded as different, even if one is an extension of the other); thus $\mathscr{H}(K)$ is the set of all functions that are defined and holomorphic on some open neighborhood of K. Call two elements f, g of $\mathscr{H}(K)$ 'equivalent' if $f = g$ on some $U \in \mathscr{U}$; write $[f]$ for the equivalence class of f, and $\mathcal{O}(K)$ for the set of all equivalence classes.

With linear operations defined in the obvious way (i.e., induced by the point-wise operations on functions), $\mathcal{O}(K)$ is a complex vector space. For each $U \in \mathscr{U}$, the correspondence $f \mapsto [f]$ defines a linear mapping $\varphi_U : \mathcal{O}(U) \to \mathcal{O}(K)$. The ranges of the φ_U form an increasingly directed family of linear subspaces with union $\mathcal{O}(K)$. Equipped with the finest locally convex topology τ that renders continuous all of the mappings φ_U (33.17), $\mathcal{O}(K)$ is a separated, locally convex TVS, called the 'space of germs of functions holomorphic in a neighborhood of K' (specifically, $[f]$ is the 'germ' of f). Note that $\mathcal{O}(K)$ is also an algebra, with the natural product induced by pointwise multiplication.

(iii) With notations as in (ii), suppose $\varphi : \mathcal{O}(K) \to F$ is a linear mapping, where F is a separated, locally convex TVS over $\mathbb{C}$. The following conditions on φ are equivalent: (a) φ is continuous for τ; (b) if U is any open neighborhood of K, then the mapping $\varphi \circ \varphi_U : \mathcal{O}(U) \to F$ is continuous; (c) if U is any open neighborhood of K, and if $f_j, f \in \mathcal{O}(U)$ are functions such that $f_j(\lambda) \to f(\lambda)$ uniformly on each compact subset of U, then $\varphi([f_j]) \to \varphi([f])$ in F.

38. TVS in duality.

The notion of duality in TVS has its spiritual roots in the point/line duality of projective geometry; practically, it has to do with linear subspaces of a TVS and the continuous linear forms that annihilate them. (For example, a primitive result in this circle of ideas is that a linear subspace of a TVS is a closed hyperplane if and only if it is the null space of some nonzero continuous linear form (22.4).) The subject of duality is very large; we limit ourselves to the rudiments.

(38.1) Definition. *The following notations are fixed for the rest of the section.* E and F denote vector spaces over $\mathbb{K}$. We assume given a non-degenerate bilinear form

$$B : E \times F \to \mathbb{K},$$

that is, the relations

   (i) $B(x_1 + x_2, y) = B(x_1, y) + B(x_2, y)$,
   (ii) $B(\lambda x, y) = \lambda B(x, y)$,
   (iii) $B(x, y_1 + y_2) = B(x, y_1) + B(x, y_2)$,
   (iv) $B(x, \lambda y) = \lambda B(x, y)$,
hold identically, and
   (v) if $B(x, y) = 0$ for all $y \in F$, then $x = \theta$,
   (vi) if $B(x, y) = 0$ for all $x \in E$, then $y = \theta$.
The spaces E and F are said to be *in duality*; the duality is said to be *defined* by the form B.

(38.2) Example. If E is any vector space over $\mathbb{K}$ and F is the vector space of all linear forms on E (with the pointwise linear operations), then the formula

$$B(x, f) = f(x) \qquad (x \in E, f \in F)$$

defines a duality, called the *canonical duality*, between E and F.

(38.3) Example. If E is any locally convex, separated TVS over $\mathbb{K}$, and F is the vector space of all continuous linear forms on E (with the pointwise linear operations), then the formula

$$B(x, f) = f(x) \qquad (x \in E, f \in F)$$

defines a duality, called the *canonical duality*, between E and F. The heart of the matter is nondegeneracy, which rests on the Hahn-Banach theorem (33.13).

(38.4) Definition. For $x \in E$ and $y \in F$ we write

$$B_x(y) = B(x, y), \qquad B^y(x) = B(x, y).$$

Evidently B_x and B^y are linear forms on F and E, respectively.

(38.5) Theorem. *The correspondence $x \mapsto B_x$ is an injective linear mapping of E into the vector space $\mathscr{V}(F, \mathbb{K})$ of all linear forms on F. Similarly, the correspondence $y \mapsto B^y$ is an injective linear mapping of F into the vector space $\mathscr{V}(E, \mathbb{K})$ of all linear forms on E.*

Proof. The proof is immediate from the definitions; in particular, injectivity results from the nondegeneracy of B. ∎

(38.6) Definition. The initial topology on E for the family of linear forms $(B^y)_{y \in F}$ is denoted $\sigma(E, F)$; it is called the topology on E *induced* by F and the given duality B. Similarly, $\sigma(F, E)$ denotes the initial topology on F for the family of linear forms $(B_x)_{x \in E}$; it is called the topology on F induced by E and the given duality.

(38.7) Theorem. *The topology $\sigma(E, F)$ is compatible with the vector space structure of E; moreover, it is separated and locally convex, and is the topology generated by the family of seminorms $(p^y)_{y \in F}$, where $p^y(x) = |B(x, y)|$. A linear form f on E is continuous for $\sigma(E, F)$ iff there exists a vector $y \in F$ such that $f = B^y$. The dual statements for $\sigma(F, E)$ also hold.*

Proof. The compatibility, local convexity and separatedness of $\sigma(E, F)$ are immediate from (33.9). For each $y \in F$, write τ_y for the initial topology on E for the mapping $B^y \colon E \to \mathbb{K}$. By definition, $\sigma(E, F)$ is the supremum of the family of topologies $(\tau_y)_{y \in F}$; since $\tau_y = \tau_{p^y}$ (37.6), it follows that $\sigma(E, F)$ coincides with the topology generated by the family of seminorms $(p^y)_{y \in F}$ (37.9).

It is trivial from the definition (38.6) that every B^y is continuous for $\sigma(E, F)$. Conversely, suppose f is any linear form on E that is continuous for

$\sigma(E, F)$. Then $p(x) = |f(x)|$ defines a $\sigma(E, F)$-continuous seminorm on E; in view of (37.17), p is in the saturation of the set of seminorms $(p^y)_{y \in F}$, i.e., there exist vectors $y_1, \ldots, y_n$ in F such that

$$p \le p_{y_1} + \cdots + p_{y_n},$$

that is,

(*) $|f(x)| \le |B(x, y_1)| + \cdots + |B(x, y_n)|$ $(x \in E)$.

It is immediate from (*) that f vanishes on the intersection of the null spaces of $B^{y_1}, \ldots, B^{y_n}$; by (21.17), f may be written as a linear combination

$$f = \lambda_1 B^{y_1} + \cdots + \lambda_n B^{y_n},$$

thus $f = B^y$ with $y = \lambda_1 y_1 + \cdots + \lambda_n y_n$.

The final assertion of the theorem follows from the foregoing ones, on considering the duality $(y, x) \mapsto B(x, y)$ between F and E. ∎

The following corollary is implicit in the preceding proof:

(38.8) Corollary. *A seminorm p on E is continuous for $\sigma(E, F)$ iff there exist vectors $y_1, \ldots, y_n$ in F such that*

$$p(x) \le |B(x, y_1)| + \cdots + |B(x, y_n)|$$

for all x in E.

(38.9) Corollary. *If F_0 is a linear subspace of F such that the relations $x \in E$, $B(x, y) = 0$ $(\forall y \in F_0)$ imply $x = \theta$, and if $F_0 \ne F$, then $\sigma(E, F_0)$ is strictly coarser than $\sigma(E, F)$, that is, $\sigma(E, F_0) \subset \sigma(E, F)$ properly.*

Proof. Observe that the restriction of B to $E \times F_0$ defines a duality between E and F_0; $\sigma(E, F_0)$ refers to this duality.

The relation $\sigma(E, F_0) \subset \sigma(E, F)$ is obvious. Assuming to the contrary that $\sigma(E, F_0) = \sigma(E, F)$, let us show that $F_0 = F$. Let $y \in F$. Since B^y is continuous for $\sigma(E, F) = \sigma(E, F_0)$, it follows from (38.7) that $B^y = B^{y_0}$ for some $y_0 \in F_0$; then $y = y_0$ by nondegeneracy, thus $y \in F_0$. ∎

(38.10) Definition. If (E, τ) is a TVS over $\mathbb{K}$, we denote by E' the vector space of all τ-continuous linear forms on E, with the pointwise linear operations. In general E' may be $\{\theta\}$, in which case a duality between E and E' is impossible (unless also $E = \{\theta\}$). However, when τ is locally convex and separated, there is a canonical duality between E and E' (38.3); in this case, $\sigma(E, E')$ is called the *weak topology* on E, and expressions such as 'weakly closed,' 'weakly continuous,' etc., refer to the weak topology.

(38.11) Theorem. *Let (E, τ) be a locally convex, separated TVS over $\mathbb{K}$. Then* (i) $\sigma(E, E') \subset \tau$; (ii) *a linear form f on E is continuous (i.e., τ-continuous) iff it is weakly continuous (i.e., $\sigma(E, E')$-continuous).*

Proof. (i) The duality in question is defined by $B(x, f) = f(x)$, thus $B^f = f$ (see (38.4) for the notations). Since $\sigma(E, E')$ is the coarsest topology on E that makes every $B^f = f\,(f \in E')$ continuous, and since every $f \in E'$ is τ-continuous, one has $\sigma(E, E') \subset \tau$.

(ii) Let f be a linear form on E. If f is τ-continuous (i.e., $f \in E'$), then $f = B^f$ shows that f is $\sigma(E, E')$-continuous. Conversely, if f is $\sigma(E, E')$-continuous, then it is immediate from $\sigma(E, E') \subset \tau$ that f is τ-continuous. ∎

(38.12) Corollary. *If (E, τ) is a locally convex, separated TVS over $\mathbb{K}$, and A is a convex subset of E, then A is closed iff it is weakly closed.*

Proof. Let us suppose $\mathbb{K} = \mathbb{C}$ (for $\mathbb{K} = \mathbb{R}$ the proof is even simpler). Both τ and $\sigma(E, E')$ are locally convex, separated topologies on E; they determine the same continuous linear forms (38.11), therefore the same continuous $\mathbb{R}$-forms (22.6), therefore the same closed convex sets (34.3). ∎

(38.13) Definition. Resume the notations in (38.1). If $S \subset E$, the *annihilator* of S (in F), denoted $S^\perp$, is the set of all $y \in F$ such that $B(s, y) = 0$ for all s in S. Dually, if $T \subset F$, the *annihilator* of T (in E), denoted $^\perp T$, is the set of all $x \in E$ such that $B(x, t) = 0$ for all t in T.

(38.14) Theorem. *Write S, T for generic subsets of E, F, respectively, where E and F are in duality (38.1).*

  (i) *$S^\perp$ is a $\sigma(F, E)$-closed linear subspace of F;*
  (i′) *$^\perp T$ is a $\sigma(E, F)$-closed linear subspace of E;*
  (ii) *$S \subset {}^\perp(S^\perp)$;*
  (ii′) *$T \subset (^\perp T)^\perp$;*
  (iii) *if $S_1 \subset S_2$ then $S_1^\perp \supset S_2^\perp$;*
  (iii′) *if $T_1 \subset T_2$ then $^\perp T_1 \supset {}^\perp T_2$;*
  (iv) *$S^\perp = ({}^\perp(S^\perp))^\perp$;*
  (iv′) *$^\perp T = {}^\perp((^\perp T)^\perp)$.*
  (v) *S is a $\sigma(E, F)$-closed linear subspace of E iff $S = {}^\perp(S^\perp)$;*
  (v′) *T is a $\sigma(F, E)$-closed linear subspace of F iff $T = (^\perp T)^\perp$.*
  (vi) *If S is any subset of E, then $^\perp(S^\perp)$ is the $\sigma(E, F)$-closed linear subspace generated by S.*
  (vi′) *If T is any subset of F, then $(^\perp T)^\perp$ is the $\sigma(F, E)$-closed linear subspace generated by T.*

Proof. (i) $S^\perp$ is the intersection of the null spaces of the $\sigma(F, E)$-continuous linear forms B_s ($s \in S$).

(ii) If $s \in S$, then for every $y \in S^\perp$ one has $B(s, y) = 0$, thus $s \in {}^\perp(S^\perp)$; this shows that $S \subset {}^\perp(S^\perp)$.

(iii) is obvious, and (i′), (ii′), (iii′) are proved dually.

(iv) $S^\perp \subset ({}^\perp(S^\perp))^\perp$ by (ii′) (with $T = S^\perp$). On the other hand, $S \subset {}^\perp(S^\perp)$, therefore $S^\perp \supset ({}^\perp(S^\perp))^\perp$ by (iii). Similarly for (iv′).

(v) If $S = {}^\perp(S^\perp)$, then S is a $\sigma(E, F)$-closed linear subspace of E by (i′). Conversely, suppose S is a $\sigma(E, F)$-closed linear subspace of E, and assume to the contrary that $S \subset {}^\perp(S^\perp)$ properly. Say $a \in {}^\perp(S^\perp)$, $a \notin S$. By (33.12) there exists a $\sigma(E, F)$-continuous linear form f on E such that $f(a) \neq 0$ and $f = 0$ on S. Choose $y \in F$ with $f = B^y$ (38.7). Since $B^y = f = 0$ on S, we have $y \in S^{\perp}$; since, moreover, $a \in {}^{\perp}(S^{\perp})$, it follows that $B(a, y) = 0$, i.e., $f(a) = 0$, a contradiction. The proof of (v') is similar.

(vi) Let M be the $\sigma(E, F)$-closed linear subspace generated by S. By (i') and (ii), ${}^{\perp}(S^{\perp})$ is a $\sigma(E, F)$-closed linear subspace containing S, therefore $M \subset {}^{\perp}(S^{\perp})$. On the other hand, since $S \subset M$ and M is $\sigma(E, F)$-closed, we have $S^{\perp} \supset M^{\perp}$, ${}^{\perp}(S^{\perp}) \subset {}^{\perp}(M^{\perp}) = M$ by (iii), (iii'), and (v). The proof of (vi') is similar. ∎

(38.15) **Corollary.** *Let (E, τ) be a locally convex, separated TVS over $\mathbb{K}$, and consider the canonical duality between E and E' (38.10). If M is a linear subspace of E, the following conditions are equivalent:* (a) *M is closed (for τ);* (b) *M is weakly closed;* (c) *$M = {}^{\perp}(M^{\perp})$.*

Proof. The equivalence of (a) and (b) follows from (38.12); (b) and (c) are equivalent by (v) of (38.14). ∎

The theme of duality will be taken up again in the context of normed spaces (Section 44).

Exercises

(38.16)   Notation as in (38.15). If M is any linear subspace of E, then the τ-closure and the weak closure of M coincide, and are equal to ${}^{\perp}(M^{\perp})$.

(38.17)   For $k = 1, 2$, let E_k, F_k be a pair of vector spaces in duality via a nondegenerate bilinear form B_k. If $T: E_1 \to E_2$ is a linear mapping that is continuous for the topologies $\sigma(E_k, F_k)$ ($k = 1, 2$), then there exists a unique mapping $T': F_2 \to F_1$ such that

$$B_1(x_1, T'y_2) = B_2(Tx_1, y_2)$$

for all $x_1 \in E_1$ and $y_2 \in F_2$. The mapping T', called the *transpose* (or *adjoint*) of T, is linear and continuous for the topologies $\sigma(F_k, E_k)$ ($k = 1, 2$).

Chapter 4

Normed Spaces, Banach Spaces, Hilbert Spaces

*39. The spaces L^p. The present section is written for the reader who objects to studying Banach spaces in a historical vacuum. The L^p spaces have motivated much of the abstract theory and are among its major applications, thus a familiarity with them is of considerable inspirational value to the student of functional analysis. In the following discussion, we rely on the formulation of integration theory in the book of P. R. Halmos [65], but the choice of integration-theoretic style is not critical; the chief action takes place in the realm of inequalities, and the reader will have no difficulty in adapting the arguments to other formulations of integration theory.

(39.1) Definition. $(X, \mathscr{S}, \mu)$ is a measure space, p is a fixed real number, $p \geq 1$. We write $\mathscr{L}^p = \mathscr{L}^p(X, \mathscr{S}, \mu)$ for the set of all measurable functions $f \colon X \to \mathbb{C}$ such that $|f|^p$ is integrable with respect to μ (i.e., $|f|^p \in \mathscr{L}^1$), and for $f \in \mathscr{L}^p$ we define

$$\|f\|_p = \left(\int |f|^p \, d\mu \right)^{1/p}.$$

If $p > 1$ we define $q = p/(p - 1)$; thus $q > 1$, and q is the unique real number such that $p^{-1} + q^{-1} = 1$, i.e., $p + q = pq$.

Our first objective is to show that $\mathscr{L}^p$ is a vector space with respect to the pointwise linear operations, and that $f \mapsto \|f\|_p$ is a seminorm on $\mathscr{L}^p$.

(39.2) Lemma. *If $p > 1$ then*

$$ab \leq \frac{a^p}{p} + \frac{b^q}{q}$$

for all $a \geq 0$, $b \geq 0$.

Proof. If $a = 0$ or $b = 0$ the conclusion is trivial; assume $a > 0$, $b > 0$. Define

$$\varphi(t) = p^{-1}t^p + q^{-1}t^{-q} \qquad (t > 0).$$

We show that φ has a minimum at $t = 1$. One has $\varphi'(t) = t^{p-1} - t^{-q-1}$, therefore $\varphi'(t) = 0$ iff $t = 1$; since $\varphi(t) \to \infty$ as $t \downarrow 0$ or $t \uparrow \infty$, it follows that φ has a minimum at $t = 1$. Thus, for all $t > 0$, one has

$$1 = \varphi(1) \leq \varphi(t) = p^{-1}t^p + q^{-1}t^{-q};$$

in particular, substituting $t = a^{1/q}b^{-1/p}$ yields

$$1 \leq p^{-1}a^{p/q}b^{-1} + q^{-1}a^{-1}b^{q/p},$$

$$ab \leq p^{-1}a^{(p/q)+1} + q^{-1}b^{(q/p)+1},$$

160

and the desired inequality is obtained on observing that $(p/q) + 1 = p$ and $(q/p) + 1 = q$. ∎

We remark that for $p = 2$, the lemma reduces to the elementary inequality $2ab \leq a^2 + b^2$, i.e., $(a - b)^2 \geq 0$.

(39.3) Theorem. (Hölder's inequality) *Assuming $p > 1$, if $f \in \mathscr{L}^p$ and $g \in \mathscr{L}^q$ then $fg \in \mathscr{L}^1$ and $\|fg\|_1 \leq \|f\|_p \|g\|_q$.*

Proof. Let $\alpha = \|f\|_p$, $\beta = \|g\|_q$. If $\alpha = 0$ or $\beta = 0$ (i.e., if $f = 0$ a.e. or $g = 0$ a.e.) then $fg = 0$ a.e. and the conclusion of the theorem is trivial. Assume $\alpha > 0$ and $\beta > 0$. If $x \in X$, application of the lemma with $a = |f(x)|/\alpha$ and $b = |g(x)|/\beta$ yields

(*) $$\frac{|f(x)g(x)|}{\alpha\beta} \leq \frac{|f(x)|^p}{p\alpha^p} + \frac{|g(x)|^q}{q\beta^q};$$

since the function on the right side of (*) is integrable, it follows that $|fg|$ is integrable, and integration of (*) yields

$$\frac{\|fg\|_1}{\alpha\beta} \leq \frac{\alpha^p}{p\alpha^p} + \frac{\beta^q}{q\beta^q} = \frac{1}{p} + \frac{1}{q} = 1,$$

thus $\|fg\|_1 \leq \alpha\beta = \|f\|_p \|g\|_q$. ∎

(39.4) Theorem. (Minkowski's inequality) *If $f, g \in \mathscr{L}^p$ then $f + g \in \mathscr{L}^p$ and $\|f + g\|_p \leq \|f\|_p + \|g\|_p$.*

Proof. If $p = 1$ the conclusion follows from the triangle inequality $|f + g| \leq |f| + |g|$.

Assume $p > 1$. Then

$$|f + g|^p \leq (|f| + |g|)^p \leq (2 \max\{|f|, |g|\})^p$$
$$= 2^p \max\{|f|^p, |g|^p\} \leq 2^p(|f|^p + |g|^p);$$

since the rightmost term is integrable, so is $|f + g|^p$, thus $f + g \in \mathscr{L}^p$.

Note that $|f + g|^{p-1} \in \mathscr{L}^q$; indeed, $(|f + g|^{p-1})^q = |f + g|^{pq-q} = |f + g|^p \in \mathscr{L}^1$ by the preceding paragraph. Since $|f| \in \mathscr{L}^p$, we have $|f + g|^{p-1}|f| \in \mathscr{L}^1$ by (39.3). Similarly $|f + g|^{p-1}|g| \in \mathscr{L}^1$. Citing Hölder's inequality,

$$(\|f + g\|_p)^p = \int |f + g|^p \, d\mu = \int |f + g|^{p-1}|f + g| \, d\mu$$

$$\leq \int |f + g|^{p-1}(|f| + |g|) \, d\mu$$

$$= \| |f + g|^{p-1}|f| \|_1 + \| |f + g|^{p-1}|g| \|_1$$

$$\leq \| |f + g|^{p-1} \|_q \|f\|_p + \| |f + g|^{p-1} \|_q \|g\|_p$$

$$= \left(\int |f + g|^{pq-q} \, d\mu \right)^{1/q} (\|f\|_p + \|g\|_p)$$

$$= \left(\int |f + g|^p \, d\mu \right)^{1/q} (\|f\|_p + \|g\|_p)$$

$$= (\|f + g\|_p)^{p/q} (\|f\|_p + \|g\|_p).$$

If $\|f + g\|_p > 0$, the preceding inequality yields (in view of $p - p/q = 1$)

$$\|f + g\|_p \le \|f\|_p + \|g\|_p,$$

and the last inequality is trivial if $\|f + g\|_p = 0$. ∎

If $f \in \mathscr{L}^p$ and $\lambda \in \mathbb{C}$, evidently $\lambda f \in \mathscr{L}^p$ and $\|\lambda f\|_p = |\lambda| \|f\|_p$. In view of (39.4), we have shown:

(39.5) Corollary. $\mathscr{L}^p$ *is a complex vector space with respect to the point-wise linear operations, and the function* $f \mapsto \|f\|_p$ *is a seminorm on* $\mathscr{L}^p$.

A key property of this seminorm is the following 'completeness theorem' that generalizes the classical Riesz-Fischer theorem:

(39.6) Theorem. *If* f_n *is a sequence in* $\mathscr{L}^p$ *such that* $\|f_m - f_n\|_p \to 0$ *as* $m, n \to \infty$, *then there exists an* f *in* $\mathscr{L}^p$ *such that* $\|f_n - f\|_p \to 0$.

Proof. To simplify the notation of the proof, let us write $\|f\|$ in place of $\|f\|_p$. Referring to such an f as a 'limit' for the sequence f_n, it clearly suffices by (39.4) to show that some subsequence of f_n possesses a limit. Therefore one can suppose, by passing to a suitable subsequence, that

$$\|f_{n+1} - f_n\| \le 2^{-n} \qquad (n = 1, 2, 3, \ldots).$$

Writing $\alpha = \sum_1^\infty \|f_{n+1} - f_n\|$, we have $0 \le \alpha < \infty$. With the convention $f_0 = 0$, define

$$g_n = |f_1| + |f_2 - f_1| + \cdots + |f_n - f_{n-1}| = \sum_1^n |f_k - f_{k-1}|.$$

By (39.4) we have $g_n \in \mathscr{L}^p$ and

$$\|g_n\| \le \sum_1^n \|f_k - f_{k-1}\| = \|f_1\| + \sum_2^n \|f_k - f_{k-1}\|$$

$$= \|f_1\| + \sum_1^{n-1} \|f_{k+1} - f_k\| \le \|f_1\| + \alpha.$$

Thus, for all n one has $(g_n)^p \in \mathscr{L}^1$ and

(*) $$\int (g_n)^p \, d\mu = (\|g_n\|)^p \le (\|f_1\| + \alpha)^p < \infty.$$

Since g_n is an increasing sequence and $p \ge 1$, the sequence $(g_n)^p$ is also increasing. It follows from (*) and the monotone convergence theorem that there exists $h \in \mathscr{L}^1$, $h \ge 0$, such that $(g_n)^p \uparrow h$ a.e. Defining $g = h^{1/p}$, we have $g \in \mathscr{L}^p$ and $g_n \uparrow g$ a.e.

Let E be a measurable set such that $\mu(E) = 0$ and $g_n(x) \uparrow g(x)$ for all $x \in X - E$. Then, for all $x \in X - E$, one has

$$g(x) = \lim_n g_n(x) = \lim_n \sum_1^n |f_k(x) - f_{k-1}(x)|,$$

thus the series

$$\sum_1^\infty |f_k(x) - f_{k-1}(x)|$$

is convergent, with sum $g(x)$; therefore (for $x \in X - E$) the series

$$\sum_1^\infty [f_k(x) - f_{k-1}(x)]$$

is also convergent, i.e., the limit

$$\lim_n \sum_1^n [f_k(x) - f_{k-1}(x)] = \lim_n f_n(x)$$

exists, and is in absolute value $\leq g(x)$. Define

$$f(x) = \begin{cases} \lim_n f_n(x) & \text{for } x \in X - E, \\ 0 & \text{for } x \in E. \end{cases}$$

Then f is measurable, $f_n \to f$ a.e., and $|f| \leq g$; therefore $|f|^p \leq g^p = h$, thus $|f|^p \in \mathscr{L}^1$, i.e., $f \in \mathscr{L}^p$.

It remains to show that $\|f_n - f\| \to 0$. Given any $\varepsilon > 0$, choose an index N such that $\|f_m - f_n\| \leq \varepsilon$ for all $m, n \geq N$. Fix $m \geq N$. Then

$$\int |f_m - f_n|^p \, d\mu \leq \varepsilon^p \qquad \text{for all } n \geq N;$$

since $|f_m - f_n|^p \in \mathscr{L}^1$ for all $n \geq N$, it follows from Fatou's Lemma that $\liminf_n |f_m - f_n|^p \in \mathscr{L}^1$, and

$$(**) \qquad \int (\liminf_n |f_m - f_n|^p) \, d\mu \leq \liminf_n \int |f_m - f_n|^p \, d\mu \leq \varepsilon^p.$$

But $|f_m - f_n|^p \to |f_m - f|^p$ a.e. as $n \to \infty$, thus

$$\liminf_n |f_m - f_n|^p = |f_m - f|^p \text{ a.e.},$$

and so $\int |f_m - f|^p \, d\mu \leq \varepsilon^p$ by $(**)$, i.e., $\|f_m - f\| \leq \varepsilon$. Since $m \geq N$ is arbitrary, the proof is complete. $\blacksquare$

We now apply a standard technique for passing from a seminorm to a norm:

(39.7) Lemma. *Let E be a vector space over $\mathbb{K}$, let r be a seminorm on E, and let $N = \{x \in E : r(x) = 0\}$. Then (i) N is a linear subspace of E, (ii) the formula $\|x + N\| = r(x)$ defines a norm on E/N, and (iii) the norm topology on E/N coincides with the quotient, by N, of the topology τ_r generated by r.*

Proof. (i) is obvious.

(ii) The formula is well-defined: if $x + N = y + N$, then $x - y \in N$, therefore $r(x) \leq r(x - y) + r(y) = r(y)$ and similarly $r(y) \leq r(x)$, thus $r(x) = r(y)$. It follows easily that $u \mapsto \|u\|$ $(u \in E/N)$ is a norm.

(iii) Let $\varepsilon > 0$. The set

$$B_\varepsilon = \{x \in E : r(x) < \varepsilon\}$$

is a basic neighborhood of θ in E for τ_r (37.5), whereas

$$U_\varepsilon = \{u \in E/N : \|u\| < \varepsilon\}$$

is a basic neighborhood of θ in E/N for the norm topology. Evidently $\pi(B_\varepsilon) = U_\varepsilon$, where $\pi: E \to E/N$ is the canonical mapping; it follows that U_ε is a basic neighborhood of θ for the quotient topology (11.8). Thus the sets U_ε are basic neighborhoods of θ for both the quotient topology and the norm topology, therefore (by additive compatibility) the two topologies are identical. ∎

We apply (39.7) to the seminorm $f \mapsto \|f\|_p$ on $\mathscr{L}^p$. Write $\mathscr{N} = \{f \in \mathscr{L}^p : \|f\|_p = 0\}$; thus $\mathscr{N}$ is the set of all measurable functions $f: X \to \mathbb{C}$ such that $f = 0$ a.e.

(39.8) Definition. $L^p = L^p(X, \mathscr{S}, \mu)$ denotes the normed space $\mathscr{L}^p/\mathscr{N}$ given by (39.7) (with $E = \mathscr{L}^p$ and $r(f) = \|f\|_p$). In view of (39.6), one sees easily:

(39.9) Theorem. L^p *is a Banach space.*

We conclude the section by working out two applications; the first is obtained by specializing p, the second by specializing the measure space.

(39.10) Example. *The Hilbert space L^2.* The case $p = 2$ is special: one then has also $q = 2$. For simplicity, let us write $\|f\|$ in place of $\|f\|_2$.

Suppose $f, g \in \mathscr{L}^2$. An elementary calculation yields

$$(1) \qquad |f + g|^2 + |f - g|^2 = 2|f|^2 + 2|g|^2.$$

{Note that $|f|^2 = ff^*$, where f^* is defined by $f^*(x) = f(x)^* = $ the complex conjugate of $f(x)$.} Integrating (1), we have

$$(2) \qquad \|f + g\|^2 + \|f - g\|^2 = 2\|f\|^2 + 2\|g\|^2;$$

on passing to quotients modulo $\mathscr{N}$ (39.8), the norm in the Banach space L^2 satisfies

$$(3) \qquad \|u + v\|^2 + \|u - v\|^2 = 2\|u\|^2 + 2\|v\|^2.$$

A Banach space whose norm satisfies (3) is called a *Hilbert space*. Abstract Hilbert spaces will be discussed formally in Section 41; we derive here some of the properties of the 'concrete' Hilbert spaces L^2.

Consider again $f, g \in \mathscr{L}^2$. In view of (39.3), each term of the identity

$$|f + g|^2 = |f|^2 + fg^* + gf^* + |g|^2$$

is integrable; integrating,

$$(4) \qquad \|f + g\|^2 = \|f\|^2 + \int fg^* \, d\mu + \int gf^* \, d\mu + \|g\|^2.$$

The middle terms on the right of (4) are analogous to the inner products in complex Euclidean spaces; accordingly, one defines

$$(f|g) = \int fg^* \, d\mu,$$

called the *inner product* of f with g (the order makes a difference). Equation (4) may then be rewritten

(4') $$\|f + g\|^2 = \|f\|^2 + (f|g) + (g|f) + \|g\|^2.$$

The following properties of inner products are readily verified (here $f, g, \ldots$ are in $\mathscr{L}^2$):

(5) $\qquad (f|f) = \|f\|^2 \geq 0, \quad \text{and} \quad (f|f) = 0 \quad \text{iff} \quad f \in \mathscr{N},$

(6) $\qquad (g|f) = (f|g)^*,$

(7) $\qquad (f_1 + f_2|g) = (f_1|g) + (f_2|g),$

(8) $\qquad (f|g_1 + g_2) = (f|g_1) + (f|g_2),$

(9) $\qquad (\lambda f|g) = \lambda(f|g),$

(10) $\qquad (f|\lambda g) = \lambda^*(f|g).$

Properties (7)–(10) are known as *sesquilinearity* (linear in f, semilinear in g) of the inner product. Clearly $(f|g)$ is unaltered if f and g are replaced by functions a.e. equal to them; thus one can introduce inner products in L^2 by defining

$$(u|v) = (f|g),$$

where $u = f + \mathscr{N}$, $v = g + \mathscr{N}$; then $(u|u) = \|u\|^2$, and the analogues of (6)–(10) hold for inner products in L^2. Also, the analogue of (4') holds in L^2:

(4'') $$\|u + v\|^2 = \|u\|^2 + (u|v) + (v|u) + \|v\|^2;$$

replacing v in (4'') successively by $-v$, iv, and $-iv$, one sees easily (cf. 41.14) that

(11) $$(u|v) = \tfrac{1}{4}\{\|u + v\|^2 - \|u - v\|^2 + i\|u + iv\|^2 - i\|u - iv\|^2\},$$

a result known as the *polarization identity*. Finally, it follows from (39.3) that

(12) $$|(u|v)| \leq \|u\|\,\|v\|,$$

a result known as the *Cauchy-Schwarz inequality*; an alternative simple proof results from (11) and (3).

(39.11) Example. *The spaces l^p.* Here X is any nonempty set, $\mathscr{S} = \mathscr{P}(X)$ is the power set of X (i.e., the σ-algebra of all subsets of X), and μ is the counting measure that assigns to each singleton $\{x\}$ the measure 1: $\mu(A) = \infty$ if A is infinite, $\mu(A) = \text{card } A$ if A is finite.

Every function $f: X \to \mathbb{C}$ is measurable; f is a simple function iff $f(X)$ is finite iff f is an integrable simple function, and in this case

$$\int f \, d\mu = \sum_{x \in X} f(x)$$

(the sum is essentially finite, since $f(x) = 0$ for all but finitely many x). Since the empty set is the only subset of X with measure 0, there is no distinction between $\mathscr{L}^p$ and L^p; a commonly used notation is $l^p(X)$, or, if the set X is explicitly understood, simply l^p. It is easy to verify that a function $f: X \to \mathbb{C}$ is in l^p iff

$$\sum_{x \in X} |f(x)|^p < \infty$$

in the sense that the finite subsums have a finite upper bound.

Of special interest are the choices $X = \mathbb{P} = \{1, 2, 3, \ldots\}$ and $X = \mathbb{Z} = \{0, \pm 1, \pm 2, \pm 3, \ldots\}$; these yield the spaces of absolutely p-summable sequences resp. bilateral sequences. The space $l^2(\mathbb{Z})$, which arises in the theory of Fourier series, is the prototypal Hilbert space.

40. Continuous linear mappings of normed spaces.

A normed space is a TVS with additional geometric structure. The concept of continuous linear mapping in normed spaces is, on the surface, purely topological-algebraic, but it turns out to have geometric overtones:

(40.1) Theorem. *If E and F are normed spaces over $\mathbb{K}$ and if $T: E \to F$ is a linear mapping, then the following conditions are equivalent:*
(a) *T is continuous;*
(b) *there exists a constant $K \geq 0$ such that $\|Tx\| \leq K\|x\|$ for all $x \in E$;*
(c) *$\{\|Tx\| : \|x\| = 1\}$ is bounded;*
(d) *$\{\|Tx\| : \|x\| < 1\}$ is bounded;*
(e) *$\{\|Tx\| : \|x\| \leq 1\}$ is bounded.*
In this case, if M_1, M_2, M_3 are the suprema of the sets described in (c), (d), (e), then $M_1 = M_2 = M_3$. Writing M for their common value, one has $\|Tx\| \leq M\|x\|$ for all $x \in E$, and M is minimal in this property.

Proof. If $E = \{\theta\}$ the situation is trivial: the only possibility is $T = 0$. Assume $E \neq \{\theta\}$, so that the set described in (c) is nonempty and therefore the definition of M_1 is unambiguous.

(a) implies (b): Assume to the contrary that no such K exists. Then, for each positive integer n, there exists a vector x_n in E such that $\|Tx_n\| > n\|x_n\|$. In particular, $x_n \neq \theta$, so we may set $y_n = n^{-1}\|x_n\|^{-1}x_n$. Then $\|y_n\| = n^{-1} \to 0$ and $\|Ty_n\| > 1$; thus y_n converges to θ but Ty_n does not, contrary to the continuity of T at θ.

(b) implies (c): If $\|x\| = 1$ then $\|Tx\| \leq K\|x\| = K$, thus $M_1 \leq K < \infty$.

(c) implies (d): By hypothesis, $M_1 < \infty$. It will suffice to show that $M_2 \leq M_1$. Thus, given $\|x\| < 1$, it is to be shown that $\|Tx\| \leq M_1$. This is trivial if $x = \theta$. If $x \neq \theta$, set $y = \|x\|^{-1}x$; then $\|y\| = 1$, therefore $\|Ty\| \leq M_1$, and so $\|Tx\| \leq \|x\|M_1 \leq M_1$.

(d) implies (e): By hypothesis, $M_2 < \infty$; let us show that $M_3 \leq M_2$. Thus, given $\|x\| \leq 1$, it is to be shown that $\|Tx\| \leq M_2$. If $\|x\| < 1$, this follows from the definition of M_2. Suppose $\|x\| = 1$. For any ε, $0 < \varepsilon < 1$, set $y_\varepsilon = \varepsilon x$; then $\|y_\varepsilon\| < 1$, therefore $\|Ty_\varepsilon\| \leq M_2$, i.e., $\|Tx\| \leq \varepsilon^{-1}M_2$, and $\|Tx\| \leq M_2$ results on letting $\varepsilon \uparrow 1$. Incidentally, $M_2 \leq M_3$ trivially, thus $M_2 = M_3$.

(e) implies (a): By hypothesis, $M_3 < \infty$. Note that $\|Tx\| \leq M_3\|x\|$ for all $x \in E$: this is trivial if $x = \theta$; if $x \neq \theta$, then $y = \|x\|^{-1}x$ has norm 1, therefore $\|Ty\| \leq M_3$, i.e., $\|Tx\| \leq M_3\|x\|$. If $x_n \in E$, $x_n \to \theta$, then $\|Tx_n\| \leq M_3\|x_n\|$ shows that $Tx_n \to \theta = T\theta$, thus T is continuous at θ; therefore T is continuous on E (5.1).

Thus, the conditions (a)–(e) are equivalent. In the course of the proof, it was shown that $M_2 \leq M_1$ and $M_3 = M_2$; since $M_1 \leq M_3$ trivially, it follows that $M_2 = M_1 = M_3$. Writing M for the common value, it was shown in the proof of '(e) implies (a)' that $\|Tx\| \leq M\|x\|$. On the other hand, if $K \geq 0$ satisfies $\|Tx\| \leq K\|x\|$ for all $x \in E$, then $M \leq K$ as shown in the proof of '(b) implies (c).' ∎

(40.2) Definition. Let E and F be normed spaces over $\mathbb{K}$ and let $T: E \to F$ be a continuous linear mapping. The *norm* of T, denoted $\|T\|$, is defined by the formula

$$\|T\| = \sup\{\|Tx\| : \|x\| \leq 1\}.$$

In view of (40.1), $0 \leq \|T\| < \infty$, $\|Tx\| \leq \|T\|\|x\|$ for all $x \in E$, and $\|T\|$ is minimal in these properties. {For linear mappings in normed spaces, the term 'bounded' is often used in place of 'continuous,' for the obvious reason; thus, one speaks of 'bounded linear mappings' or 'bounded linear operators,' 'bounded linear forms,' etc.}

The appropriateness of Definition (40.2) lies in the fact that, for fixed E and F, the set of all continuous linear mappings $T: E \to F$ may be regarded as a normed space in a natural way, as is shown in the next theorem. For clarity, we generalize the setting slightly:

(40.3) Definition. If E and F are TVS over $\mathbb{K}$, we write $\mathscr{L}(E, F)$ for the set of all continuous linear mappings $T: E \to F$. For $S, T \in \mathscr{L}(E, F)$, we write $S = T$ in case $Sx = Tx$ for all $x \in E$, i.e., $S = T$ as mappings.

(40.4) Lemma. *If E and F are TVS over $\mathbb{K}$, then $\mathscr{L}(E, F)$ is a vector space over $\mathbb{K}$ with respect to the pointwise linear operations, defined by $(S + T)x = Sx + Tx$ and $(\lambda T)x + \lambda(Tx)$ ($S, T \in \mathscr{L}(E, F)$, $x \in E$, $\lambda \in \mathbb{K}$).*

Proof. The set $\mathscr{V}(E, F)$ of *all* linear mappings $S, T, \ldots$ of E into F is a vector space with respect to the pointwise linear operations, by trivial algebra. If $S, T \in \mathscr{L}(E, F)$ and $\lambda \in \mathbb{K}$, then $S + T$ and λT are easily seen to be continuous, by the continuity of the linear operations in F; thus $\mathscr{L}(E, F)$ is a linear subspace of $\mathscr{V}(E, F)$. ∎

(40.5) Theorem. *If E and F are normed spaces over $\mathbb{K}$, then $\mathscr{L}(E, F)$ is also a normed space over $\mathbb{K}$ with respect to the pointwise linear operations and the norm $\|T\|$ defined as in (40.2).*

Proof. In view of the lemma, it is sufficient to verify that $T \mapsto \|T\|$ is a norm on $\mathscr{L}(E, F)$. Thus, assuming $S, T \in \mathscr{L}(E, F)$ and $\lambda \in \mathbb{K}$, it is to be shown that (i) if $T \neq 0$ then $\|T\| > 0$, (ii) $\|\lambda T\| = |\lambda| \|T\|$, and (iii) $\|S + T\|$ $\leq \|S\| + \|T\|$.

(i) If $T \neq 0$, say $Tx \neq \theta$, then $0 < \|Tx\| \leq \|T\| \|x\|$, therefore $\|T\| > 0$.

(ii) If $\lambda = 0$ then $\|\lambda T\| = \|0\| = 0 = |\lambda| \|T\|$. If $\lambda \neq 0$ then, since $\|(\lambda T)x\| = \|\lambda (Tx)\| = |\lambda| \|Tx\|$, one has

$$\|\lambda T\| = \sup \{\|(\lambda T)x\| : \|x\| \leq 1\}$$
$$= |\lambda| \sup \{\|Tx\| : \|x\| \leq 1\} = |\lambda| \|T\|.$$

(iii) If $\|x\| \leq 1$ then $\|(S + T)x\| = \|Sx + Tx\| \leq \|Sx\| + \|Tx\| \leq$ $\|S\| + \|T\|$, therefore

$$\|S + T\| = \sup \{\|(S + T)x\| : \|x\| \leq 1\} \leq \|S\| + \|T\|. \quad \blacksquare$$

(40.6) Definition. Notation as in (40.5). If $T_n, T \in \mathscr{L}(E, F)$, evidently $\|T_n - T\| \to 0$ if and only if $\|T_n x - Tx\| \to 0$ uniformly in $\|x\| \leq 1$, i.e., $T_n \to T$ uniformly on the closed unit ball of E. For this reason, the topology on $\mathscr{L}(E, F)$ derived from the norm $\|T\|$ is called the *uniform operator topology*. Thus, $\|T_n - T\| \to 0$ is also verbalized '$T_n \to T$ uniformly' or '$T_n \to T$ in the uniform topology.' {Two other natural topologies on $\mathscr{L}(E, F)$, the 'strong operator topology' and the 'weak operator topology,' are discussed in the exercises.} If, in addition, F is complete, then so is $\mathscr{L}(E, F)$:

(40.7) Theorem. *If E is a normed space and F is a Banach space, then the normed space $\mathscr{L}(E, F)$ described in (40.5) is in fact a Banach space.*

Proof. Suppose T_n is a sequence in $\mathscr{L}(E, F)$ such that $\|T_m - T_n\| \to 0$ as $m, n \to \infty$. For each $x \in E$, $\|T_m x - T_n x\| \leq \|T_m - T_n\| \|x\|$ shows that $T_n x$ is Cauchy in F; since F is complete, we may define $Tx = \lim T_n x$. It is to be shown that T is linear and continuous, and that $\|T_n - T\| \to 0$. The linearity of T follows from the linearity of the T_n and the continuity of the linear operations in F:

$$T(x + y) = \lim T_n(x + y) = \lim (T_n x + T_n y)$$
$$= \lim T_n x + \lim T_n y = Tx + Ty,$$

and similarly $T(\lambda x) = \lambda(Tx)$.

Given any $\varepsilon > 0$, choose an index N such that $\|T_m - T_n\| \leq \varepsilon$ for all $m, n \geq N$. Fix $n \geq N$. For each $x \in E$, $\|x\| \leq 1$, one has

(1) $\|T_m x - T_n x\| = \|(T_m - T_n)x\| \leq \varepsilon$ for all $m \geq N$;

as $m \to \infty$, one has $T_m x - T_n x \to Tx - T_n x$, therefore by the continuity of. the norm in F (16.4),

$$\|T_m x - T_n x\| \to \|Tx - T_n x\|,$$

and so $\|Tx - T_n x\| \leq \varepsilon$ by (1). Thus

(2) $\|(T - T_n)x\| \leq \varepsilon$ for all $x \in E$, $\|x\| \leq 1$.

It follows from (2) that the linear mapping $T - T_n$ is continuous, with

(2') $\|T - T_n\| \leq \varepsilon$;

therefore $T = (T - T_n) + T_n$ is also continuous, i.e., $T \in \mathscr{L}(E, F)$; since, moreover, (2') holds for every $n \geq N$, the proof is complete. ∎

An important special case is $F = \mathbb{K}$:

(40.8) Definition. If E is a normed space over $\mathbb{K}$, the *dual space* of E is the vector space $E' = \mathscr{L}(E, \mathbb{K})$ equipped with the norm

$$\|f\| = \sup\{|f(x)| : x \in E, \|x\| \leq 1\};$$

thus E' is also a normed space. Indeed, since $\mathbb{K}$ is complete, (40.7) yields:

(40.9) Corollary. *If E is any normed space, then E' is a Banach space.*

Every locally convex, separated TVS has sufficiently many continuous linear forms (33.13); in the special case of a normed space, this result can be given an extra geometric turn:

(40.10) Theorem. *Let E be a normed space and let $a \in E$, $a \neq \theta$. There exists $f \in E'$ such that $\|f\| = 1$ and $f(a) = \|a\|$.*

Proof. By (28.9), with $p(x) = \|x\|$, there exists a linear form f on E such that $f(a) = \|a\|$ and $|f(x)| \leq \|x\|$ for all $x \in E$. In particular, f is continuous and $\|f\| \leq 1$; on the other hand, $\|a\| = f(a) = |f(a)| \leq \|f\| \|a\|$, thus $\|f\| \geq 1$. ∎

(40.11) Definition. Let E be a normed space and let E' be the dual space of E (40.8). Since E' is itself a Banach space (40.9), it is susceptible to the same construct, i.e., one can form $(E')'$; this is also a Banach space, called the *bidual* of E, and denoted E''. This can go on forever: one defines $E''' = (E'')'$, which is obviously the same as $(E')''$; $E'''' = (E''')'$, etc.

It is important to inspect E'' in greater detail:

(40.12) Definition. Let E be a normed space over $\mathbb{K}$. Each $x \in E$ defines a mapping $f \mapsto f(x)$ ($f \in E'$); let us write x'' for this mapping, that is, $x'': E' \to \mathbb{K}$, $x''(f) = f(x)$. For each $x \in E$, x'' is a linear form on E' (this is immediate from the definition of the linear operations on E'); moreover, $|x''(f)| = |f(x)| \leq \|f\| \|x\|$ shows that x'' is continuous, with $\|x''\| \leq \|x\|$. In fact, $\|x''\| = \|x\|$; this is trivial if $x = \theta$, and if $x \neq \theta$ then (40.10) yields an $f \in E'$ with $\|f\| = 1$ and $f(x) = \|x\|$, therefore $\|x\| = |f(x)| = |x''(f)| \leq \|x''\| \|f\| = \|x''\|$. Thus $x'' \in E''$ and the mapping $x \mapsto x''$ preserves norms. Better yet:

(40.13) Theorem. *If E is a normed space, then the mapping $x \mapsto x''$ is an isometric linear mapping of E into its bidual E''.*

Proof. Let us verify linearity. If $x, y \in E$ and $\lambda \in \mathbb{K}$, then for all $f \in E'$ one has

$$(x + y)''(f) = f(x + y) = f(x) + f(y) = x''(f) + y''(f) = (x'' + y'')(f),$$

$$(\lambda x)''(f) = f(\lambda x) = \lambda f(x) \doteq \lambda x''(f) = (\lambda x'')(f),$$

thus $(x + y)'' = x'' + y''$ and $(\lambda x)'' = \lambda x''$. Then $\|x'' - y''\| = \|(x - y)''\| = \|x - y\|$, thus $x \mapsto x''$ is isometric, i.e., it preserves distances for the respective norm metrics. ∎

(40.14) Definition. With notation as in (40.13), $x \mapsto x''$ is called the *canonical mapping* (or 'canonical embedding') of the normed space E into its bidual E''.

The canonical embedding provides a shortcut to the completion of a normed space (cf. 24.14):

(40.15) Theorem. *Every normed space may be regarded as a dense linear subspace of a Banach space.*

Proof. Let E be a normed space and let $x \mapsto x''$ be the canonical mapping of E into E''. The image of E under this mapping is a linear subspace E_0 of E''. Let F be the closure of E_0 in E''; since E'' is a Banach space (40.9), so is F (16.11). In view of (40.13), one may identify E and E_0 as normed spaces; after the identification, E appears as a dense linear subspace of F. {We remark that, since F is a complete TVS (16.10), F may be identified with the TVS completion $\hat{E}$ of E described in (24.14).} ∎

(40.16) If E, F, G are vector spaces, and $T: E \to F$, $S: F \to G$ are linear mappings, then the composite mapping $ST: E \to G$, defined by $(ST)x = S(Tx)$, is also linear. If, moreover, E, F, G are TVS and S, T are continuous, then ST is also continuous; in the case of normed spaces, we can say a little more:

(40.17) Theorem. *If $T \in \mathscr{L}(E, F)$ and $S \in \mathscr{L}(F, G)$, where E, F, G are normed spaces over $\mathbb{K}$, then $ST \in \mathscr{L}(E, G)$ and $\|ST\| \leq \|S\| \|T\|$.*

Proof. For all $x \in E$, $\|(ST)x\| = \|S(Tx)\| \leq \|S\| \|Tx\| \leq \|S\| \|T\| \|x\|$. ∎

The next definition could obviously be generalized, but the normed space case is sufficient for our purposes:

(40.18) Definition. Let E and F be normed spaces over $\mathbb{K}$ and let $T \in \mathscr{L}(E, F)$. If g is any continuous linear form on F, then the composite $g \circ T$ is clearly a continuous linear form on E; thus, we may define a mapping $T': F' \to E'$ by $T'g = g \circ T$. Explicitly,

$$(T'g)(x) = g(Tx) \qquad (x \in E, g \in F').$$

It is clear that T' is linear. Moreover, T' is continuous: $\|T'g\| = \|g \circ T\| \leq \|g\| \|T\|$ by (40.17), thus $T' \in \mathscr{L}(F', E')$ and $\|T'\| \leq \|T\|$. (In fact $\|T'\| = \|T\|$, as we show in the next theorem.)

Guided by the finite-dimensional example, we call T' the *transpose* (or 'adjoint') of T. Since $T' \in \mathscr{L}(F', E')$ we can, in turn, form $(T')' \in \mathscr{L}(E'', F'')$; we denote it by T''. We show, in the following theorem, that T'' extends T in the same sense that E'' extends E.

(40.19) Theorem. *Let E and F be normed spaces over $\mathbb{K}$, and let $S, T \in \mathscr{L}(E, F)$ and $\lambda \in \mathbb{K}$. Then:*
  (i) $(S + T)' = S' + T'$.
  (ii) $(\lambda T)' = \lambda T'$.
  (iii) $T''x'' = (Tx)''$ *for all $x \in E$.*
  (iv) $\|T'\| = \|T\|$.
  (v) *If T is isometric and surjective, then so is T'.*

Proof. (i), (ii) For any $g \in F'$ we have, citing the linearity of g, $(S + T)'g = g \circ (S + T) = g \circ S + g \circ T = S'g + T'g = (S' + T')g$ and $(\lambda T)'g = g \circ (\lambda T) = \lambda(g \circ T) = \lambda(T'g) = (\lambda T')g$.

(iii) Let $x \in E$. Then $x'' \in E''$ and $Tx \in F$, therefore $T''x'' \in F''$ and $(Tx)'' \in F''$, thus $T''x''$ and $(Tx)''$ are eligible to be compared; for any $g \in F'$, $(T''x'')(g) = (x'' \circ T')(g) = x''(T'g) = (T'g)(x) = (g \circ T)(x) = g(Tx) = (Tx)''(g)$.

(iv) As noted in (40.18), $\|T'\| \le \|T\|$; applying this to T' in place of T, we have $\|T''\| \le \|T'\|$. On the other hand, citing (iii) and (40.13), we have

$$\|Tx\| = \|(Tx)''\| = \|T''x''\| \le \|T''\| \|x''\| = \|T''\| \|x\|$$

for all $x \in E$, thus $\|T\| \le \|T''\|$. Combining these inequalities, we have $\|T\| \le \|T''\| \le \|T'\| \le \|T\|$.

(v) Suppose T is isometric and surjective. Then $\{Tx : x \in E, \|x\| \le 1\} = \{y \in F : \|y\| \le 1\}$; therefore, for any $g \in F'$, we have

$$\|T'g\| = \sup\{|(T'g)(x)| : x \in E, \|x\| \le 1\}$$

$$= \sup\{|g(Tx)| : x \in E, \|x\| \le 1\}$$

$$= \sup\{|g(y)| : y \in F, \|y\| \le 1\} = \|g\|,$$

thus T' is isometric. Finally, if $f \in E'$ then, defining $g = f \circ T^{-1}$, clearly $g \in F'$ and $T'g = (f \circ T^{-1}) \circ T = f$, thus T' is surjective. ∎

Transposition reverses the order of operator products:

(40.20) Theorem. *If $T \in \mathscr{L}(E, F)$ and $S \in \mathscr{L}(F, G)$, then $(ST)' = T'S'$.*

Proof. Since $ST \in \mathscr{L}(E, G)$, $(ST)'$ is defined and belongs to $\mathscr{L}(G', E')$. On the other hand, $S' \in \mathscr{L}(G', F')$ and $T' \in \mathscr{L}(F', E')$, thus $T'S'$ is defined and also belongs to $\mathscr{L}(G', E')$. For any $h \in G'$, $(ST)'h = h \circ (ST) = (h \circ S) \circ T = T'(h \circ S) = T'(S'h) = (T'S')h$. ∎

Composition is simplified when $E = F = G$; we also simplify the notation:

(40.21) Definition. If E is a TVS, $\mathscr{L}(E, E)$ is denoted $\mathscr{L}(E)$.

When E is a normed space, $\mathscr{L}(E)$ provides an example of a 'normed algebra,' defined as follows:

(40.22) Definition. An *associative algebra* over $\mathbb{K}$ is a vector space A over $\mathbb{K}$, with an associative multiplication ab that is distributive with respect to addition (thus A is an associative ring) and is compatible with scalar multiplication in the sense that $\lambda(ab) = (\lambda a)b = a(\lambda b)$ for all $a, b \in A$ and $\lambda \in \mathbb{K}$. A *normed algebra* over $\mathbb{K}$ is an associative algebra A over $\mathbb{K}$ equipped with a function $a \mapsto \|a\|$ that is (i) a norm on the vector space structure of A, and is (ii) submultiplicative, i.e., $\|ab\| \leq \|a\| \|b\|$ for all $a, b \in A$. If, moreover, A is complete with respect to the norm metric (i.e., is a Banach space) then it is called a *Banach algebra*. The terms 'real normed algebra,' 'complex Banach algebra,' etc., are self-explanatory. The subject of Banach algebras will be taken up in greater depth in Chapter 6; for the present, it is sufficient to record the following:

(40.23) Theorem. *If E is a normed space, then $\mathscr{L}(E)$ is a normed algebra, with unity element the identity operator I, $\|I\| = 1$. If, moreover, E is a Banach space, then $\mathscr{L}(E)$ is a Banach algebra.*

Proof. The linear operations of $\mathscr{L}(E)$ are described in (40.4), the norm in (40.2), and the multiplication in (40.16). The identity operator is defined by $Ix = x$ $(x \in E)$. In view of (40.5), (40.17), and (40.7), it suffices to note the (easily verified) identities $S(T + U) = ST + SU$, $(S + T)U = SU + TU$, $(ST)U = S(TU)$, and $\lambda(ST) = (\lambda S)T = S(\lambda T)$. ∎

Exercises

(40.24) (i) Let E and F be TVS over $\mathbb{K}$, and write $\mathscr{L}(E, F)$ as in (40.3). The *strong operator topology* on $\mathscr{L}(E, F)$ is defined as follows. Each vector $x \in E$ defines a linear mapping $T \mapsto Tx$ of $\mathscr{L}(E, F)$ into F; thus, writing $\varphi_x(T) = Tx$, we have a family $(\varphi_x)_{x \in E}$ of linear mappings of the vector space $\mathscr{L}(E, F)$ into the TVS F. The initial topology on $\mathscr{L}(E, F)$ for the family $(\varphi_x)_{x \in E}$ is called the strong operator topology (or the 'topology of simple convergence on E,' or the 'topology of pointwise convergence on E'). Thus, the strong operator topology is compatible with the vector space structure of $\mathscr{L}(E, F)$ (19.3); a basic neighborhood of the operator 0 is given by

$$\{T : Tx_i \in V \, (i = 1, \ldots, n)\},$$

where $\{x_1, \ldots, x_n\}$ is any finite subset of E and V is a basic neighborhood of θ in F (19.6).

(ii) If F is a separated TVS, then $\mathscr{L}(E, F)$ is separated for the strong operator topology. If $T \in \mathscr{L}(E, F)$ and T_j is a net in $\mathscr{L}(E, F)$, then $T_j \to T$ for the strong operator topology (briefly, '$T_j \to T$ strongly') iff $T_j x \to Tx$ for each $x \in E$.

(iii) If F is a locally convex TVS, then the strong operator topology on $\mathscr{L}(E, F)$ is locally convex; in this case, a basic neighborhood of 0 for the strong operator topology is given by

$$\{T : p(Tx_i) \leq \varepsilon \, (i = 1, \ldots, n)\},$$

where $\{x_1, \ldots, x_n\}$ is any finite subset of E, p is any continuous seminorm on F, and $\varepsilon > 0$ (there is no loss of generality in taking $\varepsilon = 1$).

(iv) If F is a normed space over $\mathbb{K}$ then the strong operator topology on $\mathscr{L}(E, F)$ is locally convex and separated; a basic neighborhood of 0 for the strong operator topology is given by

$$\{T : \|Tx_i\| \leq \varepsilon \ (i = 1, \ldots, n)\},$$

where $\{x_1, \ldots, x_n\}$ is any finite subset of E and $\varepsilon > 0$ (there is no loss of generality in taking $\varepsilon = 1$).

(v) If E and F are both normed spaces over $\mathbb{K}$, then the uniform operator topology τ_n on $\mathscr{L}(E, F)$ is finer than the strong operator topology τ_s.

(vi) Notation as in (v). Assuming $F \neq \{\theta\}$, one has $\tau_n = \tau_s$ if and only if E is finite-dimensional.

(40.25) (i) Let E and F be TVS over $\mathbb{K}$, and write $\mathscr{L}(E, F)$ as in (40.3). The *weak operator topology* on $\mathscr{L}(E, F)$ is defined as follows. For each vector $x \in E$ and each continuous linear form g on F, the correspondence $T \mapsto g(Tx)$ defines a linear form on $\mathscr{L}(E, F)$, namely, $g \circ \varphi_x$ in the notation of (40.24). The initial topology for the family of linear forms $g \circ \varphi_x$, as x and g vary over all possibilities, is called the weak operator topology on $\mathscr{L}(E, F)$. Denote it by τ_w; it is locally convex.

(ii) τ_s is finer than τ_w.

(iii) If F is separated and locally convex, then τ_w is also separated.

(iv) If F is finite-dimensional and separated, then $\tau_s = \tau_w$. More generally, see (v).

(v) Write F' for the vector space $\mathscr{L}(F, \mathbb{K})$ of all linear forms g on F that are continuous for the given topology on F. The initial topology on F for the family of all such linear forms g is called the *weak topology* of F. If the weak topology of F coincides with the given topology on F, then $\tau_s = \tau_w$.

(vi) Assume that the weak topology on F described in (v) is separated (i.e., F has sufficiently many continuous linear forms). If $\tau_s = \tau_w$ then the given topology on F coincides with its weak topology.

(40.26) If E is a normed space over $\mathbb{K}$ and if $A \subset E$, then the following conditions are equivalent: (a) A is bounded in norm, i.e., there exists a constant $M > 0$ such that $\|x\| \leq M$ for all $x \in A$; (b) A is bounded in the sense of (26.2); (c) $f(A)$ is bounded for each $f \in E'$. The strong topology on E' (in the sense of (37.28)) coincides with the norm topology (in the sense of (40.8)).

(40.27) Let E be a vector space over $\mathbb{K}$, let $x \mapsto \|x\|_1$, $x \mapsto \|x\|_2$ be a pair of norms on E, and let τ_1, τ_2 be the topologies they generate.

(i) $\tau_1 \subset \tau_2$ iff there exists a constant $M > 0$ such that $\|x\|_1 \leq M\|x\|_2$ for all $x \in E$.

(ii) $\tau_1 = \tau_2$ iff there exist constants $M > 0$, $N > 0$ such that $N\|x\|_2 \leq \|x\|_1 \leq M\|x\|_2$ for all $x \in E$; the norms are then said to be *equivalent*.

(iii) If $x \mapsto \|x\|_1$ and $x \mapsto \|x\|_2$ are equivalent norms on E, then $(E, \| \ \|_1)$ is a Banach space iff $(E, \| \ \|_2)$ is a Banach space.

(40.28) Let E and F be normed spaces over $\mathbb{K}$ and let $T: E \to F$ be a continuous linear mapping. We say that T is *bounded below* if there exists a constant $M > 0$ such that $\|Tx\| \geq M\|x\|$ for all $x \in E$; equivalently,

$$\inf \{\|Tx\| : \|x\| = 1\} > 0.$$

(i) If T is bounded below, then T is injective.

(ii) If T is bounded below and E is a Banach space, then $T(E)$ is a complete (hence closed) linear subspace of F.

41. Hilbert spaces. One can define the concept of Hilbert space in terms of norms or in terms of inner products. We choose the former alternative; this is unconventional but not unnatural, and it has several substantial advantages, e.g., speed, and the transparency of the completion process. Another advantage is that one is obliged, as a concession to convention, to expose the equivalence of the two formulations, a perspective that deepens understanding. (For motivation for the present section, see (39.10).)

(41.1) Definition. A *Hilbert space* is a Banach space (real or complex) whose norm satisfies the identity

(P) $$\|x + y\|^2 + \|x - y\|^2 = 2\|x\|^2 + 2\|y\|^2$$

for all vectors x and y. (Guided by elementary Euclidean geometry, one says that the norm satisfies the *parallelogram law*.)

(41.2) Definition. Suppose E is a normed space and $\hat{E}$ is its completion (see (40.15)). By the continuity of the norm (16.4), it is clear that the norm of E satisfies (P) if and only if the norm of $\hat{E}$ does. Thus, the norm of E satisfies (P) if and only if its completion is a Hilbert space; accordingly, a normed space whose norm satisfies (P) is called a *pre-Hilbert space*.

Functionals satisfying an identity of the form (P) are readily constructed from 'positive bi-additive forms':

(41.3) Item. *Let E be an additively written group and let $f: E \times E \to \mathbb{K}$ be a function satisfying the following conditions:*
  (i) $f(x, x) \geq 0$,
  (ii) $f(x + y, z) = f(x, z) + f(y, z)$,
  (iii) $f(x, y + z) = f(x, y) + f(x, z)$,
for all x, y, z in E. Define $p(x) = f(x, x)^{1/2}$. Then

$$p(x + y)^2 + p(x - y)^2 = 2p(x)^2 + 2p(y)^2$$

for all x, y in E.

Proof. For fixed y, $x \mapsto f(x, y)$ is a homomorphism of E into the additive group of $\mathbb{K}$, therefore $f(-x, y) = -f(x, y)$. Similarly $f(x, -y) = -f(x, y)$. The desired identity results on adding the identities

$$f(x + y, x + y) = f(x, x) + f(y, y) + f(x, y) + f(y, x)$$
$$f(x - y, x - y) = f(x, x) + f(y, y) - f(x, y) - f(y, x)$$

(here $x - y$ corresponds to xy^{-1} in the multiplicative notation). ∎

Our aim is to show that, conversely, the norm on a pre-Hilbert space arises from a suitable bi-additive functional. This result is due, in both the real and complex cases, to P. Jordan and J. von Neumann (1935). The appropriate definitions are as follows.

(41.4) Definition. In a real pre-Hilbert space, the *inner product* of vectors x and y, denoted (x, y), is defined by the formula

(1) $$(x, y) = \tfrac{1}{4}\{\|x + y\|^2 - \|x - y\|^2\}.$$

In a complex pre-Hilbert space, the *inner product* of vectors x and y (in that order), denoted $(x|y)$, is defined by the formula

(2) $$(x|y) = \tfrac{1}{4}\{\|x + y\|^2 - \|x - y\|^2 + i\|x + iy\|^2 - i\|x - iy\|^2\}.$$

(41.5) Theorem. *In a real pre-Hilbert space, the inner product defined in (41.4) satisfies the following conditions:*

(i) $(x, x) = \|x\|^2$; *thus* $(x, x) \geq 0$ *for all* x, *and* $(x, x) = 0$ *iff* $x = \theta$;
(ii) $(y, x) = (x, y)$;
(iii) $(x + y, z) = (x, z) + (y, z)$;
(iv) $(\alpha x, y) = \alpha(x, y)$ *for all real* α.

It follows that

(v) $(x, y + z) = (x, y) + (x, z)$;
(vi) $(x, \alpha y) = \alpha(x, y)$ *for all real* α.

Proof. (i) and (ii) are obvious from the defining formula (1). It will suffice to verify (iii) and (iv). For later use, note that $(\theta, y) = 0$.

If u, v, z are any given vectors, we assert that

(*) $$(u + v, z) + (u - v, z) = 2(u, z).$$

This is derived from the parallelogram law, as follows. One has

$$\|(u + z) + v\|^2 + \|(u + z) - v\|^2 = 2\|u + z\|^2 + 2\|v\|^2,$$

that is,

(3) $$\|(u + v) + z\|^2 + \|(u - v) + z\|^2 = 2\|u + z\|^2 + 2\|v\|^2.$$

Replacing z by $-z$ in (3),

(4) $$\|(u + v) - z\|^2 + \|(u - v) - z\|^2 = 2\|u - z\|^2 + 2\|v\|^2.$$

Subtracting (4) from (3), and invoking the defining formula (1), we have

$$4(u + v, z) + 4(u - v, z) = 8(u, z),$$

which is (*). In particular (since $(\theta, z) = 0$) putting $v = u$ in (*) yields $(2u, z) = 2(u, z)$, therefore (*) may be rewritten

(**) $$(u + v, z) + (u - v, z) = (2u, z).$$

Given any vectors x, y, z, we assert that (iii) holds; this is immediate from (**) on setting $u = \tfrac{1}{2}(x + y)$, $v = \tfrac{1}{2}(x - y)$.

It remains only to verify (iv). Fix vectors x and y, and consider the function $f : \mathbb{R} \to \mathbb{R}$ defined by $f(\alpha) = (\alpha x, y)$; the problem is to show that $f(\alpha) = \alpha f(1)$ for all real α. Note that f is continuous; this is clear from the defining formula (1) and the continuity of the norm (16.4). In view of (iii), $f(\alpha + \beta) = f(\alpha) + f(\beta)$ for all real α and β; it follows that $f(\alpha) = \alpha f(1)$ for all rational α (by elementary algebra) and therefore for all real α (by continuity). ∎

The complex variant of (41.5):

(41.6) Theorem. *In a complex pre-Hilbert space, the inner product defined in* (41.4) *satisfies the following conditions:*
  (i) $(x|x) = \|x\|^2$; *thus* $(x|x) \geq 0$ *for all* x, *and* $(x|x) = 0$ *iff* $x = \theta$;
  (ii) $(y|x) = (x|y)^*$;
  (iii) $(x + y|z) = (x|z) + (y|z)$;
  (iv) $(\lambda x|y) = \lambda(x|y)$ *for all* $\lambda \in \mathbb{C}$.
It follows that
  (v) $(x|y + z) = (x|y) + (x|z)$;
  (vi) $(x|\lambda y) = \lambda^*(x|y)$ *for all* $\lambda \in \mathbb{C}$.

Proof. One can also regard the given space as a real pre-Hilbert space; in view of (41.5), the formula (1) defines an $\mathbb{R}$-bilinear form with $(x, x) = \|x\|^2$ and $(y, x) = (x, y)$. Rewriting (2) in the form

$$(2') \qquad\qquad (x|y) = (x, y) + i(x, iy),$$

it is clear that $(x|y)$ is also $\mathbb{R}$-bilinear. In particular, (iii) and (v) hold.

The key to (ii) is the observation that

$$(\dagger) \qquad\qquad (\mu x, \mu y) = (x, y) \qquad \text{for all } \mu \in \mathbb{C} \text{ with } |\mu| = 1;$$

this is immediate from the defining formula (1) and the absolute homogeneity of the norm. Applying this with $\mu = i$ at the appropriate step, (ii) results from the computation

$$(x|y)^* = (x, y) - i(x, iy) = (y, x) - i(iy, x)$$
$$= (y, x) + i(iy, -x) = (y, x) + i(iy, i^2x)$$
$$= (y, x) + i(y, ix) = (y|x).$$

In particular, $(x|x)$ is real, therefore $(2')$ yields $(x|x) = (x, x) = \|x\|^2$.

To verify (iv) it will suffice, in view of $\mathbb{R}$-bilinearity, to show that $(ix|y) = i(x|y)$; indeed, citing $(\dagger)$ twice at the appropriate step,

$$(ix|y) = (ix, y) + i(ix, iy) = (i^2x, iy) + i(x, y)$$
$$= -(x, iy) + i(x, y) = i[(x, y) + i(x, iy)] = i(x|y).$$

Finally, (vi) results from (iv) and (ii). ∎

In a pre-Hilbert space, the inner product of two vectors is dominated by the product of their norms (cf. 41.15):

(41.7) Theorem. (Cauchy-Schwarz inequality) *In a real pre-Hilbert space,*

$$(5) \qquad\qquad |(x, y)| \leq \|x\| \|y\|.$$

In a complex pre-Hilbert space,

$$(6) \qquad\qquad |(x|y)| \leq \|x\| \|y\|.$$

Proof. In either case the inequality is obvious if $x = \theta$ or $y = \theta$; assuming x and y nonzero, it is clear from homogeneity that one can suppose $\|x\| = \|y\| = 1$, in which case the problem is to show either $|(x, y)| \leq 1$ or $|(x|y)| \leq 1$, as the case may be.

Consider first the real case. Citing the defining formula (1) and the parallelogram law, we have

$$4(x, y) = \|x + y\|^2 - \|x - y\|^2,$$

$$4|(x, y)| \leq \|x + y\|^2 + \|x - y\|^2 = 2\|x\|^2 + 2\|y\|^2 = 4.$$

In the complex case, we may write $|(x|y)| = \mu(x|y)$ for suitable $|\mu| = 1$. Then $(\mu x|y) = |(x|y)|$ is real; therefore, in the notation of the proof of (41.6), (2') yields $(\mu x|y) = (\mu x, y)$. Since $\|\mu x\| = \|y\| = 1$, we have, citing the real case just proved, $|(x|y)| = |(\mu x|y)| = |(\mu x, y)| \leq 1$. ∎

In Theorems (41.5) and (41.6) it was shown that in a normed space satisfying the parallelogram law (P), one can introduce an inner product with pleasant linearity properties. We now show, conversely, that inner products (defined below) lead to norms satisfying (P). This is a convenient place to introduce some terminology, slightly more general than is strictly needed here, that will be useful later on:

(41.8) Definition. If E is a real vector space, a *bilinear form* on E is a function $\varphi: E \times E \to \mathbb{R}$ that is linear in each of its variables, i.e.,

(i) $$\varphi(x + y, z) = \varphi(x, z) + \varphi(y, z),$$

(ii) $$\varphi(\alpha x, y) = \alpha\varphi(x, y),$$

(iii) $$\varphi(x, y + z) = \varphi(x, y) + \varphi(x, z),$$

(iv) $$\varphi(x, \alpha y) = \alpha\varphi(x, y),$$

for all $x, y, z \in E$ and $\alpha \in \mathbb{R}$. The form is called *symmetric* if

(v) $$\varphi(y, x) = \varphi(x, y)$$

for all x and y (in which case, conditions (iii) and (iv) are redundant); *positive* if

(vi) $$\varphi(x, x) \geq 0$$

for all x; and *strictly positive* if

(vii) $$\varphi(x, x) > 0$$

for all $x \neq \theta$.

An *inner product* on E is a strictly positive, symmetric bilinear form on E, i.e., a function satisfying (i), (ii), (v), and (vii) (hence all the rest). A *real inner product space* is a real vector space equipped with an inner product; the inner product of x and y is usually denoted (x, y), i.e., one omits φ in the above identities.

(41.9) Definition. If E is a complex vector space, a *sesquilinear form* on E is a function $\varphi\colon E \times E \to \mathbb{C}$ that is linear in the first variable and conjugate-linear in the second, i.e.,

(i) $$\varphi(x + y, z) = \varphi(x, z) + \varphi(y, z),$$

(ii) $$\varphi(\lambda x, y) = \lambda\varphi(x, y),$$

(iii) $$\varphi(x, y + z) = \varphi(x, y) + \varphi(x, z),$$

(iv) $$\varphi(x, \lambda y) = \lambda^*\varphi(x, y),$$

for all $x, y, z \in E$ and $\lambda \in \mathbb{C}$. The form is called *Hermitian* if

(v) $$\varphi(y, x) = \varphi(x, y)^*$$

for all x and y (in which case, conditions (iii) and (iv) are redundant); *positive* if

(vi) $$\varphi(x, x) \geq 0$$

for all x; and *strictly positive* if

(vii) $$\varphi(x, x) > 0$$

for all $x \neq \theta$.

An *inner product* on E is a strictly positive, Hermitian, sesquilinear form on E, i.e., a function satisfying (i), (ii), (v), and (vii) (hence all the rest). {Incidentally, the word "Hermitian" is redundant here (cf. 41.14).} A *complex inner product space* is a complex vector space equipped with such an inner product; the inner product of x and y (in that order) will be denoted $(x\,|\,y)$.

(41.10) In the light of the preceding definitions, the gist of (41.5) and (41.6) is that every pre-Hilbert space may be regarded as an inner product space. Conversely:

(41.11) Theorem. *If E is an inner product space, then a norm on E satisfying the parallelogram law* (P) *is obtained by defining* $\|x\| = (x, x)^{1/2}$ *in the real case and* $\|x\| = (x\,|\,x)^{1/2}$ *in the complex case.*

Proof. We write out the proof for the complex case (the proof for the real case is even simpler). Evidently $\|\lambda x\| = |\lambda|\,\|x\|$ for all $x \in E$ and $\lambda \in \mathbb{C}$, and $\|x\| > 0$ when $x \neq \theta$. Moreover, (P) holds by (41.3). It remains only to verify that $\|x + y\| \leq \|x\| + \|y\|$ for all x and y.

Define $(x, y) = \mathrm{Re}\,(x\,|\,y)$ for all $x, y \in E$; evidently (x, y) is an inner product on E regarded as a real vector space, in the sense of (41.8). In particular, $(x, x) = \mathrm{Re}\,(x\,|\,x) = \mathrm{Re}\,\|x\|^2 = \|x\|^2$ for all x; it follows that

$$\|x + y\|^2 - \|x - y\|^2 = 4(x, y)$$

for all x and y, and the calculation made in (41.7) yields $|(x, y)| \leq \|x\|\,\|y\|$. Also,

$$\|x + y\|^2 = \|x\|^2 + \|y\|^2 + 2(x, y),$$

therefore, citing the triangle inequality in $\mathbb{R}$, one has

$$\|x + y\|^2 \leq \|x\|^2 + \|y\|^2 + 2|(x, y)|$$

$$\leq \|x\|^2 + \|y\|^2 + 2\|x\| \|y\| = (\|x\| + \|y\|)^2. \qquad \blacksquare$$

(41.12) Combining (41.10) and (41.11), we see that the notion of pre-Hilbert space is coextensive with that of inner product space.

More precisely, suppose E is a (say) complex pre-Hilbert space; (41.6) provides E with an inner product, defined via (2); in turn, (41.11) applied to this inner product clearly yields the original norm on E.

Conversely, suppose E is a complex inner product space, with inner product $(x|y)$; apply (41.11) to obtain a norm $\|x\| = (x|x)^{1/2}$ satisfying (P), i.e., a pre-Hilbert space structure; one verifies easily that the relation (2) holds; therefore (41.6) applied to this pre-Hilbert space yields the original inner product.

Similarly for the real case.

Thus, the terms 'pre-Hilbert space' and 'inner product space' may be used interchangeably. With a view to enjoying the best of two worlds, we make our choice as follows:

(41.13) Terminological convention. We shall use the terms *inner product space* (rather than 'pre-Hilbert space') and *Hilbert space* (rather than 'complete inner product space'). The theories of real and complex spaces are partly parallel, partly divergent. We shall restrict attention to the *complex* case, so as to avoid repetition (where the theories are parallel) and exploit the peculiar advantages of the complex field (where the theories diverge).

This is a convenient place to record some basic facts about sesquilinear forms; they are not needed until Section 57 and thereafter:

(41.14) Theorem. *Let φ be a sesquilinear form on a complex vector space E.*

(i) *(Parallelogram law) For all x, y in E,*

$$\varphi(x + y, x + y) + \varphi(x - y, x - y) = 2\varphi(x, x) + 2\varphi(y, y).$$

(ii) *(Polarization identity) For all x, y in E,*

$$\varphi(x, y) = \tfrac{1}{4}\{\varphi(x + y, x + y) - \varphi(x - y, x - y)$$

$$+ i\varphi(x + iy, x + iy) - i\varphi(x - iy, x - iy)\}.$$

(iii) *$\varphi = 0$ iff $\varphi(x, x) = 0$ for all x in E.*

(iv) *φ is Hermitian iff $\varphi(x, x)$ is real for all x in E.*

(v) *(Cauchy-Schwarz inequality) If φ is positive then*

$$|\varphi(x, y)|^2 \leq \varphi(x, x)\varphi(y, y)$$

for all x, y in E.

Proof. (i), (ii) Let $x, y \in E$. Then

(a) $\qquad \varphi(x + y, x + y) = \varphi(x, x) + \varphi(x, y) + \varphi(y, x) + \varphi(y, y);$

replacement of y by $-y$, iy, and $-iy$ yields

(b) $\qquad \varphi(x - y, x - y) = \varphi(x, x) - \varphi(x, y) - \varphi(y, x) + \varphi(y, y),$

(c) $\qquad \varphi(x + iy, x + iy) = \varphi(x, x) - i\varphi(x, y) + i\varphi(y, x) + \varphi(y, y),$

(d) $\qquad \varphi(x - iy, x - iy) = \varphi(x, x) + i\varphi(x, y) - i\varphi(y, x) + \varphi(y, y).$

Addition of (a) and (b) yields (i); and the appropriate combination of (a)–(d) yields (ii).

(iii) Immediate from (ii).

(iv) The formula $\psi(x, y) = \varphi(x, y) - \varphi(y, x)^*$ defines a sesquilinear form. Evidently φ is Hermitian iff $\psi = 0$ iff $\psi(x, x) = 0$ for all x iff $\varphi(x, x)$ is real for all x.

(v) It follows from (iv) that φ is Hermitian. Let $x, y \in E$.

Suppose first that $\varphi(x, x) = 0$; we show that $\varphi(x, y) = 0$ (whence the desired inequality holds trivially). For all complex numbers λ,

$$0 \le \varphi(\lambda x + y, \lambda x + y) = \lambda\lambda^*\varphi(x, x) + \lambda\varphi(x, y) + \lambda^*\varphi(y, x) + \varphi(y, y)$$
$$= 2 \operatorname{Re}\{\lambda\varphi(x, y)\} + \varphi(y, y);$$

in particular, if $\lambda = \alpha\varphi(x, y)^*$ with α real, then

$$0 \le 2\alpha|\varphi(x, y)|^2 + \varphi(y, y),$$

and the validity of this for all $\alpha < 0$ implies that $\varphi(x, y) = 0$.

If $\varphi(y, y) = 0$ then $\varphi(x, y) = \varphi(y, x)^* = 0$ by the foregoing.

Assume $\varphi(x, x) > 0$ and $\varphi(y, y) > 0$. By homogeneity, we can suppose without loss of generality that $\varphi(x, x) = \varphi(y, y) = 1$ (replace x and y by $\varphi(x, x)^{-1/2}x$ and $\varphi(y, y)^{-1/2}y$); it is to be shown that $|\varphi(x, y)| \le 1$. Write $|\varphi(x, y)| = \mu\varphi(x, y)$ for suitable $|\mu| = 1$; then $\varphi(\mu x, y) = \mu\varphi(x, y) = |\varphi(x, y)| \ge 0$, where $\varphi(\mu x, \mu x) = |\mu|^2\varphi(x, x) = 1$. Changing notation, we can suppose that $\varphi(x, y)$ is real; it then follows that the last two terms in the polarization identity drop out:

$$\varphi(x, y) = \tfrac{1}{4}\{\varphi(x + y, x + y) - \varphi(x - y, x - y)\},$$

whence

$$|\varphi(x, y)| \le \tfrac{1}{4}\{\varphi(x + y, x + y) + \varphi(x - y, x - y)\}$$
$$= \tfrac{1}{4}\{2\varphi(x, x) + 2\varphi(y, y)\} = 1. \qquad \blacksquare$$

Exercises

(41.15) In an inner product space, the Cauchy-Schwarz inequality (41.7) is strict if and only if x and y are linearly independent.

(41.16) If x and y are nonzero vectors in an inner product space, then $\|x + y\| = \|x\| + \|y\|$ iff $y = \alpha x$ for suitable $\alpha > 0$.

(41.17) Let E be an inner product space, A a compact convex subset of E, A_{ep} the set of extremal points of A, and A_{bp} the set of bare points of A (36.15).

(i) E is a uniformly convex normed space (36.14).

(ii) $A_{\text{bp}} \subset A_{\text{ep}}$.

(iii) A is the closed convex hull of A_{bp}.

(iv) If E is a Hilbert space, then $(A_{\mathrm{bp}})^- = (A_{\mathrm{ep}})^-$.

(v) If E is one-dimensional (i.e., $E = \mathbb{C}$), then $(A_{\mathrm{bp}})^- = A_{\mathrm{ep}}$.

(41.18) If T is a linear mapping in an inner product space (complex!), such that $(Tx|x) = 0$ for all vectors x, then $T = 0$.

42. Duality in Hilbert spaces.

The central fact about duality in Hilbert spaces is that a Hilbert space is *self-dual*; that is, if H is a Hilbert space then its dual space H' (see (40.8)) may be identified with H, in a sense to be made precise in this section (42.12).

(42.1) Definition. If E is an inner product space (41.13) each vector $y \in E$ defines a linear form y' on E via the formula $y'(x) = (x|y)$, where $(x|y)$ denotes the inner product of x and y. By the Cauchy-Schwarz inequality (41.7), $|y'(x)| = |(x|y)| \leq \|x\| \|y\|$, thus $y' \in E'$, $\|y'\| \leq \|y\|$. Better yet:

(42.2) Theorem. *If E is an inner product space, then the mapping $y \mapsto y'$ of E into E' is conjugate-linear and isometric.*

Proof. The relations $(y + z)' = y' + z'$ and $(\lambda y)' = \lambda^* y'$ follow at once from the properties of the inner product. As noted in (42.1), $\|y'\| \leq \|y\|$. On the other hand, $\|y'\| \|y\| \geq |y'(y)| = (y|y) = \|y\|^2$, therefore $\|y'\| \geq \|y\|$. ∎

If the mapping $y \mapsto y'$ maps E *onto* E', it is clear from (42.2) that E must be complete, i.e., a Hilbert space. The central miracle to be exposed in this section is that the converse is true: If H is a Hilbert space, then $y \mapsto y'$ maps H *onto* H' (42.12). The key result in this circle of ideas is the following theorem:

(42.3) Theorem. (Minimizing vector) *Let H be a Hilbert space, let A be a nonempty, closed convex subset of H, and let x be any vector in H. There exists a unique vector $y_0 \in A$ such that $\|x - y_0\| \leq \|x - y\|$ for all $y \in A$.*

Proof. The theorem is easily seen to be false in some Banach spaces; somehow, we have to make use of the parallelogram law (41.1).

Let $\delta = \inf\{\|x - y\| : y \in A\}$; we seek $y_0 \in A$ with $\delta = \|x - y_0\|$. Choose any sequence $y_n \in A$ such that $\|x - y_n\| \to \delta$. We wish to replace the distance-approximating sequence y_n by a distance-achieving vector y_0. We will show that the sequence y_n converges to a suitable vector y_0; to this end, it will suffice to show that y_n is Cauchy. If m and n are large, we know that $\|x - y_m\|$ and $\|x - y_n\|$ are near δ; we need to infer that y_m and y_n are near each other. (In two- or three-dimensional Euclidean space, this is intuitively plausible.)

Consider then $y, z \in A$. We are interested in estimating $\|z - y\|$ in terms of $\|x - z\|$ and $\|x - y\|$. This suggests looking at the equation $z - y = (x - y) - (x - z)$. By the parallelogram law,

$$\|(x - y) - (x - z)\|^2 + \|(x - y) + (x - z)\|^2 = 2\|x - y\|^2 + 2\|x - z\|^2,$$

thus

(*) $\|z - y\|^2 = 2\|x - y\|^2 + 2\|x - z\|^2 - \|2x - (y + z)\|^2.$

Since

$$\|2x - (y + z)\|^2 = 4\|x - \tfrac{1}{2}(y + z)\|^2,$$

and since $\tfrac{1}{2}(y + z) \in A$ by the convexity of A, we have

$$\|2x - (y + z)\|^2 \geq 4\delta^2$$

by the definition of δ; substituting this inequality into (*), we have

(**) $\|z - y\|^2 \leq 2\|x - y\|^2 + 2\|x - z\|^2 - 4\delta^2.$

In particular,

$$\|y_m - y_n\|^2 \leq 2\|x - y_n\|^2 + 2\|x - y_m\|^2 - 4\delta^2;$$

since the right side tends to $2\delta^2 + 2\delta^2 - 4\delta^2 = 0$ as $m, n \to \infty$, we have $\|y_m - y_n\| \to 0$.

Say $\|y_n - y_0\| \to 0$ (H is complete); $y_0 \in A$ because A is closed. Since $y_n \to y_0$, one has $x - y_n \to x - y_0$, and, by the continuity of the norm (16.4), $\|x - y_n\| \to \|x - y_0\|$. Thus $\delta = \|x - y_0\|$.

This proves existence. To see that y_0 is unique, suppose that $z_0 \in A$ also satisfies $\|x - z_0\| = \delta$. By (**),

$$\|z_0 - y_0\|^2 \leq 2\delta^2 + 2\delta^2 - 4\delta^2 = 0,$$

thus $z_0 = y_0$. ∎

To motivate the next definition, let us return to the basic problem: given $f \in H'$, H a Hilbert space, we seek a vector $y \in H$ such that $f = y'$, i.e., $f(x) = (x|y)$ for all $x \in H$. Thus, if M is the null space of f, the desired vector y must have the property that $(x|y) = 0$ for all $x \in M$. The condition $(x|y) = 0$ is reminiscent of 'orthogonality' in Euclidean spaces; so to speak, the desired vector y must be orthogonal to the null space of M. For the record:

(42.4) Definition. Vectors x, y of an inner product space E are said to be *orthogonal* if $(x|y) = 0$. Notation: $x \perp y$. A vector x is said to be orthogonal to a subset S of E if $x \perp y$ for all $y \in S$; we indicate this by writing $x \perp S$. The *annihilator* of a subset S of E is the set of all vectors x such that $x \perp S$; we denote this set by $S^\perp$. (There is a slight but harmless conflict with the notations introduced in Section 38; in the context of inner product spaces, we agree that $S^\perp$ will always mean the set defined here.)

(42.5) Orthogonality in an inner product space has the following obvious properties: $x \perp y$ iff $y \perp x$; $x \perp x$ iff $x = \theta$; if $x \perp y$ and $x \perp z$ then $x \perp \lambda y + \mu z$ for all scalars λ and μ. Another very useful property:

(42.6) Theorem. (Pythagorean relation)   *In an inner product space, $x \perp y$ implies $\|x + y\|^2 = \|x\|^2 + \|y\|^2$.*

Proof. $\|x + y\|^2 = (x|x) + (x|y) + (y|x) + (y|y) = \|x\|^2 + 0 + 0 + \|y\|^2.$ ∎

Every closed linear subspace of a Hilbert space induces an orthogonal splitting of the space:

(42.7) Theorem. (Orthogonal projection) *Let N be a closed linear subspace of a Hilbert space H. For each $x \in H$ there exists a unique representation $x = y + z$ with $y \in N$ and $z \in N^\perp$.*

Proof. Uniqueness is easy: if $x = y_1 + z_1 = y_2 + z_2$ with $y_1, y_2 \in N$ and $z_1, z_2 \in N^\perp$, then the vector $y_1 - y_2 = z_2 - z_1$ belongs both to N and to $N^\perp$, and is therefore θ (42.5).

Existence: By (42.3) there exists $y_0 \in N$ such that $\|x - y_0\| \leq \|x - y\|$ for all $y \in N$. Set $z_0 = x - y_0$; it will suffice to show that $z_0 \perp N$. Given $y \in N$, we must show that $(z_0|y) = 0$; we can suppose $\|y\| = 1$. Direct calculation shows that $z_0 - (z_0|y)y$ is orthogonal to y and therefore to $(z_0|y)y$; since $z_0 = [z_0 - (z_0|y)y] + (z_0|y)y$, it follows from (42.6) that

$$\|z_0\|^2 = \|z_0 - (z_0|y)y\|^2 + \|(z_0|y)y\|^2,$$

that is,

(†) $$\|z_0 - (z_0|y)y\|^2 = \|z_0\|^2 - |(z_0|y)|^2.$$

Since $z_0 - (z_0|y)y = (x - y_0) - (z_0|y)y = x - [y_0 + (z_0|y)y]$, where $y_0 + (z_0|y)y \in N$, it follows from the definition of y_0 that

$$\|x - y_0\| \leq \|x - [y_0 + (z_0|y)y]\|,$$

that is,

$$\|z_0\| \leq \|z_0 - (z_0|y)y\|;$$

combined with (†), this yields

$$\|z_0\|^2 \leq \|z_0\|^2 - |(z_0|y)|^2,$$

thus $(z_0|y) = 0$. ∎

The following proposition on orthogonality is analogous to (38.14):

(42.8) Theorem. *Let S and T be subsets of an inner product space E. Then:*
(1) *$S^\perp$ is a closed linear subspace of E;*
(2) *$S \subset S^{\perp\perp}$;*
(3) *if $S \subset T$ then $T^\perp \subset S^\perp$;*
(4) *$S^\perp = S^{\perp\perp\perp}$.*

Proof. The notations have the following meanings: $S^{\perp\perp} = (S^\perp)^\perp$, $S^{\perp\perp\perp} = (S^{\perp\perp})^\perp$.

(1) The fact that $S^\perp$ is a linear subspace of E is immediate from (42.5). If $x_n \in S^\perp$ and $\|x_n - x\| \to 0$ then $x \in S^\perp$ because, for any $y \in S$, the Cauchy-Schwarz inequality (41.7) yields

$$|(x|y)| \leq |(x - x_n|y)| + |(x_n|y)| \leq \|x - x_n\|\,\|y\| + 0 \to 0.$$

{More generally, it is easy to see that if $\|x_n - x\| \to 0$ and $\|y_n - y\| \to 0$ then $(x_n | y_n) \to (x | y)$.}

The proofs of (2), (3), (4) are formally the same as in (38.14). ∎

If N is a linear subspace of an inner product space, such that $N = N^{\perp\perp}$, then N is closed (42.8); in a Hilbert space, the converse is also true:

(42.9) Theorem. *If N is a closed linear subspace of a Hilbert space, then* $N^{\perp\perp} = N$.

Proof. At any rate, $N \subset N^{\perp\perp}$ (42.8). On the other hand, if $x \in N^{\perp\perp}$, write $x = y + z$ with $y \in N$ and $z \in N^{\perp}$ (42.7). Since $y \in N \subset N^{\perp\perp}$ and $x \in N^{\perp\perp}$ we have $z = x - y \in N^{\perp\perp}$, i.e., $z \perp N^{\perp}$; in particular, $z \perp z$, thus $z = \theta$ and $x = y \in N$. ∎

(42.10) Definition. If N is a closed linear subspace of a Hilbert space H, we call $N^{\perp}$ the *orthogonal complement* of N; in view of (42.9), the orthogonal complement of $N^{\perp}$ is N. The result of the projection theorem (42.7) is expressed by writing $H = N \oplus N^{\perp}$. Since $H = N + N^{\perp}$ and $N \cap N^{\perp} = \{\theta\}$, H is, in particular, the algebraic direct sum of N and $N^{\perp}$; in fact, the direct sum is topological in the sense of (20.2):

(42.11) Theorem. *If H is a Hilbert space and N is a closed linear subspace of H, then N and $N^{\perp}$ are topologically supplementary subspaces, that is, H is the topological direct sum of N and $N^{\perp}$. More precisely, if $x = y + z$ with $y \in N$ and $z \in N^{\perp}$, then $\|x\|^2 = \|y\|^2 + \|z\|^2$.*

Proof. The second statement is the Pythagorean relation (42.6). The first statement follows from the second: if $x_n = y_n + z_n$ and $x = y + z$, where $y_n, y \in N$ and $z_n, z \in N^{\perp}$, then $x_n - x = (y_n - y) + (z_n - z)$, therefore

$$\|x_n - x\|^2 = \|y_n - y\|^2 + \|z_n - z\|^2;$$

obviously $x_n \to x$ if and only if $y_n \to y$ and $z_n \to z$. {Alternatively, writing $Px = y$, one has $\|Px\| = \|y\| \le \|x\|$, thus P is continuous, therefore N and $N^{\perp}$ are topological supplements by (20.3).} ∎

The self-duality of a Hilbert space follows readily from (42.9):

(42.12) Theorem. (Fréchet-Riesz) *If f is a continuous linear form on a Hilbert space H, then there exists a unique vector $y \in H$ such that $f = y'$, that is, $f(x) = (x | y)$ for all $x \in H$. Thus, the mapping $y \mapsto y'$ of (42.2) is a conjugate-linear, isometric mapping of H onto H'.*

Proof. Uniqueness is obvious from (42.2), thus we are concerned with existence. We can suppose $f \ne 0$. The null space N of f is a closed linear subspace of H, therefore $N = N^{\perp\perp}$ by (42.9); since $N \ne H$, necessarily $N^{\perp} \ne \{\theta\}$. Choose $z \in N^{\perp}$, $z \ne \theta$; then $z \notin N$, therefore $f(z) \ne 0$; replacing z by $f(z)^{-1}z$, we can suppose $f(z) = 1$. Then, for any $x \in H$, the vector

$x - f(x)z$ is clearly annihilated by f, i.e., $x - f(x)z \in N$; since $z \in N^\perp$ it follows that $0 = (x - f(x)z \mid z)$, that is, $f(x)(z \mid z) = (x \mid z)$. Thus

$$f(x) = (x \mid \|z\|^{-2}z)$$

for all $x \in H$, so the vector $y = \|z\|^{-2}z$ meets the requirement. ∎

Exercises

(42.13) Let A be a complete, nonempty convex subset of an inner product space E. Then, given any $x \in E$, there exists a unique $y_0 \in A$ such that $\|x - y_0\| \leq \|x - y\|$ for all $y \in A$.

(42.14) The analogue of the orthogonal projection theorem (42.7) holds for a complete linear subspace of an inner product space.

(42.15) In an inner product space, $\|x + y\|^2 = \|x\|^2 + \|y\|^2$ if and only if Re $(x \mid y) = 0$. In order that $x \perp y$ it is necessary and sufficient that $\|x + y\|^2 = \|x\|^2 + \|y\|^2$ and $\|x + iy\|^2 = \|x\|^2 + \|iy\|^2$.

(42.16) The proof of the theorem on the minimizing vector (42.3) is independent of the notion of inner product, and the notion of orthogonality can also be formulated directly in terms of the norm (42.15). Give an inner-product-free proof of the theorem on orthogonal projection (42.7).

(42.17) If S is any subset of a Hilbert space H, then $S^{\perp\perp}$ is the smallest closed linear subspace of H that contains S.

43. Continuous linear mappings in Hilbert spaces.

A dividend of the self-duality of a Hilbert space H is that every closed linear subspace N of H possesses a topological supplement—in fact, a canonical one—namely, $N^\perp$ (42.11). To put it another way, the set of all closed linear subspaces of H admits a canonical bijection $N \mapsto N^\perp$ with interesting properties (it is inclusion-reversing and involutory). As we show in this section, another by-product of self-duality is that the algebra $\mathscr{L}(H)$ admits a canonical bijection $T \mapsto T^*$ possessing pleasant algebraic properties. We formulate the results slightly more generally, i.e., for a pair of Hilbert spaces H and K; this is no harder, and makes for greater clarity.

(43.1) Theorem. *If $T \in \mathscr{L}(H, K)$, where H and K are Hilbert spaces, then there exists a unique mapping $T^*: K \to H$ such that*

$$(*) \qquad (Tx \mid y) = (x \mid T^*y) \qquad \text{for all } x \in H, y \in K.$$

Moreover, $T^* \in \mathscr{L}(K, H)$.

Proof. Let $y \in K$; the problem is to define $T^*y \in H$. The mapping $x \mapsto (Tx \mid y)$ is evidently a continuous linear form on H (specifically, it is the composition $y' \circ T$), therefore by the Fréchet-Riesz theorem (42.12) there exists a unique vector $z \in H$ such that $(Tx \mid y) = (x \mid z)$ for all $x \in H$. Define $T^*y = z$. Thus $T^*: K \to H$ and the identity $(*)$ is satisfied. If also $S: K \to H$ satisfies $(*)$, then $(x \mid Sy) = (Tx \mid y) = (x \mid T^*y)$ for all $x \in H$ and $y \in K$, that is,

$(Sy)' = (T^*y)'$ for all $y \in K$, therefore $Sy = T^*y$ for all $y \in K$, in other words, $S = T^*$.

The linearity of T^* follows from the calculations

$$(T^*(y_1 + y_2))' = (y_1 + y_2)' \circ T = (y_1' + y_2') \circ T$$

$$= (y_1' \circ T) + (y_2' \circ T) = (T^*y_1)' + (T^*y_2)' = (T^*y_1 + T^*y_2)'$$

and

$$(T^*(\lambda y))' = (\lambda y)' \circ T = (\lambda^*y') \circ T$$

$$= \lambda^*(y' \circ T) = \lambda^*(T^*y)' = (\lambda(T^*y))',$$

and the injectivity of the mapping $x \mapsto x'$ (42.12).

Since $\|T^*y\| = \|(T^*y)'\| = \|y' \circ T\| \leq \|y'\| \|T\| = \|y\| \|T\|$, T^* is continuous. Thus, $T^* \in \mathcal{L}(K, H)$. {The argument shows that $\|T^*\| \leq \|T\|$; in fact $\|T^*\| = \|T\|$, as we see in the next theorem.} ∎

(43.2) Definition. With notation as in (43.1), T^* is called the *adjoint* of T.

(43.3) Theorem. *Let H and K be Hilbert spaces and let $S, T \in \mathcal{L}(H, K)$, $\lambda \in \mathbb{C}$. Then:*
  (1) $T^{**} = T$,
  (2) $(S + T)^* = S^* + T^*$,
  (3) $(\lambda T)^* = \lambda^* T^*$,
  (4) $\|T^*\| = \|T\|$,
  (5) $\|T^*T\| = \|T\|^2 = \|TT^*\|$.

Proof. We remark that complex conjugation of the identity (*) of (43.1) yields $(y|Tx) = (T^*y|x)$.

(1) T^{**} stands for $(T^*)^*$; since $T^* \in \mathcal{L}(K, H)$, one has $T^{**} \in \mathcal{L}(H, K)$, thus T and T^{**} are eligible to be compared. For all $x \in H$ and $y \in K$,

$$(T^{**}x|y) = (x|T^*y) = (Tx|y),$$

thus $T^{**} = T$.

(2), (3) For each $y \in K$ we have, by the linearity of y',

$$[(S + T)^*y]' = y' \circ (S + T) = y' \circ S + y' \circ T$$

$$= (S^*y)' + (T^*y)' = [S^*y + T^*y]' = [(S^* + T^*)y]'$$

and

$$[(\lambda T)^*y]' = y' \circ (\lambda T) = \lambda(y' \circ T)$$

$$= \lambda(T^*y)' = [\lambda^*(T^*y)]' = [(\lambda^*T^*)y]',$$

therefore $(S + T)^*y = (S^* + T^*)y$ and $(\lambda T)^*y = (\lambda^*T^*)y$.

(4) As noted in the proof of (43.1), $\|T^*\| \leq \|T\|$, therefore also $\|T\| = \|(T^*)^*\| \leq \|T^*\|$.

(5) For any $x \in H$, $\|x\| \leq 1$, citing the Cauchy-Schwarz inequality (41.7) we have

$$\|Tx\|^2 = (Tx|Tx) = (T^*Tx|x) \leq \|T^*Tx\| \|x\| \leq \|T^*Tx\| \leq \|T^*T\|,$$

therefore

$$\|T\|^2 = (\sup\{\|Tx\| : \|x\| \le 1\})^2$$
$$= \sup\{\|Tx\|^2 : \|x\| \le 1\} \le \|T^*T\|.$$

On the other hand, $\|T^*T\| \le \|T^*\|\|T\| = \|T\|^2$. Thus $\|T^*T\| = \|T\|^2$; replacement of T by T^* yields $\|TT^*\| = \|T^*\|^2 = \|T\|^2$. ∎

The effect of adjunction on a product:

(43.4) Theorem. *If $T \in \mathscr{L}(H, K)$ and $S \in \mathscr{L}(K, M)$, where H, K, M are Hilbert spaces, then $(ST)^* = T^*S^*$.*

Proof. Note that both sides of the equation belong to $\mathscr{L}(M, H)$. For all $x \in H$ and $z \in M$, $(x|(ST)^*z) = ((ST)x|z) = (S(Tx)|z) = (Tx|S^*z) = (x|T^*(S^*z)) = (x|(T^*S^*)z)$. ∎

(43.5) If H is a Hilbert space, we know that $\mathscr{L}(H)$ is a Banach algebra (40.23). Moreover, in view of (43.3) and (43.4), $T \mapsto T^*$ is a mapping of $\mathscr{L}(H)$ into itself such that (i) $T^{**} = T$, (ii) $(S + T)^* = S^* + T^*$, (iii) $(\lambda T)^* = \lambda^* T^*$, (iv) $(ST)^* = T^*S^*$, and (v) $\|T^*T\| = \|T\|^2$. It is immediate from (i) that $T \mapsto T^*$ is a bijective mapping of $\mathscr{L}(H)$ onto itself. We remark that condition (v) implies $\|T^*\| = \|T\|$ (also known from (43.3)): $\|T\|^2 = \|T^*T\| \le \|T^*\|\|T\|$ yields $\|T\| \le \|T^*\|$, and replacement of T by T^* yields the reverse inequality.

It is useful to insert here a general definition:

(43.6) Definition. A (complex) *∗-algebra* (or *involutive algebra*) is an algebra A over $\mathbb{C}$ with an *involution*, i.e., a mapping $a \mapsto a^*$ in A such that (i) $a^{**} = a$, (ii) $(a + b)^* = a^* + b^*$, (iii) $(\lambda a)^* = \lambda^* a^*$, and (iv) $(ab)^* = b^*a^*$. The terms *normed ∗-algebra* (or *involutive normed algebra*) and *Banach ∗-algebra* are self-explanatory (but the reader must be cautioned that in some contexts it may be required that the involution be continuous, or even that $\|a^*\| = \|a\|$). A Banach ∗-algebra whose involution satisfies (v) $\|a^*a\| = \|a\|^2$ for all a is called a *C∗-algebra*. {The calculation in (43.5) shows that in a C^*-algebra, $\|a^*\| = \|a\|$.}

(43.7) We may recapitulate (43.5) as follows: If H is a Hilbert space then $\mathscr{L}(H)$, with the indicated operations, is a C^*-algebra. Obviously every ∗-subalgebra of $\mathscr{L}(H)$ (i.e., a subalgebra containing adjoints) that is closed for the norm topology, is also a C^*-algebra. In a later chapter we shall prove the miraculous converse (Gel'fand-Naĭmark theorem (62.1)): If A is any C^*-algebra, there exists a Hilbert space H and a mapping $a \mapsto T_a$ of A into (but generally not onto) $\mathscr{L}(H)$ that preserves all of the structure, i.e., $T_{a+b} = T_a + T_b$, $T_{\lambda a} = \lambda T_a$, $T_{ab} = T_a T_b$, $T_{a^*} = (T_a)^*$, and $\|T_a\| = \|a\|$.

Exercises

(43.8) Suppose $T \in \mathscr{L}(H, K)$, where H and K are Hilbert spaces. We have $T^*: K \to H$ in the sense of (43.2) and $T': K' \to H'$ in the sense of (40.18). Since

K is mapped onto K' via $y \mapsto y'$, and H onto H' via $x \mapsto x'$, one expects that there will be a relation between T^* and T'. The relation is $T'y' = (T^*y)'$.

(43.9) Let H and K be Hilbert spaces. A sesquilinear form $\varphi: H \times K \to \mathbb{C}$ is said to be *bounded* if

$$\sup \{|\varphi(x, y)| : \|x\| \leq 1, \|y\| \leq 1\} < \infty,$$

in which case the supremum is denoted $\|\varphi\|$ and called the *norm* of φ. Prove:

(i) If $T \in \mathscr{L}(H, K)$ then the formula $\varphi_T(x, y) = (Tx|y)$ defines a bounded sesquilinear form φ_T on $H \times K$ with $\|\varphi_T\| = \|T\|$.

(ii) Conversely, if φ is a bounded sesquilinear form on $H \times K$, then there exists a unique $T \in \mathscr{L}(H. K)$ with $\varphi = \varphi_T$.

(iii) The set of all bounded sesquilinear forms on $H \times K$, with linear operations defined pointwise and $\|\varphi\|$ defined as above, is a Banach space, linearly isometric with $\mathscr{L}(H, K)$ via the mapping $T \mapsto \varphi_T$.

(43.10) (i) Let N be a closed linear subspace of a Hilbert space H. If $x \in H$, say $x = y + z$ with $y \in N$ and $z \in N^\perp$ (42.7), define $P_N x = y$; then $P_N \in \mathscr{L}(H)$ and $P_N{}^2 = P_N = P_N^*$.

(ii) Conversely, suppose $P \in \mathscr{L}(H)$ satisfies $P^2 = P = P^*$ (i.e., P is idempotent and self-adjoint; such an operator on H is called a *projection*). If $N = \{y \in H : Py = y\}$, then N is the range of P and $N^\perp$ is the null space of P; thus N is a closed linear subspace of H, and the relation $x = Px + (I - P)x$ shows that $P = P_N$.

(iii) The correspondence $N \mapsto P_N$ maps the set of all closed linear subspaces of H bijectively onto the set of all projections in $\mathscr{L}(H)$ (in particular, we may speak of P_N as *the* projection with range N). Under this correspondence, $P_{N^\perp} = I - P_N$; $M \subset N$ iff $P_M = P_M P_N$; $M \subset N^\perp$ iff $P_M P_N = 0$. {Accordingly, projections P, Q are said to be *orthogonal* if $PQ = 0$.}

(43.11) Let H be a Hilbert space, let $T \in \mathscr{L}(H)$, let N be a closed linear subspace of H, and let $P = P_N$ be the projection with range N (43.10). If $T(N) \subset N$, N is said to be *invariant* under T. If both N and $N^\perp$ are invariant under T, N is said to *reduce* T.

(i) N is invariant under T iff $PTP = TP$ iff $N^\perp$ is invariant under T^*.

(ii) N reduces T iff $PT = TP$ iff N reduces T^*.

(43.12) Let $T \in \mathscr{L}(H)$, H a Hilbert space, and suppose $\|T\| \leq 1$. {Such an operator on H is called a *contraction*.} Then T and T^* have the same fixed points, i.e., $Tx = x$ iff $T^*x = x$.

(43.13) Let H be a Hilbert space. The projections in $\mathscr{L}(H)$ are the idempotent contractions.

(43.14) If $T \in \mathscr{L}(H, K)$, where H and K are Hilbert spaces, then the null space of T^* is the orthogonal complement of the closure of $T(H)$ (and the null space of T is the orthogonal complement of the closure of $T^*(K)$). That is,

$$\{y : T^*y = \theta\} = \overline{T(H)}^\perp, \qquad \{x : Tx = \theta\} = \overline{T^*(K)}^\perp.$$

44. Duality in normed spaces. The following notations are fixed throughout the section: E is a normed space over $\mathbb{K}$, E' is the normed space dual of E (40.8), and E_0 is the image of E in E'' under the canonical mapping $x \mapsto x''$ (see the proof of (40.15)).

The results in the earlier part of the section deal only with the norm topologies on E, E', E''. However, the principal object of study will be the canonical duality between E and E', induced by the bilinear form

$$(x, f) \mapsto f(x) \qquad (x \in E, f \in E'),$$

as described in (38.3). For ease of reference, we record here the standard names for the topologies induced by this duality:

(44.1) Definition. The topology $\sigma(E, E')$ is called the *weak topology* on E, in conformity with (38.10). The topology $\sigma(E', E)$ is called the *weak* topology* on E'.

In the first group of results, we consider only the norm topologies. The first result is the normed space version of the Hahn-Banach theorem:

(44.2) Theorem. *If M is a linear subspace of E, then every continuous linear form g on M may be extended to a continuous linear form f on E such that $\|f\| = \|g\|$.*

Proof. Let $g \in M'$. Define $p(x) = \|g\| \|x\|$ $(x \in E)$; then p is a seminorm on E (it is a norm unless $g = 0$) and $|g(y)| \le p(y)$ for all $y \in M$. In view of (28.8), there exists a linear form f on E such that $f|M = g$ and $|f(x)| \le p(x)$ for all $x \in E$; the latter condition shows that $f \in E'$ with $\|f\| \le \|g\|$, whereas the former condition obviously implies $\|f\| \ge \|g\|$. ∎

Concerning annihilators $M^\perp$, see (38.13), (38.14).

(44.3) Theorem. *If M is a linear subspace of E, then (suggestively, though inexactly) $M' = E'/M^\perp$, in the sense that there exists a natural isometric linear mapping of $E'/M^\perp$ onto M'.*

Proof. First we construct a mapping $T: E' \to M'$ as follows. If $f \in E'$ then the restriction of f to M is a continuous linear form on M; thus, writing $Tf = f|M$, we have $Tf \in M'$. Clearly T is linear, and, since obviously $\|Tf\| \le \|f\|$, T is continuous. Moreover, T is surjective: if $g \in M'$ then there exists $f \in E'$ with $Tf = g$ (44.2). The null space of T is $M^\perp$: $Tf = 0$ iff $f|M = 0$ iff $f \in M^\perp$. In particular, since T is continuous, $M^\perp$ is a closed linear subspace of E'. {In fact, $M^\perp$ is even weak* closed, by part (i) of (38.14).}

Let $\pi: E' \to E'/M^\perp$ be the canonical mapping, and equip $E'/M^\perp$ with the quotient norm (16.8). Since T vanishes on $M^\perp$, it induces a mapping $S: E'/M^\perp \to M'$ via the formula $S(\pi(f)) = Tf$; S is well-defined, since $\pi(f) = \pi(g)$ means $f - g \in M^\perp$ and therefore $T(f - g) = 0$, i.e., $Tf = Tg$. Thus, $S \circ \pi = T$. It is easy to see that S is a linear mapping of $E'/M^\perp$ onto M'; moreover, since the null space of T is precisely $M^\perp$, it follows that S is injective, and therefore bijective.

Let $u \in E'/M^\perp$. The proof will be concluded by showing that $\|Su\| = \|u\|$. For any $f \in u$, $Su = S(\pi(f)) = Tf$, therefore $\|Su\| = \|Tf\| \le \|f\|$; then

$$\|Su\| \le \inf\{\|f\| : f \in u\} = \|u\|.$$

To establish the reverse inequality, fix $f_0 \in u$ and write $g_0 = f_0|M$, i.e., $g_0 = Tf_0 = Su$. By (44.2), there exists $f \in E'$ with $f|M = g_0$ and $\|f\| = \|g_0\|$. Then $f|M = g_0 = f_0|M$ shows that $f - f_0 \in M^\perp$; thus $\pi(f) = \pi(f_0) = u$ and therefore $\|u\| = \|\pi(f)\| \leq \|f\| = \|g_0\| = \|Su\|$. ∎

(44.4) Theorem. *If M is a closed linear subspace of E, then* (*suggestively, though inexactly*) *$M^\perp = (E/M)'$, in the sense that there exists a natural isometric linear mapping of $(E/M)'$ onto $M^\perp$.*

Proof. Let $\pi: E \to E/M$ be the canonical mapping, and equip E/M with the quotient norm (16.8). {We are assuming that M is closed for the norm topology; in fact, M is also weakly closed (38.12). We remark that a closed linear subspace of E' may fail to be weak* closed, thus the canonical duality is in general not perfectly symmetric. However, symmetry reigns in 'reflexive' spaces, to be discussed below.}

Define a mapping $T: (E/M)' \to M^\perp$ as follows. If $\xi \in (E/M)'$, clearly $\xi \circ \pi$ is a continuous linear form on E that vanishes on M, thus $\xi \circ \pi \in M^\perp$; define $T\xi = \xi \circ \pi$. Clearly T is linear, and $\|T\xi\| \leq \|\xi\| \|\pi\| \leq \|\xi\|$. Moreover, T is injective: if $T\xi = 0$ then, for all $x \in E$, $0 = (T\xi)(x) = (\xi \circ \pi)(x) = \xi(\pi(x))$, thus $\xi = 0$ on E/M, i.e., $\xi = 0$. {Remark: $T = \pi'$, the transpose of π in the sense of (40.18).}

Let $f \in M^\perp$. The proof will be concluded by showing that there exists (a necessarily unique) $\xi \in (E/M)'$ with $T\xi = f$ and $\|\xi\| \leq \|T\xi\|$ (this will show that T is bijective and, in view of the earlier inequality, isometric). Given $u \in E/M$, the problem is to define $\xi(u)$. If $u = \pi(x) = \pi(y)$, then $x - y \in M$ and so $f(x - y) = 0$, i.e., $f(x) = f(y)$; thus, the definition $\xi(u) = f(x)$ is unambiguous. Clearly $f = \xi \circ \pi$; since f is linear, so is ξ. Also, ξ is continuous; this follows from the fact that E/M bears the final topology for π (11.8), but we shall verify it directly to get extra geometric information. Let $u \in E/M$. For any $x \in u$ we have $|\xi(u)| = |f(x)| \leq \|f\| \|x\|$, therefore

$$|\xi(u)| \leq \|f\| \inf \{\|x\| : x \in u\} = \|f\| \|u\|;$$

thus ξ is continuous, with $\|\xi\| \leq \|f\|$. Since $T\xi = \xi \circ \pi = f$, the proof is complete. ∎

The foregoing results lack symmetry in E and E'. The reason is that, in general, E cannot be canonically identified with the dual $(E')' = E''$ of E'. To put the matter another way, the canonical image E_0 of E is, in general, a proper linear subspace of E''; their coincidence is a special event:

(44.5) Definition. E is said to be *reflexive* if $E_0 = E''$, i.e., if the canonical embedding $x \mapsto x''$ maps E onto E''. We remark that, since the canonical mapping is isometric (40.13) and E'' is a Banach space (40.9), a reflexive normed space is necessarily a Banach space; accordingly, we apply the term only to Banach spaces, i.e., we speak of 'reflexive Banach spaces' rather than 'reflexive normed spaces.'

The phenomenon of reflexivity will be explored more deeply in the next section, but it is appropriate to pause here to record two important examples.

(44.6) Example. *The following conditions on E are equivalent: (a) E is finite-dimensional; (b) E' is finite-dimensional; (c) E" is finite-dimensional. In this case* dim E = dim E' = dim $E"$ *and E is reflexive.*

Proof. (a) implies (b): Let $x_1, \ldots, x_n$ be a basis of E. For each $j = 1, \ldots, n$, let f_j be the linear form on E such that $f_j(x_k) = \delta_{jk}$ for $k = 1, \ldots, n$. The f_j are continuous by (23.9); we show that they are a basis of E'.

If $f \in E'$ then, defining $\lambda_j = f(x_j)$, one has $f = \sum_1^n \lambda_j f_j$; indeed, for any k,

$$\left(\sum_{j=1}^n \lambda_j f_j \right)(x_k) = \sum_{j=1}^n \lambda_j \delta_{jk} = \lambda_k = f(x_k),$$

therefore $\sum_1^n \lambda_j f_j = f$ on the linear span of the x_k, i.e., on E. Thus the f_j span E'.

The f_j are linearly independent. For, if $\sum_1^n \lambda_j f_j = 0$ then, for all k, $0 = \sum_{j=1}^n \lambda_j f_j(x_k) = \lambda_k$, thus the coefficients all vanish.

Summarizing: If E is finite-dimensional, then E' is also finite-dimensional and dim E' = dim E.

(b) implies (c): Apply the preceding sentence to E' in place of E. In particular, dim $E"$ = dim E'.

(c) implies (a): Since $x \mapsto x"$ is a vector space isomorphism of E onto E_0, we have dim E = dim $E_0 \leq$ dim $E" < \infty$. It follows that dim E_0 = dim E = dim E' = dim $E" < \infty$, and therefore $E_0 = E"$. ∎

(44.7) Example. *Every Hilbert space is reflexive.*

Proof. Let H be a Hilbert space (41.13). Each $y \in H$ determines a continuous linear form y' on H via the formula $y'(x) = (x|y)$, and $y \mapsto y'$ is an isometric, conjugate-linear mapping of H onto H' (42.12). Given any $\varphi \in H"$, we seek $x \in H$ such that $\varphi = x"$; that is, we require that for all $y \in H$, $\varphi(y') = x"(y') = y'(x) = (x|y) = (y|x)^* = (x'(y))^*$, i.e., $x'(y) = (\varphi(y'))^*$. The existence of such an x is immediate from the fact that $y \mapsto (\varphi(y'))^*$ is a continuous linear form on H. ∎

We now consider the relative strength of the various topologies on E and E'.

(44.8) Theorem. *Let τ be the norm topology on E. Then:*
  (i) $\tau \supset \sigma(E, E')$;
  (ii) $\tau = \sigma(E, E')$ *if and only E is finite-dimensional.*
Proof. (i) The relation $\tau \supset \sigma(E, E')$ was noted in (38.11).

(ii) Suppose first that $\tau = \sigma(E, E')$. Consider the τ-neighborhood $V = \{x \in E : \|x\| \leq 1\}$ of θ. By hypothesis, V is also a weak neighborhood of θ, therefore there exists a finite subset $\{f_1, \ldots, f_n\}$ of E' such that

(*) $\qquad \{x : |f_j(x)| \leq 1 \text{ for } j = 1, \ldots, n\} \subset V$

(cf. (38.6), (19.6)). In particular, if $f_j(x) = 0$ for $j = 1, \ldots, n$ then $x = \theta$; indeed, $f_j(kx) = kf_j(x) = 0$ for $j = 1, \ldots, n$ and for all positive integers k, therefore by (*) one has $\|kx\| \leq 1$, i.e., $\|x\| \leq 1/k$, for all k. Thus, the intersection of the null spaces of $f_1, \ldots, f_n$ is $\{\theta\}$; since every $f \in E'$ vanishes on $\{\theta\}$, it follows that f is a linear combination of $f_1, \ldots, f_n$ (21.17). Thus E' is spanned by $f_1, \ldots, f_n$; it follows from (44.6) that E is also finite-dimensional.

Conversely, suppose E is finite-dimensional. Since E is a separated TVS for both τ and $\sigma(E, E')$, it follows from (23.3) that $\tau = \sigma(E, E')$. ∎

On E' there are three natural topologies: the norm topology, the weak topology $\sigma(E', E'')$, and the weak* topology $\sigma(E', E)$. Before discussing the relationships between them, it is convenient to reformulate the weak* topology:

(44.9) Definition. As above, E_0 denotes the image of E in E'' under the canonical embedding $x \mapsto x''$. The canonical bilinear form

$$(f, x'') \mapsto x''(f) = f(x) \qquad (f \in E', \, x'' \in E_0)$$

effects a duality (in the sense of (38.1)) between E' and E_0 (by the same argument that shows that E and E' are in duality via the bilinear form $(x, f) \mapsto f(x)$).

(44.10) Theorem. *With notation as in* (44.9),

   (i) $\sigma(E', E_0) = \sigma(E', E)$;

   (ii) *the vector space isomorphism $x \mapsto x''$ of E onto E_0 is bicontinuous for the topologies $\sigma(E, E')$ and $\sigma(E_0, E')$;*

   (iii) *the relative topology on E_0 induced by $\sigma(E'', E')$ (i.e., by the weak* topology of E'') is $\sigma(E_0, E')$;*

   (iv) *the canonical mapping $x \mapsto x''$ is a continuous mapping of E into E'', for the topologies $\sigma(E, E')$ and $\sigma(E'', E')$, respectively.*

Proof. (i) If $\{x_1, \ldots, x_n\}$ is any finite subset of E then, since $x_k''(f) = f(x_k)$ for all $f \in E'$,

$$\{f : |x_k''(f)| \leq 1 \text{ for } k = 1, \ldots, n\} = \{f : |f(x_k)| \leq 1 \text{ for } k = 1, \ldots, n\};$$

the left and right sides are basic neighborhoods of θ for $\sigma(E', E_0)$ and $\sigma(E', E)$, respectively.

   (ii) For convenience, let us write $T: E \to E_0$, $Tx = x''$. If $\{f_1, \ldots, f_n\}$ is any finite subset of E', then

$$U = \{x : |f_k(x)| \leq 1 \quad \text{for } k = 1, \ldots, n\}$$

is a basic neighborhood of θ for $\sigma(E, E')$, whereas

$$V = \{x'' : |x''(f_k)| \leq 1 \quad \text{for } k = 1, \ldots, n\}$$

is a basic neighborhood of θ for $\sigma(E_0, E')$. Since $x''(f_k) = f_k(x)$, clearly $T(U) = V$, $U = T^{-1}(V)$; this shows that T and T^{-1} are continuous at θ for the indicated topologies, therefore T is a homeomorphism.

   (iii) If $\{f_1, \ldots, f_n\}$ is any finite subset of E', then

$$W = \{\varphi \in E'' : |\varphi(f_k)| \leq 1 \quad \text{for } k = 1, \ldots, n\}$$

is a basic neighborhood of θ for $\sigma(E'', E')$; thus, a basic neighborhood of θ for the relative topology on E_0 induced by $\sigma(E'', E')$ is

$$W \cap E_0 = \{x'' \in E_0 : |x''(f_k)| \leq 1 \quad \text{for } k = 1, \ldots, n\},$$

which is also a basic neighborhood of θ for $\sigma(E_0, E')$.

(iv) This is immediate from (ii) and (iii). ∎

The comparison of the various topologies on E' is as follows:

(44.11) Theorem. *Let τ' be the norm topology on E'. Then:*
(i) $\tau' \supset \sigma(E', E'') \supset \sigma(E', E)$;
(ii) $\sigma(E', E'') = \sigma(E', E)$ *if and only if E is reflexive;*
(iii) $\tau' = \sigma(E', E'')$ *if and only if E is finite-dimensional.*

Proof. (i) $\tau' \supset \sigma(E', E'')$ by the first assertion of (44.8). We have $\sigma(E', E) = \sigma(E', E_0)$ by (44.10), and $\sigma(E', E_0) \subset \sigma(E', E'')$ because $E_0 \subset E''$ (cf. the proof of (38.9)), therefore $\sigma(E', E) \subset \sigma(E', E'')$.

(ii) E is reflexive iff $E_0 = E''$ iff $\sigma(E', E_0) = \sigma(E', E'')$ (see (38.9)); but $\sigma(E', E_0) = \sigma(E', E)$.

(iii) Applying the second assertion of (44.8) to E' in place of E, we have $\tau' = \sigma(E', E'')$ iff E' is finite-dimensional; but E' is finite-dimensional iff E is finite-dimensional (44.6). In this case all topologies in sight on E' coincide.

{Note that if $\tau' = \sigma(E', E)$ then $\tau' = \sigma(E', E'')$ by (i), therefore E is finite-dimensional by (iii).} ∎

The following result, known as the Alaoglu-Bourbaki theorem, brings the power of compactness to bear in the theory of duality. It has many key applications, several of which will be taken up in the sequel. {Notably, in connection with reflexive Banach spaces in the next section (45.4); the Gel'fand topology on the maximal ideal space of a commutative Banach algebra (52.10); and, in tandem with the Kreĭn-Mil'man theorem, the sufficiency of the irreducible representations of certain ∗-algebras (67.27).}

(44.12) Theorem. *The closed unit ball of E' is weak* compact; that is, the set*

$$B = \{f \in E' : \|f\| \leq 1\}$$

is compact for the topology $\sigma(E', E)$.

Proof. If $f \in B$ then $|f(x)| \leq \|x\|$ for all $x \in E$. Thus, writing

$$D_x = \{\lambda \in \mathbb{K} : |\lambda| \leq \|x\|\}$$

(a closed disc or a closed interval, according as $\mathbb{K} = \mathbb{C}$ or $\mathbb{K} = \mathbb{R}$), we have $f(x) \in D_x$ for all $x \in E$. This attracts attention to the space

$$D = \prod_{x \in E} D_x,$$

equipped with the product topology; by Tihonov's theorem, D is compact. We write $\pi_x : D \to D_x$ for the canonical projection of D onto D_x.

Equip B with the relative topology induced by $\sigma(E', E)$. The theorem will be proved by showing that B is homeomorphic with a certain closed subset C of D.

Define a mapping $T: B \to D$ as follows. If $f \in B$ then $f(x) \in D_x$ for all $x \in E$, thus we may define $Tf = (f(x))_{x \in E}$. The element $Tf = (f(x))_{x \in E}$ of D has the special property that the xth coordinate is a linear function of the index x; thus, writing C for the set of all $(\lambda_x)_{x \in E}$ in D such that

$$\lambda_{y+z} = \lambda_y + \lambda_z, \qquad \lambda_{\mu y} = \mu \lambda_y$$

for all $y, z \in E$ and $\mu \in \mathbb{K}$, we have $T(B) \subset C$. In fact,

$$T(B) = C.$$

Indeed, if $(\lambda_x)_{x \in E} \in C$ it is clear that the function $f: E \to \mathbb{K}$ defined by $f(x) = \lambda_x$ is a linear form on E; moreover, $|f(x)| = |\lambda_x| \le \|x\|$ shows that $f \in B$, and $Tf = (f(x))_{x \in E} = (\lambda_x)_{x \in E}$.

The proof will be carried out with a series of three observations: (i) T is injective; (ii) C is a closed subset of D (and is therefore compact for the relative topology); (iii) T maps B bicontinuously onto $T(B) = C$.

(i) If $f, g \in B$ and $Tf = Tg$, that is, if $(f(x))_{x \in E} = (g(x))_{x \in E}$, then $f(x) = g(x)$ for all $x \in E$, i.e., $f = g$.

(ii) Fix $y, z \in E$ and consider the mapping $\varphi: D \to \mathbb{K}$ defined by

$$\varphi((\lambda_x)_{x \in E}) = \lambda_{y+z} - \lambda_y - \lambda_z.$$

Thus, writing $u = (\lambda_x)_{x \in E}$, we have

$$\varphi(u) = \pi_{y+z}(u) - \pi_y(u) - \pi_z(u) \qquad (u \in D).$$

Obviously φ is continuous, therefore the set

$$\varphi^{-1}(\{0\}) = \{(\lambda_x)_{x \in E} \in D : \lambda_{y+z} = \lambda_y + \lambda_z\}$$

is closed in D; denote it by $C_{y,z}$. Similarly, for fixed $w \in E$ and $\mu \in \mathbb{K}$, the set

$$\{(\lambda_x)_{x \in E} \in D : \lambda_{\mu w} = \mu \lambda_w\}$$

is closed in D; denote it by $C_{\mu,w}$. Evidently

$$C = \left(\bigcap C_{y,z} \right) \cap \left(\bigcap C_{\mu,w} \right),$$

where y, z, w vary over E and μ varies over $\mathbb{K}$, thus C is closed in D.

(iii) In view of (i), T maps B bijectively onto $T(B) = C$. Fix $f_0 \in B$. A subbasic open neighborhood of f_0 for the relative weak* topology on B has the form

$$V = \{f \in B : |f(z) - f_0(z)| < \varepsilon\},$$

where $z \in E$ and $\varepsilon > 0$. Then

$$\begin{aligned} T(V) &= \{(f(x))_{x \in E} : f \in V\} \\ &= \{(f(x))_{x \in E} : f \in B \ \& \ |f(z) - f_0(z)| < \varepsilon\} \\ &= \{(f(x))_{x \in E} : f \in B \ \& \ |\pi_z(Tf) - \pi_z(Tf_0)| < \varepsilon\}, \end{aligned}$$

which is clearly a subbasic open neighborhood of Tf_0 for the relative topology induced on $C = T(B)$ by the product topology on D. ∎

Exercises

(44.13) If M is a linear subspace of the normed space E and if y_0 is a vector in E such that

$$\inf \{\|y_0 + z\| : z \in M\} = d > 0,$$

then there exists $f \in E'$ such that $f(y_0) = 1$, $f = 0$ on M, and $\|f\| = 1/d$.

(44.14) If $1 < p < \infty$ then, in the notation of Section 39, the Banach spaces L^p are reflexive. The space l^1 of (39.11) is not reflexive.

(44.15) If E is any normed space, there exist a compact space $\mathscr{X}$ and a linear isometry $E \to \mathscr{C}(\mathscr{X})$.

(44.16) (Kreĭn-Šmul'jan) Let E be a Banach space and let K be a convex subset of E'. The following conditions on K are equivalent: (a) K is weak* closed; (b) for every $r > 0$, the set $K \cap B_r$ is weak* closed, where

$$B_r = \{f \in E' : \|f\| \le r\}.$$

45. The bidual of a normed space; reflexive Banach spaces.

Let E be a normed space over $\mathbb{K}$ and let E_0 be the image of E in E'' under the canonical embedding $x \mapsto x''$ (40.14). We show in this section that E is a reflexive Banach space (i.e., $E_0 = E''$) if and only if its closed unit ball is compact for the weak topology $\sigma(E, E')$. The proof is based on a preliminary theorem concerning the relation of an arbitrary normed space to its bidual. We motivate the preliminary theorem with two remarks.

(45.1) Item. *If M is a dense linear subspace of E, then the ball $\{y \in M : \|y\| \le 1\}$ is dense in the ball $\{x \in E : \|x\| \le 1\}$. More precisely, if $x \in E$ then there exists a sequence $y_n \in M$ such that $\|y_n - x\| \to 0$ and $\|y_n\| = \|x\|$ for all n.*

Proof. The statement of this result refers to the norm topology. {We remark that a linear subspace of E is dense for the norm topology if and only if it is dense for the weak topology $\sigma(E, E')$ (38.16).}

It is sufficient to prove the second assertion. Suppose $x \in E$. If $x = \theta$ take $y_n = \theta$ for all n. Assume $x \ne \theta$. By hypothesis, there exists a sequence $x_n \in M$ such that $x_n \to x$, i.e., $\|x_n - x\| \to 0$; then $\|x_n\| \to \|x\| > 0$ (16.4), therefore (note that at most finitely many x_n can be θ) $\|x_n\|^{-1}x_n \to \|x\|^{-1}x$, $\|x\|\|x_n\|^{-1}x_n \to x$, thus the sequence $y_n = \|x\|\|x_n\|^{-1}x_n$ meets the requirements. ∎

(45.2) Item. *E_0 is dense in E'' for the weak* topology $\sigma(E'', E')$.*

Proof. Let T be the weak* closure of E_0 in E''; by (38.14), $T = (^\perp T)^\perp$. To show that $T = E''$ it will suffice to show that $^\perp T = \{0\}$. Suppose $f \in {}^\perp T$, that is, $\varphi(f) = 0$ for all $\varphi \in T$; in particular, $x''(f) = 0$ for all $x \in E$, i.e., $f(x) = 0$ for all $x \in E$, i.e., $f = 0$. ∎

In general, the norm closure of E_0 in E'' falls short of E'' (e.g., if E is a nonreflexive Banach space then E_0 is a closed, proper linear subspace of E''), thus (45.1) is generally irrelevant for the subspace E_0 of E''. However, in view of (45.2) it is natural to ask: is the closed unit ball of E_0 weak* dense in the closed unit ball of E''? The answer is affirmative:

(45.3) Theorem. *The closed unit ball of E_0 is weak* dense in the closed unit ball of E''; that is, the closure of $\{x'' : x \in E, \|x\| \leq 1\}$ for the topology $\sigma(E'', E')$ coincides with $\{\varphi \in E'' : \|\varphi\| \leq 1\}$.*

Proof. Let $A = \{x'' : x \in E, \|x\| \leq 1\}$, the image of the closed unit ball of E under the canonical embedding of E in E''; let $B = \{\varphi \in E'' : \|\varphi\| \leq 1\}$. Obviously $B \supset A$; it is to be shown that B is the weak* closure of A.

First, we note that B is weak* closed. This is immediate from the weak* compactness of B (44.12), but we prefer to give a more elementary direct proof. Suppose φ_0 is in the weak* closure of B. For each fixed $f \in E'$ with $\|f\| \leq 1$, the mapping $\varphi \mapsto \varphi(f)$ $(\varphi \in E'')$ is weak* continuous and maps B into the closed set $D = \{\lambda : |\lambda| \leq 1\}$, therefore it maps the weak* closure of B into D, and in particular $|\varphi_0(f)| \leq 1$; varying f, $\|\varphi_0\| \leq 1$, i.e., $\varphi_0 \in B$.

Write C for the weak* closure of A. Since B is weak* closed and contains A, we have $B \supset C$; it is to be shown that $B = C$. {We write out the proof for $\mathbb{K} = \mathbb{C}$; for the case that $\mathbb{K} = \mathbb{R}$, one need only suppress 'Re' in the following argument.}

Assuming $\varphi \in E''$, $\varphi \notin C$, it will suffice to show that $\|\varphi\| > 1$. Since the weak* continuous linear forms on E'' are given by elements of E' (38.7), it results from (34.4) that there exist $f \in E'$ and a real number β such that (i) $\operatorname{Re} \psi(f) < \beta$ for all $\psi \in C$, and (ii) $\operatorname{Re} \varphi(f) > \beta$. Since $A \subset C$, we infer from (i) that $\operatorname{Re} f(x) < \beta$ whenever $\|x\| \leq 1$. It follows that $\|f\| \leq \beta$. {Proof: For each x one can write $|f(x)| = f(\mu x)$ for suitable $|\mu| = 1$.} Then (ii) yields $\beta < |\varphi(f)| \leq \|\varphi\| \|f\| \leq \|\varphi\|\beta$, thus $\|\varphi\| > 1$. ∎

The following characterization of reflexive spaces is an easy consequence of the above theorem and the Alaoglu-Bourbaki theorem:

(45.4) Theorem. *Let E be a normed space and let $S = \{x \in E : \|x\| \leq 1\}$ be the closed unit ball of E. In order that E be a reflexive Banach space, it is necessary and sufficient that S be weakly compact, i.e., compact for the topology $\sigma(E, E')$.*

Proof. Write $T: E \to E''$ for the canonical embedding $Tx = x''$, and let $B = \{\varphi \in E'' : \|\varphi\| \leq 1\}$.

Suppose first that E is reflexive, i.e., $T(E) = E''$; since T is isometric, $T(S) = B$. Equip E with the weak topology $\sigma(E, E')$, and E'' with the weak* topology $\sigma(E'', E')$; by (ii) of (44.10), T is a homeomorphism for these topologies. Since B is weak* compact (44.12), it follows that $S = T^{-1}(B)$ is weakly compact.

Conversely, suppose S is weakly compact. Equip E and E'' with the weak and weak* topologies, respectively; by (iv) of (44.10), T is a continuous mapping of E into E'' for these topologies. By hypothesis, S is weakly compact, therefore $T(S)$ is weak* compact and hence is weak* closed in E''. Thus $T(S)$ coincides with its weak* closure; in view of (45.3), this means that $T(S) = B$, from which it is immediate that $T(E) = E''$. ∎

Exercises

(45.5) Let E be a normed space and let M be a closed linear subspace of E. In order that E be a reflexive Banach space, it is necessary and sufficient that M and E/M be reflexive Banach spaces.

(45.6) If M is a linear subspace of a reflexive Banach space E, the following conditions are equivalent: (a) M is closed; (b) M is weakly closed; (c) $M = {}^\perp(M^\perp)$; (d) if M_0 is the image of M under the canonical mapping $x \mapsto x''$ (of E onto E''), then $M_0 = (M^\perp)^\perp$.

(45.7) A Banach space E is reflexive if and only if E' is reflexive.

(45.8) The following conditions on a normed space E are equivalent: (a) E_0 is norm-dense in E''; (b) the completion of E is reflexive; (c) E' is reflexive.

*(45.9) (D. P. Mil'man)   Every uniformly convex Banach space is reflexive.

Chapter 5

Category

The notion of category is one of the major themes of functional analysis, on the order of magnitude of such themes as linearity, completeness, compactness, and convexity. Historically, it is most closely allied to the notion of completeness, having entered functional analysis via complete metric spaces (in connection, specifically, with the classical examples of metrizable, complete TVS); by the time of Banach's book (1932), its key role in functional analysis was securely established.

The classical terminology, according to which a space (or a subset of a space) is either 'of the first category' or 'of the second category' appears to be on the way out, the word 'category' itself having been conscripted for higher service. We follow here the widely used terminology of Bourbaki, noting parenthetically the earlier nomenclature.

Nowadays the concepts are formulated in general topological spaces, and are meaningful in large classes of special topological spaces; in the following section we expose this material—a brief interlude in general topology, slightly more general than is needed in this book—for the reader's convenience (and because it exceeds the topological prerequisites set forth in the Preface).

46. Baire spaces. Throughout this section, X denotes a topological space.

(46.1) Definition. A subset A of X is said to be *rare* (in X) if its closure has no interior points, i.e., int $(\overline{A}) = \varnothing$. {The classical term: '$A$ is nowhere dense in X.'} A subset B of X is said to be *meager* (in X) if it is the union of a countable family of rare subsets of X, i.e., $B = \bigcup_1^\infty A_n$ with A_n rare for all n ($n = 1, 2, 3, \ldots$); otherwise, B is said to be a *nonmeager* subset of X. The definitions apply, in particular, to X regarded as a subset of itself (though, of course, a nonempty topological space cannot be rare in itself). {The classical term for X meager: 'X is a space of the first category.' The classical term for X nonmeager: 'X is a space of the second category.'}

(46.2) It is important to emphasize that the definitions (46.1) are *relative to X*. For example, a nonempty rare subset A of X is not a rare subset of itself in the relative topology. The same remark applies to the notion of meager. {To put it another way, one must distinguish between the notion of 'meager subset' (in the sense of (46.1)) and 'meager subspace' (i.e., a subset which is a meager subset of itself in the relative topology).}

The following theorem embodies the key classical result:

(46.3) Theorem. (Baire category theorem) *If (X, d) is a complete metric space and A is a meager subset of X, then $X - A$ is dense in X.*

Proof. In topological terms, X is a metrizable topological space whose topology can be defined by a complete metric d.

By hypothesis, $A = \bigcup_1^\infty A_n$, where int $(\overline{A}_n) = \varnothing$ for all n. Replacing A by the possibly larger set $B = \bigcup_1^\infty \overline{A}_n$ (which is also meager), it will suffice to show that $X - B$ is dense; thus, changing notation, we can suppose that the A_n are closed sets, int $(A_n) = \varnothing$. Note that if U is any nonempty open set, then the open sets $U \cap (X - A_n)$ are nonempty; indeed, $U \cap (X - A_n) = \varnothing$ would imply $U \subset A_n$, contrary to int $(A_n) = \varnothing$.

Let V be any nonempty open set in X; it is to be shown that $V \cap (X - A) \neq \varnothing$, i.e., $\bigcap_1^\infty V \cap (X - A_n) \neq \varnothing$. To this end, we shall construct a sequence U_n of nonempty open sets such that (i) $\overline{U}_{n+1} \subset U_n \cap (X - A_n)$, and (ii) diam $(\overline{U}_n) < 1/n$.

Let U_1 be any nonempty open set with $\overline{U}_1 \subset V$ and diam $(\overline{U}_1) < 1$ (e.g., U_1 can be a suitable open ball). Since $U_1 \cap (X - A_1)$ is a nonempty open set, there exists a nonempty open set U_2 with $\overline{U}_2 \subset U_1 \cap (X - A_1)$ and diam $(\overline{U}_2) < \frac{1}{2}$. Suppose, inductively, that $U_1, U_2, \ldots, U_n$ have already been constructed with the desired properties. Since $U_n \cap (X - A_n)$ is a nonempty open set, there exists a nonempty open set U_{n+1} with $\overline{U}_{n+1} \subset U_n \cap (X - A_n)$ and diam $(\overline{U}_{n+1}) < 1/(n + 1)$.

Since $U_n \supset \overline{U}_{n+1}$, and therefore $\overline{U}_n \supset \overline{U}_{n+1}$, and since diam $(\overline{U}_n) \to 0$, it follows from a well-known property of complete metric spaces that $\bigcap_1^\infty \overline{U}_n = \{a\}$ for suitable $a \in X$. Then $a \in \overline{U}_1 \subset V$ and $a \in \overline{U}_{n+1} \subset X - A_n$ for all n, hence $a \in V \cap (X - A_n)$ for all n. This completes the proof that $X - A$ is dense in X. ∎

The condition in (46.3) can be reformulated in various ways, and, in practice, it is useful to do so:

(46.4) Theorem. *The following conditions on a topological space X are equivalent:*

(a) *if A is any meager subset of X then $X - A$ is dense in X;*

(b) *if U is a nonempty open subset of X then U is nonmeager;*

(c) *if $A = \bigcup_1^\infty A_n$, where the A_n are closed sets with int $(A_n) = \varnothing$ for all n, then int $(A) = \varnothing$;*

(d) *if $A = \bigcap_1^\infty U_n$, where the U_n are open sets with $\overline{U}_n = X$, then $\overline{A} = X$.*

Proof. (a) implies (b): If U is a nonempty open set in X, then $X - U$ is a closed, proper subset of X, and is therefore not dense in X; in view of (a), U is nonmeager.

(b) implies (c): Obviously A is meager, and int $(A) \subset A$, therefore int (A) is also meager; in view of (b), int $(A) = \varnothing$.

(c) implies (d): Suppose $A = \bigcap_1^\infty U_n$, where the U_n are dense open sets in X. Since $X - A = \bigcup_1^\infty (X - U_n)$, where $X - U_n$ is closed and int $(X - U_n) = X - \overline{U}_n = \varnothing$, we have $X - \overline{A} = $ int $(X - A) = \varnothing$ by (c), i.e., $\overline{A} = X$.

(d) implies (a): Let A be a meager subset of X; it is to be shown that $X - A$ is dense. As argued in the proof of (46.3), we can suppose $A = \bigcup_1^\infty A_n$, where the A_n are closed and int $(A_n) = \varnothing$. Then $X - A = \bigcap_1^\infty (X - A_n)$, where the $X - A_n$ are dense open sets, therefore $X - A$ is dense by (d). ∎

(46.5) Definition. A *Baire space* is a topological space satisfying one (hence all) of the conditions in (46.4).

(46.6) The Baire category theorem (46.3) can now be put as follows: *Every complete metric space is a Baire space.*

For our purposes, the key property of Baire spaces is the following:

(46.7) Theorem. *If X is a Baire space and $(f_\iota)_{\iota \in I}$ is a pointwise bounded family of continuous real-valued functions on X, then the family is uniformly bounded on some nonempty open subset of X.*

Proof. By assumption, for each $x \in X$ there exists a constant M_x such that $|f_\iota(x)| \leq M_x$ for all $\iota \in I$; we seek a nonempty open set U and a constant M such that $|f_\iota(x)| \leq M$ for all $x \in U$ and $\iota \in I$.

For each $\iota \in I$ and each positive integer n, the set

$$A_{\iota n} = \{x \in X : |f_\iota(x)| \leq n\}$$

is closed, by the continuity of f_ι; therefore, the set A_n, defined for each n by

$$A_n = \bigcap_{\iota \in I} A_{\iota n} = \{x : |f_\iota(x)| \leq n \quad \text{for all } \iota \in I\},$$

is also closed. The pointwise boundedness assumption means that $X = \bigcup_1^\infty A_n$; since X is a Baire space, int $(A_m) \neq \varnothing$ for some index m, by (c) of (46.4). Then $U = \text{int}(A_m)$ and $M = m$ meet the requirements. ∎

(46.8) The value space of the f_ι in (46.7) is not critical; for instance, 'complex-valued' works just as well as 'real-valued.' More generally, the theorem is true for continuous mappings $f_\iota : X \to E_\iota$, where, for each ι, E_ι is a normed space; one simply replaces the given functions $x \mapsto f_\iota(x)$ by the real-valued functions $x \mapsto \|f_\iota(x)\|$, which are also continuous (16.4), and to which (46.7) is applicable. {Still more generally, the E_ι can be valued groups, in the sense of (8.2).}

Exercises

(46.9) (i) If X is a Baire space and U is a nonempty open subset of X, then U is a Baire space in the relative topology. (ii) Every compact space is a Baire space; better yet, (iii) every locally compact space is a Baire space.

(46.10) If X is a Baire space, $(f_\iota)_{\iota \in I}$ is a pointwise bounded family of continuous, real-valued functions on X, and U is any nonempty open set in X, then there exists a nonempty open set V in X such that $V \subset U$ and the f_ι are uniformly bounded on V.

(46.11) If X is a Baire space and A is a meager subset of X, then $X - A$ is a Baire space in the relative topology.

47. Uniform boundedness principle.

The main result of this section is an application of the Baire category theorem to Banach spaces; it is also known as the *Banach-Steinhaus theorem*:

(47.1) Theorem. (Uniform boundedness principle) *Let E be a Banach space and $(E_\iota)_{\iota \in I}$ a family of normed spaces over $\mathbb{K}$, and let $T_\iota \colon E \to E_\iota$ ($\iota \in I$) be a family of continuous linear mappings. If, for each $x \in E$, the family $(\|T_\iota x\|)_{\iota \in I}$ is bounded, then $\|T_\iota\|$ is bounded.*

Proof. By assumption, the family $(T_\iota)_{\iota \in I}$ is pointwise bounded; we wish to show that the family is uniformly bounded on the closed unit ball of E.

For each $\iota \in I$ define $f_\iota \colon E \to \mathbb{R}$ by the formula $f_\iota(x) = \|T_\iota x\|$; f_ι is continuous, by the continuity of the norm in E_ι (16.4). By (46.7) there exist a nonempty open set U in E and a constant M, such that $f_\iota(x) \leq M$ for all $\iota \in I$ and $x \in U$. We can suppose U is an open ball, say $U = \{x : \|x - a\| < r\}$ for suitable $a \in E$ and $r > 0$. Then

(*) $\|x - a\| < r$ implies $\|T_\iota x\| \leq M$ for all $\iota \in I$.

The proof will be concluded by showing that $\|T_\iota\| \leq 2M/r$ for all $\iota \in I$. Fix $\iota \in I$. Given any $y \in E$, $\|y\| < 1$, it is to be shown that $\|T_\iota y\| \leq 2M/r$. Set $x = a + ry$; clearly $x \in U$, therefore $\|T_\iota x\| \leq M$ by (*), i.e., $\|T_\iota a + r(T_\iota y)\| \leq M$. On the other hand $a \in U$, therefore $\|T_\iota a\| \leq M$. Then

$$\|r(T_\iota y)\| \leq \|r(T_\iota y) + T_\iota a\| + \|-(T_\iota a)\| \leq M + M,$$

thus $\|T_\iota y\| \leq 2M/r$. ∎

(47.2) Corollary. *If E is a Banach space over $\mathbb{K}$ and $(f_\iota)_{\iota \in I}$ is a pointwise bounded family of continuous linear forms on E, then $\|f_\iota\|$ is bounded.*

Proof. Put $E_\iota = \mathbb{K}$ for all $\iota \in I$ and quote (47.1). ∎

(47.3) Corollary. *If E is a normed space over $\mathbb{K}$ and $(x_\iota)_{\iota \in I}$ is a family of vectors in E such that, for each $f \in E'$, the family $(f(x_\iota))_{\iota \in I}$ is bounded, then $\|x_\iota\|$ is bounded.*

Proof. {So to speak, every 'weakly bounded' set in a normed space is bounded. Incidentally, the converse of this proposition is trivial.}

Let $x \mapsto x''$ be the canonical embedding of E in E'' (40.14). For each $f \in E'$, the family $(x_\iota''(f))_{\iota \in I} = (f(x_\iota))_{\iota \in I}$ is bounded by hypothesis; thus, (x_ι'') is a pointwise bounded family of continuous linear forms on E'. Since E' is a Banach space (40.9), it follows from (47.2) that $\|x_\iota''\|$ is bounded; but $\|x_\iota''\| = \|x_\iota\|$ (40.13). ∎

We conclude the section with two applications to Hilbert space.

(47.4) Corollary. *If H is a Hilbert space and $(x_\iota)_{\iota \in I}$ is a family of vectors in H such that, for each $y \in H$, the family $((x_\iota | y))_{\iota \in I}$ is bounded, then $\|x_\iota\|$ is bounded.*

Proof. This is immediate from (47.3) and the Fréchet-Riesz theorem (42.12). ∎

(47.5) Corollary. *If H and K are Hilbert spaces and if $T \colon H \to K$ and $S \colon K \to H$ are mappings such that*

$$(Tx | y) = (x | Sy)$$

for all $x \in H$ and $y \in K$, then T and S are continuous linear mappings, $S = T^*$.

Proof. In view of (43.1), it will suffice to show that T is linear and continuous.

If $x_1, x_2 \in H$ then, for all $y \in K$,

$$(T(x_1 + x_2)|y) = (x_1 + x_2|Sy) = (x_1|Sy) + (x_2|Sy)$$
$$= (Tx_1|y) + (Tx_2|y) = (Tx_1 + Tx_2|y),$$

thus $T(x_1 + x_2) = Tx_1 + Tx_2$. Similarly $T(\lambda x) = \lambda(Tx)$. Thus T is linear.

It remains to show that the set of vectors $\{Tx : \|x\| \le 1\}$ is bounded. Let $y \in K$; by (47.4) it is sufficient to show that the set $\{(Tx|y) : \|x\| \le 1\}$ is bounded. Indeed, $\|x\| \le 1$ implies $|(Tx|y)| = |(x|Sy)| \le \|x\| \|Sy\| \le \|Sy\|$. ∎

Exercises

(47.6) The completeness of E is essential in (47.1) and (47.2); i.e., 'Banach space' cannot be weakened to 'normed space.'

(47.7) Let E and F be Banach spaces over $\mathbb{K}$, and suppose $T: E \to F$ and $S: F' \to E'$ are mappings such that $g(Tx) = (Sg)(x)$ for all $x \in E$ and $g \in F'$. Then S and T are continuous linear mappings, and $S = T'$ in the sense of (40.18).

(47.8) Let E and F be normed spaces over $\mathbb{K}$, let $T_n: E \to F$ $(n = 1, 2, 3, \ldots)$ be a sequence of continuous linear mappings such that $\|T_n\|$ is bounded, and let

$$M = \{x \in E : T_n x \text{ is Cauchy in } F\}.$$

Then M is a closed linear subspace of E. If, moreover, F is a Banach space, then the formula

$$Tx = \lim T_n x \qquad (x \in M)$$

defines a continuous linear mapping $T: M \to F$. {One can replace 'sequence' by 'net.'}

48. Banach's open mapping and closed graph theorems.

This section is an application of the Baire category theorem to metrizable, complete TVS. There is no loss of generality in restricting attention to real scalars (see the remarks at the end of the section). The principal result is as follows:

(48.1) Theorem. (Open mapping theorem) *If $T: E \to F$ is a surjective, continuous linear mapping, where E and F are metrizable complete TVS, then T is open.*

The assertion is that if U is an open set in E then $T(U)$ is open in F. In view of the additivity of T, it is sufficient to prove that if V is any neighborhood of θ in E then $T(V)$ is a neighborhood of $T\theta = \theta$ in F, i.e., there exists an open neighborhood W of θ in F such that $W \subset T(V)$. To show that a proposed W satisfies this inclusion amounts to solving a system of linear equations: for each $y \in W$ one seeks $x \in V$ with $Tx = y$. A category argument shows that, in a sense, approximate solutions exist:

(48.2) Lemma. *Suppose $T: E \to F$ is a surjective, continuous linear mapping, where E and F are TVS, and F is a Baire space. If V is any neighborhood of θ in E, then $\overline{T(V)}$ is a neighborhood of θ in F.*

A second lemma shows, in effect, how to pass from approximate solutions to solutions (by a method of successive approximations):

(48.3) Lemma. *Suppose $T: E \to F$ is a continuous mapping, where E is a complete metric space and F is a metric space. Assume that for each $r > 0$ there exists $\rho > 0$ such that*

(*) $$B_\rho(Tx) \subset \overline{T(B_r(x))} \qquad \text{for all } x \in E$$

(here $B_r(x)$ denotes the closed ball of radius r and center x). Then:

(i) *If r and ρ satisfy* (*), *then for each $a > r$ one has*

(**) $$B_\rho(Tx) \subset T(B_a(x)) \qquad \text{for all } x \in E.$$

(ii) *T is an open mapping.*

We turn to the proofs.

Proof of Lemma (48.3). (i) Suppose r and ρ satisfy (*), and let $a > r$. Write $a = \sum_1^\infty r_n$ with $r_n > 0$ for all n, and $r_1 = r$. {E.g., let $r_{n+1} = 2^{-n}(a - r)$ for $n = 1, 2, 3, \ldots$.} For each n there exists, by hypothesis, a $\rho_n > 0$ such that

(*)$_n$ $$B_{\rho_n}(Tx) \subset \overline{T(B_{r_n}(x))} \qquad \text{for all } x \in E.$$

We can suppose $\rho_1 = \rho$ and $\rho_n \to 0$.

Given any $x_0 \in E$ and $y \in B_\rho(Tx_0)$, the problem is to show that $y \in T(B_a(x_0))$; thus, we seek $x \in B_a(x_0)$ with $y = Tx$. The desired point x will be obtained as the limit of a suitable sequence x_n.

We construct a sequence $x_1, x_2, x_3, \ldots$ in E such that

(†) $$x_n \in B_{r_n}(x_{n-1}) \quad \text{and} \quad Tx_n \in B_{\rho_{n+1}}(y)$$

for $n = 1, 2, 3, \ldots$. The construction is inductive. We have $y \in B_\rho(Tx_0) = B_{\rho_1}(Tx_0)$; by (*)$_1$ with $x = x_0$, y can be approximated, as nearly as we like, by elements Tx with $x \in B_{r_1}(x_0)$; thus we may choose $x_1 \in B_{r_1}(x_0)$ with $Tx_1 \in B_{\rho_2}(y)$. Then $y \in B_{\rho_2}(Tx_1)$; by (*)$_2$ with $x = x_1$, we may choose $x_2 \in B_{r_2}(x_1)$ with $Tx_2 \in B_{\rho_3}(y)$, etc.

Say d is the metric on E; thus $x_n \in B_{r_n}(x_{n-1})$ means that $d(x_n, x_{n-1}) \le r_n$. Since the r_n are summable, it follows that the sequence x_n is Cauchy, therefore $x_n \to x$ for suitable $x \in E$. Moreover,

$$d(x_n, x_0) \le \sum_{k=1}^n d(x_k, x_{k-1}) \le \sum_1^n r_k;$$

letting $n \to \infty$ we have $d(x, x_0) \le a$ by the continuity of d, thus $x \in B_a(x_0)$.

It remains to show that $Tx = y$. Indeed, since $\rho_n \to 0$ it is clear from (†) that $Tx_n \to y$, whereas $Tx_n \to Tx$ by the continuity of T.

(ii) Let $x_0 \in E$, $a > 0$; thus $B_a(x_0)$ is a basic neighborhood of x_0. It is to be shown that $T(B_a(x_0))$ is a neighborhood of Tx_0. Choose any r with $0 < r < a$. By hypothesis, there exists $\rho > 0$ satisfying (*). Since $a > r$, we know from (i) that (**) holds; thus it is clear that $T(B_a(x))$ is a neighborhood of Tx for each x in E. ∎

Proof of Lemma (48.2). Let V be a neighborhood of θ in E. Choose a balanced neighborhood W of θ such that $W + W \subset V$, and fix a scalar λ with $|\lambda| > 1$ (e.g., $\lambda = 2$). We show that

$$E = \bigcup_{n=1}^{\infty} \lambda^n W.$$

Let $x \in E$. Then $\mu x \in W$ for a suitable nonzero scalar μ (17.11). Choose n so that $|\lambda^{-n}\mu^{-1}| \leq 1$; since W is balanced,

$$\lambda^{-n}x = (\lambda^{-n}\mu^{-1})(\mu x) \in \lambda^{-n}\mu^{-1}W \subset W,$$

thus $x \in \lambda^n W$.

Since T is surjective and linear, we have

$$F = T(E) = \bigcup_1^{\infty} T(\lambda^n W) = \bigcup_1^{\infty} \lambda^n T(W);$$

since F is a Baire space, there exists an index m such that $\overline{\lambda^m T(W)}$ has non-empty interior. It follows that $\overline{T(W)}$ has nonempty interior (because $y \mapsto \lambda^m y$ is a homeomorphism of F). Let $A = \text{int} \,(\overline{T(W)})$. Since $-T(W) = T(-W) = T(W)$, it follows that $-A = A$. Then, invoking the continuity of addition at the second inclusion sign, we have

$$\theta \in A + A \subset \overline{T(W)} + \overline{T(W)} \subset \overline{T(W) + T(W)} = \overline{T(W + W)} \subset \overline{T(V)};$$

since $A + A$ is open (2.10), it follows that $\overline{T(V)}$ is a neighborhood of θ. ∎

Proof of Theorem (48.1). E and F may be metrized with additively invariant metrics, relative to which they are complete metric spaces (7.9). To show that T is open, we verify the hypotheses of Lemma (48.3). Let $r > 0$, and let $B_r(\theta)$ be the closed ball with radius r and center θ. Since F is a Baire space (46.6), it follows from Lemma (48.2) that the closure of $T(B_r(\theta))$ is a neighborhood of θ in F, thus

$$B_\rho(\theta) \subset \overline{T(B_r(\theta))}$$

for suitable $\rho > 0$. It follows easily from the additivity of T and the invariance of the metrics that

$$B_\rho(Tx) \subset \overline{T(B_r(x))}$$

for all $x \in E$. By (48.3), T is open. ∎

The open mapping theorem is applicable, in particular, to Banach spaces (16.10), as are the following corollaries.

(48.4) Corollary. *If E and F are metrizable, complete TVS, and $T: E \to F$ is a continuous vector space isomorphism, then T is bicontinuous.*

(48.5) Corollary. *Let E be a vector space and let τ_1, τ_2 be topologies on E, relative to each of which E is a metrizable, complete TVS. If τ_1 and τ_2 are comparable (i.e., if $\tau_2 \supset \tau_1$ or $\tau_1 \supset \tau_2$) then $\tau_1 = \tau_2$.*

Proof. If $\tau_2 \supset \tau_1$, apply (48.4) to the identity mapping $(E, \tau_2) \to (E, \tau_1)$. ∎

(48.6) Corollary. (Closed graph theorem) *Let* $T: E \to F$ *be a linear mapping, where* E *and* F *are metrizable, complete TVS, and let*

$$G_T = \{(x, Tx) : x \in E\}$$

be the graph of T. *In order that* T *be continuous, it is necessary and sufficient that* G_T *be a closed subset of the product space* $E \times F$.

Proof. Equipped with the product vector space structure and the product topology (11.9), $E \times F$ is also a metrizable, complete TVS. {If E and F are metrized by the additively invariant metrics d, δ, then the metric

$$D((x, y), (x', y')) = d(x, x') + \delta(y, y')$$

induces the product topology on $E \times F$ and is clearly invariant; since d and δ are complete metrics, so is D, thus $E \times F$ is complete (7.8).}

Clearly G_T is a linear subspace of $E \times F$, and the mapping $S: G_T \to E$ defined by $S(x, Tx) = x$ is a vector space isomorphism. It is trivial that S is continuous.

If G_T is closed in $E \times F$, then G_T is also a metrizable, complete TVS (12.1), therefore S is bicontinuous by (48.4); composing with the projection of $E \times F$ onto F, we see that $x \mapsto S^{-1}x = (x, Tx) \mapsto Tx$ is continuous.

If, conversely, T is continuous, then G_T is closed in $E \times F$ by elementary general topology. {No linear or metric structure is needed for this; it is sufficient that E and F be separated topological spaces.} ∎

(48.7) Corollary. *Let* E *be a metrizable, complete TVS, and let* M *and* N *be closed linear subspaces of* E *such that* $E = M + N$ *and* $M \cap N = \{\theta\}$ *Then* E *is the topological direct sum of* M *and* N.

Proof. By the arguments in the proof of (48.6), M, N and $M \times N$ are metrizable, complete TVS. The mapping $(y, z) \mapsto y + z$ is a continuous vector space isomorphism of $M \times N$ onto E; by (48.4) it is bicontinuous, thus E is the topological direct sum of M and N (20.2). ∎

(48.8) Corollary. *Let* $T: E \to F$ *be a continuous linear mapping, where* E *and* F *are metrizable, complete TVS. In order that* T, *regarded as a mapping of* E *onto* $T(E)$, *be open for the relative topology induced on* $T(E)$ *by* F, *it is necessary and sufficient that* $T(E)$ *be closed in* F.

Proof. Sufficiency: If $T(E)$ is closed in F, then it is a metrizable, complete TVS for the relative topology, therefore $T: E \to T(E)$ is open by (48.1).

Necessity: Assume $T: E \to T(E)$ is open. Let M be the null space of T. Since T is continuous, M is a closed linear subspace of E, therefore the quotient TVS E/M is complete and metrizable (12.1). Let $\pi: E \to E/M$ be the canonical mapping.

Let $S: E/M \to T(E)$ be the unique mapping such that $T = S \circ \pi$. Since M is the kernel of T, S is a vector space isomorphism. We assert that S is bicontinuous. If V is an open set in $T(E)$, then $T^{-1}(V)$ is open in E by the continuity of T; since $T^{-1}(V) = \pi^{-1}(S^{-1}(V))$ and since E/M bears the final topology for π, it follows that $S^{-1}(V)$ is open in E/M. On the other hand, if U is an open set in E/M, then $\pi^{-1}(U)$ is open in E; thus

$$S(U) = S(\pi(\pi^{-1}(U))) = T(\pi^{-1}(U))$$

is open in $T(E)$ by the hypothesis on T.

In particular, S is a bicontinuous isomorphism for the additive topological group structures of E/M and $T(E)$; since E/M is complete, it follows easily from (5.12) that $T(E)$ is also complete, therefore $T(E)$ is closed in F by elementary general topology. {More concretely, let d be an additively invariant metric generating the topology of F and therefore of $T(E)$. Since $T(E)$ is complete, it follows from (7.8) that $(T(E), d)$ is a complete metric space, therefore $T(E)$ is closed in F.} ∎

(48.9) The foregoing results are applicable also to complex TVS, since the underlying real TVS (obtained by restricting scalar multiplication) has the same additive topological group structure (and therefore the same uniform structure). Moreover, for the proof of Lemma (48.2) the mapping T need only be assumed real-linear; it follows that the above results hold with T either linear or conjugate-linear.

Exercises

(48.10) Let $T: E \to F$ be a linear mapping, where E and F are metrizable TVS over $\mathbb{K}$, and E is complete. Let d and δ be additively invariant metrics generating the topologies of E and F, and define

$$d'(x, y) = d(x, y) + \delta(Tx, Ty) \qquad (x, y \in E).$$

Then T is continuous for the given topologies if and only if (E, d') is a complete metric space. (If $\mathbb{K} = \mathbb{C}$, the same result holds with T conjugate-linear.)

(48.11) Let E be a vector space over $\mathbb{K}$, let $x \mapsto \|x\|_1$, $x \mapsto \|x\|_2$ be two Banach space norms on E, and let τ_1, τ_2 be the topologies they generate. If $\tau_1 \subset \tau_2$ then $\tau_1 = \tau_2$; so to speak, if two Banach space norms on E are comparable, then they are equivalent (40.27).

(48.12) If E and F are Banach spaces and $T: E \to F$ is a continuous linear mapping, the following conditions on T are equivalent: (a) T is bounded below (40.28); (b) T is injective and $T(E)$ is a closed linear subspace of F.

Chapter 6

Banach Algebras

Normed algebras and Banach algebras were defined in (40.22); some examples have been given in (40.23)—and in Section 0. For the rest of the book, Banach algebra means complex Banach algebra. {Much of what follows is clearly valid for real Banach algebras as well, but complex function-theoretic methods make a crucial appearance in Section 51; from that point on, the study of real Banach algebras becomes a distinct, somewhat more complicated enterprise [cf. **116**, p. 40, Th. 1.7.6]. Although it is true that every real Banach algebra may be realized as, loosely speaking, the real part of a suitable complex Banach algebra [**116**, p. 8, Th. 1.3.2], this fact is by no means a prescription for routinizing the study of the real theory by means of the complex.}

Banach algebras dominate the rest of the book. They made their debut in 1941 in a spectacular paper of I. M. Gel'fand [**52**]. Gel'fand's exposition is still unsurpassed in elegance; the present chapter draws heavily on it. As an application of the theory, Gel'fand's proof (in a paper of the same vintage [**53**]) of a classical theorem of Wiener remains a showpiece for functional-analytic techniques. (An exposition of Gel'fand's proof, begun in Section 0, is completed in Section 63 on the basis of the theory developed in the present chapter.)

49. Preliminaries. For ready reference, we repeat an earlier definition (40.22):

(49.1) Definition. A (complex) *normed algebra* is an associative algebra A over $\mathbb{C}$, whose underlying vector space is a normed space with a norm satisfying $\|xy\| \leq \|x\| \|y\|$. {Alternatively, A is a complex normed space equipped with an associative, bilinear product satisfying $\|xy\| \leq \|x\| \|y\|$.} A *Banach algebra* is a complete normed algebra—informally, an associative algebra superimposed on a Banach space, with a submultiplicative norm.

A dividend of submultiplicativity in a normed algebra is that the topological and uniform structures are compatible with multiplication:

(49.2) Theorem. *In a normed algebra A, (i) the multiplication $(x, y) \mapsto xy$ is jointly continuous in its factors, and (ii) the product of two Cauchy sequences is Cauchy.*

Proof. (i) The assertion is that multiplication is a continuous mapping $A \times A \to A$, where $A \times A$ bears the product topology. Continuity at $(0, 0)$ is obvious from $\|xy\| \leq \|x\| \|y\|$, so one may quote (18.2). A more direct proof results on applying the norm to the identity

$$xy - x_0 y_0 = (x - x_0)(y - y_0) + x_0(y - y_0) + (x - x_0)y_0.$$

(ii) Suppose x_n and y_n are Cauchy sequences in A. Since $\|x_n\|$ and $\|y_n\|$ are bounded (16.4), it follows from

$$\|x_m y_m - x_n y_n\| \leq \|x_m - x_n\|\,\|y_m\| + \|x_n\|\,\|y_m - y_n\|$$

that $x_n y_n$ is Cauchy. ∎

Every normed algebra may be regarded as a dense subalgebra of a Banach algebra:

(49.3) Theorem. (Completion) *If A is a normed algebra, there exists a Banach algebra B and an algebra monomorphism $\varphi\colon A \to B$ such that φ is isometric and $\varphi(A)$ is dense in B. If A has a unity element e, then $\varphi(e)$ is a unity element for B.*

Proof. Viewing A as a normed space, one first constructs a Banach space B and an isometric linear mapping $\varphi\colon A \to B$ such that $\varphi(A)$ is dense in B. {A fast proof is available via the Hahn-Banach theorem (40.15); a more elementary proof is indicated in (16.18).}

Identifying A with $\varphi(A)$, i.e., viewing A as a dense linear subspace of the Banach space B, there remains the task of extending the multiplication of A to all of B. {A sophisticated proof is available via (24.13), but the following argument is straightforward and brief.} Assuming $u, v \in B$, we are to define uv. Choose sequences x_n, y_n in A with $x_n \to u$ and $y_n \to v$. Since $x_n y_n$ is Cauchy (49.2) it converges to a limit w in B. It is proposed to define $uv = w$; it must be checked that w is independent of the particular sequences used to approximate u and v. If also $x_n', y_n' \in A$ with $x_n' \to u$ and $y_n' \to v$, then the identity

$$x_n y_n - x_n' y_n' = (x_n - x_n')y_n + x_n'(y_n - y_n')$$

shows that $\lim x_n' y_n' = \lim x_n y_n = w$; thus the definition $uv = w$ is unambiguous. Moreover, since

$$\|x_n y_n\| \leq \|x_n\|\,\|y_n\|$$

for all n, passage to the limit yields $\|w\| \leq \|u\|\,\|v\|$ by (16.4), thus $\|uv\| \leq \|u\|\,\|v\|$.

The required algebraic identities for products in B are inherited from A by straightforward limiting arguments; selected samples follow. One shows first that $(u, v) \mapsto uv$ is bilinear. {For example, to prove that $u(v + w) = uv + uw$, choose sequences x_n, y_n, z_n in A converging to u, v, w, respectively, and pass to the limit in the equations $x_n(y_n + z_n) = x_n y_n + x_n z_n$.} Since $\|uv\| \leq \|u\|\,\|v\|$, the calculation in (49.2) shows that $(u, v) \mapsto uv$ is a continuous mapping. To verify the associative law, consider the mapping $f\colon B \times B \times B \to B$ defined by

$$f(u, v, w) = (uv)w - u(vw);$$

f is continuous and vanishes on the dense subset $A \times A \times A$, therefore f is identically zero. {Explicitly, since A is separated, $\{0\}$ is a closed subset, whence $f^{-1}(\{0\})$ is a closed set containing $A \times A \times A$.} ∎

(49.4) A normed algebra without a unity element can be equipped with one (49.5); we review here the algebraic preliminaries. Suppose A is an associative algebra over $\mathbb{C}$ (with or without a unity element). The Cartesian product $A_1 = A \times \mathbb{C}$ becomes an associative algebra on defining

$$(x, \lambda) + (y, \mu) = (x + y, \lambda + \mu)$$

$$\mu(x, \lambda) = (\mu x, \mu \lambda)$$

$$(x, \lambda)(y, \mu) = (xy + \mu x + \lambda y, \lambda \mu).$$

The element $u = (0, 1)$ is a unity element for A_1. Note that $(x, \lambda) = (x, 0) + \lambda u$. The mapping $x \mapsto (x, 0)$ is an algebra monomorphism $A \to A_1$; identifying A with its image, we may regard A as a subalgebra of A_1, and the elements of A_1 are uniquely representable in the form $x + \lambda u$ with $x \in A$ and $\lambda \in \mathbb{C}$. The mapping $x + \lambda u \mapsto \lambda$ is an algebra epimorphism $A_1 \to \mathbb{C}$ with kernel A, thus A is an ideal of A_1 and $A_1/A \cong \mathbb{C}$; since $\mathbb{C}$ is a field, it follows that A is a maximal ideal of A_1. The algebra A_1 is called the *unitification* of A.

(49.5) Theorem. (Adjunction of unity) *Let A be a normed algebra (with or without unity element), let A_1 be the algebra unitification of A (49.4), and define*

$$\|x + \lambda u\| = \|x\| + |\lambda|.$$

Then A_1 is a normed algebra, $\|u\| = 1$, and A is a closed, maximal ideal of A_1; moreover, A_1 is a Banach algebra if and only if A is a Banach algebra.

Proof. It is straightforward to check that A_1 is a normed algebra and $\|u\| = 1$. From the relation

$$\|(x + \lambda u) - (y + \mu u)\| = \|x - y\| + |\lambda - \mu|,$$

we see that $x_n + \lambda_n u \to x + \lambda u$ iff $x_n \to x$ and $\lambda_n \to \lambda$; in particular, it is clear that A is a closed subset of A_1. The same relation shows that a sequence $x_n + \lambda_n u$ is Cauchy iff x_n and λ_n are Cauchy; it follows easily that A_1 is complete iff A is complete. ∎

If $(A, \| \ \|)$ is a normed algebra with unity element u, the relation $u = uu$ implies that $\|u\| \le \|u\|^2$, thus $\|u\| \ge 1$. The formula $|x| = \|u\|^{-1}\|x\|$ defines an equivalent norm on A (40.27) for which $|u| = 1$, but in general the new norm is not submultiplicative (49.9). It is a corollary of the next theorem that one can have the best of both worlds; the result itself is not of great practical importance, but its proof is too beautiful to be omitted:

(49.6) Theorem. *If A is an associative algebra over $\mathbb{C}$, with unity element u, and if A has a Banach space norm $x \mapsto \|x\|$ such that $(x, y) \mapsto xy$ is separately continuous in each factor, then there exists an equivalent norm $x \mapsto |x|$ on A such that $(A, |\ |)$ is a Banach algebra (in particular, multiplication is jointly continuous in its factors) and such that $|u| = 1$.*

Proof. A Banach algebra is ready at hand: the Banach algebra $\mathscr{B} = \mathscr{L}(A)$ of all continuous linear mappings $T: A \to A$, with

$$\|T\| = \sup \{\|Tx\| : \|x\| \le 1\}$$

(see (40.23)). Moreover, $\mathscr{B}$ has unity element I, $\|I\| = 1$. The strategy is to suitably embed A in $\mathscr{B}$.

Each x in A determines a pair of linear mappings U_x, V_x on A via the formulas

$$U_x y = xy, \qquad V_x y = yx.$$

By hypothesis, U_x and V_x are continuous, thus U_x, V_x are in $\mathscr{B}$.

It is routine to verify that $x \mapsto U_x$ is an algebra monomorphism $A \to \mathscr{B}$ with $U_u = I$. {For example, $U_{xy} = U_x U_y$ results from the associative law $(xy)z = x(yz)$; and $U_x u = x$ shows that $x \mapsto U_x$ is injective.} It follows trivially that the formula

$$|x| = \|U_x\|$$

defines a submultiplicative norm on A with $|u| = 1$; it remains to show that this new norm on A is equivalent to the old one.

Let us write $\mathscr{U} = \{U_x : x \in A\}$ for the image of A under the mapping $x \mapsto U_x$; thus $\mathscr{U}$ is a subalgebra of $\mathscr{B}$. We now show that

(*) $\mathscr{U} = \{T \in \mathscr{B} : TV_y = V_y T \quad \text{for all } y \in A\}.$

Since $U_x V_y = V_y U_x$ for all x and y (by the associative law), $\mathscr{U}$ is contained in the right side of (*). Conversely, suppose $T \in \mathscr{B}$ commutes with every V_y; setting $x = Tu$, we have

$$U_x y = xy = (Tu)y = V_y Tu = TV_y u = T(uy) = Ty$$

for all $y \in A$, thus $T = U_x$.

It follows at once from (*) that $\mathscr{U}$ is a norm-closed subalgebra of $\mathscr{B}$, hence is itself a Banach algebra; thus $x \mapsto |x|$ is in fact a Banach algebra norm on A.

The mapping $x \mapsto U_x$ is a vector space isomorphism $(A, \| \ \|) \to (\mathscr{U}, \| \ \|)$ between a pair of Banach spaces. We assert that this mapping is bicontinuous; this will show that the new norm on A is equivalent to the old one, thus ending the proof. By the open mapping theorem, it will suffice to show that the mapping is continuous (48.4). Writing

$$S = \{x \in A : \|x\| \le 1\},$$

it is to be shown that the set $\{\|U_x\| : x \in S\}$ is bounded (40.1). By the uniform boundedness principle (47.1) it is enough to show that for each $y \in A$ the set $\{\|U_x y\| : x \in S\}$ is bounded; but

$$\{\|U_x y\| : x \in S\} = \{\|xy\| : x \in S\} = \{\|V_y x\| : \|x\| \le 1\}$$

is indeed bounded (with supremum $\|V_y\|$) by the assumed continuity of V_y. ∎

(49.7) **Corollary.** *If $(A, \| \ \|)$ is a Banach algebra with unity u, then there exists an equivalent Banach algebra norm $x \mapsto |x|$ with $|u| = 1$.*

If A is a Banach algebra with a unity element u of norm one, the mapping $\lambda \mapsto \lambda u$ identifies $\mathbb{C}$ in A both algebraically and metrically; no confusion can result from writing 1 in place of u.

Exercises

(49.8) A pair (B, φ) with the properties in (49.3) is called a *completion* of A. If (B_1, φ_1) and (B_2, φ_2) are completions of A, then there exists an isometric isomorphism $\psi: B_1 \to B_2$ such that $\psi \circ \varphi_1 = \varphi_2$. In this sense, the completion of A is unique.

(49.9) (i) Let $(B, \| \ \|)$ be a normed algebra with unity element u. The equivalent norm $|x| = \|u\|^{-1}\|x\|$ is submultiplicative iff $\|u\| \|xy\| \leq \|x\| \|y\|$ for all x, y in B.

(ii) Let $(A, \| \ \|)$ be a normed algebra without unity, in which not all products are zero. Say $ab \neq 0$. Choose a positive real number r such that $r\|ab\| > \|a\| \|b\|$ (necessarily $r > 1$). Let A_1 be the algebra unitification of A (49.4) and define

$$\|x + \lambda u\| = \|x\| + r|\lambda|$$

for all $x \in A$, $\lambda \in \mathbb{C}$. The result is that A_1 becomes a normed algebra and $\|u\| = r$. The norm $|x + \lambda u| = \|u\|^{-1}\|x + \lambda u\|$ is not submultiplicative.

(49.10) If $(A, \| \ \|)$ is a normed space with a bilinear mapping $\varphi: A \times A \to A$ that is continuous at $(0, 0)$, then there exists an equivalent norm $x \mapsto |x|$ on A such that $|\varphi(x, y)| \leq |x| |y|$ for all $x, y \in A$.

(49.11) Let A be an associative ring. If S is a nonempty subset of A, we write

$$S' = \{x \in A : xs = sx \quad \text{for all } s \in S\},$$

called the *commutant* of S in A. In particular, A' is the center of A. {This competes (unconfusingly) with the notation for dual space (40.8); there is never enough notation to go around.} The set $(S')'$ is called the *bicommutant* of S in A, denoted briefly S''. One has $(S'')' = (S')'' = ((S')')'$, denoted briefly S'''—but see (iii) below.

  (i) $S \subset S''$.
  (ii) If $S \subset T \subset A$ then $S' \supset T'$.
  (iii) $S''' = S'$.
  (iv) S' is a subring of A.
  (v) S is commutative (i.e., $st = ts$ for all $s, t \in S$) iff $S \subset S'$ iff S'' is commutative.
  (vi) If A is an algebra then S' is a subalgebra of A.
  (vii) If A is a ring with involution $x \mapsto x^*$ and if S is a $*$-subset of A (i.e., $s \in S$ implies $s^* \in S$), then S' is a $*$-subring of A.
  (viii) If A is a ring with unity, then S' is a full subring of A. {A subring B of A is said to be *full* if B contains the unity element of A and if, whenever $b \in B$ has an inverse in A, b^{-1} is in B; so to speak, 'B contains inverses.'}
  (ix) If A has a separated topology such that, for each $x \in A$, the mappings $y \mapsto xy$ and $y \mapsto yx$ are continuous, then S' is a closed subset of A. In particular, if A is a normed algebra then S' is a closed subalgebra of A.

(49.12) Notation as in (49.6) and its proof. One has $V_{xy} = V_y V_x$ for all x, y in A, thus $x \mapsto V_x$ is an algebra anti-isomorphism of A into $\mathscr{B}$. Write $\mathscr{V} = \{V_x : x \in A\}$ for its range. Then, in the notation of (49.11), one has $\mathscr{U} = \mathscr{V}'$, $\mathscr{V} = \mathscr{U}'$ and therefore $\mathscr{U} = \mathscr{U}''$, $\mathscr{V} = \mathscr{V}''$.

(49.13) In a sense, (49.6) is a theorem about Banach spaces with a multiplication; the result may fail when norms are replaced by values in the sense of

(12.1). Specifically, let A be an associative algebra over $\mathbb{C}$, equipped with a topology relative to which A is a metrizable complete TVS. In particular, the topology of A may be generated by a value $x \mapsto |x|$ (12.1). Assume, moreover, that the multiplication $(x, y) \mapsto xy$ is jointly continuous in its factors. It may be impossible to choose the value so that $|xy| \leq |x| |y|$ for all x, y (15.13).

(49.14) Let A be an additively written (but not necessarily abelian) group, with another operation $A \times A \to A$, written $(x, y) \mapsto xy$, satisfying

$$x(y + z) = xy + xz, \qquad (x + y)z = xz + yz$$

for all x, y, z. Assume A has a topology τ such that (1) τ is compatible with the additive group structure of A, and (2) for each $x \in A$, the mappings $y \mapsto xy$ and $y \mapsto yx$ are continuous (it suffices to assume continuity at 0).

(i) If $(x, y) \mapsto xy$ is continuous at $(0, 0)$, then it is continuous on $A \times A$.

(ii) If A is a Baire space (46.5) and is metrizable (cf. 6.3), then $(x, y) \mapsto xy$ is continuous (on $A \times A$).

(iii) If, in (49.6), A is assumed to be a metrizable, complete TVS (instead of a Banach space), the geometrical part of the conclusion may be lost (49.13); but (ii) shows that the topological part survives (cf. 46.6).

50. Invertibility in a Banach algebra with unity.

The example in Section 0 makes clear the importance of 'reciprocals' in a particular Banach algebra; in this section, we look at the concept in a general setting.

Throughout this section, A denotes a fixed (complex) Banach algebra with unity element 1, $\|1\| = 1$.

(50.1) As in any ring with unity, an element x of A is said to be *left-invertible* if there exists an element y in A such that $yx = 1$ (such an element y is called a *left inverse* of x); the terms *right-invertible* and *right inverse* are defined dually. If x is both left- and right-invertible, say $yx = 1 = xz$, then

$$y = y1 = y(xz) = (yx)z = 1z = z;$$

such an element x is said to be *invertible*, and the unique element with which all left and right inverses coincide is called the *inverse* of x and is denoted x^{-1}. Thus $xx^{-1} = x^{-1}x = 1$ for an invertible element x. {The invertible elements of A are also called its 'units.'} We write U for the set of all invertible elements of A; it is elementary that U is a group, with multiplication as law of composition and 1 as neutral element.

Obviously invertible are the nonzero scalar multiples of 1. When A contains nonscalar elements (i.e., when A has dimension greater than one), it is not obvious at first glance that it has any nonscalar invertible elements; in fact, they exist in surprising abundance:

(50.2) Theorem. *If $z \in A$ and $\|z\| < 1$, then $1 - z$ is invertible. Explicitly, the sequence*

$$y_n = 1 + z + z^2 + \cdots + z^{n-1}$$

converges to a limit y—formally, one writes $y = \sum_{n=0}^{\infty} z^n$ (the convention is that

$z^0 = 1$)—*and one has* $(1 - z)y = y(1 - z) = 1$. *Thus* $1 - z$ *is invertible,*
with inverse

$$(1 - z)^{-1} = \sum_{n=0}^{\infty} z^n.$$

Moreover,

$$\|(1 - z)^{-1}\| \le \frac{1}{1 - \|z\|}.$$

Proof. Since $\|z\| < 1$, the following computation shows that the
sequence y_n is Cauchy (therefore convergent):

$$\|y_{n+p} - y_n\| = \left\|\sum_{k=n}^{n+p-1} z^k\right\| \le \sum_{k=n}^{n+p-1} \|z^k\| \le \sum_{k=n}^{n+p-1} \|z\|^k$$

$$= \|z\|^n \sum_{k=0}^{p-1} \|z\|^k \le \|z\|^n \sum_{k=0}^{\infty} \|z\|^k = \frac{\|z\|^n}{1 - \|z\|}.$$

Let $y = \lim y_n$. Then $\|y\| = \lim \|y_n\|$ (16.4); since

$$\|y_n\| \le 1 + \|z\| + \cdots + \|z\|^{n-1} \le \sum_{k=0}^{\infty} \|z\|^k = \frac{1}{1 - \|z\|}$$

for all n, it follows that $\|y\| \le (1 - \|z\|)^{-1}$. Also,

$$zy_n = y_n z = z + z^2 + \cdots + z^n = y_{n+1} - 1;$$

passage to the limit yields $zy = yz = y - 1$, thus $y - zy = y - yz = 1$,
that is, $(1 - z)y = y(1 - z) = 1$. ∎

(50.3) Corollary. *If* $x \in A$ *and* $\|1 - x\| < 1$, *then* x *is invertible,*

$$x^{-1} = \sum_{n=0}^{\infty} (1 - x)^n,$$

and

$$\|x^{-1}\| \le \frac{1}{1 - \|1 - x\|}.$$

Proof. Apply (50.2) to the element $z = 1 - x$. ∎

(50.4) Corollary. *If* $x \in A$ *and* λ *is a complex number such that* $|\lambda| > \|x\|$,
then $\lambda 1 - x$ *is invertible,*

$$(\lambda 1 - x)^{-1} = \sum_{n=0}^{\infty} \lambda^{-n-1} x^n,$$

and

$$\|(\lambda 1 - x)^{-1}\| \le \frac{1}{|\lambda| - \|x\|}.$$

Proof. Write $z = \lambda^{-1} x$; thus $\lambda 1 - x = \lambda(1 - z)$ and $\|z\| < 1$. By
(50.2), $1 - z$ is invertible and

$$(1 - z)^{-1} = \sum_{n=0}^{\infty} z^n, \qquad \|(1 - z)^{-1}\| \le \frac{1}{1 - \|z\|};$$

then $\lambda 1 - x = \lambda(1 - z)$ is also invertible, and

$$(\lambda 1 - x)^{-1} = \lambda^{-1}(1 - z)^{-1} = \sum_{n=0}^{\infty} \lambda^{-1} z^n = \sum_{n=0}^{\infty} \lambda^{-n-1} x^n,$$

$$\|(\lambda 1 - x)^{-1}\| = |\lambda|^{-1} \|(1 - z)^{-1}\| \leq \frac{|\lambda|^{-1}}{1 - \|z\|} = \frac{1}{|\lambda| - \|x\|}. \quad \blacksquare$$

(50.5) Corollary. *The group U of invertible elements is an open subset of A. Specifically, if $x \in U$ then*

$$\{y : \|y - x\| < \|x^{-1}\|^{-1}\} \subset U.$$

Proof. If $\|y - x\| < \|x^{-1}\|^{-1}$ then

$$\|1 - yx^{-1}\| = \|(x - y)x^{-1}\| \leq \|x - y\| \|x^{-1}\| < 1$$

shows that $yx^{-1} \in U$ (50.3), therefore $y = (yx^{-1})x \in UU \subset U$. $\quad \blacksquare$

(50.6) An element of A is called *singular* if it is not invertible. The set of singular elements of A will be denoted S; thus $S = \complement U$, a closed subset of A (50.5).

(50.7) Theorem. *The mapping $x \mapsto x^{-1}$ $(x \in U)$ is bicontinuous (hence U is a topological group).*

Proof. The mapping is involutory $((x^{-1})^{-1} = x)$, so it will suffice to prove that it is continuous. Assuming x_n, x in U with $\|x_n - x\| \to 0$, it is to be shown that $\|x_n^{-1} - x^{-1}\| \to 0$. We have

$$x_n^{-1} - x^{-1} = x_n^{-1}(x - x_n)x^{-1},$$

therefore

$$\|x_n^{-1} - x^{-1}\| \leq \|x_n^{-1}\| \|x - x_n\| \|x^{-1}\|;$$

the proof will be concluded by showing that $\|y^{-1}\|$ remains bounded when y is sufficiently near x. Indeed, if $\|y - x\| \leq \frac{1}{2}\|x^{-1}\|^{-1}$ then y is invertible (50.5) and

$$\|1 - x^{-1}y\| = \|x^{-1}(x - y)\| \leq \|x^{-1}\| \|x - y\| \leq \frac{1}{2};$$

citing (50.3), we have $x^{-1}y$ invertible and

$$\|(x^{-1}y)^{-1}\| \leq \frac{1}{1 - \|1 - x^{-1}y\|} \leq \frac{1}{1 - \frac{1}{2}} = 2,$$

thus $\|y^{-1}x\| \leq 2$, and therefore

$$\|y^{-1}\| = \|(y^{-1}x)x^{-1}\| \leq \|y^{-1}x\| \|x^{-1}\| \leq 2\|x^{-1}\|.$$

Thus $\|y^{-1}\| \leq 2\|x^{-1}\|$ whenever $\|y - x\| \leq \frac{1}{2}\|x^{-1}\|^{-1}$. $\quad \blacksquare$

The following improvement on (50.7) is a toe in the door for complex function theory:

(50.8) Theorem. *The mapping $x \mapsto x^{-1}$ $(x \in U)$ is differentiable at every $x \in U$ in the following sense: if h is a nonzero complex variable, then the limit*

$$\lim_{h \to 0} (1/h)[(x + h1)^{-1} - x^{-1}]$$

exists and is equal to $-(x^{-1})^2$.

Proof. Let $x \in U$. If $0 < |h| < \|x^{-1}\|^{-1}$ then

$$\|(x + h1) - x\| = |h| < \|x^{-1}\|^{-1},$$

thus $x + h1$ is invertible (50.5); moreover,

$$(x + h1)^{-1} - x^{-1} = (x + h1)^{-1}[x - (x + h1)]x^{-1}$$
$$= -h(x + h1)^{-1}x^{-1},$$

thus

$$(1/h)[(x + h1)^{-1} - x^{-1}] = -(x + h1)^{-1}x^{-1},$$

the right side of which tends to $-(x^{-1})^2$ as $h \to 0$ (50.7). {Tally a point for elementary calculus in Banach algebras: the derivative of x^{-1} is $-x^{-2}$.} ∎

One way of proceeding would be to expand (50.8) into a full-blown theory of vector-valued analytic functions. For our purposes, it turns out to be simpler to reduce matters to the ordinary theory of scalar-valued analytic functions. The philosophy is that if $\lambda \mapsto F(\lambda)$ is a pleasant A-valued function of a complex variable λ, then, for each continuous linear form f on A, $\lambda \mapsto f(F(\lambda))$ is a pleasant complex-valued function of λ; the strategy succeeds because there are enough continuous linear forms (Hahn-Banach theorem).

Exercises

(50.9) Let A be an associative ring with unity. If $x \in A$ is left-invertible and has a unique left inverse, then x is invertible.

(50.10) In an associative ring (with product $(x, y) \mapsto xy$) one defines an auxiliary operation

$$x \circ y = x + y - xy,$$

called the *circle composition*. This operation is associative. An element x is said to be *right quasiregular* (RQR) if there exists an element y such that $x \circ y = 0$; the term *left quasiregular* (LQR) is defined dually; and x is said to be *quasiregular* (QR) if it is both LQR and RQR (the unique element x' with $x \circ x' = x' \circ x = 0$ is called the *adverse* of x).

(i) The set Q of all QR elements is a group, with $x \circ y$ as law of composition, 0 as neutral element, and x' as the 'inverse' of x.

(ii) In a ring with unity element 1, the mapping $x \mapsto 1 - x$ ($x \in Q$) is a group isomorphism of Q onto the group U of invertible elements.

(iii) In a Banach algebra (with or without unity element), every element x satisfying $\|x\| < 1$ is QR. Moreover, Q is an open set; thus every Banach algebra is a Q-ring (cf. 15.13).

(50.11) Let E be a normed space and consider the normed algebra $\mathscr{B} = \mathscr{L}(E)$.

(i) Every left-invertible element of $\mathscr{B}$ is bounded below (cf. 40.28).

(ii) Every right-invertible element of $\mathscr{B}$ is surjective.

(iii) Assume E is a Banach space. A continuous linear mapping $T: E \to E$ is bounded below iff T is injective and its range is a closed linear subspace of E.

(iv) Assume E is a Hilbert space. An element of the Banach algebra $\mathscr{B}$ is left-invertible iff it is bounded below; and right-invertible iff it is surjective.

(50.12) If A is a Banach algebra with unity and if B is a full subring of A (49.11), then the closure $\bar{B}$ is also a full subring of A.

(50.13) Let A be an associative ring with unity and let T be a nonempty subset of A.

(i) T'' is a full subring of A containing T (49.11).

(ii) The intersection of a family of full subrings of A is a full subring.

(iii) There exists a smallest full subring containing T; it is contained in T''.

(50.14) Let A be an associative algebra with unity and let T be a nonempty subset of A.

(i) There exists a smallest full subalgebra B containing T; one has $B \subset T''$.

(ii) If, in addition, A is a Banach algebra, then $\bar{B}$ is the smallest closed, full subalgebra of A containing T.

51. Resolvent and spectrum.

We continue with the notations of the preceding section: A is a Banach algebra with unity element 1, $\|1\| = 1$, U is the group of invertible elements and $S = \complement U$ is the set of singular elements of A.

(51.1) Definition. If $x \in A$, the *resolvent set* of x is the set of all $\lambda \in \mathbb{C}$ such that $\lambda 1 - x$ is invertible; we denote it by $\rho(x)$ (or by $\rho_A(x)$ when it is necessary to indicate its dependence on the containing algebra A). Thus $\rho(x) = \{\lambda \in \mathbb{C} : \lambda 1 - x \in U\}$.

(51.2) Theorem. *For every $x \in A$, $\rho(x)$ is an open subset of $\mathbb{C}$ and*

$$\{\lambda \in \mathbb{C} : |\lambda| > \|x\|\} \subset \rho(x).$$

Proof. $\rho(x)$ is the inverse image of the open set U under the continuous mapping $\lambda \mapsto \lambda 1 - x$ $(\lambda \in \mathbb{C})$, therefore it is open in $\mathbb{C}$. The asserted inclusion holds by (50.4). {So to speak, $\rho(x)$ is a deleted neighborhood of ∞ on the Riemann sphere.} ∎

(51.3) Definition. If $x \in A$, the *resolvent function* of x is the function $R_x : \rho(x) \to A$ defined by the formula

$$R_x(\lambda) = (\lambda 1 - x)^{-1} \qquad (\lambda \in \rho(x)).$$

The properties of the resolvent function flow from the properties of inversion and the following algebraic identity:

(51.4) Lemma. (Resolvent identity) *If $x \in A$ and $R = R_x$, then*

$$R(\lambda) - R(\mu) = -(\lambda - \mu)R(\lambda)R(\mu)$$

for all λ, μ in $\rho(x)$.

Proof. For all λ, μ in $\rho(x)$,

$$(\lambda 1 - x)^{-1} - (\mu 1 - x)^{-1} = (\lambda 1 - x)^{-1}[(\mu 1 - x) - (\lambda 1 - x)](\mu 1 - x)^{-1}$$
$$= -(\lambda - \mu)(\lambda 1 - x)^{-1}(\mu 1 - x)^{-1}. \qquad ∎$$

The key property of the resolvent function is analyticity:

(51.5) Theorem. *If $x \in A$ and $R = R_x$ is the resolvent function of x, then*

(i) $\lim\limits_{\lambda \to \infty} R(\lambda) = 0$;

(ii) *R is differentiable at every $\lambda \in \rho(x)$, in the sense that the limit*

$$\lim_{h \to 0} (1/h)[R(\lambda + h) - R(\lambda)]$$

exists; the limit is $-(\lambda 1 - x)^{-2}$.

Proof. (i) In view of (51.2) it is permissible to contemplate the limit. Assuming $\lambda_n \in \rho(x)$, $|\lambda_n| \to \infty$, it is to be shown that $\|R(\lambda_n)\| \to 0$. We can suppose that the λ_n are nonzero. Since $\lambda_n^{-1} \to 0$, we have $1 - \lambda_n^{-1}x \to 1$; citing (50.7) we have $(1 - \lambda_n^{-1}x)^{-1} \to 1$, therefore

$$R(\lambda_n) = (\lambda_n 1 - x)^{-1} = \lambda_n^{-1}(1 - \lambda_n^{-1}x)^{-1} \to 0 \cdot 1 = 0.$$

(ii) At any rate, R is continuous (it is the composite of continuous mappings $\lambda \mapsto \lambda 1 - x$ and $y \mapsto y^{-1}$). Let $\lambda \in \rho(x)$ and let h be a nonzero complex variable; for h sufficiently small, we have $\lambda + h \in \rho(x)$ and, by the resolvent identity,

$$R(\lambda + h) - R(\lambda) = -hR(\lambda + h)R(\lambda).$$

Thus

$$(1/h)[R(\lambda + h) - R(\lambda)] = -R(\lambda + h)R(\lambda)$$

for all sufficiently small $h \neq 0$; by continuity, the right side tends to $-R(\lambda)^2 = -(\lambda 1 - x)^{-2}$ as $h \to 0$. ∎

Application of continuous linear forms to the resolvent function produces analytic functions of the garden variety:

(51.6) Theorem. *Let $x \in A$, let $R = R_x$ be the resolvent function of x, and let f be any continuous linear form on A. Then*

(i) $\lim\limits_{\lambda \to \infty} f(R(\lambda)) = 0$;

(ii) *the complex-valued function $\lambda \mapsto f(R(\lambda))$ ($\lambda \in \rho(x)$) is differentiable at each $\lambda \in \rho(x)$, with derivative $-f((\lambda 1 - x)^{-2})$.*

Proof. (i) As $\lambda \to \infty$, we have $R(\lambda) \to 0$ (51.5), therefore $f(R(\lambda)) \to f(0) = 0$.

(ii) Write $F(\lambda) = f(R(\lambda))$ ($\lambda \in \rho(x)$). Fix $\lambda \in \rho(x)$ and let h be a nonzero complex variable. As $h \to 0$, we have

$$(1/h)[R(\lambda + h) - R(\lambda)] \to -R(\lambda)^2$$

by (51.5), therefore

$$(1/h)[f(R(\lambda + h)) - f(R(\lambda))] \to -f(R(\lambda)^2)$$

by the continuity and linearity of f; in other words,

$$\frac{F(\lambda + h) - F(\lambda)}{h} \to -f((\lambda 1 - x)^{-2})$$

as $h \to 0$. ∎

The resolvent set of an element is large (51.2) but not too large:

(51.7) Theorem.   *For each x in A, $\rho(x)$ is a proper subset of $\mathbb{C}$.*

Proof.   Assume to the contrary that $\rho(x) = \mathbb{C}$ for some element x; then the resolvent function R_x is defined on all of $\mathbb{C}$. Let f be any continuous linear form on A and define $F(\lambda) = f(R_x(\lambda))$ ($\lambda \in \mathbb{C}$). In view of (51.6), F is a bounded entire function, therefore F is constant by Liouville's theorem; moreover, the constant must be zero, thus $F(\lambda) = 0$ for all λ. In particular, $0 = F(0) = -f(x^{-1})$; since f is arbitrary, the Hahn-Banach theorem yields the absurdity $x^{-1} = 0$ (40.10). ∎

A striking dividend of analyticity is that the theory of complex Banach algebras contains no new fields:

(51.8) Theorem. (Gel'fand-Mazur theorem)   *If A is a division ring then it is one-dimensional over $\mathbb{C}$, and $A = \mathbb{C}$ in the sense that the mapping $\lambda \mapsto \lambda 1$ is an isometric algebra isomorphism of $\mathbb{C}$ onto A.*

Proof.   The hypothesis is that every nonzero element of A is invertible, that is, $S = \{0\}$. Let $x \in A$; we are to show that x is a scalar multiple of 1. According to (51.7), $\mathbb{C}\rho(x)$ contains a complex number λ; this means that $\lambda 1 - x$ is singular, therefore $\lambda 1 - x = 0$. ∎

(51.9) Definition.   If $x \in A$, the complement of $\rho(x)$ is called the *spectrum* of x (relative to A) and is denoted by $\sigma(x)$ (or by $\sigma_A(x)$ when it is necessary to indicate its dependence on A); thus,

$$\sigma(x) = \mathbb{C}\rho(x) = \{\lambda \in \mathbb{C} : \lambda 1 - x \text{ is singular}\}.$$

The spectrum of an element is small, but not too small:

(51.10) Theorem.   *For each x in A, $\sigma(x)$ is a nonempty compact subset of $\mathbb{C}$, and $\sigma(x) \subset \{\lambda : |\lambda| \le \|x\|\}$.*

Proof.   By (51.2), $\sigma(x)$ is a closed subset of the disc $\{\lambda : |\lambda| \le \|x\|\}$, thus $\sigma(x)$ is compact; it is nonempty by (51.7). ∎

(51.11) Definition.   If $x \in A$, the *spectral radius* of x (with respect to A), denoted $r(x)$ (or $r_A(x)$), is defined by the formula

$$r(x) = \sup \{|\lambda| : \lambda \in \sigma(x)\}.$$

In view of (51.10) we have $0 \le r(x) \le \|x\|$.

The following definition will not be needed until Section 56, but this is a good place to record it:

(51.12) Definition.   Let $x \in A$. The resolvent set $\rho(x)$ is an open subset of $\mathbb{C}$ (51.2); its connected components are also open sets (because $\mathbb{C}$ is locally connected) and there can be only countably many of them (because $\mathbb{C}$ is separable). One of the components of $\rho(x)$ contains the (connected) set $\{\lambda \in \mathbb{C} : |\lambda| > r(x)\}$; the other components of $\rho(x)$ are contained in $\{\lambda \in \mathbb{C} : |\lambda| \leq r(x)\}$. The former is called the *unbounded component* of $\rho(x)$; the bounded components of $\rho(x)$ are called the *holes* of $\sigma(x)$.

Exercises

(51.13) Let $\mathscr{A} = \mathscr{L}(H)$, where H is a separable, infinite-dimensional Hilbert space, let $(x_n)_{n \in \mathbb{Z}}$ be an orthonormal basis of H indexed by the set of all integers, and let U be the element of $\mathscr{A}$ such that $Ux_n = x_{n+1}$ for all $n \in \mathbb{Z}$. (Such an operator U is called a *bilateral shift*.) Let $\mathscr{B}$ be the closed subalgebra of $\mathscr{A}$ generated by I and U, that is, the closed linear span of the set $\{I, U, U^2, U^3, \ldots\}$. Then U is invertible in $\mathscr{A}$ but $U^{-1} \notin \mathscr{B}$.

(51.14) Let A be a Banach algebra with unity element 1, $\|1\| = 1$, and let B be a closed subalgebra of A such that $1 \in B$.
  (i) If $x \in B$ then $\sigma_A(x) \subseteq \sigma_B(x)$. The inclusion may be proper.
  (ii) If $x \in B$ then $r_A(x) = r_B(x)$.

*(51.15) The only real Banach algebras that are division rings are the reals, the complexes, and the quaternions.

(51.16) If A is an associative algebra with unity element 1, if $x, y \in A$ and if λ is a nonzero scalar, then $\lambda 1 - xy$ is invertible iff $\lambda 1 - yx$ is invertible. It follows that $\sigma_A(xy) - \{0\} = \sigma_A(yx) - \{0\}$ (cf. 53.2).

(51.17) Let A be a Banach algebra with unity.
  (i) The algebras $\mathscr{U} = \{U_x : x \in A\}$, $\mathscr{V} = \{V_x : x \in A\}$ described in (49.12) are full subalgebras of $\mathscr{L}(A)$.
  (ii) If $x \in A$ then $\sigma_A(x) = \sigma_{\mathscr{L}(A)}(U_x) = \sigma_{\mathscr{L}(A)}(V_x)$.

(51.18) Let A be a complex algebra without unity element. If $x \in A$, the *spectrum* of x in A, denoted $\sigma_A(x)$, is defined by the formula

$$\sigma_A(x) = \{0\} \cup \{\lambda \in \mathbb{C} : \lambda \neq 0 \text{ and } \lambda^{-1}x \text{ is not QR}\}$$

(see (50.10) for the definition of QR). {Thus, in an arbitrary algebra, 0 belongs to the spectrum of an element unless the algebra has a unity and the element is invertible.} Let A_1 be the unitification of A (cf. 49.4), with A regarded as an ideal of A_1.

  (i) As a subalgebra of A_1, A is 'full' in the following sense: if $x \in A$, $y \in A_1$ and either $x \circ y = 0$ or $y \circ x = 0$, then $y \in A$. In particular, if $x \in A$ then x is QR in A iff it is QR in A_1.
  (ii) $\sigma_A(x) = \sigma_{A_1}(x)$ for all x in A.
  (iii) $\sigma_A(xy) = \sigma_A(yx)$ for all x, y in A (cf. 51.16).
  (iv) If, in particular, A is a Banach algebra, then, for every $x \in A$, $\sigma(x)$ is a compact (and trivially nonempty) subset of $\mathbb{C}$; as in (51.11), the number $\sup \{|\lambda| : \lambda \in \sigma_A(x)\}$ is called the *spectral radius* of x in A, denoted $r_A(x)$.

(51.19) Let A be an associative ring. For each $x \in A$, write

$$I_x = \{yx - y : y \in A\}.$$

  (i) I_x is a left ideal of A.

(ii) x is LQR iff $x \in I_x$ iff $I_x = A$.

(iii) If x is not LQR, then there exists a modular maximal left ideal M of A such that $x \notin M$. {A left ideal I of A is said to be *modular* if there exists an element $e \in A$ such that $a - ae \in I$ for all $a \in A$; such an element e is called a 'right unity mod I.'}

(iv) Every element of the radical of A is QR. {The *radical* R of A is defined to be the intersection of all the modular maximal left ideals; R coincides with the intersection of all the modular maximal right ideals (defined dually), hence is a two-sided ideal of A.}

(v) If A is a Banach algebra, then every modular maximal left ideal is a closed linear subspace of A (hence so is the radical R of A). Every element of R has spectral radius 0.

52. Gel'fand representation theorem.

The Gel'fand theory, applicable to a commutative Banach algebra, is a technique for exploiting the ring structure via the set of maximal ideals (the incentive for doing so is illustrated in Section 0). The following definition is appropriate and concise:

(52.1) Definition. A commutative Banach algebra with a unity element of norm one will be called a *Gel'fand algebra*.

For the rest of the section, A denotes a fixed Gel'fand algebra (but the first two lemmas do not require commutativity); U is the group of invertible elements of A, and $S = \complement U$ the set of singular elements. It is worth noting that the commutativity of A implies that S is closed under multiplication.

(52.2) Example. Let T be a compact space and let $\mathscr{C}(T)$ be the set of all continuous functions $x: T \to \mathbb{C}$, equipped with the pointwise operations

$$(x + y)(t) = x(t) + y(t), \qquad (\lambda x)(t) = \lambda x(t), \qquad (xy)(t) = x(t)y(t)$$

and the norm

$$\|x\|_\infty = \sup \{|x(t)| : t \in T\};$$

then $\mathscr{C}(T)$ is a Gel'fand algebra (cf. 16.16).

The gist of the Gel'fand representation theorem is that for every Gel'fand algebra A, there exists a compact space T and a pleasant algebra homomorphism $A \to \mathscr{C}(T)$ sending the unity element of A onto the constant function 1; if so, then each $t \in T$ determines an algebra epimorphism $x \mapsto x(t)$ of A onto $\mathbb{C}$ (the 'evaluation mapping' at t), whose kernel is a maximal ideal of A. This draws attention to the maximal ideals of the Gel'fand algebra A; the crux of the Gel'fand theorem is to characterize the maximal ideals of A and to organize them into a topological space.

(52.3) Lemma. (i) *If I is a proper ideal of A, then so is $\bar{I}$.*
(ii) *If M is a maximal ideal of A, then M is closed.*

Proof. (i) From the continuity of the operations, it is clear that $\bar{I}$ is also an ideal. Since I excludes 1, evidently $I \subset S$; but S is closed (50.6), therefore $\bar{I} \subset S$, thus $\bar{I}$ is also proper.

(ii) Let M be a maximal (in particular, proper) ideal of A; in view of (i) we have $M \subset \overline{M} \subset A$ with $\overline{M}$ a proper ideal of A, therefore $M = \overline{M}$ by maximality. ∎

If I is a proper ideal of A, then A/I is an algebra with operations

$$(x + I) + (y + I) = (x + y) + I, \quad \lambda(x + I) = \lambda x + I, \quad (x + I)(y + I) = xy + I$$

and with unity element $1 + I$. Thus, writing $\pi: A \to A/I$ for the canonical mapping $\pi(x) = x + I$, π is an algebra epimorphism. If, in addition, I is closed, and if, for each u in A/I, one defines

$$\|u\| = \inf\{\|x\| : x \in u\},$$

then A/I is a Banach space (16.11); better yet:

(52.4) Lemma. *If I is a closed, proper ideal of A, then A/I is a Banach algebra with unity element $1 + I$, $\|1 + I\| = 1$. The canonical mapping $\pi: A \to A/I$ is continuous, with $\|\pi\| = 1$.*

Proof. Suppose $u, v \in A/I$. If $x \in u$ and $y \in v$ then $xy \in uv$, therefore

$$\|uv\| \le \|xy\| \le \|x\| \, \|y\|;$$

varying x and y, it results that $\|uv\| \le \|u\| \, \|v\|$. Combining this with the preliminary remarks, we see that A/I is a Banach algebra with unity element $\pi(1) = 1 + I$.

The inequality $\|\pi(x)\| \le \|x\|$ is immediate from $x \in \pi(x)$, thus π is continuous with $\|\pi\| \le 1$. In particular, $\|\pi(1)\| \le \|1\| = 1$. But $\|\pi(1)\| \ge 1$ results from $\pi(1) = \pi(1)\pi(1)$ and submultiplicativity of the norm. Thus $\|\pi(1)\| = 1 = \|1\|$ and therefore $\|\pi\| = 1$. {We remark that the commutativity of A implies that of A/I, thus A/I is also a Gel'fand algebra.} ∎

(52.5) Lemma. *If M is a maximal ideal of the Gel'fand algebra A, then A/M is the field of complex numbers; more precisely, A/M is one-dimensional and the mapping $\lambda \mapsto \lambda(1 + I)$ is an isometric algebra isomorphism of $\mathbb{C}$ onto A/M.*

Proof. Since M is closed (52.3), A/M is also a Gel'fand algebra (52.4), and A/M is a field by elementary commutative ring theory; quote the Gel'fand-Mazur theorem (51.8). ∎

(52.6) Definition. With notations as in (52.5), we write $\pi_M: A \to A/M$ for the canonical mapping. For each $x \in A$, there exists a unique complex number $f_M(x)$ such that $\pi_M(x) = f_M(x)(1 + M)$; thus $f_M: A \to \mathbb{C}$ is an algebra epimorphism with kernel M.

The multiplicative linear forms f_M described above are continuous and have norm one. {Proof: $|f_M(x)| = \|f_M(x)(1 + M)\| = \|\pi_M(x)\| \le \|x\|$ for all x in A, and $|f_M(1)| = \|\pi_M(1)\| = 1$.} This is not an accident:

(52.7) Lemma. *Every algebra epimorphism $f: A \to \mathbb{C}$ is continuous, with $\|f\| = 1$; in fact, if M is the kernel of f, then $f = f_M$.*

Proof. Let M be the kernel of f; since M is a proper ideal, all of its elements are singular. If $x \in A$ then $f(x - f(x)1) = 0$, thus $x - f(x)1$ belongs to M and is therefore singular; it results from (50.4) that $|f(x)| \leq \|x\|$. Thus f is a continuous linear form, $\|f\| \leq 1$; since $|f(1)| = 1 = \|1\|$, we conclude that $\|f\| = 1$.

It follows that M is a closed hyperplane and A/M is one-dimensional (22.4), thus A/M is the set of all scalar multiples of $1 + M$. We therefore have the set-up of (52.5) (without having used the Gel'fand-Mazur theorem); writing f_M as in (52.6), we have $f_M(x - f(x)1) = 0$ for all x in A, thus $f_M = f$.

{A brisker, though less elementary proof: Since A/M is isomorphic to the field $\mathbb{C}$, M is a maximal ideal of A; thus f_M exists in the sense of (52.6), and $f_M = f$ as above.} ∎

(52.8) Definition. A *character* of A is an algebra epimorphism $f: A \to \mathbb{C}$; we write $\mathcal{X}$ for the set of all characters of A. According to (52.7), $\mathcal{X}$ is a subset of the closed unit ball of A' (the dual space of A); when $\mathcal{X}$ is equipped with the relative weak* topology, we call it the *character space* of A.

(52.9) Lemma. (i) *The character space $\mathcal{X}$ of A is compact.*
(ii) *The mapping $M \mapsto f_M$ (52.6) is a bijection of the set of all maximal ideals onto $\mathcal{X}$.*

Proof. (i) Writing $\mathcal{Y}$ for the closed unit ball of A', we have $\mathcal{X} \subset \mathcal{Y}$; since $\mathcal{Y}$ is weak* compact by the Alaoglu-Bourbaki theorem (44.12), it will suffice to show that $\mathcal{X}$ is weak* closed in A'. Suppose f_j is a net in $\mathcal{X}$ weak* convergent to $f \in A'$; thus $f_j(x) \to f(x)$ for each x in A. In particular, $f(1) = \lim f_j(1) = \lim 1 = 1$; moreover, for each pair x, y in A we have

$$f(xy) = \lim f_j(xy) = \lim \{f_j(x)f_j(y)\} = \lim f_j(x) \lim f_j(y) = f(x)f(y),$$

thus $f: A \to \mathbb{C}$ is an algebra epimorphism, that is, $f \in \mathcal{X}$.
(ii) The mapping $M \mapsto f_M$ is onto $\mathcal{X}$ by (52.7); if $f_{M_1} = f_{M_2}$ then $M_1 = M_2$ (because M is the kernel of f_M), thus the mapping is injective. ∎

(52.10) Definition. We write $\mathcal{M}$ for the set of all maximal ideals of A. The *Gel'fand topology* on $\mathcal{M}$ is the topology that makes the bijection $\mathcal{M} \mapsto \mathcal{X}$ described in (52.9) a homeomorphism. Equipped with the Gel'fand topology, $\mathcal{M}$ is called the *maximal ideal space* of A; it is, of course, compact.

(52.11) Definition. Each x in A determines a complex-valued function $\hat{x}: \mathcal{M} \to \mathbb{C}$ via the formula

$$\hat{x}(M) = f_M(x) \qquad (M \in \mathcal{M});$$

the function $\hat{x}$ is called the *Gel'fand transform* of x. {In the notation of (52.6), $\hat{x}(M)$ is the unique complex number such that $\pi_M(x) = \hat{x}(M)(1 + M)$.}

(52.12) Lemma. *For each x in A, the function $\hat{x}: \mathcal{M} \to \mathbb{C}$ is continuous; in fact, the Gel'fand topology is the coarsest topology on $\mathcal{M}$ that renders the functions $\hat{x}$ ($x \in A$) continuous.*

Proof. It obviously suffices to prove the second statement. Let $M_0 \in \mathcal{M}$. Working back through (52.10) and the definition of the weak* topology (44.1), we see that a basic neighborhood $\mathcal{N}$ of M_0 for the Gel'fand topology is given by

$$\mathcal{N} = \{M : |f_M(x_k) - f_{M_0}(x_k)| < \varepsilon \quad \text{for } k = 1, \ldots, n\},$$

where $\varepsilon > 0$ and $x_1, \ldots, x_n$ are fixed elements of A; but

$$\mathcal{N} = \{M : |\hat{x}_k(M) - \hat{x}_k(M_0)| < \varepsilon \quad \text{for } k = 1, \ldots, n\},$$

and such sets $\mathcal{N}$ are a neighborhood base at M_0 for the coarsest topology rendering the functions $\hat{x}$ continuous. ∎

All the pieces are ready for assembly:

(52.13) Theorem. (Gel'fand representation theorem) *Let A be a Gel'fand algebra and let $\mathcal{M}$ be the maximal ideal space of A (52.10).*

(i) *The mapping $x \mapsto \hat{x}$ ($x \in A$) described in (52.11) is an algebra homomorphism $A \to \mathcal{C}(\mathcal{M})$ such that $\hat{1}$ is the constant function 1;*

(ii) *for each $x \in A$, the range of the function $\hat{x}$ is $\sigma(x)$;*

(iii) *$\|\hat{x}\|_\infty = r(x) \le \|x\|$ for all $x \in A$, thus the mapping $x \mapsto \hat{x}$ is continuous (with norm 1);*

(iv) *the kernel of the homomorphism $x \mapsto \hat{x}$ is the intersection of all the maximal ideas of A; it is also the set of all $x \in A$ such that $\sigma(x) = \{0\}$.*

The mapping $x \mapsto \hat{x}$ is called the **Gel'fand transformation** *and its range is denoted $\hat{A}$.*

(v) *$\hat{A}$ is a full subalgebra of $\mathcal{C}(\mathcal{M})$.*

The algebra of functions $\hat{A}$ is separating, in the following sense:

(vi) *if M_1, M_2 are maximal ideals with $M_1 \ne M_2$, then there exists an element x of A such that $\hat{x}(M_1) \ne \hat{x}(M_2)$.*

Proof. (i) The details are straightforward. For example, the computation

$$(xy)\hat{\ }(M) = f_M(xy) = f_M(x)f_M(y) = \hat{x}(M)\hat{y}(M)$$

shows that $(xy)\hat{\ } = \hat{x}\hat{y}$; and $\hat{1}(M) = f_M(1) = 1$ for all M, thus $\hat{1} = 1$.

(ii) If $\lambda \in \sigma(x)$, that is, if $\lambda 1 - x$ is singular, then the principal ideal $(\lambda 1 - x)A$ is proper; if M is a maximal ideal of A that contains $(\lambda 1 - x)A$ (Zorn's lemma) then $f_M(\lambda 1 - x) = 0$, thus $\hat{x}(M) = \lambda$. This shows that $\sigma(x)$ is contained in the range of $\hat{x}$. The steps of the argument are reversible.

(iii) In view of (ii) and (51.11), we have

$$\|\hat{x}\|_\infty = \sup\{|\hat{x}(M)| : M \in \mathcal{M}\}$$
$$= \sup\{|\lambda| : \lambda \in \sigma(x)\} = r(x) \le \|x\|$$

for each x in A.

(iv) In view of (ii), $\hat{x} = 0$ iff $\sigma(x) = \{0\}$; also, $\hat{x} = 0$ iff $\hat{x}(M) = 0$ for all M iff $f_M(x) = 0$ for all M iff $x \in M$ for all M.

(v) Assuming $\hat{x}$ is invertible in $\mathcal{C}(\mathcal{M})$, it is to be shown that $\hat{x}^{-1} \in \hat{A}$ (49.11). Obviously $\hat{x}$ vanishes at no point of $\mathcal{M}$, thus $0 \notin \hat{x}(\mathcal{M}) = \sigma(x)$ and therefore x is invertible in A. Then $1 = (xx^{-1})^\wedge = \hat{x}(x^{-1})^\wedge$, thus $(\hat{x})^{-1} = (x^{-1})^\wedge \in \hat{A}$.

(vi) This is immediate from (52.12) and the fact that the Gel'fand topology is separated, but it is instructive to prove a little more: A is a 2-fold transitive algebra of functions, in the sense that if $M_1 \neq M_2$ and if λ_1, λ_2 is any specified pair of complex numbers, then there exists $x \in A$ such that $\hat{x}(M_1) = \lambda_1$ and $\hat{x}(M_2) = \lambda_2$. Note that $M_2 \not\subset M_1$. $\{M_2 \subset M_1$ would imply $M_2 = M_1$ by the maximality of M_2.$\}$ Choose $x_1 \in M_2$ with $x_1 \notin M_1$; then $f_{M_2}(x_1) = 0$ and $f_{M_1}(x_1) \neq 0$, thus $\hat{x}_1(M_2) = 0$ and $\hat{x}_1(M_1) \neq 0$. $\{$This, too, settles (vi).$\}$ Multiplying x_1 by a scalar, we can suppose $\hat{x}_1(M_1) = 1$. Similarly, choose x_2 so that $\hat{x}_2(M_1) = 0$ and $\hat{x}_2(M_2) = 1$. The element $x = \lambda_1 x_1 + \lambda_2 x_2$ meets the requirements. $\{$Better yet, see (52.15).$\}$ ∎

The formulation of the Gel'fand theorem in terms of functions on the maximal ideal space $\mathcal{M}$ is the traditional one due to Gel'fand. It is often congenial to notate the theorem in terms of the character space $\mathcal{X}$, by regarding the Gel'fand transformation $x \mapsto \hat{x}$ as an algebra homomorphism $A \to \mathscr{C}(\mathcal{X})$, the Gel'fand transform of x being defined by the formula

$$\hat{x}(f) = f(x) \qquad (f \in \mathcal{X}).$$

$\{$So viewed, $\hat{x}$ is the restriction to $\mathcal{X}$ of the functional x'' described in (40.12).$\}$

Exercises

(52.14) (i) In a Banach algebra with unity, the closure of a proper left ideal is a proper left ideal.

(ii) In any Banach algebra, if I is a proper, modular left ideal (51.19), then so is $\bar{I}$.

(52.15) With notation as in the Gel'fand representation theorem (52.13), the algebra $\hat{A}$ is n-fold transitive for every positive integer n; that is, if $M_1, \ldots, M_n$ are distinct maximal ideals and $\lambda_1, \ldots, \lambda_n$ are any specified complex numbers, then there exists an element x of A such that $\hat{x}(M_k) = \lambda_k$ for $k = 1, \ldots, n$.

(52.16) Let B be a Banach algebra with unity element 1, $\|1\| = 1$, let $x \in B$ and let A be the closed subalgebra generated by 1 and x; explicitly, A is the closed linear span of the set $\{1, x, x^2, x^3, \ldots\}$. Then A is a Gel'fand algebra whose maximal ideal space is homeomorphic with $\sigma_A(x)$.

(52.17) If A is any Banach algebra with unity and if x, y are elements of A such that $xy = yx$, then $r(x + y) \leq r(x) + r(y)$ and $r(xy) \leq r(x)r(y)$.

(52.18) (i) Let T be any nonempty set and let $\mathscr{B}(T)$ be the set of all bounded functions $x: T \to \mathbb{C}$, equipped with the pointwise algebra operations and the norm $\|x\|_\infty = \sup\{|x(t)| : t \in T\}$ (cf. 16.15). Then $\mathscr{B}(T)$ is a Gel'fand algebra.

(ii) Let T be a topological space and let $\mathscr{C}^\infty(T)$ be the set of all $x \in \mathscr{B}(T)$ that are continuous. Then $\mathscr{C}^\infty(T)$ is a closed subalgebra of $\mathscr{B}(T)$ containing the constant function 1, hence is a Gel'fand algebra. If T is completely regular, then $\mathscr{C}^\infty(T)$ separates the points of T. The example in (i) is a special case (take T to be discrete).

(iii) Let T be a locally compact space and let $\mathscr{C}^0(T)$ be the set of all continuous $x: T \to \mathbb{C}$ that 'vanish at infinity' (in the sense that, for each $\varepsilon > 0$, the set

$\{t \in T : |x(t)| \geq \varepsilon\}$ is compact). Then $\mathscr{C}^0(T)$ is a closed subalgebra of $\mathscr{C}^\infty(T)$ (hence is a commutative Banach algebra) that separates the points of T; it has a unity element iff T is compact.

(52.19) Let A be a commutative Banach algebra without unity. {Note that the unitification A_1 of A is a Gel'fand algebra (49.5).}

(i) If $f: A \to \mathbb{C}$ is an algebra epimorphism, then f is a continuous linear form, $\|f\| \leq 1$, and the kernel of f is a (closed) modular maximal ideal of A.

(ii) If M is a modular maximal ideal of A, then M is a closed linear subspace of A, and A/M is one-dimensional. Let $\pi_M: A \to A/M$ be the canonical mapping, let u_M be the unity element of A/M, and, for each $x \in A$, let $f_M(x)$ be the unique complex number such that $\pi_M(x) = f_M(x)u_M$. Then $f_M: A \to \mathbb{C}$ is a continuous algebra epimorphism of norm ≤ 1, with kernel M.

(iii) Let $\mathscr{M}$ be the set of all modular maximal ideals M of A, and let $\mathscr{X}$ be the set of all algebra epimorphisms $f: A \to \mathbb{C}$. The correspondence $M \mapsto f_M$ described in (ii) maps $\mathscr{M}$ bijectively onto $\mathscr{X}$.

(iv) $\mathscr{X} \cup \{0\}$ is the set of all continuous linear forms f on A such that $f(xy) = f(x)f(y)$ for all $x, y \in A$; it is a weak* compact subset of A' (the dual space of A). Therefore $\mathscr{X}$ is locally compact for the weak* topology. {It can happen that $\mathscr{X}$ is compact, even though A has no unity element.} The set $\mathscr{M}$, equipped with the topology induced by $\mathscr{X}$, is called the *modular maximal ideal space* of A.

(v) For each $x \in A$, the function $\hat{x}: \mathscr{M} \to \mathbb{C}$ defined by $\hat{x}(M) = f_M(x)$ is continuous and vanishes at infinity, i.e., $\hat{x} \in \mathscr{C}^0(\mathscr{M})$ (cf. 52.18). The nonzero range of $\hat{x}$ coincides with the nonzero spectrum of x, thus

$$\sigma_A(x) = \hat{x}(\mathscr{M}) \cup \{0\}.$$

In particular, $\|\hat{x}\|_\infty = r_A(x) \leq \|x\|$ (cf. 51.18).

(vi) The mapping $x \mapsto \hat{x}$ is an algebra homomorphism $A \to \mathscr{C}^0(\mathscr{M})$, called the *Gel'fand transformation*; its kernel, the intersection of all the modular maximal ideals, is the radical of A.

53. The rational functional calculus.

The objective is to define $f(x)$, where x is an element of an algebra with unity and f is a suitable rational function. In defining expressions of the form $f(x)$, it is a general principle that the more restrictions one places on x, the fewer need be imposed on f. {For example, if x belongs to a Banach algebra with unity, then f can be any complex-valued function that is analytic in a neighborhood of the spectrum of x (see (53.10)). If x is any continuous linear mapping (briefly, 'operator') on a Hilbert space, then f can be any function that is σ-analytic on some 'spectral set' σ of x (see Section 66). If x is a normal operator on a Hilbert space, then f can be any continuous function on the spectrum of x (see Section 65).} In all cases, it is a desideratum that the 'spectral mapping formula'

$$\sigma(f(x)) = f(\sigma(x))$$

hold. In the present section we impose the minimal restriction on x, paying the penalty that the class of admissible functions f is small; the resulting spectral mapping formula is utterly elementary, yet exceedingly useful.

For the rest of the section, A is an associative algebra, with unity element 1, over the field $\mathbb{C}$ of complex numbers (actually, any algebraically complete field would do). We write $\mathbb{C}[t]$ for the algebra of all polynomial forms $p = \sum_{k=0}^{n} a_k t^k$ in an indeterminate t, with complex coefficients a_k (we distinguish between a polynomial form and the polynomial functions it may define via substitution).

(53.1) Lemma. *Let $x \in A$. There exists a unique algebra homomorphism $\varphi: \mathbb{C}[t] \to A$ such that $\varphi(1) = 1$ and $\varphi(t) = x$.*

Notation: If $p \in \mathbb{C}[t]$ we define $p(x) = \varphi(f)$; thus if $p = \sum_{k=0}^{n} a_k t^k$, then $p(x) = \sum_{k=0}^{n} a_k x^k$.

The range of φ is the smallest subalgebra of A containing x and 1, i.e., it is the linear span of the set $\{1, x, x^2, x^3, \ldots\}$.

Proof. Since the set $\{1, t, t^2, t^3, \ldots\}$ is a basis for $\mathbb{C}[t]$ as a complex vector space, there exists a unique linear mapping $\varphi: \mathbb{C}[t] \to A$ such that $\varphi(t^n) = x^n$ for $n = 0, 1, 2, 3, \ldots$ (with the conventions that $t^0 = 1$ and $x^0 = 1$). The fact that φ preserves products follows from the bilinearity of the mapping

$$(p, q) \mapsto \varphi(pq) \qquad (p, q \in \mathbb{C}[t])$$

and the fact that

$$\varphi(t^m t^n) = \varphi(t^{m+n}) = x^{m+n} = x^m x^n = \varphi(t^m)\varphi(t^n)$$

for all nonnegative integers m and n. Thus φ has the desired properties—and is clearly uniquely determined by them. ∎

The concepts of invertibility and spectrum defined in Sections 50 and 51 make sense in any algebra with unity:

(53.2) Definition. If $x \in A$, the *spectrum* of x (in A) is defined by the formula

$$\sigma(x) = \{\lambda \in \mathbb{C} : \lambda 1 - x \text{ is not invertible in } A\}.$$

The notation $\sigma_A(x)$ is used when it is necessary to call attention to the containing algebra A (cf. 53.7).

The spectrum of an element can be empty (53.7); when it is nonempty, there is a 'spectral mapping theorem' for polynomial functions:

(53.3) Lemma. *If $x \in A$ and $\sigma(x) \neq \varnothing$, then*

$$\sigma(p(x)) = p(\sigma(x))$$

for all p in $\mathbb{C}[t]$. For nonconstant p, the formula holds even if $\sigma(x) = \varnothing$.

Proof. Note that the choice $A = \mathbb{C}$ in (53.1) is permissible; the result is that every polynomial form $p \in \mathbb{C}[t]$ yields a function $p: \mathbb{C} \to \mathbb{C}$. The expression

$$p(\sigma(x)) = \{p(\lambda) : \lambda \in \sigma(x)\}$$

is to be understood in this sense.

If p is a constant, say $p = a_0 \in \mathbb{C}$, then

$$\sigma(p(x)) = \sigma(a_0 1) = \{a_0\} = p(\sigma(x)).$$

{It is only in showing the last equality that the nonemptiness of $\sigma(x)$ is used.}

Assume p nonconstant, say $\deg p = n \geq 1$. Let $\mu \in \mathbb{C}$. Factoring $p - \mu$ into linear factors, say

$$p - \mu = a_0(t - \lambda_1)\cdots(t - \lambda_n),$$

where $a_0 \neq 0$, we have

(*) $\qquad\qquad p(x) - \mu 1 = a_0(x - \lambda_1 1)\cdots(x - \lambda_n 1);$

since the factors on the right side of (*) commute with each other, it is clear that $p(x) - \mu 1$ is invertible iff $x - \lambda_k 1$ is invertible for all k. In other words, $p(x) - \mu 1$ is singular iff $x - \lambda_k 1$ is singular for some k, thus the following statements are equivalent: $\mu \in \sigma(p(x))$; $\lambda_k \in \sigma(x)$ for some k; $\sigma(x)$ contains one of the zeros of $p - \mu$; the function $\lambda \mapsto p(\lambda) - \mu$ vanishes at some point of $\sigma(x)$; $\mu \in p(\sigma(x))$. ∎

The next order of business is to generalize from polynomial forms to rational forms. We write $\mathbb{C}(t)$ for the field of fractions of the integral domain $\mathbb{C}[t]$; thus $\mathbb{C}(t)$ is the set of all rational forms $f = p/q$, where $p, q \in \mathbb{C}[t]$ and $q \neq 0$. The rational form f is in *reduced form* if (as can be arranged) p and q are relatively prime in $\mathbb{C}[t]$, i.e., when they possess no nonconstant common factor. When f is itself a polynomial, the polynomial q of the reduced form is a constant, conventionally taken to be 1.

Suppose $x \in A$. If $f = p/q \in \mathbb{C}(t)$ in reduced form, how should $f(x)$ be defined? Yearning to be used is the expression $p(x)q(x)^{-1}$; to use it, we will have to arrange that $q(x)$ be invertible, i.e., that $0 \notin \sigma(q(x))$. But $\sigma(q(x)) = q(\sigma(x))$ (usually). Thus we should avoid rational forms whose denominator (in the reduced form) has a zero in $\sigma(x)$; in the language of complex function theory, we should stick to rational forms f that have no poles in $\sigma(x)$. The Riemann sphere can be detained in the waiting room a little longer; for the present, the following simple notion will suffice:

(53.4) Definition. Let $x \in A$ and $f \in \mathbb{C}(t)$. We say that f is *nonsingular on* $\sigma(x)$ if, in the reduced form $f = p/q$, q has no zeros in $\sigma(x)$. We write $\mathbb{C}(t; \sigma(x))$ for the set of all such f; it is clearly a subalgebra of $\mathbb{C}(t)$, and

$$\mathbb{C}[t] \subset \mathbb{C}(t; \sigma(x)) \subset \mathbb{C}(t).$$

{Note that $\mathbb{C}[t] = \mathbb{C}(t; \sigma(x))$ iff $\sigma(x) = \mathbb{C}$; and $\mathbb{C}(t; \sigma(x)) = \mathbb{C}(t)$ iff $\sigma(x) = \varnothing$.}

The term 'rational functional calculus' refers to the following result and its consequences (including the next section):

(53.5) Theorem. *Let $x \in A$. There exists a unique algebra homomorphism* $\Phi: \mathbb{C}(t; \sigma(x)) \to A$ *such that* $\Phi(1) = 1$ *and* $\Phi(t) = x$.

Notation: If $f \in \mathbb{C}(t; \sigma(x))$ we define $f(x) = \Phi(f)$; writing $f = p/q$, where $p, q \in \mathbb{C}[t]$ and q has no zeros in $\sigma(x)$, we have $f(x) = p(x)q(x)^{-1} = q(x)^{-1}p(x)$. The range of Φ is the smallest full subalgebra of A containing x.

Proof. *Uniqueness:* Such a Φ must extend the mapping φ of (53.1) (by the uniqueness of φ). Suppose $f = p/q \in \mathbb{C}(t; \sigma(x))$ in reduced form. Since q and $1/q$ are in $\mathbb{C}(t; \sigma(x))$, it results from $q(1/q) = (1/q)q = 1$ that $\Phi(q) = \varphi(q)$ is invertible, with inverse $\Phi(1/q)$. Then $f = p(1/q)$ yields

$$\Phi(f) = \Phi(p)\Phi(1/q) = \varphi(p)\varphi(q)^{-1},$$

which shows that Φ is just as unique as φ.

Existence: Let $f \in \mathbb{C}(t; \sigma(x))$; we are to define $\Phi(f)$. Write $f = p/q$, where p and q are polynomials and q has no zeros in $\sigma(x)$. (Liberally, it is not assumed that f is in reduced form; this simplifies some of the later algebra.) Note that $q(x)$ $(= \varphi(q))$ is invertible in A. {This is obvious if q is a constant. Otherwise, citing (53.3) we have $0 \notin q(\sigma(x)) = \sigma(q(x))$.} Note also that $p(x)$ commutes with $q(x)$ and with $q(x)^{-1}$. {The relation $p(x)q(x) = q(x)p(x)$ is obvious; left- and right-multiplying by $q(x)^{-1}$, we have also $q(x)^{-1}p(x) = p(x)q(x)^{-1}$.} We define

$$\Phi(f) = p(x)q(x)^{-1};$$

this is permissible because if $f = p_1/q_1$ is another such representation with q_1 nonvanishing on $\sigma(x)$, then $p(x)q(x)^{-1} = p_1(x)q_1(x)^{-1}$ results from $q_1 p = p_1 q$ and (53.1). It is straightforward to check that Φ is an algebra homomorphism with $\Phi(1) = 1$ and $\Phi(t) = x$.

Finally, assuming $f \in \mathbb{C}(t; \sigma(x))$ and $\Phi(f)$ invertible in A, it is to be shown that $\Phi(f)^{-1}$ is in the range of Φ (cf. 49.11). Write $f = p/q$ in reduced form. By hypothesis, $\Phi(f) = p(x)q(x)^{-1}$ is invertible, therefore so is $p(x)$, that is, $0 \notin \sigma(p(x))$; it follows from (53.3) that $0 \notin p(\sigma(x))$ (a triviality if $\sigma(x) = \varnothing$), that is, p has no zeros in $\sigma(x)$. Thus $1/f = q/p \in \mathbb{C}(t; \sigma(x))$, and it follows from $f(1/f) = 1$ that $\Phi(f)^{-1} = \Phi(1/f)$. ∎

The spectral mapping formula is valid for the rational functional calculus:

(53.6) Theorem. (Spectral mapping theorem)   *If $x \in A$ and $\sigma(x) \neq \varnothing$, then*

$$\sigma(f(x)) = f(\sigma(x))$$

for all f in $\mathbb{C}(t; \sigma(x))$. For nonconstant f, the formula holds even if $\sigma(x) = \varnothing$.

Proof. Let $f \in \mathbb{C}(t; \sigma(x))$, say $f = p/q$ in reduced form; when $\sigma(x) = \varnothing$, we assume in addition that f is not a constant. Let $\mu \in \mathbb{C}$. Then

$$f(x) - \mu 1 = p(x)q(x)^{-1} - \mu 1 = [p(x) - \mu q(x)]q(x)^{-1}$$
$$= (p - \mu q)(x)q(x)^{-1},$$

thus $f(x) - \mu 1$ is singular iff $(p - \mu q)(x)$ is singular; let us analyze the latter condition. If $\sigma(x) \neq \varnothing$, then

(*) $$\sigma((p - \mu q)(x)) = (p - \mu q)(\sigma(x))$$

by (53.3), thus $(p - \mu q)(x)$ is singular iff $p - \mu q$ has a zero in $\sigma(x)$. If $\sigma(x) = \varnothing$ then f is nonconstant by hypothesis, therefore $p - \mu q \neq 0$; then either $p - \mu q$ is a (nonzero) constant a (in which case $(p - \mu q)(x) = a1$ cannot be singular) or $p - \mu q$ is not a constant (in which case (*) holds, showing that $(p - \mu q)(x)$ cannot be singular). We conclude that $(p - \mu q)(x)$ is singular iff $p - \mu q$ has a zero in $\sigma(x)$. Thus we have the following chain of equivalent statements: $\mu \in \sigma(f(x)); f(x) - \mu 1$ singular; $(p - \mu q)(x)$ singular; $p - \mu q$ has a zero in $\sigma(x)$; $[p(\lambda) - \mu q(\lambda)]q(\lambda)^{-1} = 0$ for some $\lambda \in \sigma(x)$ (recall that q has no zeros in $\sigma(x)$); $f(\lambda) - \mu = 0$ for some $\lambda \in \sigma(x)$; $\mu \in f(\sigma(x))$. ∎

Exercises

(53.7) If $A = \mathbb{C}(t)$ and $B = \mathbb{C}[t]$, then $\sigma_A(t) = \varnothing$ whereas $\sigma_B(t) = \mathbb{C}$.

(53.8) If B is a full subalgebra of A (49.11) and if $x \in B$, then $\sigma_B(x) = \sigma_A(x)$.

(53.9) Let A be a Banach algebra with unity, let $x \in A$, let B be the smallest full subalgebra containing x (53.5), and let $C = \bar{B}$. Then $\sigma_A(x) = \sigma_B(x) = \sigma_C(x)$.

*(53.10) (Holomorphic functional calculus) Let A be a Banach algebra with unity and let $x \in A$. As in (37.30), write $\mathcal{O}(\sigma(x))$ for the algebra of germs of functions holomorphic in a neighborhood of $\sigma(x)$.

There exists a unique algebra homomorphism $\Phi: \mathcal{O}(\sigma(x)) \to A$ such that Φ 'extends' the polynomial functional calculus (i.e., $\Phi([1]) = 1$ and $\Phi([\lambda]) = x$, where $[1]$ is the germ of the constant function 1, and $[\lambda]$ is the germ of the identity function $\lambda \mapsto \lambda$) and Φ is continuous for the topology on $\mathcal{O}(\sigma(x))$ described in (37.30).

If $f \in \mathcal{O}(U)$, where U is an open neighborhood of $\sigma(x)$, one defines $f(x) = \Phi([f])$; the spectral mapping formula $\sigma(f(x)) = f(\sigma(x))$ holds.

*54. $(f \circ g)(x) = f(g(x))$. The formula in the title is a plausible substitution property of the rational functional calculus; this section is devoted to its (surprisingly sticky) verification. {For an application to the theory of 'spectral sets,' see the proof of (66.12).}

As in the preceding section, A is an associative algebra over $\mathbb{C}$ with unity element 1. It is useful to have a formal notation for the result of (53.5):

(54.1) **Definition.** Let $x \in A$. The unique algebra homomorphism $\mathbb{C}(t; \sigma(x)) \to A$ such that $1 \mapsto 1$ and $t \mapsto x$ will be denoted Φ_x (or $\Phi_x{}^A$, when it is necessary to emphasize the algebra A). Thus (53.5),

$$f(x) = \Phi_x(f)$$

for all f in $\mathbb{C}(t; \sigma(x))$. We call Φ_x the *evaluation mapping* at x.

Suppose, in particular, that $A = \mathbb{C}$ and $x = \lambda \in \mathbb{C}$. Then $\sigma_{\mathbb{C}}(\lambda) = \{\lambda\}$, $\mathbb{C}(t; \sigma_{\mathbb{C}}(\lambda)) = \mathbb{C}(t; \{\lambda\})$. A rational form f belongs to $\mathbb{C}(t; \{\lambda\})$ iff it is non-singular at λ; i.e., writing $f = p/q$ in reduced form, $f \in \mathbb{C}(t; \{\lambda\})$ iff $q(\lambda) \neq 0$. The mapping

$$f \mapsto p(\lambda)/q(\lambda) \qquad (f \in \mathbb{C}(t; \{\lambda\}))$$

is evidently an algebra homomorphism $\mathbb{C}(t; \{\lambda\}) \to \mathbb{C}$ such that $1 \mapsto 1$ and $t \mapsto \lambda$, thus, in the notation of (54.1), it must be the mapping $\Phi_\lambda{}^{\mathbb{C}}$. In other words, for $f \in \mathbb{C}(t; \{\lambda\})$ the complex number $f(\lambda) = \Phi_\lambda{}^{\mathbb{C}}(f)$ defined by (54.1) coincides with the result of the elementary calculation $p(\lambda)/q(\lambda)$. It is useful to have a concise notation for this special case:

(54.2) Definition. For each $\lambda \in \mathbb{C}$, we write $\varphi_\lambda = \Phi_\lambda{}^{\mathbb{C}}$. Thus

$$\varphi_\lambda \colon \mathbb{C}(t; \{\lambda\}) \to \mathbb{C}$$

is the unique algebra homomorphism such that $\varphi_\lambda(1) = 1$ and $\varphi_\lambda(t) = \lambda$. For all f in $\mathbb{C}(t; \{\lambda\})$, the definition $f(\lambda) = \varphi_\lambda(f)$ coincides with the usual interpretation of $f(\lambda)$.

Our objective is to show that the formula $(f \circ g)(x) = f(g(x))$ is valid whenever it makes sense. Making sense of $f(g(x))$ is easy: one requires that $g \in \mathbb{C}(t; \sigma(x))$ and that $f \in \mathbb{C}(t; \sigma(g(x)))$; the main technical point is in the interpretation of $f \circ g$. The situation is clarified by viewing it in the context of the Riemann sphere:

(54.3) Definition. We write $\mathfrak{S} = \mathbb{C} \cup \{\infty\}$ for the Riemann sphere, regarded as the one-point compactification of $\mathbb{C}$. {No attempt is made to include the point at infinity in any algebraic operations.} Each rational form $f \in \mathbb{C}(t)$ defines a continuous function $f \colon \mathfrak{S} \to \mathfrak{S}$ as follows. Write $f = p/q$ in reduced form and let

$$\mathscr{D}_f = \{\lambda \in \mathbb{C} : f \in \mathbb{C}(t; \{\lambda\})\} = \{\lambda \in \mathbb{C} : q(\lambda) \neq 0\}.$$

The set $\mathbb{C} - \mathscr{D}_f$ is finite (and, unless f is a polynomial, nonempty); its points are called the *finite poles* of f. We call $\mathscr{D}_f$ the *natural domain* (or 'algebraic domain') of f; the formula

(i) $$\lambda \mapsto p(\lambda)/q(\lambda) \qquad (\lambda \in \mathscr{D}_f)$$

defines a continuous, complex-valued function $\mathscr{D}_f \to \mathbb{C}$. {In the notations of (54.2), it is the function $\lambda \mapsto \varphi_\lambda(f) = f(\lambda)$ $(\lambda \in \mathscr{D}_f)$.} Note that $\mathscr{D}_f$ is dense in $\mathbb{C}$ and therefore in $\mathfrak{S}$; thus the function (i) can have at most one continuous extension to $\mathfrak{S}$. On the other hand, if $\mu \in \mathbb{C} - \mathscr{D}_f$ then

(ii) $$\lim_{\lambda \to \mu} [p(\lambda)/q(\lambda)] = \infty.$$

Also,

(iii) $$\lim_{\lambda \to \infty} [p(\lambda)/q(\lambda)]$$

exists: it is ∞ if $\deg p > \deg q$; 0 if $\deg p < \deg q$; and the ratio of the leading coefficients of p and q if $\deg p = \deg q$. Defining

$$f \colon \mathfrak{S} \to \mathfrak{S}$$

by the formulas

$$f(\lambda) = p(\lambda)/q(\lambda) \qquad \text{for } \lambda \in \mathscr{D}_f,$$

$$f(\lambda) = \infty \qquad \text{for } \lambda \in \mathbb{C} - \mathscr{D}_f,$$

$$f(\infty) = \lim_{\lambda \to \infty} [p(\lambda)/q(\lambda)],$$

the result is that f may be regarded as a continuous function $\mathfrak{S} \to \mathfrak{S}$. The points of the finite set

$$\{\xi \in \mathfrak{S} : f(\xi) = \infty\}$$

are called the *poles* of f; these are the points of $\mathbb{C} - \mathscr{D}_f$, augmented by ∞ when $\deg p > \deg q$. {Note: Since every subset of $\mathfrak{S}$ with finite complement is dense in $\mathfrak{S}$, f may be described as the unique continuous function $\mathfrak{S} \to \mathfrak{S}$ such that $f(\lambda) = p(\lambda)/q(\lambda)$ for λ in some subset $\mathscr{D}$ of $\mathscr{D}_f$ with $\mathscr{D}_f - \mathscr{D}$ finite.} A function $\mathfrak{S} \to \mathfrak{S}$ is called a *rational function* if it arises from some rational form in the foregoing way.

If $f: \mathfrak{S} \to \mathfrak{S}$ is a rational function, then its values are finite at all but finitely many points of $\mathfrak{S}$, that is, the set

$$f^{-1}(\mathbb{C}) = \{\xi \in \mathfrak{S} : f(\xi) \in \mathbb{C}\}$$

is cofinite in $\mathfrak{S}$; f has no poles iff $f^{-1}(\mathbb{C}) = \mathfrak{S}$ iff f is a constant (fundamental theorem of algebra). Explicitly ruled out is the constant function $\mathfrak{S} \to \{\infty\}$.

(54.4) Lemma. *If $g: \mathfrak{S} \to \mathfrak{S}$ is a nonconstant rational function and if $\eta \in \mathfrak{S}$, then the equation $g(\xi) = \eta$ can have at most finitely many solutions ξ in $\mathfrak{S}$.*

Proof. This is clear if $\eta = \infty$ (f has only finitely many poles).

Assuming $\eta \in \mathbb{C}$, suppose to the contrary that the set $g^{-1}(\{\eta\})$ is infinite. Then $S = \mathbb{C} \cap g^{-1}(\{\eta\})$ is an infinite set of complex numbers such that $g(\lambda) = \eta$ for all $\lambda \in S$. Write $g = p/q$ in reduced form; since $\eta \in \mathbb{C}$ we have $S \subset \mathscr{D}_g$, thus

$$\eta = g(\lambda) = p(\lambda)/q(\lambda)$$

for all λ in S. Thus $p - \eta q$ is a polynomial such that $(p - \eta q)(\lambda) = 0$ for all $\lambda \in S$; since S is infinite, we conclude that $p - \eta q = 0$. This implies that g is the constant function η, contrary to hypothesis. ∎

If $f, g: \mathfrak{S} \to \mathfrak{S}$ are rational functions, then the composite function $f \circ g: \mathfrak{S} \to \mathfrak{S}$ is continuous. However, if g is a constant equal to some finite pole of f, then $f \circ g$ is the constant function $\mathfrak{S} \to \{\infty\}$, hence is not rational; in all other cases, all is well:

(54.5) Lemma. *Let $f, g: \mathfrak{S} \to \mathfrak{S}$ be rational functions, such that g is not a constant equal to some finite pole of f. Then the composite function $f \circ g: \mathfrak{S} \to \mathfrak{S}$ is also a rational function.*

Proof. If g is a constant, say $g = \eta$, $\eta \in \mathscr{D}_f$, then $(f \circ g)(\lambda) = f(\eta)$ for all $\lambda \in \mathbb{C}$, hence $f \circ g = f(\eta)$ on $\mathfrak{S}$ by continuity.

Suppose g is not a constant. Consider the set

$$\mathscr{D} = \{\lambda \in \mathbb{C} : \lambda \in \mathscr{D}_g \text{ and } g(\lambda) \in \mathscr{D}_f\}.$$

We assert that $\mathbb{C} - \mathscr{D}$ is finite. Indeed,

$$\mathbb{C} - \mathscr{D} = (\mathbb{C} - \mathscr{D}_g) \cup \{\lambda \in \mathbb{C} : \lambda \in \mathscr{D}_g \text{ and } g(\lambda) \in \mathbb{C} - \mathscr{D}_f\}.$$

The first term on the right is certainly finite; to see that the second term is finite, note that $\mathbb{C} - \mathcal{D}_f$ is finite, and, for each μ in $\mathbb{C} - \mathcal{D}_f$, the equation $g(\lambda) = \mu$ can have at most finitely many solutions (54.4).

Define a function $h_0 \colon \mathcal{D} \to \mathbb{C}$ by the formula

$$h_0(\lambda) = f(g(\lambda)) \qquad (\lambda \in \mathcal{D}).$$

It is obvious by elementary algebra that there exists a rational function h such that $h(\lambda) = h_0(\lambda)$ for all λ in $\mathcal{D}$. Thus h and $f \circ g$ are continuous functions on $\mathfrak{S}$ such that $h|\mathcal{D} = f \circ g|\mathcal{D}$; since $\mathcal{D}$ is dense in $\mathfrak{S}$, we conclude that $h = f \circ g$. In particular, $f \circ g$ is a rational function. $\blacksquare$

Note in particular that the element t of $\mathbb{C}(t)$ induces the identity mapping on $\mathfrak{S}$, i.e., $t(\xi) = \xi$ for all $\xi \in \mathfrak{S}$. It follows that $t \circ g = g$ (as functions on $\mathfrak{S}$) for every $g \in \mathbb{C}(t)$. {Proof: For all $\lambda \in \mathcal{D}_g$, $(t \circ g)(\lambda) = t(g(\lambda)) = g(\lambda)$.} Similarly the constant polynomial 1 satisfies $1 \circ g = 1$ for all $g \in \mathbb{C}(t)$.

It is expedient to cast the notion of composition in more algebraic terms. To this end, we consider another special case of (54.1):

(54.6) Definition. In (54.1) let $\mathbb{C}(t)$ play the role of A and let $g \in \mathbb{C}(t)$ play the role of x. If g is nonconstant then $\sigma(g) = \varnothing$ and therefore $\mathbb{C}(t; \sigma(g)) = \mathbb{C}(t)$; if $g = \lambda \in \mathbb{C}$ then $\sigma(g) = \{\lambda\}$ and so $\mathbb{C}(t; \sigma(g)) = \mathbb{C}(t; \{\lambda\})$. In either case, (54.1) names an algebra homomorphism $\Phi_g{}^{\mathbb{C}(t)}$; we write briefly $\Phi_g = \Phi_g{}^{\mathbb{C}(t)}$. Thus

$$\Phi_g \colon \mathbb{C}(t; \sigma(g)) \to \mathbb{C}(t)$$

is the unique algebra homomorphism such that $\Phi_g(1) = 1$ and $\Phi_g(t) = g$. Note the meaning of the relation $f \in \mathbb{C}(t; \sigma(g))$: g is not a constant equal to a pole of f. {We shall shortly see that the ugly notation $f(g) = \Phi_g(f)$ can be dispensed with (54.7).}

When $g = \lambda \in \mathbb{C}$, there is no essential difference between the mapping $\Phi_\lambda = \Phi_\lambda{}^{\mathbb{C}(t)}$ of (54.6) and the mapping φ_λ of (54.2): the composition of $\varphi_\lambda \colon \mathbb{C}(t; \{\lambda\}) \to \mathbb{C}$ with the inclusion mapping $\mathbb{C} \to \mathbb{C}(t)$ has the properties that characterize Φ_λ. Thus the main interest of (54.6) is in the case that g is nonconstant:

(54.7) Lemma. *If $g \in \mathbb{C}(t)$ is nonconstant and*

$$\Phi_g \colon \mathbb{C}(t) \to \mathbb{C}(t)$$

is the mapping described in (54.6), then

$$\Phi_g(f) = f \circ g$$

for all $f \in \mathbb{C}(t)$.

Proof. Fix $f \in \mathbb{C}(t)$ and write $h = \Phi_g(f)$. From (54.5) and its proof, we know that $f \circ g$ is a rational function, and, writing

$$\mathcal{D} = \{\lambda \in \mathbb{C} : \lambda \in \mathcal{D}_g \text{ and } g(\lambda) \in \mathcal{D}_f\}$$

$$= \{\lambda \in \mathbb{C} : \lambda \text{ is not a pole of } g, \text{ and } g(\lambda) \text{ is not a pole of } f\},$$

$\mathbb{C} - \mathscr{D}$ is finite. On the other hand, $h = \Phi_g(f)$ is a rational function, thus $\mathbb{C} - \mathscr{D}_h$ is finite. Defining

$$\mathscr{E} = \mathscr{D} \cap \mathscr{D}_h,$$

$\mathbb{C} - \mathscr{E}$ is finite; for $\lambda \in \mathscr{E}$, all three of $g(\lambda), f(g(\lambda)), h(\lambda)$ are in $\mathbb{C}$. We are to show that $h = f \circ g$; since $\mathscr{E}$ is dense in $\mathfrak{S}$, it will suffice to show that $h(\lambda) = f(g(\lambda))$ for all λ in $\mathscr{E}$.

Fix $\lambda \in \mathscr{E}$ and write $f = p/q$ in reduced form. For brevity, let us write $\Phi = \Phi_g$. Then

$$h = \Phi(f) = \Phi(pq^{-1}) = \Phi(p)\Phi(q)^{-1}.$$

Say $p = \sum_{j=0}^{m} a_j t^j$, $q = \sum_{k=0}^{n} b_k t^k$; then

(*) $$\Phi(p) = \sum_{j=0}^{m} a_j g^j, \qquad \Phi(q) = \sum_{k=0}^{n} b_k g^k.$$

Since $g \in \mathbb{C}(t; \{\lambda\})$ (because $\lambda \in \mathscr{D}_g$), it is clear from (*) that $\Phi(p), \Phi(q) \in \mathbb{C}(t; \{\lambda\})$; moreover,

$$\varphi_\lambda(\Phi(p)) = \sum_{j=0}^{m} a_j \varphi_\lambda(g)^j = \sum_{j=0}^{m} a_j g(\lambda)^j = p(g(\lambda)),$$

and similarly $\varphi_\lambda(\Phi(q)) = q(g(\lambda))$. Since $g(\lambda)$ is not a pole of f—i.e., $g(\lambda)$ is not a zero of q—we have $\varphi_\lambda(\Phi(q)) = q(g(\lambda)) \neq 0$, that is, $(\Phi(q))(\lambda)$ is a nonzero complex number; from this it is clear that λ is not a pole of $\Phi(q)^{-1}$, thus $\Phi(q)^{-1} \in \mathbb{C}(t; \{\lambda\})$. Since $\Phi(p), \Phi(q)$, and $\Phi(q)^{-1}$ belong to $\mathbb{C}(t; \{\lambda\})$, it results from $h = \Phi(p)\Phi(q)^{-1}$ that

$$h(\lambda) = \varphi_\lambda(h) = \varphi_\lambda(\Phi(p))\varphi_\lambda(\Phi(q))^{-1} = p(g(\lambda))q(g(\lambda))^{-1} = f(g(\lambda)). \quad \blacksquare$$

(54.8) Theorem. *Let A be an associative algebra with unity and let $x \in A$. If f and g are rational forms such that $g(x)$ and $f(g(x))$ exist, then $(f \circ g)(x)$ exists and $(f \circ g)(x) = f(g(x))$.*

Proof. The hypothesis is that $g \in \mathbb{C}(t; \sigma(x))$ and $f \in \mathbb{C}(t; \sigma(g(x)))$; in other words, g has no poles in $\sigma(x)$, and f has no poles in $\sigma(g(x))$. We must show that $f \circ g$ is a rational function with no poles in $\sigma(x)$.

First, $f \circ g$ is a rational function. {Proof: If g is not a constant, then $f \circ g$ is rational by (54.5). If $g = \mu \in \mathbb{C}$ then $g(x) = \mu 1$, $\sigma(g(x)) = \{\mu\}$; since $f \in \mathbb{C}(t; \sigma(g(x))) = \mathbb{C}(t; \{\mu\})$, we conclude that μ is not a pole of f, therefore $f \circ g$ is rational (54.5)—in fact, it is the constant $f(\mu)$.}

Next, $f \circ g$ has no poles in $\sigma(x)$. {Proof: Suppose to the contrary that $f \circ g$ has a pole λ in $\sigma(x)$ (in particular, $\sigma(x)$ is nonempty). Then $f(g(\lambda)) = (f \circ g)(\lambda) = \infty$ shows that $g(\lambda)$ is a pole of f, thus f has a pole in $g(\sigma(x))$; since $g(\sigma(x)) = \sigma(g(x))$ (53.6), this contradicts the hypothesis on f.}

Summarizing, both $f(g(x))$ and $(f \circ g)(x)$ are defined; it remains to show that they are equal.

Let us first dispose of the case that g is a constant, say $g = \mu \in \mathbb{C}$. Then $g(x) = \mu 1$ and $f \in \mathbb{C}(t; \{\mu\})$, that is, $\mu \in \mathscr{D}_f$. On the one hand, $f(g(x)) = f(\mu 1)$. On the other hand, $f \circ g$ is the constant function $f(\mu)$, thus $(f \circ g)(x) = f(\mu)1$. Finally, $f(\mu 1) = f(\mu)1$. {Proof: Writing $f = p/q$ in reduced form, we have $f(\mu 1) = p(\mu 1)q(\mu 1)^{-1} = (p(\mu)1)(q(\mu)1)^{-1} = (p(\mu)/q(\mu))1 = f(\mu)1.$}

Assume now that g is not a constant. Let us regard $g \in \mathbb{C}(t; \sigma(x))$ as fixed and $f \in \mathbb{C}(t; \sigma(g(x)))$ as variable. We have shown that $f \circ g \in \mathbb{C}(t; \sigma(x))$ for all such f. We may therefore define a mapping

$$\Phi \colon \mathbb{C}(t; \sigma(g(x))) \to A$$

by the formula $\Phi(f) = (f \circ g)(x)$. Since $1 \circ g = 1$ and $t \circ g = g$, we have $\Phi(1) = 1$ and $\Phi(t) = g(x)$. Moreover, Φ is an algebra homomorphism; indeed, it is clear from (54.7) and (54.1) that Φ is the composite of two homomorphisms,

$$f \mapsto f \circ g \mapsto (f \circ g)(x)$$

(note that $\mathbb{C}(t; \sigma(g(x))) \subset \mathbb{C}(t) =$ the domain of Φ_g). These properties of Φ are characteristic of $\Phi_{g(x)}$; thus, for all $f \in \mathbb{C}(t; \sigma(g(x)))$, we have

$$\Phi_{g(x)}(f) = \Phi(f),$$

that is, $f(g(x)) = (f \circ g)(x)$. ∎

55. Gel'fand's formula for the spectral radius.

Let A be a Banach algebra with unity element 1, $\|1\| = 1$. For $x \in A$, the spectral radius of x in A is defined by the formula

$$r(x) = \sup \{|\lambda| : \lambda \in \sigma(x)\},$$

and one has $r(x) \le \|x\|$ (51.11).

(55.1) Theorem. *If $x \in A$ then $r(x) = \lim_{n \to \infty} \|x^n\|^{1/n}$.*

Proof. {The existence of the limit can be given an elementary proof (55.3), but the proof that it equals $r(x)$ is nonelementary—indeed, the definition of $r(x)$ already entails the fact that the spectrum is nonempty.}

The existence of the limit, and the validity of the formula, will be verified simultaneously by proving the following two inequalities:

(1) $$\limsup \|x^n\|^{1/n} \le r(x),$$

(2) $$r(x) \le \liminf \|x^n\|^{1/n}.$$

Proof of (1): Let $R = R_x \colon \rho(x) \to A$ be the resolvent function of x (51.3), that is,

$$R(\lambda) = (\lambda 1 - x)^{-1} \qquad (\lambda \in \rho(x)),$$

where $\rho(x) = \mathbb{C} - \sigma(x)$ is the resolvent set of x. We know that R is differentiable at every point of $\rho(x)$ (51.5), and $\rho(x)$ includes the set $\{\lambda \in \mathbb{C} : |\lambda| > \|x\|\}$ (51.2).

Fix a continuous linear form f on A. We define a complex-valued function G on the open disc

$$D_1 = \{\mu \in \mathbb{C} : |\mu| < 1/r(x)\}$$

(the convention is that $1/r(x) = +\infty$ when $r(x) = 0$) as follows: define $G(0) = 0$; if $0 < |\mu| < 1/r(x)$ then $|\mu^{-1}| > r(x)$ and we define

$$G(\mu) = f[R(\mu^{-1})].$$

It follows from (51.6) that G is differentiable at every point of the punctured disc $D_1 - \{0\}$. {On the punctured disc, G is the composite of the differentiable functions $\mu \mapsto \mu^{-1}$ and $\lambda \mapsto f[R(\lambda)]$.}

Consider the (possibly smaller) disc

$$D_2 = \{\mu \in \mathbb{C} : |\mu| < 1/\|x\|\}$$

(with the convention that $1/\|x\| = +\infty$ when $x = 0$). If $0 < |\mu| < 1/\|x\|$ then $|\mu^{-1}| > \|x\|$ and therefore (50.4)

$$R(\mu^{-1}) = \sum_{n=0}^{\infty} \mu^{n+1} x^n;$$

since f is continuous and linear, we conclude that

(*) $$G(\mu) = \sum_{n=0}^{\infty} f(x^n)\mu^{n+1}.$$

The representation (*) is trivially valid at $\mu = 0$ (recall that $G(0) = 0$ by definition). Thus the series representation (*) of G is valid on the entire open disc D_2, and in particular G is analytic on D_2. But G is known to be analytic on the (possibly larger) disc D_1, hence is representable on D_1 by a power series; by the theorem on uniqueness of coefficients, the series (*) must also represent G on D_1. In particular, for each $\mu \in D_1$ the sequence $f(x^n)\mu^{n+1}$ is bounded (indeed, it is a null sequence).

In summary, for each $\mu \in D_1$ and for each continuous linear form f on A, the sequence $f(x^n)\mu^{n+1} = f(\mu^{n+1}x^n)$ is bounded. Citing the uniform boundedness principle (47.3), we conclude that for each $\mu \in D_1$, the sequence $\|\mu^{n+1}x^n\|$ is bounded. Thus, for each $\mu \in D_1$ there exists a constant $M_\mu > 0$ such that

(**) $$|\mu^{n+1}|\, \|x^n\| \le M_\mu \qquad (n = 1, 2, 3, \ldots).$$

Fix $\lambda \in \mathbb{C}$ with $|\lambda| > r(x)$. Since $\mu = \lambda^{-1}$ is in D_1, it results from (**) that

$$\|x^n\| \le |\lambda|^n(|\lambda| M_\mu),$$

thus $\|x^n\|^{1/n} \le |\lambda|(|\lambda| M_\mu)^{1/n}$ for all n. Then

$$\limsup \|x^n\|^{1/n} \le |\lambda| \lim (|\lambda| M_\mu)^{1/n} = |\lambda|;$$

since this is true for every λ satisfying $|\lambda| > r(x)$, we conclude that

$$\limsup \|x^n\|^{1/n} \le r(x).$$

Proof of (2): From the spectral mapping theorem (53.3) we have $\sigma(x^n) = \{\lambda^n : \lambda \in \sigma(x)\}$, therefore $r(x^n) = r(x)^n$ for every positive integer n. Since $r(x^n) \le \|x^n\|$, we have

$$r(x) = r(x^n)^{1/n} \le \|x^n\|^{1/n}$$

for all n, therefore (2) holds.

{Postlude: Since $r(x) \le \|x^n\|^{1/n}$ for all n, and the right side is known to converge to $r(x)$, we infer that $\lim \|x^n\|^{1/n} = \inf \|x^n\|^{1/n}$.} ∎

If $x \in B \subset A$, where B is a closed subalgebra of A that contains 1, the evident inclusion $\sigma_A(x) \subset \sigma_B(x)$ implies that $r_A(x) \le r_B(x)$; although the inclusion may be proper (51.14), the inequality is never strict:

(55.2) Corollary. *If $x \in B \subset A$, where B is a closed subalgebra of A such that $1 \in B$, then $r_B(x) = r_A(x)$.*

Proof. According to (55.1), the spectral radius depends only on the sequence of norms $\|x^n\|$. ∎

So to speak, when the algebra B containing x is enlarged to A, the spectrum of x in general shrinks but its spectral radius does not; the crux of the matter is *how* the spectrum shrinks, as will be explained in the next section.

Exercises

(55.3) (i) In (55.1) the hypothesis $\|1\| = 1$ is inessential: the spectral radius formula holds even if $\|1\| > 1$.

(ii) The formula $\lim \|x^n\|^{1/n} = \inf \|x^n\|^{1/n}$ is valid in any normed algebra.

(55.4) With notations as in (55.1), the following conditions on x are equivalent: (a) $r(x) = \|x\|$; (a′) there exists $\lambda \in \sigma(x)$ with $|\lambda| = \|x\|$; (b) $\|x^n\| = \|x\|^n$ for all positive integers n; (b′) $\|x^n\| = \|x\|^n$ for infinitely many positive integers n.

(55.5) If A is a Gel′fand algebra, then $\lim \|x^n\|^{1/n} = \|\hat{x}\|_\infty$ for all x in A (here $\hat{x}$ is the Gel′fand transform of x (52.13)); the radical of A is the set of all x such that $\|x^n\|^{1/n} \to 0$.

(55.6) The formula $r_A(x) = \lim \|x^n\|^{1/n}$ holds in every Banach algebra A (with or without unity).

56. Topological divisors of zero; boundary of the spectrum.

Let A be a Banach algebra with unity element 1, $\|1\| = 1$.

If B is a closed subalgebra of A containing 1, and if $x \in B$, then $\sigma_B(x) \supset \sigma_A(x)$. So to speak, in passing from B to the larger algebra A, the spectrum of x can only shrink. How much of $\sigma_B(x)$ survives? The present section is devoted to some useful partial answers. For $\lambda \in \sigma_B(x)$ to 'survive,' it must be the case that $\lambda 1 - x$, known to be singular in B, remains singular in A; if $\lambda 1 - x$ is 'drastically singular' in some sense, then its singularness may be 'permanent,' i.e., not ameliorated by enlargement of the algebra. For example, if $\lambda 1 - x$ is a left divisor of zero in B, say $(\lambda 1 - x)y = 0$ with y a nonzero element of B (one could even suppose $\|y\| = 1$), then clearly $\lambda 1 - x$ can have no inverse (nor even a left inverse) in A. The property of being a divisor of zero is a severe restriction (therefore difficult to achieve); a less restrictive condition (therefore easier to achieve) is available in a normed algebra:

(56.1) Definition. An element x of A is called a *topological divisor of zero* (briefly, TDZ) in A if there exists a sequence $y_n \in A$ with $\|y_n\| = 1$ for all n, such that either $xy_n \to 0$ (in which case x is called a *left TDZ*) or $y_n x \to 0$ (in which case x is called a *right TDZ*). If there exists a sequence $y_n \in A$ with $\|y_n\| = 1$, such that both $xy_n \to 0$ and $y_n x \to 0$, then x is called a *two-sided TDZ*.

(56.2) Lemma. *Every TDZ is singular. More precisely, if x is a left [right] TDZ, then x has no left [right] inverse.*

Proof. Assuming x is left-invertible, say $yx = 1$, it is to be shown that x is not a left TDZ. Indeed, if y_n is any sequence in A with $\|y_n\| = 1$, it results from

$$1 = \|y_n\| = \|(yx)y_n\| = \|y(xy_n)\| \le \|y\|\,\|xy_n\|$$

that $xy_n \nrightarrow 0$. ∎

(56.3) Lemma. *Let S be the set of singular elements of A and let ∂S be the boundary of S. Every element of ∂S is a two-sided TDZ in A.*

Proof. Let $x \in \partial S = \bar{S} \cap \overline{\complement S}$. Since $S = \complement U$ is closed (50.6), we have $\partial S = S \cap \bar{U}$. Thus x is singular, but there exists a sequence of invertible elements x_n such that $\|x_n - x\| \to 0$. Let $y_n = x_n^{-1}$ and define $z_n = \|y_n\|^{-1}y_n$; it will be shown that $xz_n \to 0$ and $z_n x \to 0$. At any rate,

$$xz_n = (x - x_n)z_n + x_n z_n = (x - x_n)z_n + \|y_n\|^{-1}1,$$

thus

$$\|xz_n\| \le \|x - x_n\| + \|y_n\|^{-1},$$

and similarly

$$\|z_n x\| \le \|x - x_n\| + \|y_n\|^{-1};$$

it will therefore suffice to show that $\|y_n\| \to \infty$.

For all n, we have $1 = y_n x_n = y_n(x_n - x) + y_n x$, therefore

(*) $$\|1 - y_n x\| \le \|y_n\|\,\|x_n - x\|.$$

Since y_n is invertible but x is not, $y_n x$ cannot be invertible; in view of (50.3), $\|1 - y_n x\| \ge 1$ for all n. Since $\|x_n - x\| \to 0$, it results from (*) that $\|y_n\| \to \infty$. ∎

(56.4) Lemma. *Let $x \in A$ and let λ be a boundary point of $\sigma_A(x)$. Then $\lambda 1 - x$ is a two-sided TDZ in A.*

Proof. The hypothesis is that $\lambda \in \partial\sigma_A(x)$. Since $\sigma_A(x)$ is closed (51.10), we have

$$\partial\sigma_A(x) = \overline{\sigma_A(x)} \cap \overline{\complement\sigma_A(x)} = \sigma_A(x) \cap \overline{\rho_A(x)};$$

thus $\lambda \in \sigma_A(x)$, and there exists a sequence $\lambda_n \in \rho_A(x)$ such that $\lambda_n \to \lambda$. This means that $\lambda 1 - x$ is singular but is the limit of the sequence of invertible elements $\lambda_n 1 - x$, therefore $\lambda 1 - x$ is a two-sided TDZ (56.3). ∎

(56.5) Theorem. *Let B be a closed subalgebra of A containing 1 and let $x \in B$. Then*

$$\partial \sigma_B(x) \subset \partial \sigma_A(x).$$

In particular, if $\sigma_B(x)$ has empty interior then $\sigma_B(x) = \sigma_A(x)$.

Proof. If $\lambda \in \partial \sigma_B(x)$, then $\lambda 1 - x$ is a TDZ in B (56.4); a fortiori, $\lambda 1 - x$ is a TDZ in A, therefore $\lambda 1 - x$ is singular in A (56.2), thus $\lambda \in \sigma_A(x)$. Since $\sigma_B(x) \supset \sigma_A(x)$, it follows that λ is also a boundary point of $\sigma_A(x)$.

If int $\sigma_B(x) = \varnothing$ then $\sigma_B(x) = \partial \sigma_B(x) \subset \sigma_A(x) \subset \sigma_B(x)$. ∎

In the notation of (56.5), $\sigma_B(x) - \sigma_A(x)$ is trivially open in $\sigma_B(x)$; surprisingly, it is even open in $\mathbb{C}$:

(56.6) Corollary. *If $x \in B \subset A$, where B is a closed subalgebra of A containing 1, then*

$$\sigma_B(x) = \sigma_A(x) \cup W,$$

where W is the union of certain of the holes of $\sigma_A(x)$; thus $\sigma_B(x)$ is the union of $\sigma_A(x)$ and the holes of $\sigma_A(x)$ that it intersects. Moreover, $\rho_A(x)$ and $\rho_B(x)$ have the same unbounded component.

Proof. {For the terminology, see (51.12). To put the matter suggestively, the passage from $\sigma_A(x)$ to $\sigma_B(x)$ is a question of filling in certain of its holes.}

We have $\sigma_B(x) \supset \sigma_A(x)$ and $\sigma_B(x) - \sigma_A(x) = \sigma_B(x) \cap \rho_A(x)$; to prove the first assertion of the corollary, it will suffice to show that $\sigma_B(x)$ contains every component of $\rho_A(x)$ that it intersects.

Let C be a component (bounded or unbounded) of $\rho_A(x)$. We wish to show that either $C \cap \sigma_B(x) = \varnothing$ or $C \cap \sigma_B(x) = C$. Write $D = C \cap \sigma_B(x)$. We know that D is a closed subset of the connected space C; to show that $D = \varnothing$ or $D = C$, it will suffice to show that the boundary of D relative to the space C is empty, i.e., $\partial_C D = \varnothing$. Assume to the contrary; say

$$\lambda \in \partial_C D = C \cap \bar{D} \cap \overline{C - D}.$$

It follows that λ is also a boundary point of $\sigma_B(x)$ in $\mathbb{C}$. {Proof: One has $D \subset \sigma_B(x)$ and $C - D = C - \sigma_B(x) \subset \complement \sigma_B(x)$, thus $\lambda \in \bar{D} \subset \sigma_B(x)$ and $\lambda \in \overline{C - D} \subset \complement \sigma_B(x)$.} Thus $\lambda \in \partial \sigma_B(x) \subset \sigma_A(x)$ (56.5), whereas $\lambda \in C \subset \rho_A(x) = \complement \sigma_A(x)$, a contradiction. This completes the proof of the first assertion.

Suppose, in particular, that C is the unbounded component of $\rho_A(x)$. From the foregoing, we know that $C \cap \sigma_B(x) = \varnothing$, thus $C \subset \rho_B(x)$; this shows that C is contained in the unbounded component of $\rho_B(x)$. On the other hand, it results from $\rho_B(x) \subset \rho_A(x)$ that the unbounded component of $\rho_B(x)$ is contained in C. Thus C is also the unbounded component of $\rho_B(x)$. ∎

A useful application of (56.5) to involutive algebras:

(56.7) Corollary. *Let A be a Banach $*$-algebra with unity element 1, and let B be a closed $*$-subalgebra of A containing 1, such that $\sigma_B(y)$ is real for every self-adjoint element y of B. Then B is a full subalgebra of A, thus $\sigma_B(x) = \sigma_A(x)$ for all x in B.*

Proof. Assuming $x \in B$ and $0 \in \sigma_B(x)$, it is to be shown that $0 \in \sigma_A(x)$. Since x is not invertible in B, either x^*x or xx^* must fail to be invertible in B. Suppose, for example, that $y = x^*x$ is singular in B, thus $0 \in \sigma_B(y)$. Since $y^* = y$, we have $\sigma_B(y) \subset \mathbb{R}$ by hypothesis, thus 0 must be a boundary point of $\sigma_B(y)$; citing (56.5), we have $0 \in \sigma_A(y)$. Thus x^*x is not invertible in A; a fortiori, x is not invertible in A, that is, $0 \in \sigma_A(x)$. This shows that B is a full subalgebra of A, and therefore $\sigma_B(x) = \sigma_A(x)$ for all x in B (53.8). {Concerning the hypothesis on B, see (56.12).}

Exercises

(56.8) Let A be a normed algebra, let $x \in A$, and let U_x, V_x be the continuous linear mappings on A defined by $U_x y = xy$, $V_x y = yx$ $(y \in A)$.

(i) The following conditions on x are equivalent: (a) x is a left TDZ in A; (b) U_x is not bounded below; (c) U_x is a left TDZ in $\mathscr{L}(A)$. So are the conditions (a') x is a right TDZ in A; (b') V_x is not bounded below; (c') V_x is a left TDZ in $\mathscr{L}(A)$.

(ii) Assume A is a Banach algebra with unity. The following conditions on x are equivalent: (a) x is a left TDZ in A; (b) either x is a left divisor of zero, or the right ideal xA is not closed. Similarly with 'right' in place of 'left.'

(56.9) Let A be a finite-dimensional, associative algebra over $\mathbb{C}$, with unity element 1, and let $x \in A$. The following conditions on x are equivalent: (a) x is singular (i.e., not invertible); (b) x has no left inverse; (b') x has no right inverse; (c) x is a right divisor of zero; (c') x is a left divisor of zero; (d) x is a two-sided divisor of zero, that is, $xy = yx = 0$ for some nonzero element y. In particular, in a finite-dimensional Banach algebra with unity, all variants of the concepts 'singular,' 'divisor of zero,' and 'topological divisor of zero' merge.

(56.10) Let A be a Banach algebra with unity and let $x \in A$. If x is not left-invertible, but x is the limit of a sequence of left-invertible elements, then x is a right TDZ.

(56.11) The proof of (56.6) is a proposition in elementary topology: If X is a locally connected topological space, and if S and T are closed subsets of X such that $S \subset T$ and $\partial T \subset S$, then $T = S \cup W$, where W is the union of those components of $\complement S$ that T intersects.

(56.12) Let A be a Banach algebra with unity element 1, let B be a closed subalgebra of A, and let $y \in B$. Then $\sigma_B(y)$ is real iff $\sigma_A(y)$ is real (in which case the two spectra are equal).

*57. Spectrum in $\mathscr{L}(E)$. In special Banach algebras, the theory of the spectrum may be expected to have special features. In this section, we explore spectrum in the context of the Banach algebra $\mathscr{L}(E)$, where E is a nonzero complex Banach space (40.23). A few of the preliminary results are valid with E a normed space; the central results require E to be a Banach space; reflexivity produces some extra overtones; and the case of a Hilbert space offers the most detailed results.

If E is a normed space (always assumed complex and $\neq \{\theta\}$ in this section) then $\mathscr{L}(E)$ is a normed algebra with unity element I, $\|I\| = 1$ (40.23); if E is a Banach space then $\mathscr{L}(E)$ is a Banach algebra; if E is a reflexive Banach space, then the mapping $T \mapsto T'$ (40.18) is a norm-preserving vector space isomorphism of $\mathscr{L}(E)$ onto $\mathscr{L}(E')$ such that $(ST)' = T'S'$ for all S, T in $\mathscr{L}(E)$ (cf. (40.19), (40.20)); and if E is a Hilbert space then $\mathscr{L}(E)$ is a C^*-algebra (43.7).

If $T \in \mathscr{L}(E)$ we write briefly $\sigma(T)$ instead of $\sigma_{\mathscr{L}(E)}(T)$. The core idea in the following analysis is that there is a variety of ways in which a continuous linear mapping may fail to be invertible.

Among the singular elements of $\mathscr{L}(E)$ are the divisors of zero; they may be characterized as follows:

(57.1) Theorem. *Let E be a normed space and let $T \in \mathscr{L}(E)$.*
(i) *T is a left divisor of zero in $\mathscr{L}(E)$ iff T is not injective.*
(ii) *T is a right divisor of zero in $\mathscr{L}(E)$ iff $T(E)$ is not dense in E.*

Proof. (i) Suppose T is not injective. Let y be a nonzero vector in E such that $Ty = \theta$, let f be a nonzero continuous linear form on E (40.10), and define $S \in \mathscr{L}(E)$ by the formula

$$Sx = f(x)y \qquad (x \in E);$$

evidently $TS = 0$, thus T is a left divisor of zero in $\mathscr{L}(E)$. The converse is obvious.

(ii) Suppose $T(E)$ is not dense in E, say $y \notin \overline{T(E)}$. Let f be a continuous linear form on E such that $f(y) \neq 0$ and $f = 0$ on $T(E)$ (33.12), and let z be a nonzero vector in E. Define $S \in \mathscr{L}(E)$ by the formula

$$Sx = f(x)z \qquad (x \in E);$$

then $STx = f(Tx)z = \theta$ for all $x \in E$, whereas $Sy = f(y)z \neq \theta$, thus T is a right divisor of zero. The converse is obvious. ∎

Each mode of noninvertibility invites naming a corresponding subset of the spectrum:

(57.2) Definition. Let E be a normed space and let $T \in \mathscr{L}(E)$. A complex number λ is said to be an *eigenvalue* of T if $T - \lambda I$ is not injective, i.e., there exists a nonzero vector x such that $Tx = \lambda x$; the set of all such λ is called the *point spectrum* of T, denoted $\sigma_p(T)$. The set of all complex numbers λ such that the range of $T - \lambda I$ is not dense is called the *compression spectrum* of T, denoted $\sigma_{com}(T)$. {There is no universally lovely notation to cover all the subsets that have to be named.}

(57.3) Corollary. *If E is a normed space and $T \in \mathscr{L}(E)$, then*

$$\sigma_p(T) = \{\lambda \in \mathbb{C} : T - \lambda I \text{ is a left divisor of zero in } \mathscr{L}(E)\},$$

$$\sigma_{com}(T) = \{\lambda \in \mathbb{C} : T - \lambda I \text{ is a right divisor of zero in } \mathscr{L}(E)\}.$$

Obviously $\sigma_p(T) \subset \sigma(T)$ and $\sigma_{\text{com}}(T) \subset \sigma(T)$; none of these sets is guaranteed to be nonempty. For $\tau = \sigma_p$, σ_{com}, or σ, we have

$$\tau(aT + bI) = \{a\lambda + b : \lambda \in \tau(T)\},$$

provided that $\tau(T)$ is nonempty, $a, b \in \mathbb{C}$ and $a \neq 0$.

Another way to be singular is to be a topological divisor of zero (56.2); left TDZ's are characterized as follows:

(57.4) Theorem. *Let E be a normed space and let $T \in \mathscr{L}(E)$. The following conditions on T are equivalent:*
(a) *T is a left TDZ in $\mathscr{L}(E)$;*
(b) *there exists a sequence of vectors x_n in E such that $\|x_n\| = 1$ and $Tx_n \to \theta$;*
(c) *T is not bounded below (40.28).*

Proof. The equivalence of (b) and (c) is obvious from the definitions.

(b) implies (a): Let x_n be a sequence of unit vectors such that $Tx_n \to \theta$, let $f \in E'$ with $\|f\| = 1$ (40.10), and, for each n, define $S_n \in \mathscr{L}(E)$ by the formula

$$S_n x = f(x) x_n \qquad (x \in E).$$

Clearly $\|S_n\| = \|f\| \|x_n\| = 1$. Moreover,

$$TS_n x = f(x) Tx_n \qquad (x \in E),$$

therefore $\|TS_n\| = \|f\| \|Tx_n\| \to 0$. Thus T is a left TDZ in $\mathscr{L}(E)$.

(a) implies (b): Let S_n be a sequence in $\mathscr{L}(E)$ such that $\|S_n\| = 1$ and $\|TS_n\| \to 0$. For each n, let y_n be a unit vector such that $\|S_n y_n\| \geq \frac{1}{2}$. Then $x_n = \|S_n y_n\|^{-1} S_n y_n$ is a unit vector, and

$$\|Tx_n\| = \|S_n y_n\|^{-1} \|TS_n y_n\| \leq 2\|TS_n\| \to 0. \qquad \blacksquare$$

This names another subset of the spectrum:

(57.5) Definition. Let E be a normed space and let $T \in \mathscr{L}(E)$. A complex number λ is said to be an *approximate eigenvalue* of T if there exists a sequence of vectors x_n in E such that $\|x_n\| = 1$ and $Tx_n - \lambda x_n \to \theta$; the set of all such λ is called the *approximate point spectrum* of T, denoted $\sigma_{\text{ap}}(T)$.

(57.6) Corollary. *If E is a normed space and $T \in \mathscr{L}(E)$, then*

$$\sigma_{\text{ap}}(T) = \{\lambda \in \mathbb{C} : T - \lambda I \text{ is a left TDZ in } \mathscr{L}(E)\}$$

$$= \{\lambda \in \mathbb{C} : T - \lambda I \text{ is not bounded below}\}.$$

Evidently $\sigma_{\text{ap}}(T) \subset \sigma(T)$, and

$$\sigma_{\text{ap}}(aT + bI) = \{a\lambda + b : \lambda \in \sigma_{\text{ap}}(T)\}$$

provided that $a \neq 0$ and $\sigma_{\text{ap}}(T) \neq \varnothing$. When E is a Banach space, the approximate point spectrum is definitely nonempty:

(57.7) Theorem.   *If E is a Banach space and $T \in \mathcal{L}(E)$, then $\partial\sigma(T) \subseteq \sigma_{\text{ap}}(T)$.*

Proof.   If $\lambda \in \partial\sigma(T)$ then $T - \lambda I$ is a left (even two-sided) TDZ in $\mathcal{L}(E)$ (56.4), therefore $\lambda \in \sigma_{\text{ap}}(T)$ (57.6). {In particular, $\sigma_{\text{ap}}(T)$ contains every $\lambda \in \sigma(T)$ such that $|\lambda| = r(T)$ (51.11).} ∎

The characterization of right TDZ's is less transparent (though the situation is improved when E is a Banach space (57.17)):

(57.8) Theorem.   *Let E be a normed space and let $T \in \mathcal{L}(E)$. The following conditions on T are equivalent:*

  (a) *T is a right TDZ in $\mathcal{L}(E)$;*
  (b) *T' is a left TDZ in $\mathcal{L}(E')$;*
  (c) *there exists a sequence $f_n \in E'$ such that $\|f_n\| = 1$ and $f_n \to 0$ uniformly on $T(E_1)$, where $E_1 = \{x \in E : \|x\| \leq 1\}$.*

Proof.   (a) implies (b): If $S_n \in \mathcal{L}(E)$, $\|S_n\| = 1$, and $\|S_n T\| \to 0$, then $\|S_n'\| = 1$ and $\|T'S_n'\| = \|(S_n T)'\| = \|S_n T\| \to 0$.

(b) implies (a): Assuming T' is a left TDZ in $\mathcal{L}(E')$, by (57.4) there exists a sequence $f_n \in E'$ such that $\|f_n\| = 1$ and $\|T'f_n\| \to 0$. Fix a vector $z \in E$ with $\|z\| = 1$ and, for each n, define $S_n \in \mathcal{L}(E)$ by the formula

$$S_n x = f_n(x) z \qquad (x \in E);$$

then $\|S_n\| = \|f_n\| \|z\| = 1$, and it results from the calculation

$$S_n T x = f_n(Tx) z = [(f_n \circ T)(x)] z = [(T'f_n)(x)] z$$

that $\|S_n T\| = \|T'f_n\| \to 0$.

The equivalence of (b) and (c) follows at once from the observation that if $f_n \in E'$, $\|f_n\| = 1$, then $\|T'f_n\| = \|f_n \circ T\| = \sup\{|f_n(Tx)| : x \in E_1\}$. ∎

In view of (57.8) and (57.6), the set of all $\lambda \in \mathbb{C}$ such that $T - \lambda I$ is a right TDZ is precisely the set $\sigma_{\text{ap}}(T')$; we refrain from giving a name to this subset of $\sigma(T)$, thus leaving a dissymmetry in the nomenclature. The situation is inherently dissymmetric:

(57.9) Theorem.   *Let E be a Banach space and let $T \in \mathcal{L}(E)$. The following conditions on T are equivalent:*

  (a) *T is singular;*
  (b) *T is either a right divisor of zero or a left TDZ in $\mathcal{L}(E)$.*

Proof.   Suppose T is neither a right divisor of zero nor a left TDZ. Then $T(E)$ is dense in E (57.1) and T is bounded below (57.4). The latter condition implies that T is injective and $T(E)$ is closed (40.28). Thus T is bijective, and continuity of the inverse comes free of charge with boundedness below (no need to quote the open mapping theorem), thus T is invertible. This proves that (a) implies (b), and the converse is trivial. ∎

(57.10) Corollary.   *If E is a Banach space and $T \in \mathcal{L}(E)$, then*

$$\sigma(T) = \sigma_{\text{com}}(T) \cup \sigma_{\text{ap}}(T).$$

Proof.   Immediate from (57.9), (57.3), and (57.6). ∎

(57.11) Corollary. *Let E be a reflexive Banach space and let $T \in \mathscr{L}(E)$. If T is singular but is not a divisor of zero in $\mathscr{L}(E)$, then T is both a left TDZ and a right TDZ.*

Proof. By (57.9), T is a left TDZ in $\mathscr{L}(E)$. Since E is reflexive, T' satisfies the same hypotheses relative to $\mathscr{L}(E')$, therefore T' is a left TDZ in $\mathscr{L}(E')$— whence T is a right TDZ in $\mathscr{L}(E)$ (by either (57.8) or reflexivity). ∎

(57.12) Corollary. *Let E be a reflexive Banach space and let $T \in \mathscr{L}(E)$. If T is injective and $T(E)$ is a dense, proper subspace of E, then T is both a left TDZ and a right TDZ.*

Proof. T is not a divisor or zero in $\mathscr{L}(E)$ (57.1) but is obviously singular; quote (57.11). ∎

The hypothesis of (57.12) ties in with one of the two standard concepts that round out our analysis of the spectrum:

(57.13) Definition. Let E be a normed space and let $T \in \mathscr{L}(E)$. The *residual spectrum* of T, denoted $\sigma_r(T)$, is the set of all complex numbers λ such that $T - \lambda I$ is injective but its range is not dense in E; thus

$$\sigma_r(T) = \sigma_{\text{com}}(T) - \sigma_p(T).$$

The *continuous spectrum* of T, denoted $\sigma_c(T)$, is the set of all complex numbers λ such that $T - \lambda I$ is injective, has dense range, but is singular; thus

$$\sigma_c(T) = \sigma(T) - \{\sigma_p(T) \cup \sigma_{\text{com}}(T)\}.$$

It is obvious from the formulas that

$$\sigma(T) = \sigma_p(T) \cup \sigma_c(T) \cup \sigma_r(T),$$

the terms on the right being mutually disjoint.

(57.14) Remark. Notation as in (57.13). In view of (57.3), a complex number λ belongs to $\sigma_p(T)$ iff $T - \lambda I$ is a left divisor of zero; to $\sigma_r(T)$ iff $T - \lambda I$ is a right divisor of zero but not a left divisor of zero; and to $\sigma_c(T)$ iff $T - \lambda I$ is singular but is not a divisor of zero. Exercise (57.34) sheds further light on the concept of continuous spectrum in the case that E is a Banach space.

In order that $T \in \mathscr{L}(E)$ be invertible (E a normed space), it is obviously necessary and sufficient that T be bounded below and surjective. In other words, T is singular iff (i) T is not bounded below, or (ii) T is not surjective. Condition (i) is equivalent to T being a left TDZ in $\mathscr{L}(E)$ (57.4). Our next goal is to show that, when E is a Banach space, condition (ii) is equivalent to T being a right TDZ in $\mathscr{L}(E)$ (57.17).

(57.15) Lemma. *Let E be a normed space, let $T \in \mathscr{L}(E)$, and suppose that T' is bounded below. Then $\overline{T(E_1)}$ is a neighborhood of zero, where $E_1 = \{x \in E : \|x\| \le 1\}$.*

Proof. Let $M > 0$ with $\|T'f\| \geq M\|f\|$ for all $f \in E'$; it will be shown that $\overline{T(E_1)}$ contains the ball $\{y \in E : \|y\| \leq M\}$.

Assuming $y \notin \overline{T(E_1)}$, it will suffice to show that $\|y\| > M$. Since $\overline{T(E_1)}$ is closed and convex, by the Hahn-Banach theorem there exist $f \in E'$ and a real number β such that $\mathrm{Re}\,f(y) > \beta$ and $\mathrm{Re}\,f(z) < \beta$ for all $z \in \overline{T(E_1)}$ (34.4). It follows that

(*) $|(T'f)(x)| < \beta$ for all $x \in E_1$.

{Proof: Suppose $x \in E_1$. Write $|f(Tx)| = \mu f(Tx)$ for suitable $|\mu| = 1$. Then $|(T'f)(x)| = \mu f(Tx) = f(T(\mu x)) = \mathrm{Re}\,f(T(\mu x)) < \beta$ because $\mu x \in E_1$.} From (*) we have $\|T'f\| \leq \beta$; thus

$$M\|f\| \leq \|T'f\| \leq \beta < \mathrm{Re}\,f(y) \leq |f(y)| \leq \|f\|\,\|y\|,$$

therefore $M < \|y\|$. ∎

(57.16) Theorem. *If E is a Banach space and $T \in \mathscr{L}(E)$, the following conditions on T are equivalent:*

(a) *T is surjective;*

(b) *T' is bounded below.*

Proof. (b) implies (a): Suppose T' is bounded below. By the lemma, $\overline{T(E_1)}$ is a neighborhood of zero, where $E_1 = \{x : \|x\| \leq 1\}$. For each $r > 0$ let $E_r = \{x : \|x\| \leq r\} = rE_1$; then

$$\overline{T(E_r)} = \overline{T(rE_1)} = \overline{rT(E_1)} = r\overline{T(E_1)}$$

shows that $\overline{T(E_r)}$ is a neighborhood of zero for every $r > 0$. It follows that T is an open mapping (48.3); in particular, $T(E)$ is an open linear subspace of E, thus $T(E) = E$.

(a) implies (b): Suppose $T(E) = E$. Since $T'f = f \circ T$ for all $f \in E'$, it follows at once that T' is injective. {More generally, it is an immediate consequence of the Hahn-Banach theorem (33.12) that T' is injective iff T has dense range.} Let

$$\mathscr{S} = \{f \in E' : \|T'f\| \leq 1\}.$$

For each $x \in E$, the set $\{f(x) : f \in \mathscr{S}\}$ is bounded; for, writing $x = Ty$ for suitable y, we have

$$|f(x)| = |f(Ty)| = |(T'f)(y)| \leq \|T'f\|\,\|y\| \leq \|y\|$$

for all f in $\mathscr{S}$. Thus, the set $\mathscr{S}$ is pointwise bounded on the Banach space E; by the uniform boundedness principle (47.2), $\mathscr{S}$ is bounded in norm. Say $\|f\| \leq M$ for all $f \in \mathscr{S}$, where $0 < M < \infty$. The proof will be concluded by showing that $\|T'g\| \geq (1/M)\|g\|$ for all g in E'. This inequality holds trivially when $T'g = 0$ (because T' is injective). Assuming $T'g \neq 0$, set $f = \|T'g\|^{-1}g$; since $\|T'f\| = 1$ we have $f \in \mathscr{S}$, therefore $\|f\| \leq M$, i.e., $\|g\| \leq M\|T'g\|$. ∎

(57.17) Collorary. *If E is a Banach space and $T \in \mathscr{L}(E)$, the following conditions on T are equivalent:*

(a) T is not surjective;
(b) T is a right TDZ in $\mathscr{L}(E)$;
(c) T' is not bounded below.

Proof. The equivalence of (b) and (c) is valid for E a normed space: T is a right TDZ in $\mathscr{L}(E)$ iff T' is a left TDZ in $\mathscr{L}(E')$ (57.8) iff T' is not bounded below (57.4). By (57.16), (a) and (c) are also equivalent. ∎

It follows at once from (57.17) that if $T \in \mathscr{L}(E)$, E a Banach space, then

$$\sigma_{\mathrm{ap}}(T') = \{\lambda \in \mathbb{C} : T - \lambda I \text{ is not surjective}\}.$$

The dual result,

$$\sigma_{\mathrm{ap}}(T) = \{\lambda \in \mathbb{C} : T' - \lambda I \text{ is not surjective}\},$$

is valid for an arbitrary normed space E:

(57.18) Theorem. *If E is a normed space and $T \in \mathscr{L}(E)$, the following conditions on T are equivalent;*
(a) *T is bounded below;*
(b) *T' is surjective.*

Proof. (a) implies (b): Let $M > 0$ be a constant such that $\|Tx\| \geq M\|x\|$ for all $x \in E$. It follows that T is injective, and

(*) $$Tx \mapsto x$$

is a well-defined continuous linear mapping of $T(E)$ onto E. Assuming $g \in E'$, we seek $f \in E'$ such that $T'f = g$. Define $f_0 : T(E) \to \mathbb{C}$ by the formula $f_0(Tx) = g(x)$; thus f_0 is the composite of the mapping (*) with g, therefore f_0 is a continuous linear form on $T(E)$. By the Hahn-Banach theorem, f_0 has an extension f in E' (44.2). Then $g = f_0 \circ T = f \circ T = T'f$.

(b) implies (a): If T' is surjective then, since E' is a Banach space (40.9), T'' is bounded below (57.16); it follows from the properties of the canonical embedding $E \to E''$ that T is also bounded below (cf. (40.13), (40.19)). ∎

We conclude the section with various results that exploit the special character of Hilbert space.

(57.19) Theorem. *If H is a Hilbert space and $T \in \mathscr{L}(H)$, the following conditions on T are equivalent:*
(a) *T is left-invertible in $\mathscr{L}(H)$;*
(b) *T is bounded below;*
(c) *T is not a left TDZ in $\mathscr{L}(H)$.*

Proof. (a) implies (b): This is true for H a normed space: if $S \in \mathscr{L}(H)$ with $ST = I$, then $\|x\| = \|S(Tx)\| \leq \|S\| \|Tx\|$ for all $x \in H$.

(b) implies (a): If T is bounded below, then T is injective, $T(H)$ is a closed linear subspace of H, and T maps H bicontinuously onto $T(H)$ (40.28). If $S : H \to H$ is the unique linear mapping such that $S(Tx) = x$ for all $x \in H$ and $S = 0$ on $T(H)^{\perp}$ (cf. 42.7), then $S \in \mathscr{L}(H)$ and $ST = I$.

The equivalence of (b) and (c) requires only that H be a normed space (57.4). ∎

(57.20) Corollary. *If H is a Hilbert space and $T \in \mathscr{L}(H)$, the following conditions on T are equivalent:*

(a) *T is right-invertible in $\mathscr{L}(H)$;*

(b) *T is surjective;*

(c) *T is not a right TDZ in $\mathscr{L}(H)$.*

Proof. It is obvious from the properties of adjunction that T is a left [right] divisor of zero iff T^* is a right [left] divisor of zero, and similarly for TDZ's. Thus (a) and (c) are equivalent by the parallel assertion of (57.19).

The equivalence of (b) and (c) requires only that H be a Banach space (57.17); but the following argument that (a) and (b) are equivalent in the Hilbert space setting is simpler.

(a) implies (b): Obvious.

(b) implies (a): Let N be the null space of T, and let T_0 be the restriction of T to $N^\perp$. Since $H = N \oplus N^\perp$, it is clear that $T_0 : N^\perp \to H$ is a continuous vector space isomorphism; by the open mapping theorem, it is bicontinuous (48.1). The formula $Sx = T_0^{-1}x$ $(x \in H)$ defines an element of $\mathscr{L}(H)$ such that $TS = I$. ∎

The following result is a useful characterization of invertibility in the Hilbert space setting; it is a special case of a number of the earlier results in the section (cf. 43.8), but the elementary indigenous proof is clearer:

(57.21) Theorem. *Let H be a Hilbert space and let $T \in \mathscr{L}(H)$. The following conditions on T are equivalent:*

(a) *T is invertible;*

(b) *T and T^* are bounded below;*

(c) *T is bounded below and T^* is injective;*

(d) *T is bounded below and $T(H)$ is dense.*

Proof. The implications (a) $\Rightarrow$ (b) $\Rightarrow$ (c) are obvious.

(c) implies (d): More generally, it follows from the relation $(T^*y | x) = (y | Tx)$ that the null space of T^* coincides with $T(H)^\perp = \overline{T(H)}^\perp$, therefore T^* is injective iff $T(H)$ is dense (42.9).

(d) implies (a): Immediate from (40.28). ∎

The adjunction $T \mapsto T^*$ for Hilbert space operators is reminiscent of complex conjugation in $\mathbb{C}$. The analogy is pursued to consummation in von Neumann's theory of spectral sets (Section 66); for the present, we are interested in the operatorial analogue of 'real number':

(57.22) Definition. Let $T \in \mathscr{L}(H)$, H a Hilbert space. T is said to be *self-adjoint* (or 'Hermitian') if $T^* = T$.

For arbitrary $T \in \mathscr{L}(H)$, the formula $\varphi_T(x, y) = (Tx | y)$ defines a sesquilinear form on H (cf. 43.9), such that

$$(\varphi_T(y, x))^* = (Ty | x)^* = (x | Ty) = (T^*x | y) = \varphi_{T^*}(x, y);$$

clearly T is self-adjoint iff φ_T is Hermitian iff $(Tx | x)$ is real for all x (cf. 41.14).

(57.23) Lemma. *Let H be a Hilbert space and let $T \in \mathscr{L}(H)$.*
(i) *If $(Tx|x) \geq 0$ for all $x \in H$, then $T + I$ is invertible.*
(ii) *$T^*T + I$ is invertible for every T.*

Proof. (i) In particular, $(Tx|x)$ is real for all x, thus $T^* = T$. The calculation

$$\|(T + I)x\|^2 = \|Tx\|^2 + (Tx|x) + (x|Tx) + \|x\|^2$$
$$= \|Tx\|^2 + 2(Tx|x) + \|x\|^2 \geq \|x\|^2$$

shows that $T + I$ is bounded below; being self-adjoint, it must be invertible (57.21).
(ii) $(T^*Tx|x) = (Tx|Tx) \geq 0$ for all x. ∎

(57.24) Lemma. *If H is a Hilbert space and $T \in \mathscr{L}(H)$ is self-adjoint, then $\sigma(T) \subset \mathbb{R}$.*

Proof. Assuming $\mu \in \mathbb{C}$ has Cartesian form $\mu = \alpha + i\beta$ with $\beta \neq 0$, it is to be shown that $T - \mu I$ is invertible. Since

$$T - \mu I = (T - \alpha I) - i\beta I = -\beta[-\beta^{-1}(T - \alpha I) + iI],$$

where $-\beta^{-1}(T - \alpha I)$ is also self-adjoint, it clearly suffices (after a change of notation) to show that $T + iI$ is invertible. Indeed,

$$(T + iI)^*(T + iI) = (T + iI)(T + iI)^* = T^2 + I$$

is invertible by (57.23). ∎

(57.25) Lemma. *Let H be a Hilbert space and suppose $T \in \mathscr{L}(H)$ is self-adjoint. Define*

$$m = \inf \{(Tx|x) : \|x\| = 1\},$$
$$M = \sup \{(Tx|x) : \|x\| = 1\}.$$

Then $m, M \in \sigma(T)$.

Proof. We show that $m \in \sigma(T)$. {The proof that $M \in \sigma(T)$ is similar; alternatively, one can consider $-T$.} This will be done by finding a sequence of unit vectors x_n such that $\|(T - mI)x_n\| \to 0$. Define a sesquilinear form φ on H by the formula

$$\varphi(x, y) = ((T - mI)x|y) \qquad (x, y \in H).$$

Evidently $\varphi(x, x) \geq 0$ for all $x \in H$ with $\|x\| = 1$; it follows that $\varphi(x, x) \geq 0$ for all $x \in H$, therefore φ satisfies the Cauchy-Schwarz inequality:

$$|\varphi(x, y)|^2 \leq \varphi(x, x)\varphi(y, y)$$

for all x, y in H (41.14). Note also that

$$|\varphi(x, y)| \leq \|T - mI\| \|x\| \|y\|$$

for all x, y in H. It follows that for every vector x,

$$\|(T - mI)x\|^4 = ((T - mI)x|(T - mI)x)^2$$
$$= [\varphi(x, (T - mI)x)]^2$$
$$\leq \varphi(x, x)\varphi((T - mI)x, (T - mI)x)$$
$$\leq \varphi(x, x)\|T - mI\| \|(T - mI)x\|^2,$$

therefore

(*) $$\|(T - mI)x\|^2 \leq \varphi(x, x)\|T - mI\|.$$

By the definition of m, there exists a sequence of vectors x_n such that $\|x_n\| = 1$ and $\varphi(x_n, x_n) \to 0$; in view of (*), we have $\|(T - mI)x_n\| \to 0$. ∎

(57.26) Lemma. *Let $T \in \mathcal{L}(H)$, H a Hilbert space. The following conditions on T are equivalent:*
(a) $(Tx|x) \geq 0$ *for all $x \in H$;*
(b) $T^* = T$ *and $\sigma(T) \subset [0, \infty)$.*

Proof. (b) implies (a): In the notation of (57.25) we have $m \in \sigma(T)$, therefore $m \geq 0$.

(a) implies (b): In particular, $(Tx|x)$ is real for all x, thus $T^* = T$ and therefore $\sigma(T) \subset \mathbb{R}$ (57.24). Assuming $\beta \in \mathbb{R}$, $\beta < 0$, it will suffice to show that $T - \beta I$ is invertible. But

$$T - \beta I = -\beta(-\beta^{-1}T + I),$$

where $-\beta^{-1}T$ also satisfies (a); quote (57.23). ∎

(57.27) Theorem. *Let H be a Hilbert space and suppose $T \in \mathcal{L}(H)$ is self-adjoint. Define*

$$m = \inf\{(Tx|x) : \|x\| = 1\},$$

$$M = \sup\{(Tx|x) : \|x\| = 1\}.$$

Then $\{m, M\} \subset \sigma(T) \subset [m, M]$, thus $[m, M]$ is the smallest interval containing $\sigma(T)$.

Proof. We have $((T - mI)x|x) \geq 0$ for all unit vectors x, hence for all vectors x, thus $\sigma(T - mI) \subset [0, \infty)$ by (57.26). Since

$$\sigma(T - mI) = \{\lambda - m : \lambda \in \sigma(T)\},$$

we conclude that $\sigma(T) \subset [m, \infty)$. Similarly $\sigma(T) \subset (-\infty, M]$, thus $\sigma(T) \subset [m, M]$. Finally, $\{m, M\} \subset \sigma(T)$ by (57.25). ∎

These results motivate the following definition:

(57.28) Definition. Let H be a Hilbert space. If $S, T \in \mathcal{L}(H)$ are self-adjoint and if $(Sx|x) \leq (Tx|x)$ for all $x \in H$, we write $S \leq T$ (or $T \geq S$).

It is routine to check that this defines a partial ordering of the set of all self-adjoint operators on H. {In particular, antisymmetry results from the fact that if $(Tx|x) = 0$ for all $x \in H$ then $T = 0$ (cf. 41.14).} The message of (57.26): For a self-adjoint operator T, $T \geq 0$ iff $\sigma(T) \geq 0$.

Exercises

(57.29) If $T \in \mathscr{L}(E)$, E a normed space, then
$$\sigma_{\text{com}}(T) = \sigma_{\text{p}}(T');$$
if E is a Hilbert space, then $\sigma_{\text{com}}(T) = (\sigma_{\text{p}}(T^*))^*$.

(57.30) If $T \in \mathscr{L}(E)$, E a Banach space, then
$$\sigma(T) = \sigma_{\text{ap}}(T) \cup \sigma_{\text{p}}(T');$$
if E is a Hilbert space, then $\sigma(T) = \sigma_{\text{ap}}(T) \cup (\sigma_{\text{p}}(T^*))^*$.

(57.31) If $T \in \mathscr{L}(E)$, E a reflexive Banach space, then
$$\sigma(T) = \sigma_{\text{p}}(T) \cup \sigma_{\text{p}}(T') \cup [\sigma_{\text{ap}}(T) \cap \sigma_{\text{ap}}(T')];$$
if E is a Hilbert space, then
$$\sigma(T) = \sigma_{\text{p}}(T) \cup (\sigma_{\text{p}}(T^*))^* \cup [\sigma_{\text{ap}}(T) \cap (\sigma_{\text{ap}}(T^*))^*].$$

(57.32) Let E be a normed space, $T \in \mathscr{L}(E)$, $E_1 = \{x \in E : \|x\| \leq 1\}$. Then T is invertible in $\mathscr{L}(E)$ iff T is injective and $T(E_1)$ is a neighborhood of zero.

(57.33) If $T \in \mathscr{L}(E)$, E a normed space, then $\sigma_{\text{ap}}(T)$ is a closed subset of $\mathbb{C}$.

(57.34) Let E be a Banach space. If $T \in \mathscr{L}(E)$ is injective and has dense range, then the following conditions on T are equivalent: (a) T is invertible; (b) T is bounded below; (c) T is surjective. It follows that $\sigma_{\text{c}}(T) \subset \sigma_{\text{ap}}(T)$ for every T in $\mathscr{L}(E)$.

(57.35) If H is a Hilbert space and if $T \in \mathscr{L}(H)$ is *normal* ($T^*T = TT^*$), then $\sigma(T) = \sigma_{\text{ap}}(T)$.

(57.36) Let H be a Hilbert space and let $T \in \mathscr{L}(H)$. Then T is bounded below iff there exists a number $c > 0$ such that $T^*T \geq cI$.

(57.37) Let H be a Hilbert space and let $T \in \mathscr{L}(H)$. The set
$$W(T) = \{(Tx|x) : \|x\| = 1\}$$
is called the *numerical range* of T.

(i) $W(T)$ is a convex subset of $\mathbb{C}$.

(ii) $\sigma(T) \subset \overline{W(T)}$.

(iii) If T is normal then conv $\sigma(T) = \overline{W(T)}$; this includes (57.27) as a special case.

(57.38) Let H be a Hilbert space and let P, $Q \in \mathscr{L}(H)$ be projections (43.10). Then $P(H) \subset Q(H)$ iff $P \leq Q$ in the sense of (57.28).

Chapter 7

C*-Algebras

C*-algebras are the Banach algebras most closely related to Hilbert space; their special importance among Banach algebras reflects the special importance of Hilbert space among Banach spaces. Continuing in this vein, with allusions to their many applications, one can build a strong case for the urgency of studying C*-algebras on the grounds that they are important and useful; it is equally plausible that mathematicians study them because they are so beautiful. The aim of this chapter is to persuade the reader that C*-algebras are beautiful; the proof that they are useful is deferred until the next chapter.

58. Preliminaries. For convenient reference, we repeat some earlier definitions (43.6):

(58.1) Definition. A normed algebra [Banach algebra] with an involution $x \mapsto x^*$ is called a *normed $*$-algebra* [*Banach $*$-algebra*]; its norm is said to be *$*$-quadratic* if $\|x^*x\| = \|x\|^2$ for all x. A *C*-algebra* is a Banach $*$-algebra whose norm is $*$-quadratic.

An obvious operatorial example of C*-algebra: any norm-closed $*$-sub-algebra of $\mathscr{L}(H)$, H a Hilbert space (43.7). The climax of the present chapter is a proof of the converse: *Every C*-algebra may be identified with one of the operatorial examples* (Gel'fand-Naĭmark theorem (62.1)). To put it mildly, the $*$-quadratic property of the norm has striking geometric and algebraic consequences.

The first three theorems are very useful and quite near the surface.

(58.2) Theorem. *If A is a normed $*$-algebra with $*$-quadratic norm, then the involution is isometric, that is, $\|x^*\| = \|x\|$ for all $x \in A$.*

Proof. Cf. (43.5). ∎

In general, spectral radius satisfies $r(x) \leq \|x\|$ (51.11); equality is a special event:

(58.3) Theorem. *If A is a C*-algebra with unity and if $x \in A$ satisfies $x^*x = xx^*$, then $r_A(x) = \|x\|$.*

Proof. {In any ring with involution, an element x satisfying $x^*x = xx^*$ is called *normal*. An element y satisfying $y^* = y$ is called *self-adjoint*.}

Since $y = x^*x$ is self-adjoint, we have $\|y\|^2 = \|y^*y\| = \|y^2\|$. {In particular, $\|1\| = 1$ (put $y = 1$).} Induction yields $\|y\|^{2^k} = \|y^{2^k}\|$, thus $\|y\| = \|y^n\|^{1/n}$ for $n = 2^k$ ($k = 1, 2, 3, \ldots$); it follows from (55.1) that $\|y\| = r(y)$. Thus $r(x^*x) = \|x^*x\| = \|x\|^2$.

250

On the other hand, since x commutes with x^*, we have $(x^*x)^n = (x^n)^*(x^n)$, thus $\|(x^*x)^n\| = \|(x^n)^*(x^n)\| = \|x^n\|^2$ for all positive integers n; then

$$\|(x^*x)^n\|^{1/n} = (\|x^n\|^{1/n})^2$$

for all n, and passage to the limit yields $r(x^*x) = r(x)^2$. Thus $r(x)^2 = r(x^*x) = \|x\|^2$. ∎

(58.4) Theorem. *If A is a C*-algebra with unity and if $x \in A$ is self-adjoint, then $\sigma_A(x) \subset \mathbb{R}$.*

Proof. By an elementary algebraic maneuver, it suffices to show that $x - i1$ is invertible (cf. the proof of (57.24)). Assume to the contrary that $i \in \sigma(x)$. Then, for every complex number λ, we have

$$\lambda + 1 = \lambda - i^2 \in \sigma(\lambda 1 - ix),$$

therefore $|\lambda + 1| \le \|\lambda 1 - ix\|$. In particular, if λ is real then

$$(\lambda + 1)^2 \le \|\lambda 1 - ix\|^2 = \|(\lambda 1 - ix)^*(\lambda 1 - ix)\|$$

$$= \|\lambda^2 1 + x^2\| \le \lambda^2 + \|x\|^2;$$

thus $1 + 2\lambda \le \|x\|^2$ for all real λ, which is absurd. ∎

(58.5) Corollary. *If A is a C*-algebra with unity and if B is a closed *-subalgebra of A containing 1, then B is a full subalgebra of A, thus $\sigma_B(x) = \sigma_A(x)$ for all x in B.*

Proof. Quote (56.7). ∎

The final two theorems of the section are 'complements,' i.e., they are not used in the rest of the text. The first, which is more or less useless, shows that the *-quadratic property extends to the completion. The second asserts that the *-quadratic property is extendible to the unitification; we take this as license to restrict attention to C*-algebras with unity.

(58.6) Theorem. *If A is a normed *-algebra with *-quadratic norm and if B is the Banach algebra completion of A (49.3), then the involution of A can be extended to make B a C*-algebra.*

Proof. We regard A as a dense subalgebra of B. It is immediate from (58.2) that the mapping $x \mapsto x^*$ ($x \in A$) extends to an isometric mapping of B onto B, and it is routine to check that the extended mapping is an involution of B. Passage to the limit in the identity $\|x^*x\| = \|x\|^2$ ($x \in A$) shows that the identity remains true for all x in B. ∎

(58.7) Theorem. *A C*-algebra without unity may be embedded in a C*-algebra with unity.*

More precisely, let $(A, \|\ \|)$ be a C*-algebra without unity, let $(A_1, \|\ \|)$ be its Banach algebra unitification with the natural norm $\|a + \lambda 1\| = \|a\| + |\lambda|$ (49.5), and define an involution on A_1 by the formula $(a + \lambda 1)^* = a^* + \lambda^* 1$. There exists a norm $a + \lambda 1 \mapsto |a + \lambda 1|$ on A_1, equivalent to the natural norm, *such that* (i) $(A_1, |\ |)$ *is a C*-algebra, and* (ii) *the embedding* $(A, \|\ \|) \to$ $(A_1, |\ |)$ *is isometric.*

Proof. As in (49.5), we regard A as a subalgebra of A_1. We write $a, b, \ldots$ for elements of A, and $x, y, \ldots$ for elements of A_1. Each $x \in A_1$ defines a left-multiplication mapping $y \mapsto xy$ $(y \in A_1)$; since A is an ideal of A_1, it is invariant under this mapping. Thus, for each $x \in A_1$, we may define a linear mapping

$$T_x : A \to A$$

by the formula $T_x b = xb$ $(b \in A)$. Clearly $T_x \in \mathscr{L}(A)$ and $\|T_x\| \le \|x\|$.

It is straightforward to check that the mapping $x \mapsto T_x$ is an algebra homomorphism $A_1 \to \mathscr{L}(A)$. We assert that this mapping is injective. Indeed, if $T_x = 0$, where $x = a + \lambda 1$, then $ab + \lambda b = 0$ for all $b \in A$. Since A has no unity element, it follows that $\lambda = 0$. {Proof: If $\lambda \ne 0$, then $(-\lambda^{-1}a)b = b$ for all $b \in A$, thus $-\lambda^{-1}a$ is a left unity element for A; in a ring with involution, a left unity element is a unity element.} Then $ab = 0$ for all $b \in A$. In particular $aa^* = 0$, and $a = 0$ results from $\|a\|^2 = \|aa^*\|$. Thus $x = 0$ as asserted.

It follows that the formula

$$|x| = \|T_x\| \qquad (x \in A_1)$$

defines a normed algebra norm on A_1. The burden of the proof is to show that (i) this new norm on A_1 is equivalent to the natural norm, (ii) $|x^*x| = |x|^2$ for all x in A_1, and (iii) $|a| = \|a\|$ for all a in A. We prove these three assertions in reverse order.

(iii) Let $a \in A$. The problem is to show that $\|T_a\| = \|a\|$. We already know that $\|T_a\| \le \|a\|$. On the other hand,

$$\|a\|^2 = \|aa^*\| = \|T_a a^*\| \le \|T_a\| \|a^*\| = \|T_a\| \|a\|,$$

whence $\|a\| \le \|T_a\|$.

(ii) Let $x = a + \lambda 1 \in A_1$. It is to be shown that $\|T_{x^*x}\| = \|T_x\|^2$. For all $b \in A$, we have $T_x b = xb \in A$, therefore

$$\|T_x b\|^2 = \|xb\|^2 = \|(xb)^*(xb)\| = \|b^*x^*xb\| \le \|b^*\| \|x^*xb\|$$

$$= \|b\| \|T_{x^*x} b\| \le \|b\|^2 \|T_{x^*x}\|,$$

whence $\|T_x\|^2 \le \|T_{x^*x}\|$. On the other hand,

$$\|T_{x^*x}\| = \|T_{x^*} T_x\| \le \|T_{x^*}\| \|T_x\|,$$

thus

(*) $$\|T_x\|^2 \le \|T_{x^*x}\| \le \|T_{x^*}\| \|T_x\|.$$

It follows from (*) that $\|T_x\| \le \|T_{x^*}\|$ and (replacing x by x^*) $\|T_{x^*}\| \le \|T_x\|$, thus $\|T_{x^*}\| = \|T_x\|$; substituting this into (*) yields $\|T_x\|^2 = \|T_{x^*x}\|$.

(i) It remains to show that the two norms on A_1 are equivalent; since A_1 is complete for the norm $\|x\|$, and since $|x| = \|T_x\| \le \|x\|$ shows that the norms are comparable, it will suffice by the open mapping theorem to show that A_1 is complete for the norm $|x|$ (48.5). In view of (iii), $(A, |\ |)$ is a complete and therefore closed linear subspace of $(A_1, |\ |)$; the quotient normed space A_1/A induced by the norm $|x|$ (16.7) is one-dimensional, hence is trivially complete; therefore $(A_1, |\ |)$ is also complete (16.12). {Alternatively, the image of A in $\mathscr{L}(A)$ is complete, hence closed; therefore the image of A_1 in $\mathscr{L}(A)$ is closed (23.6), hence complete.} ∎

Exercises

(58.8) Theorems (58.3) and (58.4) also hold for C^*-algebras without unity (cf. 51.18).

*(58.9) Every normed ∗-algebra with ∗-quadratic norm is semisimple. {A ring is called *semisimple* if its radical is zero (cf. 51.19).}

59. Commutative C^*-algebras.

(59.1) **Example.** Let T be a compact space and let $\mathscr{C}(T)$ be the set of all continuous functions $x: T \to \mathbb{C}$, equipped with the pointwise algebra operations, the pointwise involution $x^*(t) = (x(t))^*$ ($=$ the complex conjugate of $x(t)$) and the norm $\|x\|_\infty = \sup\{|x(t)| : t \in T\}$. Then $\mathscr{C}(T)$ is a commutative C^*-algebra with unity.

The converse, due to I. M. Gel'fand and M. A. Naĭmark [54], is the basis for much 'spectral theory' (cf. (65.1), Section 70):

(59.2) **Theorem.** (Commutative Gel'fand-Naĭmark theorem) *If A is a commutative C^*-algebra with unity and if $\mathscr{M}$ is the maximal ideal space of A, then the Gel'fand transformation $x \mapsto \hat{x}$ (52.13) maps A isometrically and ∗-isomorphically onto $\mathscr{C}(\mathscr{M})$. Suggestively, $A = \mathscr{C}(\mathscr{M})$.*

Proof. Since every $x \in A$ is normal and therefore satisfies $\|\hat{x}\|_\infty = r(x) = \|x\|$ (58.3), the Gel'fand transformation maps A onto a complete— hence closed—subalgebra $\hat{A}$ of $\mathscr{C}(\mathscr{M})$. The algebra $\hat{A}$ contains the constant functions and separates the points of $\mathscr{M}$ (52.13). To conclude that $\hat{A} = \mathscr{C}(\mathscr{M})$, we need only show that $\hat{A}$ is a ∗-subalgebra of $\mathscr{C}(\mathscr{M})$ (Weierstrass-Stone theorem); it will suffice to show that $(\hat{x})^* = (x^*)^\smallfrown$ for every $x \in A$. Let $x \in A$ and write $x = y + iz$ with y and z self-adjoint. {Such a 'Cartesian decomposition' is possible—and uniquely so—in every complex ∗-algebra: $y = (1/2)(x + x^*)$, $z = (1/2i)(x - x^*)$.} Since $\sigma(y)$ and $\sigma(z)$ are real (58.4), the functions $\hat{y}$ and $\hat{z}$ are real-valued (52.13), therefore $(x^*)^\smallfrown = (y - iz)^\smallfrown = \hat{y} - i\hat{z} = (\hat{y} + i\hat{z})^* = (\hat{x})^*$. ∎

Exercises

(59.3) The Banach algebras $\mathscr{B}(T)$, $\mathscr{C}^\infty(T)$, $\mathscr{C}^0(T)$ (cf. 52.18), equipped with the pointwise involution $x^*(t) = (x(t))^*$, are commutative C^*-algebras.

(59.4) If A is a commutative C^*-algebra without unity and $\mathscr{M}$ is the space of modular maximal ideals of A (52.19), then the Gel'fand transformation maps A isometrically and ∗-isomorphically onto $\mathscr{C}^0(\mathscr{M})$.

60. *-Representations. The definition of C^*-algebra is a judicious mixture of algebra, geometry, and analysis (the $*$-algebra structure, the $*$-quadratic norm, and completeness). These aspects are so tightly bound together in the $*$-quadratic property of the norm that a statement about one aspect has reverberations for the other aspects. For example, the following theorem shows that algebraic morphisms automatically have desirable geometric and topological features (in particular, $*$-isomorphisms are automatically isometric):

(60.1) Theorem. *If A and B are C^*-algebras with unity, and if $\varphi: A \to B$ is a $*$-homomorphism such that $\varphi(1) = 1$, then $\|\varphi(a)\| \leq \|a\|$ for all a in A (in particular, φ is continuous).*

Proof. {By '$*$-homomorphism' is meant an algebra homomorphism φ such that $\varphi(a^*) = \varphi(a)^*$ for all a.} It is elementary that $\sigma_B(\varphi(a)) \subset \sigma_A(a)$ for all a in A (if $a - \lambda 1$ is invertible in A, then $\varphi(a) - \lambda 1 = \varphi(a - \lambda 1)$ is invertible in B), therefore $r_B(\varphi(a)) \leq r_A(a)$. If a is self-adjoint then so is $\varphi(a)$, and (58.3) yields

$$\|\varphi(a)\| = r_B(\varphi(a)) \leq r_A(a) = \|a\|.$$

If $a \in A$ is arbitrary then a^*a is self-adjoint, therefore

$$\|\varphi(a)\|^2 = \|\varphi(a)^*\varphi(a)\| = \|\varphi(a^*a)\| \leq \|a^*a\| = \|a\|^2,$$

thus $\|\varphi(a)\| \leq \|a\|$. ∎

The case that $B = \mathscr{L}(H)$, H a Hilbert space, is of special importance:

(60.2) Definition. If A is a $*$-algebra and H is a Hilbert space, a $*$-homomorphism $\varphi: A \to \mathscr{L}(H)$ is called a *$*$-representation* of A on H. If A has a unity element 1 it is desirable, but not obligatory, that $\varphi(1) = 1_H$ (the identity operator on H); such $*$-representations will be called *unital*. A $*$-representation is said to be *faithful* if it is injective (cf. 60.11).

(60.3) Theorem. *If A is a C^*-algebra with unity, $(H_\iota)_{\iota \in I}$ is a family of Hilbert spaces, and if, for each $\iota \in I$, $\varphi_\iota: A \to \mathscr{L}(H_\iota)$ is a unital $*$-representation of A on H_ι, then there exists a unital $*$-representation $\varphi: A \to \mathscr{L}(H)$ such that*

$$\|\varphi(a)\| = \sup_{\iota \in I} \|\varphi_\iota(a)\|$$

for all $a \in A$.

The heart of the matter is the construction of the Hilbert space H. The details are as follows.

(60.4) Suppose $(H_\iota)_{\iota \in I}$ is any family of Hilbert spaces. Let H be the set of all families $x = (x_\iota)$, with $x_\iota \in H_\iota$ for all $\iota \in I$, such that $\sum_{\iota \in I} \|x_\iota\|^2 < \infty$. {The latter condition means that, as J runs over all finite subsets of I, the ordinary sums $\sum_{\iota \in J} \|x_\iota\|^2$ remain bounded; the infinite sum is then defined to be the supremum of the finite subsums. It is elementary that $\|x_\iota\| = 0$ for all but countably many $\iota \in I$.}

If $x = (x_\iota)$ and $y = (y_\iota)$ are in H, then the coordinatewise sum $(x_\iota + y_\iota)$ is also in H; this follows at once from the relations

$$\|x_\iota + y_\iota\|^2 + \|x_\iota - y_\iota\|^2 = 2\|x_\iota\|^2 + 2\|y_\iota\|^2.$$

Also, if $x = (x_\iota) \in H$ and $\lambda \in \mathbb{C}$, then $(\lambda x_\iota) \in H$ results from the relations $\|\lambda x_\iota\| = |\lambda|\,\|x_\iota\|$. Clearly H is a vector space relative to the coordinatewise linear operations

$$x + y = (x_\iota + y_\iota), \qquad \lambda x = (\lambda x_\iota).$$

From the relations

$$(x_\iota | y_\iota) = \tfrac{1}{4}\{\|x_\iota + y_\iota\|^2 - \|x_\iota - y_\iota\|^2 + i\|x_\iota + iy_\iota\|^2 - i\|x_\iota - iy_\iota\|^2\}$$

it follows that $\sum_{\iota \in I} |(x_\iota | y_\iota)| < \infty$; the formula

$$(x|y) = \sum_{\iota \in I} (x_\iota | y_\iota)$$

(the limit of the net of finite subsums) evidently defines a sesquilinear form on H such that

$$(x|x) = \sum_{\iota \in I} \|x_\iota\|^2.$$

In particular, $(x|x) > 0$ when $x \neq \theta$, thus H is an inner product space (41.9); the norm derived from the inner product is given by

$$\|x\| = (x|x)^{1/2} = \left(\sum_{\iota \in I} \|x_\iota\|^2\right)^{1/2}.$$

It remains to show that H is complete relative to this norm. Suppose x_n is a Cauchy sequence in H, say $x_n = (x_{n\iota})$ for $n = 1, 2, 3, \ldots$. For each $\iota \in I$

$$\|x_{m\iota} - x_{n\iota}\|^2 \leq \|x_m - x_n\|^2$$

for all m and n, thus the sequence $x_{1\iota}, x_{2\iota}, x_{3\iota}, \ldots$ is Cauchy in H_ι; we define

$$x_\iota = \lim_{n \to \infty} x_{n\iota} \qquad (\iota \in I).$$

We show simultaneously that (x_ι) belongs to H and that x_n converges to it. Let $\varepsilon > 0$ and choose an index N such that $\|x_m - x_n\| \leq \varepsilon$ whenever $m, n \geq N$. If $m, n \geq N$ and J is a finite subset of I, then

$$\sum_{\iota \in J} \|x_{m\iota} - x_{n\iota}\|^2 \leq \varepsilon^2;$$

keeping n and J fixed and letting $m \to \infty$, we have

$$\sum_{\iota \in J} \|x_\iota - x_{n\iota}\|^2 \leq \varepsilon^2;$$

keeping n fixed and varying J, we see that the family $(x_\iota - x_{n\iota})$ is in H and has norm $\leq \varepsilon$, therefore the family $(x_\iota - x_{n\iota}) + (x_{n\iota}) = (x_\iota)$ is also in H, and, setting $x = (x_\iota)$, we have

$$\|x - x_n\|^2 \leq \varepsilon^2;$$

the validity of the inequality for all $n \geq N$ shows that $x_n \to x$. This completes the proof that H is a Hilbert space.

(60.5) Definition. With notation as in (60.4), the Hilbert space H is called the *orthogonal sum* of the family $(H_\iota)_{\iota \in I}$. Notation: $H = \bigoplus_{\iota \in I} H_\iota$.

Proof of Theorem (60.3). Let $H = \bigoplus_{\iota \in I} H_\iota$ as described above. We note first that if $(T_\iota)_{\iota \in I}$ is a family of operators, with $T_\iota \in \mathscr{L}(H_\iota)$ for all $\iota \in I$, and if $\sup_{\iota \in I} \|T_\iota\| < \infty$, then the formula

$$Tx = (T_\iota x_\iota) \qquad (x = (x_\iota) \in H)$$

defines an operator $T \in \mathscr{L}(H)$; moreover, writing $M = \sup_{\iota \in I} \|T_\iota\|$, it is clear from the relations

$$\|T_\varkappa x_\varkappa\|^2 \leq \|Tx\|^2 = \sum_{\iota \in I} \|T_\iota x_\iota\|^2 \leq M^2 \sum_{\iota \in I} \|x_\iota\|^2 = M^2 \|x\|^2$$

that $\|T\| = M$.

We are ready to define the mapping $\varphi: A \to \mathscr{L}(H)$. If $a \in A$ then $\|\varphi_\iota(a)\| \leq \|a\|$ for all $\iota \in I$ (60.1), thus the family of operators $(\varphi_\iota(a))_{\iota \in I}$ is bounded in norm (this is the only use we make of the assumption that A is a C^*-algebra); let $\varphi(a)$ be the operator on H such that

$$\varphi(a)x = (\varphi_\iota(a)x_\iota)$$

for all $x = (x_\iota)$ in H. As noted above,

$$\|\varphi(a)\| = \sup_{\iota \in I} \|\varphi_\iota(a)\|.$$

It is straightforward to check that φ is a $*$-homomorphism and that $\varphi(1) = 1_H$. ∎

We remark that the kernel of φ is the intersection of the kernels of the φ_ι.

(60.6) Definition. The $*$-representation φ constructed in the proof of (60.3) is called the *direct sum* of the family $(\varphi_\iota)_{\iota \in I}$.

Exercises

(60.7) If A and B are C^*-algebras with unity and if $\varphi: A \to B$ is a unital algebra homomorphism such that $\|\varphi(a)\| \leq \|a\|$ for all $a \in A$, then φ is a $*$-homomorphism.

(60.8) Let $\varphi: A \to \mathscr{L}(H)$ be a $*$-representation and let

$$N = \{x \in H : \varphi(a)x = 0 \quad \text{for all } a \in A\}.$$

Then N is a closed linear subspace of H such that both N and $N^\perp$ are invariant under all of the operators $\varphi(a)$. If $N = \{\theta\}$, φ is said to be *nondegenerate*. In general, writing $\psi(a) = \varphi(a)|N^\perp$ (the restriction of $\varphi(a)$ to $N^\perp$), the $*$-representation $\psi: A \to \mathscr{L}(N^\perp)$ is nondegenerate, thus the vectors $\psi(a)x$ ($a \in A$, $x \in N^\perp$) have closed linear span $N^\perp$.

(60.9) If A and B are arbitrary C^*-algebras (with or without unity) and $\varphi: A \to B$ is a $*$-homomorphism, then $\|\varphi(a)\| \leq \|a\|$ for all $a \in A$.

*(60.10) If A is a C^*-algebra and I is a closed ideal of A, then I is a $*$-ideal and A/I is a C^*-algebra with respect to the quotient norm (cf. 52.4).

*(60.11)   Let A and B be C^*-algebras and let $\varphi: A \to B$ be a $*$-homomorphism.

(i) If φ is injective, then $\|\varphi(a)\| = \|a\|$ for all $a \in A$.

(ii) For every φ, $\varphi(A)$ is a closed $*$-subalgebra of B.

(60.12)   (i) Let $(A_\iota)_{\iota \in I}$ be a family of C^*-algebras, let A be the set of all families $a = (a_\iota)_{\iota \in I}$, with $a_\iota \in A_\iota$ for all $\iota \in I$, such that $\sup_{\iota \in I} \|a_\iota\| < \infty$. Equipped with the coordinatewise $*$-algebra operations and the norm $\|a\| = \sup_{\iota \in I} \|a_\iota\|$, A is a C^*-algebra, called the C^*-*sum* of the family. Notation: $A = \bigoplus_{\iota \in I} A_\iota$. {The term 'product' and the notation $\prod_{\iota \in I} A_\iota$ are also used.}

(ii) If $(A_\iota)_{\iota \in I}$ and $(B_\iota)_{\iota \in I}$ are families of C^*-algebras and if, for each $\iota \in I$, $\varphi_\iota: A_\iota \to B_\iota$ is a $*$-homomorphism, then the formula $\varphi((a_\iota)) = (\varphi_\iota(a_\iota))$ defines a $*$-homomorphism $\varphi: \bigoplus_{\iota \in I} A_\iota \to \bigoplus_{\iota \in I} B_\iota$.

(iii) If $(H_\iota)_{\iota \in I}$ is a family of Hilbert spaces, then there exists a natural faithful $*$-representation $\bigoplus_{\iota \in I} \mathscr{L}(H_\iota) \to \mathscr{L}(\bigoplus_{\iota \in I} H_\iota)$.

(iv) If $(A_\iota)_{\iota \in I}$ is a family of C^*-algebras, $(H_\iota)_{\iota \in I}$ is a family of Hilbert spaces, and if, for each $\iota \in I$, φ_ι is a $*$-representation of A_ι on H_ι, then the family (φ_ι) defines in a natural way a $*$-representation of $\bigoplus_{\iota \in I} A_\iota$ on $\bigoplus_{\iota \in I} H_\iota$.

(60.13)   If A is a Banach $*$-algebra with continuous involution, then every $*$-representation $\varphi: A \to \mathscr{L}(H)$ is continuous.

61. States on a C^*-algebra.

The main target (reached in the next section) is the Gel'fand-Naĭmark theorem: Given a C^*-algebra A, one has to produce a Hilbert space H and an isometric $*$-representation $\varphi: A \to \mathscr{L}(H)$. What intrinsic aspects of A reflect the vectors of the extrinsic Hilbert space H that has to be produced? Answer: a special family of linear forms. Namely, if $\varphi: A \to \mathscr{L}(H)$ is a $*$-representation and u is a vector in H, then the formula $f(a) = (\varphi(a)u|u)$ defines a continuous linear form on A such that $f(a^*a) = (\varphi(a)^*\varphi(a)u|u) = \|\varphi(a)u\|^2 \geq 0$ for all $a \in A$. Let us formalize this idea:

(61.1) Definition.   Let A be a (complex) $*$-algebra and let $f: A \to \mathbb{C}$ be a linear form; f is said to be a *state* on A if $f(a^*a) \geq 0$ for all $a \in A$. If A has a unity element and $f(1) = 1$, the state f is said to be *normalized*.

The strategy for proving the Gel'fand-Naĭmark theorem involves three steps: (1) show that the given C^*-algebra has sufficiently many states (the hardest technical problem), (2) show that each state f may be used to construct a $*$-representation φ_f (the key intuition), and (3) sum up the representations φ_f via (60.3) to get the desired faithful representation φ. Our objective in this section is to prove (1) and (2); the pieces are put together in the next section.

(61.2)   The following remarks are offered as motivation; they can be omitted without loss of continuity. Let A be a C^*-algebra with unity and suppose it were possible (as promised by the Gel'fand-Naĭmark theorem) to construct an isometric, unital $*$-representation $\varphi: A \to \mathscr{L}(H)$. It would then follow that $\sigma(a^*a) \subset [0, \infty)$ for all $a \in A$. {Proof: Let $\mathscr{A} = \varphi(A)$; $\mathscr{A}$ is a $*$-subalgebra of $\mathscr{L}(H)$ with $1_H \in \mathscr{A}$. Since A is complete and φ is isometric,

$\mathscr{A}$ is norm-closed in $\mathscr{L}(H)$. Let $a \in A$ and write $S = \varphi(a)$; thus $\varphi(a^*a) = S^*S$ and $\sigma(a^*a) = \sigma_{\mathscr{A}}(S^*S)$. We know that $\sigma_{\mathscr{L}(H)}(S^*S) \subset [0, \infty)$ (put $T = S^*S$ in (57.26)); since $\mathscr{A}$ is a full subalgebra of $\mathscr{L}(H)$ (58.5), we conclude that $\sigma(a^*a) = \sigma_{\mathscr{A}}(S^*S) = \sigma_{\mathscr{L}(H)}(S^*S) \subset [0, \infty).\}$ The proof of the Gel'fand-Naĭmark theorem requires that this spectral property of C^*-algebras (called *symmetry*) be established in advance. The details are as follows.

(61.3) Definition. Let A be a C^*-algebra with unity. We say that $a \in A$ is *positive* (relative to A) if $a^* = a$ and $\sigma(a) \subset [0, \infty)$. Notation: $a \geq 0$.

(61.4) Lemma. *If $a \in B \subset A$, where A is a C^*-algebra with unity and B is a closed $*$-subalgebra of A containing the unity element, then $a \geq 0$ relative to B iff $a \geq 0$ relative to A.*

Proof. Immediate from the fact that B is a full subalgebra of A (58.5). ∎

In a C^*-algebra with unity, every positive element has a unique positive square root:

(61.5) Theorem. *Let A be a C^*-algebra with unity and let $a \in A$, $a \geq 0$. There exists a unique element $b \in A$ such that $b \geq 0$ and $b^2 = a$; moreover, one has $b \in \{a\}''$. Notation: $b = a^{1/2}$.*

Proof. Let $B = \{a\}''$ (the bicommutant of a in A) and write $B = \mathscr{C}(T)$, T compact (59.2). Since $a \geq 0$ relative to B (61.4), it is clear from the functional representation that there exists $b \in B$ with $b \geq 0$ and $b^2 = a$.

If also $c \in A$ with $c \geq 0$ and $c^2 = a$, we are to show that $c = b$. Clearly c commutes with a, therefore b commutes with c. Then $\{b, c\}''$ is a commutative C^*-algebra with unity, in which b and c are positive elements such that $b^2 = a = c^2$; clearly $b = c$ by the functional representation for $\{b, c\}''$. ∎

(61.6) Lemma. *In a C^*-algebra with unity, the sum of positive elements is positive.*

Proof. Suppose $a \geq 0$ and $b \geq 0$; we are to show that the self-adjoint element $a + b$ has nonnegative spectrum. At any rate, $\sigma(a + b) \subset \mathbb{R}$ by (58.4).

Let $\alpha = \|a\|$, $\beta = \|b\|$. From $\sigma(a) \subset [0, \alpha]$ it is clear that $\sigma(\alpha 1 - a) \subset [0, \alpha]$; citing (58.3), we have $\|\alpha 1 - a\| = r(\alpha 1 - a) \leq \alpha$. Similarly $\|\beta 1 - b\| \leq \beta$, thus

$$(*) \qquad \|(\alpha + \beta)1 - (a + b)\| \leq \|\alpha 1 - a\| + \|\beta 1 - b\| \leq \alpha + \beta.$$

If $\lambda \in \sigma(a + b)$ then $(\alpha + \beta) - \lambda \in \sigma((\alpha + \beta)1 - (a + b))$, therefore

$$|(\alpha + \beta) - \lambda| \leq r((\alpha + \beta)1 - (a + b)) = \|(\alpha + \beta)1 - (a + b)\|;$$

citing (*), we conclude that $|(\alpha + \beta) - \lambda| \leq \alpha + \beta$, which implies that $\lambda \geq 0$. ∎

(61.7) Lemma. *In a C^*-algebra with unity, if a is an element such that $-a^*a \geq 0$, then $a = 0$.*

Proof. Since aa^* and a^*a have the same nonzero spectra (51.16), one also has $-aa^* \geq 0$. Write $a = b + ic$ with b and c self-adjoint. Elementary algebra yields

$$a^*a + aa^* = 2b^2 + 2c^2.$$

Since $b^* = b$, it is clear from (59.2) (or from (53.3)) that $2b^2 \geq 0$. Similarly $2c^2 \geq 0$, thus

$$a^*a = 2b^2 + 2c^2 - aa^* \geq 0$$

by (61.6). From $-a^*a \geq 0$ and $a^*a \geq 0$ we infer that $\sigma(a^*a) = \{0\}$, whence $a^*a = 0$ (58.3) and therefore $a = 0$. ∎

(61.8) Theorem. *If A is a C^*-algebra with unity, then $a^*a \geq 0$ for every $a \in A$.*

Proof. Let $B = \{a^*a\}''$ (the bicommutant of a^*a in A) and write $B = \mathscr{C}(T)$, T compact (59.2). Since a^*a is self-adjoint, it may be viewed as a real-valued continuous function on T; one may therefore write

$$a^*a = b - c,$$

with $b, c \in \{a^*a\}''$, $b \geq 0$, $c \geq 0$, and $bc = 0$. It will suffice to show that $c = 0$. Set $x = ac$; then

$$x^*x = c(a^*a)c = c(b - c)c = -c^3,$$

thus $-x^*x = c^3 \geq 0$ (by the functional representation); then $x = 0$ by (61.7), whence $c^3 = 0$ and therefore $c = 0$. ∎

A key role in the construction of states on C^*-algebras is played by the following simple geometrical characterization:

(61.9) Theorem. *Let A be a C^*-algebra with unity and let f be a linear form on A. The following conditions on f are equivalent: (a) f is a state; (b) f is continuous and $\|f\| = f(1)$.*

Proof. (a) implies (b): By the hypothesis on f, the mapping $(a, b) \mapsto f(ab^*)$ is a positive sesquilinear form on A (41.9). It follows from (41.14) that the sesquilinear form is Hermitian and satisfies the Cauchy-Schwarz inequality; expressed in terms of f, this means that

$$f(ba^*) = (f(ab^*))^*, \qquad |f(ab^*)|^2 \leq f(aa^*)f(bb^*)$$

for all $a, b \in A$. Setting $b = 1$, we have

(i) $$f(a^*) = f(a)^*,$$

(ii) $$|f(a)|^2 \leq f(aa^*)f(1),$$

for all $a \in A$. The first of these relations shows that $f(a)$ is real when a is self-adjoint; the continuity of f will be derived from the second. We observe that for every self-adjoint $b \in A$, we have

(iii) $$|f(b)| \leq \|b\|f(1).$$

{Proof: Writing $B = \{b\}''$, the bicommutant of b in A, we have $B = \mathscr{C}(T)$ for a suitable compact space T (59.2); then b is a real-valued function on T with $-\|b\| \leq b(t) \leq \|b\|$ for all $t \in T$. Since $\|b\|1 - b$ is a continuous function on T with nonnegative values, we may define $c = (\|b\|1 - b)^{1/2} \in \mathscr{C}(T) = B$; then

$$f(\|b\|1 - b) = f(c^2) = f(c^*c) \geq 0,$$

thus $f(b) \leq \|b\|f(1)$. Similarly $f(b) \geq -\|b\|f(1)$.} Combining (ii) and (iii), we have, for every a in A,

$$|f(a)|^2 \leq f(aa^*)f(1) \leq \|a^*a\|f(1)^2 = \|a\|^2f(1)^2,$$

thus $|f(a)| \leq \|a\|f(1)$. This shows that f is continuous and $\|f\| \leq f(1)$. On the other hand, $\|f\| \geq f(1)$ results from $\|1\| = 1$.

(b) implies (a): Suppose f is a continuous linear form on A such that $f(1) = \|f\| > 0$. Replacing f by $f(1)^{-1}f$, we can suppose that $\|f\| = f(1) = 1$.

Let $a \in A$; it is to be shown that $f(a^*a) \geq 0$. Writing I for the closed interval $[0, \|a^*a\|]$, we know from (61.8) that $\sigma(a^*a) \subset I$; it will suffice to show that $f(a^*a) \in I$. Let D be any closed disc containing I, say

$$D = \{\lambda \in \mathbb{C} : |\lambda - \lambda_0| \leq \alpha\};$$

since I is the intersection of all such discs, it will suffice to show that $f(a^*a) \in D$. Since

$$\sigma(a^*a - \lambda_0 1) = \sigma(a^*a) - \lambda_0 \subset D - \lambda_0 = \{\mu \in \mathbb{C} : |\mu| \leq \alpha\},$$

it results from (58.3) that $\|a^*a - \lambda_0 1\| = r(a^*a - \lambda_0 1) \leq \alpha$; then

$$|f(a^*a) - \lambda_0| = |f(a^*a) - \lambda_0 f(1)| = |f(a^*a - \lambda_0 1)| \leq \|f\|\,\|a^*a - \lambda_0 1\| \leq \alpha,$$

thus $f(a^*a) \in D$. ∎

It follows from (61.9) that every C^*-algebra with unity has an abundance of normalized states:

(61.10) Theorem. *Let A be a C^*-algebra with unity. If a is any element of A, there exists a normalized state f on A such that $f(a^*a) = \|a^*a\|$.*

Proof. Consider the commutative $*$-subalgebra $B = \{a^*a\}''$, the bicommutant of a^*a in A; since B is a full subalgebra of A (49.11), $\sigma_B(a^*a) = \sigma_A(a^*a)$. {One could also quote (58.5).} We write briefly $\sigma(a^*a)$.

Suppose $\alpha \in \sigma(a^*a)$. Let f_0 be a character of B such that $f_0(a^*a) = \alpha$ (52.13). One has $\|f_0\| = 1 = f_0(1)$ (52.7). By the Hahn-Banach theorem, f_0 may be extended to a continuous linear form f on A such that $\|f\| = \|f_0\| = 1$ (44.2). Since $f(1) = f_0(1) = 1 = \|f\|$, f is a state on A (61.9). Since $\|a^*a\| = r(a^*a)$ (58.3), the role of α can be played by $\|a^*a\|$ (cf. 61.8). ∎

The preceding theorem assures an adequate supply of states on a C^*-algebra; the next theorem shows that each state generates a $*$-representation that reproduces it:

(61.11) Theorem. *If A is a C^*-algebra with unity and if f is a state on A, then there exist a unital $*$-representation $\varphi: A \to \mathcal{L}(H)$ and a vector $u \in H$ such that $f(a) = (\varphi(a)u|u)$ for all $a \in A$.*

Proof. In particular, we seek a Hilbert space H. Where is the inner product to come from? Answer: from the function $(x, y) \mapsto f(y^*x)$. The details are as follows. The formula

(1) $$\langle x, y \rangle = f(y^*x) \qquad (x, y \in A)$$

defines a positive sesquilinear form on A (41.9). {The choice of $f(y^*x)$ instead of $f(xy^*)$ reflects a preference for using the left regular representation of A rather than the right regular antirepresentation.} Applying (41.14), we have the Cauchy-Schwarz inequality

(2) $$|\langle x, y \rangle|^2 \le \langle x, x \rangle \langle y, y \rangle,$$

as well as the Hermitian property

(3) $$\langle x, y \rangle^* = \langle y, x \rangle.$$

The form $\langle x, y \rangle$ is in general not strictly positive; to get strict positivity, we must pass to quotients modulo a suitable linear subspace N of A. Namely, let

$$N = \{x \in A : \langle x, x \rangle = 0\};$$

it is clear from (2) that

(4) $$N = \{x \in A : \langle x, y \rangle = 0 \quad \text{for all } y \in A\}$$
$$= \{y \in A : \langle x, y \rangle = 0 \quad \text{for all } x \in A\},$$

therefore N is obviously a linear subspace of A. Write $E = A/N$ for the quotient vector space and $x \mapsto \tilde{x} = x + N$ for the canonical linear surjection $A \to E$. We define an inner product on E by the formula

(5) $$(\tilde{x}|\tilde{y}) = \langle x, y \rangle;$$

this is permissible since the relations $\tilde{x}_1 = \tilde{x}_2$ and $\tilde{y}_1 = \tilde{y}_2$ (i.e., $x_1 - x_2 \in N$ and $y_1 - y_2 \in N$) imply, in view of (4), that

$$\langle x_1, y_1 \rangle - \langle x_2, y_2 \rangle = \langle x_1 - x_2, y_1 \rangle + \langle x_2, y_1 - y_2 \rangle = 0.$$

It is straightforward to check that (5) defines a strictly positive sesquilinear form on E, i.e., an inner product; in other words, E is a pre-Hilbert space (41.11). Let H be the completion of E (41.2); we regard E as a dense linear subspace of the Hilbert space H.

This takes care of the Hilbert space. To get a $*$-representation, we must exploit further the multiplicative structure of A. From the relation $f(y^*(ax)) = f((a^*y)^*x)$ we have

(6) $$\langle ax, y \rangle = \langle x, a^*y \rangle$$

for all a, x, y in A. It follows at once from (4) and (6) that N is a left ideal of A. Each $a \in A$ defines a linear mapping $\varphi(a): E \to E$ by the formula

(7) $$\varphi(a)\tilde{x} = (ax)^\sim \qquad (x \in A).$$

{The mapping is well-defined because if $\tilde{x} = \tilde{y}$, i.e., if $x - y \in N$, then $a(x - y) \in N$, $(ax)^\sim = (ay)^\sim$.} Routine calculations show that $a \mapsto \varphi(a)$ is an algebra homomorphism of A into the algebra of all linear mappings on E:

(8) $$\varphi(a + b) = \varphi(a) + \varphi(b),$$

(9) $$\varphi(\lambda a) = \lambda \varphi(a),$$

(10) $$\varphi(ab) = \varphi(a)\varphi(b),$$

for all $a, b \in A$ and $\lambda \in \mathbb{C}$. Moreover,

(11) $$(\varphi(a)\tilde{x} \mid \tilde{y}) = (\tilde{x} \mid \varphi(a^*)\tilde{y})$$

for all $a, x, y \in A$. {Proof: Citing (6) at the appropriate step, we have $(\varphi(a)\tilde{x} \mid \tilde{y}) = ((ax)^\sim \mid \tilde{y}) = \langle ax, y \rangle = \langle x, a^*y \rangle = (\tilde{x} \mid \varphi(a^*)\tilde{y}).$}

We have constructed, so to speak, a '∗-representation' $\varphi: A \to \mathscr{V}(E)$, where $\mathscr{V}(E)$ is the algebra of *all* linear mappings on E (cf. the proof of (40.4)). {We remark that the construction of φ satisfying (8)–(10) can be performed for any positive sesquilinear form on any ∗-algebra; and (11) holds iff (6) does.}

To produce a ∗-representation $A \to \mathscr{L}(H)$, we wish to extend each $\varphi(a)$ to a continuous linear mapping on H. To this end, we observe that each $\varphi(a)$ is a continuous linear mapping on E; indeed,

(12) $$\|\varphi(a)\tilde{x}\| \leq \|a\| \|\tilde{x}\|$$

for all $x \in A$. {Proof: Fix $a, x \in A$. For all $y \in A$,

$$f(x^*(y^*y)x) = f((yx)^*(yx)) \geq 0,$$

therefore the formula $f_x(y) = f(x^*yx)$ defines a state on A. Applying to f_x the inequality (iii) in the proof of (61.9), we have

$$f_x(a^*a) \leq \|a^*a\| f_x(1),$$

thus

$$f(x^*a^*ax) \leq \|a^*a\| f(x^*x),$$

which is (12) in disguise.} Each $\varphi(a)$ has a unique continuous (linear) extension to H; we denote the extension also by $\varphi(a)$. Taking into account (8)–(11), we see that $\varphi: A \to \mathscr{L}(H)$ is a ∗-representation of A on H. The calculation $\varphi(1)\tilde{x} = (1x)^\sim = \tilde{x}$ ($x \in A$) shows that $\varphi(1) = 1_H$.

Finally, writing $u = \tilde{1} = 1 + N$, we have

$$(\varphi(a)u \mid u) = ((a1)^\sim \mid \tilde{1}) = \langle a1, 1 \rangle = f(1^*a1) = f(a)$$

for all $a \in A$. ∎

At this point, the reader may advance directly to the Gel'fand-Naĭmark theorem (62.1).

The rest of the section is devoted to complementary material that gives additional perspective on what has gone before and opens several new themes of general importance.

(61.12) Definition. Let A be a C^*-algebra with unity. If a and b are self-adjoint elements of A such that $b - a \geq 0$ in the sense of (61.3), we write $a \leq b$ (or $b \geq a$).

(61.13) Theorem. *Let A be a C^*-algebra with unity and let A_s be the set of all self-adjoint elements of A.*

(i) *With respect to the relation $a \leq b$ defined in (61.12), A_s is an ordered real vector space in the sense of (31.1).*

(ii) *A linear form $f: A \to \mathbb{C}$ is a state on A iff the restriction of f to A_s is a positive linear form in the sense of (32.1).*

Proof. (i) It is obvious that A_s is a real-linear subspace of A. Let $P = \{a \in A_s : a \geq 0\}$. Obviously $\alpha P \subset P$ for all nonnegative real numbers α, and $P + P \subset P$ by (61.6). Also, $P \cap (-P) = \{0\}$. {Proof: If $a \geq 0$ and $-a \geq 0$, then $\sigma(a) = \{0\}$, therefore $a = 0$ by (58.3).} Thus P is a salient, pointed, convex cone; quote (31.5) and (31.4).

(ii) Suppose f is a state on A. As noted in the proof of (61.9), f is real-valued on A_s. Moreover, if $a \in A_s$, $a \geq 0$, then $a = b^*b$ for suitable $b \in A$ (61.5), therefore $f(a) = f(b^*b) \geq 0$.

Conversely, suppose f is a linear form on A such that f is real-valued on A_s and $f|A_s$ is a positive linear form in the sense of (32.1). For all $a \in A$ we have $a^*a \geq 0$ by (61.8), therefore $f(a^*a) \geq 0$; thus f is a state. ∎

(61.14) Definition. Let A be a $*$-algebra with unity. We write Σ for the set (conceivably empty) of all normalized states on A, and, for each $a \in A$, we write

$$\Sigma(a) = \{f(a) : f \in \Sigma\}.$$

We call $\Sigma(a)$ the *numerical status* of a in A. {The notations Σ_A and $\Sigma_A(a)$ are used when it is necessary to emphasize the algebra A.}

(61.15) Theorem. *If A is a C^*-algebra with unity, then Σ is a nonempty, weak* compact, convex subset of A' (the dual space of A). {So topologized, Σ is called the* **state space** *of A.}*

For each $a \in A$, $\Sigma(a)$ is a nonempty, compact, convex subset of $\mathbb{C}$.

Proof. The nonemptiness of Σ is obvious from (61.10), and its convexity is obvious from the definition of normalized state (61.1). Every $f \in \Sigma$ is a continuous linear form of norm one (61.9), thus Σ is contained in the closed unit ball of A'. The closed unit ball of A' is weak* compact by the Alaoglu-Bourbaki theorem (44.12); to prove that Σ is weak* compact, it suffices to note that Σ is weak* closed in A'. {If $f_j \in \Sigma$ is a net that is weak* convergent to $f \in A'$, then $f(1) = \lim f_j(1) = 1$ and $f(a^*a) = \lim f_j(a^*a) \geq 0$ for all $a \in A$.}

The assertion concerning $\Sigma(a)$ follows at once from the linearity and weak* continuity of the mapping $f \mapsto f(a)$ $(f \in A')$. ∎

The numerical status of an element is independent of the containing algebra:

(61.16) Theorem. *If A is a C^*-algebra with unity and if B is a closed *-subalgebra of A containing the unity element, then $\Sigma_B(b) = \Sigma_A(b)$ for all $b \in B$.*

Proof. In view of (61.9) and the Hahn-Banach theorem (44.2), $\Sigma_B = \{f | B : f \in \Sigma_A\}$. ∎

The following result can be improved upon (61.21) but is adequate for present purposes:

(61.17) Theorem. *If A is a C^*-algebra with unity and if $a \in A$ is self-adjoint, then $\Sigma(a) = $ conv $\sigma(a)$.*

Proof. We know that $\sigma(a)$ is real (58.4); let m be the smallest element of $\sigma(a)$, M the largest. Thus $[m, M] = $ conv $\sigma(a)$. Evidently $m1 \le a \le M1$ in the sense of (61.12); citing (ii) of (61.13), we have $m \le f(a) \le M$ for all $f \in \Sigma$, thus $\Sigma(a) \subset [m, M]$. On the other hand, $\sigma(a) \subset \Sigma(a)$ by the proof of (61.10), therefore conv $\sigma(a) \subset \Sigma(a)$ by the convexity of $\Sigma(a)$. ∎

(61.18) Corollary. *Let A be a C^*-algebra with unity and let $a \in A$ be self-adjoint. Then $a \ge 0$ iff $\Sigma(a) \subset [0, \infty)$.*

Exercises

(61.19) Let A be a C^*-algebra with unity.

(i) The *-representation constructed in (61.11) is cyclic. {In general, a *-representation $\varphi: A \to \mathscr{L}(H)$ is said to be *cyclic* if there exists a vector $u \in H$ such that the linear subspace $\{\varphi(a)u : a \in A\}$ is dense in H; such a vector u is then called a *cyclic vector* for the representation.}

(ii) If $\varphi: A \to \mathscr{L}(H)$ and $\psi: A \to \mathscr{L}(K)$ are cyclic representations of A, with cyclic vectors u and v, respectively, and if $(\varphi(a)u | u) = (\psi(a)v | v)$ for all $a \in A$, then there exists a unitary operator $W: H \to K$ (i.e., a bijective linear isometry) such that $\varphi(a) = W^* \psi(a) W$ for all $a \in A$; in this sense, the representation of a state via a cyclic *-representation is unique up to unitary equivalence.

(61.20) Let A be a C^*-algebra with unity.

(i) If B is a C^*-algebra with unity and $\varphi: A \to B$ is a unital *-monomorphism, then $\Sigma_A(a) = \Sigma_B(\varphi(a))$ for all $a \in A$.

(ii) If $\varphi: A \to \mathscr{L}(H)$ is a faithful unital *-representation of A, then, for all $a \in A$, $\Sigma(a)$ is the closure of the numerical range (57.37) of $\varphi(a)$, i.e., $\Sigma(a) = \overline{W(\varphi(a))}$.

(61.21) Let A be a C^*-algebra with unity and let $a \in A$.

(i) conv $\sigma(a) \subset \Sigma(a)$.

(ii) If Σ_{ep} is the set of all extremal points of Σ, and if $\Sigma_{\mathrm{ep}}(a) = \{f(a) : f \in \Sigma_{\mathrm{ep}}\}$, then $\Sigma(a)$ is the closed convex hull of $\Sigma_{\mathrm{ep}}(a)$, that is,

$$\Sigma(a) = (\mathrm{conv}\, \Sigma_{\mathrm{ep}}(a))^-.$$

(iii) If a is normal, then conv $\sigma(a) = \Sigma(a)$.

(61.22) With notation as in the proof of (61.11), ker $\varphi = \{a \in A : aA \subseteq N\} = \{a \in A : f(x^*a^*ax) = 0$ for all $x \in A\}$.

(61.23) If f is a linear form on a $*$-algebra A, define

$$f^*(a) = (f(a^*))^* \qquad (a \in A);$$

f^* is also a linear form on A, called the *adjoint* of f. If $f^* = f$, that is, if $f(a^*) = f(a)^*$ for all $a \in A$, f is said to be *self-adjoint*; an equivalent condition is that f be real-valued on the self-adjoint elements of A. The correspondence $f \mapsto f^*$ is conjugate-linear and involutory:

$$(f + g)^* = f^* + g^*, \qquad (\lambda f)^* = \lambda^* f^*, \qquad f^{**} = f.$$

Every linear form f has a unique representation $f = g + ih$ with g and h self-adjoint (called the *Cartesian decomposition* of f).

62. Gel'fand-Naĭmark representation theorem.

(62.1) Theorem. *Let A be a C^*-algebra with unity. There exist a Hilbert space H and an isometric unital $*$-representation $\varphi: A \to \mathscr{L}(H)$.*

Proof. Let Σ be the set of all normalized states on A (61.1). For each $f \in \Sigma$, write

$$\varphi_f : A \to \mathscr{L}(H_f)$$

for the unital $*$-representation constructed in (61.11), and let $\varphi: A \to \mathscr{L}(H)$ be the direct sum of the family $(\varphi_f)_{f \in \Sigma}$ (60.6). In particular,

$$\|\varphi(a)\| = \sup \{\|\varphi_f(a)\| : f \in \Sigma\}.$$

It remains to show that φ is isometric.

From (60.1) we know that $\|\varphi(a)\| \leq \|a\|$ for all $a \in A$. On the other hand, if $a \in A$ then there exists $f \in \Sigma$ such that $f(a^*a) = \|a\|^2$ (61.10); writing $u \in H_f$ for the unit vector constructed in (61.11), we have

$$\|a\|^2 = f(a^*a) = (\varphi_f(a^*a)u|u) = (\varphi_f(a)^*\varphi_f(a)u|u)$$
$$= \|\varphi_f(a)u\|^2 \leq \|\varphi_f(a)\|^2,$$

therefore $\|a\| \leq \|\varphi_f(a)\| \leq \|\varphi(a)\|$. ∎

In the above proof, one could replace Σ by any subset $\Sigma_0 \subseteq \Sigma$ for which the direct sum of the representations φ_f $(f \in \Sigma_0)$ is faithful (60.11). For this to happen, the kernels of the φ_f $(f \in \Sigma_0)$ must have intersection $\{0\}$. The kernel of φ_f is described in (61.22). In order that the representation obtained from Σ_0 be faithful, the following condition is necessary and sufficient: if $a \in A$, $a \neq 0$, then there exist $x \in A$ and $f \in \Sigma_0$ such that $f(x^*a^*ax) > 0$. A sufficient condition is

$$\{a \in A : f(a^*a) = 0 \quad \text{for all } f \in \Sigma_0\} = \{0\};$$

this condition is satisfied, for example, by the set Σ_{ep} of all extremal points of Σ (cf. 61.21).

The particular representation constructed in the proof of (62.1) is called the *universal representation* of A.

Exercises

(62.2) The assumption of a unity element in (62.1) can be dropped.

(62.3) Let A be a C^*-algebra and let n be a positive integer. Let A_n be the *-algebra of all $n \times n$ matrices (a_{ij}) over A, with the usual algebra operations and with *-transposition as involution: $(a_{ij})^* = (b_{ij})$ with $b_{ij} = a_{ji}^*$ for all i, j. Suitably normed, A_n is also a C^*-algebra.

(62.4) Let A be a C^*-algebra with unity. An alternate proof of the Gel'fand-Naĭmark theorem may be given in which the extension procedure in the proof of (61.10) is replaced by an extension theorem for positive linear forms (32.2). The three steps of the proof are as follows.

   (i) With notation as in (61.13), A_s is an Archimedean ordered real vector space, with order unit 1 (31.6).

   (ii) If N is a maximal left ideal of A, then there exists a normalized state f on A such that $N = \{x \in A : f(x^*x) = 0\}$.

   (iii) The Gel'fand-Naĭmark theorem follows from (ii), the semisimplicity of A (58.9), and (60.11).

Chapter 8

Miscellaneous Applications

63. Wiener's theorem (the punchline). Here is the theorem in condensed form (for a detailed statement, see Section 0):

(63.1) Theorem. (N. Wiener) *The reciprocal of a nowhere vanishing absolutely convergent trigonometric series is also an absolutely convergent trigonometric series.*

Gel'fand's proof, to be given here, depends on the representation theory of Section 52; to avoid some notational conflicts between Sections 0 and 52, let us start more or less from scratch.

As in Section 0, let $l^1(\mathbb{Z})$ be the algebra of all functions $x\colon \mathbb{Z} \to \mathbb{C}$ such that

$$\sum_{n=-\infty}^{\infty} |x(n)| < \infty,$$

with the pointwise linear operations and the convolution product

$$(xy)(m) = \sum_{n=-\infty}^{\infty} x(m-n)y(n).$$

Equipped with the norm

$$\|x\|_1 = \sum_{n=-\infty}^{\infty} |x(n)|,$$

$l^1(\mathbb{Z})$ is a commutative normed algebra with unity (see Remark 3, formula (18), and Remark 9 of Section 0). Writing

$$e_n(m) = \delta_{mn} \qquad (m, n \in \mathbb{Z}),$$

the linear span of the e_n is a dense subalgebra $c_{00}(\mathbb{Z})$ of $l^1(\mathbb{Z})$; explicitly,

$$e_m e_n = e_{m+n} \qquad (m, n \in \mathbb{Z});$$

e_0 is the unity element of $l^1(\mathbb{Z})$; and $e_n = (e_1)^n$ for all $n \in \mathbb{Z}$ (with the understanding that $(e_1)^0 = e_0$ and $(e_1)^n = (e_1^{-1})^{-n} = (e_{-1})^{-n}$ when $n < 0$).

(63.2) Lemma. $l^1(\mathbb{Z})$ *is a Gel'fand algebra.*

Proof. Obviously $\|e_0\|_1 = 1$. All that remains to be proved is that $l^1(\mathbb{Z})$ is complete for the norm $\|\ \|_1$; the general format of the proof is the same as for the completeness proof in (60.4) (alternatively, see (39.11)). ∎

Let us calculate the characters of $l^1(\mathbb{Z})$:

(63.3) Lemma. (i) *If μ is a complex number with $|\mu| = 1$, the formula*

$$(*) \qquad f_\mu(x) = \sum_{n=-\infty}^{\infty} x(n)\mu^n \qquad (x \in l^1(\mathbb{Z}))$$

defines a character f_μ of $l^1(\mathbb{Z})$.

(ii) *The correspondence* $\mu \mapsto f_\mu$ *is a homeomorphism of the unit circle* $\mathbb{T} = \{\mu \in \mathbb{C} : |\mu| = 1\}$ *with the character space* $\mathscr{X}$ *of* $l^1(\mathbb{Z})$.

Proof. (i) From the inequality

$$\left| \sum_{n=-\infty}^{\infty} x(n)\mu^n \right| \leq \sum_{n=-\infty}^{\infty} |x(n)\mu^n| = \sum_{n=-\infty}^{\infty} |x(n)| = \|x\|_1,$$

it is clear that (*) defines a continuous linear form on $l^1(\mathbb{Z})$. In particular,

$$f_\mu(e_n) = \mu^n \qquad (n \in \mathbb{Z}),$$

therefore

$$f_\mu(e_m e_n) = f_\mu(e_{m+n}) = \mu^{m+n} = \mu^m \mu^n = f_\mu(e_m) f_\mu(e_n).$$

This shows that f_μ is multiplicative on the linear span of the e_n, that is, on $c_{00}(\mathbb{Z})$; since $c_{00}(\mathbb{Z})$ is dense in $l^1(\mathbb{Z})$, it follows from elementary continuity and linearity arguments that f_μ is multiplicative on $l^1(\mathbb{Z})$. Thus $f_\mu \in \mathscr{X}$, the character space of $l^1(\mathbb{Z})$ (52.8).

(ii) Suppose f is any character of $l^1(\mathbb{Z})$. Let $\mu = f(e_1)$. Since $\|f\| = 1$ (52.7), we have

$$|\mu| = |f(e_1)| \leq \|e_1\| = 1;$$

also $f(e_{-1}) = f((e_1)^{-1}) = f(e_1)^{-1} = \mu^{-1}$, therefore

$$|\mu^{-1}| = |f(e_{-1})| \leq \|e_{-1}\| = 1,$$

thus $|\mu| = 1$. For all $n \in \mathbb{Z}$ we have

$$f(e_n) = f((e_1)^n) = f(e_1)^n = \mu^n = f_\mu(e_n);$$

this shows that f and f_μ agree on the dense linear subspace $c_{00}(\mathbb{Z})$, therefore $f = f_\mu$ on $l^1(\mathbb{Z})$ by continuity.

Summarizing, the mapping $\mu \mapsto f_\mu$ is a surjection $\mathbb{T} \to \mathscr{X}$. From the formula $\mu = f_\mu(e_1)$, it is clear that the mapping is injective.

In other words, the mapping $f \mapsto f(e_1)$ is a bijection $\mathscr{X} \to \mathbb{T}$. To prove that it is a homeomorphism, it suffices (by the compactness of $\mathscr{X}$ and the separatedness of $\mathbb{T}$) to show that it is continuous; but this is obvious from the definition of the topology on $\mathscr{X}$ (52.8). ∎

In view of (52.10), the maximal ideal space of $l^1(\mathbb{Z})$ may be identified with $\mathbb{T}$. When this is done, the Gel'fand transform of $x \in l^1(\mathbb{Z})$ is the function $\hat{x}$ on $\mathbb{T}$ given by the formula

$$\hat{x}(\mu) = f_\mu(x) = \sum_{n=-\infty}^{\infty} x(n)\mu^n \qquad (\mu \in \mathbb{T});$$

in other words, for all $t \in \mathbb{R}$,

$$\hat{x}(e^{it}) = \sum_{n=-\infty}^{\infty} x(n)e^{int}.$$

Thus $\hat{x}$ may be interpreted either as a continuous function on $\mathbb{T}$ or as a continuous periodic function on $\mathbb{R}$ of period 2π. The range of the Gel'fand transformation,

$$(l^1(\mathbb{Z}))^\wedge = \{\hat{x} : x \in l^1(\mathbb{Z})\},$$

may thus be identified as the algebra (operations pointwise) of all continuous periodic functions $\mathbb{R} \to \mathbb{C}$ of period 2π whose Fourier series are absolutely convergent.

If such a function, say $\hat{x}$, never vanishes, then $1/\hat{x} = \hat{y}$ for suitable $y \in l^1(\mathbb{Z})$ (by (v) of (52.13)). This is Wiener's theorem.

Exercise

(63.4) With notation as in the proof of (63.3), $\sigma(e_1) = \mathbb{T}$. If $U \in \mathscr{L}(l^1(\mathbb{Z}))$ is the multiplication operator $Ux = e_1 x$, then $\sigma(U) = \mathbb{T}$. In view of the formulas $Ue_n = e_{n+1}$, $U^{-1}e_n = e_{n-1}$, U is called a *bilateral shift operator.*

64. Stone-Čech compactification. A topological space T is called *completely regular* if it is uniformizable and separated [**21**, Ch. IX, §1, Def. 4]. {A topological space T is uniformizable iff it has the following property: if $t \in U \subset T$, U open, there exists a continuous function $x: T \to [0, 1]$ such that $x(t) = 1$ and $x = 0$ on $T - U$ [**21**, Ch. IX, §1, Th. 2]. Thus the completely regular spaces are the $T_{3\frac{1}{2}}$ spaces according to a classical nomenclature.} Since every compact space is uniformizable [**19**, Ch. II, §4, Th. 1], it follows that every subspace of a compact space is completely regular. The Stone-Čech theorem is a converse: Every completely regular space may be embedded in a compact space. More precisely:

(64.1) Theorem. *Let T be a completely regular space. There exists a pair $(\mathscr{X}, \varphi)$, consisting of a compact space $\mathscr{X}$ and a mapping $\varphi: T \to \mathscr{X}$, with the following properties:*

(1) φ is injective and maps T bicontinuously onto $\varphi(T)$, where $\varphi(T)$ is equipped with the relative topology induced by $\mathscr{X}$;

(2) $\varphi(T)$ is dense in $\mathscr{X}$;

(3) if $x: T \to \mathbb{R}$ is any bounded continuous function, there exists a continuous function $F: \mathscr{X} \to \mathbb{R}$ such that $x = F \circ \varphi$. Diagrammatically:

The gist of (1) and (2) is that T may be identified with a dense subset of $\mathscr{X}$; this done, (3) asserts that every bounded continuous real function on T has a continuous extension to $\mathscr{X}$.

There exist straightforward proofs of (64.1) based on the Tihonov theorem, with $\mathscr{X}$ taken to be a closed subset of a suitable Cartesian product of closed intervals. The following functional-analytic proof is especially tidy: the desired space $\mathscr{X}$ turns up as the character space of a suitable Gel'fand algebra (but secretly the Tihonov theorem has done its work in the way the character space is topologized—via the Alaoglu-Bourbaki theorem (44.12)).

The functional-analytic proof begins spectacularly, by producing $\mathscr{X}$ at once: As noted in (59.3), the algebra $\mathscr{C}^{\infty}(T)$ of all bounded continuous functions $x: T \to \mathbb{C}$ is a commutative C^*-algebra with unity, relative to the pointwise $*$-algebra operations and the norm

$$\|x\|_{\infty} = \sup\{|x(t)| : t \in T\};$$

we take $\mathscr{X}$ to be the character space of $\mathscr{C}^{\infty}(T)$ (52.8). Thus $\mathscr{X}$ is the set of all algebra epimorphisms $f: \mathscr{C}^{\infty}(T) \to \mathbb{C}$, equipped with the relative weak* topology it acquires as a subset of (the closed unit ball of) the dual space of $\mathscr{C}^{\infty}(T)$.

The mapping $\varphi: T \to \mathscr{X}$ is produced equally swiftly. Each $t \in T$ defines a character $f_t \in \mathscr{X}$ by the formula

$$f_t(x) = x(t) \qquad (x \in \mathscr{C}^{\infty}(T));$$

we define

$$\varphi(t) = f_t \qquad (t \in T).$$

It remains to be shown that the pair $(\mathscr{X}, \varphi)$ has the required properties (1)–(3); this is accomplished in the following lemmas.

(64.2) Lemma. φ *is injective.*

Proof.   Suppose $s, t \in T$, $s \neq t$. By complete regularity, there exists $x \in \mathscr{C}^{\infty}(T)$ such that $x(s) = 0$ and $x(t) = 1$, thus $f_s(x) \neq f_t(x)$.   ∎

(64.3) Lemma. φ *maps T bicontinuously onto $\varphi(T)$.*

Proof.   Let t_j be a net in T and let $t \in T$. The problem is to show that $t_j \to t$ (for the given topology on T) iff $f_{t_j} \to f_t$ (for the weak* topology). The latter condition means that $f_{t_j}(x) \to f_t(x)$—that is, $x(t_j) \to x(t)$—for all $x \in \mathscr{C}^{\infty}(T)$.

If $t_j \to t$ then, for every $x \in \mathscr{C}^{\infty}(T)$, $x(t_j) \to x(t)$ by the continuity of x.

Conversely, suppose $x(t_j) \to x(t)$ for all x. Given any neighborhood V of t, we are to show that $t_j \in V$ ultimately. Choose $x \in \mathscr{C}^{\infty}(T)$ so that $x(t) = 1$ and $x = 0$ on $T - V$. By supposition, $x(t_j) \to x(t) = 1$, thus there exists an index k such that $|x(t_j) - 1| < 1$ for all $j \geq k$; it follows that if $j \geq k$ then $x(t_j) \neq 0$, thus $t_j \in V$.   ∎

(64.4) Lemma. $\varphi(T)$ *is dense in $\mathscr{X}$.*

Proof.   For the first time in the proof, we make use of the representation theorem for commutative C^*-algebras (59.2): the Gel'fand transformation $x \mapsto \hat{x}$ maps $\mathscr{C}^{\infty}(T)$ onto $\mathscr{C}(\mathscr{X})$ (isometrically and $*$-isomorphically).

Assume to the contrary that $\overline{\varphi(T)}$ is a proper subset of $\mathscr{X}$. Say $f \in \mathscr{X}$, $f \notin \overline{\varphi(T)}$. By the complete regularity of $\mathscr{X}$, there exists $F \in \mathscr{C}(\mathscr{X})$ such that

$F(f) = 1$ and $F = 0$ on $\overline{\varphi(T)}$. In particular, $F = 0$ on $\varphi(T)$ and $F \neq 0$. Say $F = \hat{x}$, $x \in \mathscr{C}^\infty(T)$. Then

$$0 = F(\varphi(t)) = \hat{x}(f_t) = f_t(x) = x(t)$$

for all $t \in T$, thus $x = 0$, whence $F = \hat{x} = 0$, a contradiction. ∎

(64.5) **Lemma.** *The pair* $(\mathscr{X}, \varphi)$ *satisfies condition* (3) *of* (64.1).

Proof. Suppose $x: T \to \mathbb{R}$ is bounded and continuous. In particular, $x \in \mathscr{C}^\infty(T)$. Define $F = \hat{x}$, the Gel'fand transform of x. For all $t \in T$,

$$F(\varphi(t)) = \hat{x}(f_t) = f_t(x) = x(t),$$

thus $F \circ \varphi = x$; in particular, F is real-valued on $\varphi(T)$, therefore F is real-valued on $\mathscr{X}$ (64.4). ∎

This completes the proof of (64.1). Such a pair $(\mathscr{X}, \varphi)$ is called a *Stone-Čech compactification* of T. {It is not difficult to show, but inappropriate to do it here, that such a pair is essentially unique. See, e.g., [**85**, p. 153], [**134**, p. 331, Th. A].}

65. The continuous functional calculus (spectral theorem for a normal operator).

For general remarks on functional calculi, see the introduction of Section 53. The continuous functional calculus is applicable to any normal element of any C^*-algebra A with unity; applied to the C^*-algebra $A = \mathscr{L}(H)$, H a Hilbert space, this theory is the core of what is known as the spectral theory of normal operators.

(65.1) **Theorem.** *Let* A *be a* C^*-*algebra with unity, suppose* $a \in A$ *is normal* $(a^*a = aa^*)$, *and write* u *for the identity function* $u(\lambda) \equiv \lambda$ *on* $\sigma(a)$. *There exists a unique* $*$-*homomorphism*

$$\Phi: \mathscr{C}(\sigma(a)) \to A$$

such that $\Phi(1) = 1$ *and* $\Phi(u) = a$. *Moreover,* Φ *is isometric and its range is the closed subalgebra* C *of* A *generated by* 1, a, *and* a^*.

Proof. Let B be any closed commutative $*$-subalgebra of A that contains 1 and a. {For example, $B = \{a, a^*\}''$, the bicommutant of $\{a, a^*\}$ in A; or $B = C$.} One has $\sigma_B(a) = \sigma_A(a)$. {Cf. (58.5), (53.8).} Thus we may write $\sigma(a)$ unambiguously.

Let $\mathscr{M}$ be the maximal ideal space of B and let $x \mapsto \hat{x}$ be the Gel'fand transformation $B \to \mathscr{C}(\mathscr{M})$; by the commutative Gel'fand-Naĭmark theorem (59.2), we know that $x \mapsto \hat{x}$ is an isometric $*$-isomorphism of B onto $\mathscr{C}(\mathscr{M})$.

The mapping $\Phi: \mathscr{C}(\sigma(a)) \to A$ is defined as follows. Suppose $f \in \mathscr{C}(\sigma(a))$. Since $\hat{a}: \mathscr{M} \to \sigma(a)$ is continuous (and surjective), the composite function $f \circ \hat{a}$ is in $\mathscr{C}(\mathscr{M})$. We define $\Phi(f)$ to be the unique element of B such that

$$(\Phi(f))^\wedge = f \circ \hat{a}.$$

It is straightforward to check that Φ is a $*$-homomorphism with $\Phi(1) = 1$ and $\Phi(u) = a$. Moreover,

$$\|\Phi(f)\| = \|(\Phi(f))^\wedge\|_\infty = \|f \circ \hat{a}\|_\infty$$

$$= \sup\{|f(\hat{a}(M))| : M \in \mathscr{M}\}$$

$$= \sup\{|f(\lambda)| : \lambda \in \sigma(a)\} = \|f\|_\infty,$$

thus Φ is isometric.

Assuming $\Psi : \mathscr{C}(\sigma(a)) \to A$ is a $*$-homomorphism with $\Psi(1) = 1$ and $\Psi(u) = a$, let us shown that $\Psi = \Phi$. From (60.1) we know that Ψ is continuous. It follows that the set

$$\mathscr{A} = \{f \in \mathscr{C}(\sigma(a)) : \Psi(f) = \Phi(f)\}$$

is a closed $*$-subalgebra of $\mathscr{C}(\sigma(a))$ containing 1 and u; by the Weierstrass-Stone theorem (the function u already separates the points of $\sigma(a)$) we have $\mathscr{A} = \mathscr{C}(\sigma(a))$, that is, $\Psi = \Phi$.

It remains to be shown that the range of Φ is C. From the properties of Φ, we know that its range is a closed $*$-subalgebra of A containing 1 and a, therefore $\Phi(\mathscr{C}(\sigma(a))) \supset C$. On the other hand, $\Phi^{-1}(C)$ is a closed $*$-subalgebra of $\mathscr{C}(\sigma(a))$ containing 1 and u, therefore $\Phi^{-1}(C) = \mathscr{C}(\sigma(a))$, thus $\Phi(\mathscr{C}(\sigma(a))) \subset C$. ∎

(65.2) Definition. With notation as in (65.1), we define $f(a) = \Phi(f)$ for all $f \in \mathscr{C}(\sigma(a))$.

With this notation, (65.1) yields the following formulas: $(f + g)(a) = f(a) + g(a)$, $(\alpha f)(a) = \alpha f(a)$, $(fg)(a) = f(a)g(a)$, $f^*(a) = (f(a))^*$, $1(a) = 1$, $u(a) = a$, and $\|f(a)\| = \|f\|_\infty$, for all $f, g \in \mathscr{C}(\sigma(a))$ and $\alpha \in \mathbb{C}$.

(65.3) Theorem. (Spectral mapping theorem) *If A is a C^*-algebra with unity and if $a \in A$ is normal, then*

$$\sigma(f(a)) = f(\sigma(a))$$

for all $f \in \mathscr{C}(\sigma(a))$.

Proof. Let $f \in \mathscr{C}(\sigma(a))$. With notation as in the proof of (65.1), we have $f(a) = \Phi(f) \in B$; thus $\sigma(f(a))$ is the range of the function $(\Phi(f))^\wedge = f \circ \hat{a}$, namely, the set

$$\{(f \circ \hat{a})(M) : M \in \mathscr{M}\} = \{f(\hat{a}(M)) : M \in \mathscr{M}\} = \{f(\lambda) : \lambda \in \sigma(a)\}. \quad ∎$$

(65.4) Theorem. *Let A be a C^*-algebra with unity and suppose $a \in A$ is normal. Then*

$$(f \circ g)(a) = f(g(a))$$

whenever $g \in \mathscr{C}(\sigma(a))$ and $f \in \mathscr{C}(\sigma(g(a)))$.

Proof. Fix $g \in \mathscr{C}(\sigma(a))$. Then $g(a)$ is normal and $\sigma(g(a)) = g(\sigma(a))$ (65.3). It follows that if $f \in \mathscr{C}(\sigma(g(a)))$ then the composite function $f \circ g$ is defined and belongs to $\mathscr{C}(\sigma(a))$, thus $(f \circ g)(a)$ exists. We may therefore define a mapping

$$\Phi : \mathscr{C}(\sigma(g(a))) \to A$$

by the formula $\Phi(f) = (f \circ g)(a)$. Note that $\Phi(u) = g(a)$, where $u(\lambda) \equiv \lambda$ is the identity function on $\sigma(a)$. {Proof: $u \circ g = g$.} Also, $\Phi(1) = 1$. {Proof: $1 \circ g = 1$, suitably interpreted.} It is straightforward to check that Φ is a *-homomorphism. Thus Φ has the properties that characterize the mapping provided by (65.1) for the normal element $g(a)$; in other words,

$$\Phi(f) = f(g(a))$$

for all $f \in \mathscr{C}(\sigma(g(a)))$, which is precisely the assertion of the theorem. ∎

(65.5) **Theorem.** *Let A and B be C*-algebras with unity, let $\varphi : A \to B$ be a unital *-homomorphism, and let $a \in A$ be normal. Then*

$$\varphi(f(a)) = f(\varphi(a))$$

for all $f \in \mathscr{C}(\sigma(a))$.

Proof. As noted in the proof of (60.1), $\sigma(\varphi(a)) \subset \sigma(a)$; thus if $f \in \mathscr{C}(\sigma(a))$, the f that appears on the right side of the desired formula is to be interpreted as the restriction $f|\sigma(\varphi(a))$. Write u for the identity function $u(\lambda) \equiv \lambda$ on $\sigma(a)$.

The mappings $f \mapsto \varphi(f(a))$ and $f \mapsto f(\varphi(a))$ are *-homomorphisms $\mathscr{C}(\sigma(a)) \to B$, each of which satisfies $1 \mapsto 1$ and $u \mapsto \varphi(a)$, therefore the mappings are identical by the argument in the proof of (65.1). ∎

(65.6) **Application.** (*Spectral theorem for a normal operator*) Let T be a normal operator on a Hilbert space H. Regarding T as an element of the C*-algebra $\mathscr{L}(H)$, the unique mapping

$$f \mapsto f(T) \qquad (f \in \mathscr{C}(\sigma(T)))$$

defined in (65.2) is one formulation of the 'spectral theorem' for a normal operator (the so-called functional representation formulation). Using measure-theoretic techniques, the class of functions can be greatly enlarged (though injectivity and the spectral mapping formula fall by the wayside); the basic idea is that each vector $x \in H$ determines a positive linear form (i.e., a positive Radon measure) μ_x on $\mathscr{C}(\sigma(T))$ via the formula $\mu_x(f) = (f(T)x|x)$. {For the details, cf. [11], [35, Ch. XV, §11], [39, p. 112, Th. 4.71], [66, §44, Th. 1], [123, p. 309, Sect. 12.24], [133, p. 249, Cor. 9.2.2].} Another strategy for gaining flexibility in the function class is to opt for the subalgebra $\{T, T^*\}''$ in the proof of (65.1); the resulting maximal ideal space $\mathscr{M}$ is then a Stonian (or 'extremally disconnected') space—the closure of every open set is clopen (closed and open)—and the characteristic functions of the clopen sets provide the supply of projection operators E ($E^* = E = E^2$) required for other formulations of the spectral theorem. {For an exposition of the details, cf. [40].}

Exercises

(65.7) Notation as in the proof of (65.1). When $B = C$, the function $\hat{a}: \mathcal{M} \to \sigma(a)$ is injective, hence is a homeomorphism.

(65.8) Deduce another proof of the spectral mapping theorem (65.3) by specializing (65.5).

(65.9) Let A be a C^*-algebra with unity, let τ be a nonempty compact subset of the complex plane, and suppose $\varphi: \mathscr{C}(\tau) \to A$ is an algebra homomorphism such that $\varphi(1) = 1$ and $\|\varphi(f)\| \le \|f\|_\infty$ for all $f \in \mathscr{C}(\tau)$. Write u for the identity function $u(\lambda) \equiv \lambda$ on τ, and let $a = \varphi(u)$. Then $\tau \supset \sigma(a)$, φ is a *-homomorphism, and a is normal.

***(65.10)** Let $T \in \mathscr{L}(H)$, H a Hilbert space, and suppose there exists an algebra homomorphism $\varphi: \mathscr{C}(\sigma(T) \to \mathscr{L}(H)$ such that $\varphi(1) = I$ and $\varphi(u) = T$, where u is the identity function $u(\lambda) \equiv \lambda$ on $\sigma(T)$. Then φ is continuous and T is similar to a normal operator (i.e., there exists an invertible operator S such that $S^{-1}TS$ is normal).

***(65.11)** (Fuglede-Putnam theorem) If A is a C^*-algebra with unity, if $a, b \in A$ are normal, and if $x \in A$ satisfies $xa = bx$, then also $xa^* = b^*x$; in particular, $\{a\}'$—the commutant of a in A—is a *-subalgebra of A, thus $\{a\}'' = \{a, a^*\}''$. It follows that, in the notation of (65.2), if $x \in A$ satisfies $xa = ax$, then also $xf(a) = f(a)x$.

***(65.12)** (P. R. Halmos)   If T is a normal operator on a Hilbert space H, then there exists a self-adjoint operator S on H, and a continuous function $f: \sigma(S) \to \mathbb{C}$, such that $T = f(S)$.

(65.13)   Let A be a C^*-algebra with unity.
(i) If $a \in A$, $a \ge 0$, then there exists a unique $b \in A$ such that $b \ge 0$ and $b^2 = a$. Notation: $a^{1/2} = b$. Necessarily $b \in \{a\}''$, the bicommutant of a in A.
(ii) Every element of A is a linear combination of unitary elements.

*66. Spectral sets.

The theory of spectral sets, due to J. von Neumann [110], is a theory of 'functional calculus' type for continuous linear mappings (briefly, *operators*) on a Hilbert space (cf. the introduction of Section 53). The basic theory can easily be formulated so as to apply (with variable success) to an element of an arbitrary Banach algebra. However, in a sense the *raison d'être* of the theory is to explain why operator theory in Hilbert spaces is different from operator theory in Banach spaces; accordingly, we restrict the discussion to Hilbert space operators from the start. The reader will readily see which arguments generalize and which do not.

For the rest of the section, H is a fixed Hilbert space and T is an operator on H, i.e., $T \in \mathscr{L}(H)$ (of course, the hypotheses on T are in a state of perpetual flux).

As in Sections 53 and 54, $\mathfrak{S} = \mathbb{C} \cup \{\infty\}$ denotes the Riemann sphere, and, for every subset σ of $\mathfrak{S}$, $\mathbb{C}(t; \sigma)$ is the algebra of all rational forms $f \in \mathbb{C}(t)$ that have no poles in σ. Recall that every $f \in \mathbb{C}(t)$ may be regarded as a continuous function $f: \mathfrak{S} \to \mathfrak{S}$ (54.3). A subset σ of $\mathfrak{S}$ is said to be *bounded* if $\sigma \subset \mathbb{C}$ and σ is bounded in the ordinary sense; thus, σ is unbounded iff ∞ belongs to the closure of σ (i.e., either $\infty \in \sigma$, or there exists a sequence of complex numbers $\lambda_n \in \sigma$ with $|\lambda_n| \to \infty$, or both).

(66.1) Definition. If σ is a nonempty subset of $\mathfrak{S}$ and if $f \in \mathbb{C}(t; \sigma)$, then, regarding f as a function on $\mathfrak{S}$, $f|\sigma$ is a finite-valued function, i.e., $f|\sigma: \sigma \to \mathbb{C}$; we write $\mathscr{A}(\sigma)$ for the set of all such restrictions $f|\sigma$. {To avoid endless repetition, we assume for the rest of the section that all subsets $\sigma, \tau, \ldots$ of $\mathfrak{S}$ considered are *nonempty*.}

(66.2) Proposition. (i) $\mathscr{A}(\sigma)$ *is an algebra of functions, for every* $\sigma \subset \mathfrak{S}$.
(ii) *If σ is a closed (i.e., compact) subset of $\mathfrak{S}$, then* $\mathscr{A}(\sigma) \subset \mathscr{C}(\sigma)$.

Proof. (i) Clear.
(ii) For compact σ, $\mathscr{C}(\sigma)$ is the commutative C^*-algebra described in (59.1); the inclusion $\mathscr{A}(\alpha) \subset \mathscr{C}(\sigma)$ is obvious since the functions in $\mathscr{A}(\sigma)$ are continuous and $\mathbb{C}$-valued. ∎

(66.3) Definition. If $\sigma \subset \mathfrak{S}$, f is a function whose domain of definition includes σ, and f is $\mathfrak{S}$-valued on σ, we write

$$\|f\|_\sigma = \sup\{|f(\lambda)| : \lambda \in \sigma\},$$

with the convention that $|\infty| = \infty$. Thus $\|f\|_\sigma < \infty$ means that f is bounded on σ, i.e., $f(\sigma)$ is a bounded subset of $\mathfrak{S}$. In particular, if σ is closed and $f \in \mathbb{C}(t; \sigma)$, then $\|f\|_\sigma$ is a substitute notation for $\|f|\sigma\|_\infty$.

(66.4) Suppose $\sigma \supset \sigma(T)$. Obviously $\mathbb{C}(t; \sigma) \subset \mathbb{C}(t; \sigma(T))$; thus, if $f \in \mathbb{C}(t; \sigma)$ then $f(T)$ exists in the sense of the rational functional calculus (53.5). By the notation

$$f \mapsto f(T) \qquad (f \in \mathbb{C}(t; \sigma)),$$

we mean the restriction, to $\mathbb{C}(t; \sigma)$, of the rational functional calculus $\mathbb{C}(t; \sigma(T)) \to \mathscr{L}(H)$ for T given by (53.5). Note: *If $\infty \in \sigma$ then $\mathbb{C}(t; \sigma)$ excludes all rational forms $f = p/q$ with $\deg p > \deg q$ (in particular, it excludes all nonconstant polynomials).*

(66.5) Definition. A set $\sigma \subset \mathfrak{S}$ is called a *spectral set* for T if (i) $\sigma \supset \sigma(T)$, and (ii) $\|f(T)\| \le \|f\|_\sigma$ for all $f \in \mathbb{C}(t; \sigma)$.

It is obvious that $\mathbb{C}$ is a spectral set for T. {Let $f \in \mathbb{C}(t; \mathbb{C}) = \mathbb{C}[t]$. If f is a constant, $f = c1$, then $\|f(T)\| = \|cI\| = |c| = \|f\|_{\mathbb{C}}$; if f is a nonconstant polynomial, then $\|f(T)\| < \infty = \|f\|_{\mathbb{C}}$.} It is nonobvious, but true, that T has bounded spectral sets; explicitly, it is a fundamental theorem of von Neumann that the closed disc $\{\lambda : |\lambda| \le \|T\|\}$ is a spectral set for T. A sketch of von Neumann's theorem is given at the end of the section. {C. Foiaş has shown that von Neumann's result characterizes Hilbert spaces among Banach spaces, i.e., it is false for all other Banach spaces [46].}

It turns out that one need only consider spectral sets that are closed subsets of $\mathfrak{S}$ (but it's nice to be flexible). The next propositions clarify this point.

(66.6) Proposition. *Let $\sigma(T) \subset \sigma \subset \mathfrak{S}$. The following conditions on σ are equivalent:* (a) *σ is a spectral set for T;* (b) *if $f \in \mathbb{C}(t)$ and $\|f\|_\sigma \leq 1$ then $\|f(T)\| \leq 1$.*

Proof. (a) implies (b): If $f \in \mathbb{C}(t)$ and $\|f\|_\sigma \leq 1$ then f can have no poles in σ, i.e., $f \in \mathbb{C}(t; \sigma)$, thus $\|f(T)\| \leq \|f\|_\sigma \leq 1$.

(b) implies (a): Assuming $f \in \mathbb{C}(t; \sigma)$, it is to be shown that $\|f(T)\| \leq \|f\|_\sigma$. The inequality holds trivially if $\|f\|_\sigma = \infty$ (which means that f has a pole in $\bar{\sigma}$); it also holds if $\|f\|_\sigma = 0$, since application of (b) to nf ($n = 1, 2, 3, \ldots$) yields $\|f(T)\| = 0$; and if $0 < \|f\|_\sigma < \infty$, the desired inequality results on applying (b) to $\|f\|_\sigma^{-1} f$. ∎

Every superset of a spectral set is a spectral set:

(66.7) Proposition. *If σ is a spectral set for T and if $\tau \supset \sigma$, then τ is also a spectral set for T.*

Proof. If $f \in \mathbb{C}(t; \tau)$ then $f \in \mathbb{C}(t; \sigma)$ and $\|f(T)\| \leq \|f\|_\sigma \leq \|f\|_\tau$. ∎

(66.8) Proposition. *Let $\sigma \supset \sigma(T)$. Then σ is a spectral set for T iff $\bar{\sigma}$ is a spectral set for T.*

Proof. Write $\tau = \bar{\sigma}$ and note that $\|f\|_\sigma = \|f\|_\tau$ for every $f \in \mathbb{C}(t)$, by the continuity of f as a function on $\mathfrak{S}$.

Suppose τ is a spectral set for T. To show that σ is, we invoke the criterion of (66.6): if $f \in \mathbb{C}(t)$ and $\|f\|_\sigma \leq 1$, then $\|f\|_\tau = \|f\|_\sigma \leq 1$, therefore $\|f(T)\| \leq 1$ (because τ is a spectral set for T).

The converse is immediate from (66.7). ∎

Spectral sets are always supersets of the spectrum; it is a special event when the spectrum is itself a spectral set:

(66.9) Proposition. *The following conditions on T are equivalent:* (a) *$\sigma(T)$ is a spectral set for T;* (b) *$\|f(T)\| = r(f(T))$ for all $f \in \mathbb{C}(t; \sigma(T))$.*

Proof. Here r denotes spectral radius (51.11). One knows that $r(S) \leq \|S\|$ for every operator S. {Operators for which equality holds are called *normaloid*.} If $f \in \mathbb{C}(t; \sigma(T))$ then, from the spectral mapping formula (53.6)

$$\sigma(f(T)) = f(\sigma(T)),$$

it follows that $r(f(T)) = \|f\|_{\sigma(T)}$.

(a) implies (b): For all $f \in \mathbb{C}(t; \sigma(T))$, we have

$$r(f(T)) \leq \|f(T)\| \leq \|f\|_{\sigma(T)} = r(f(T)).$$

(b) implies (a): For all $f \in \mathbb{C}(t; \sigma(T))$, we have

$$\|f(T)\| = r(f(T)) = \|f\|_{\sigma(T)},$$

therefore $\sigma(T)$ is a spectral set for T. ∎

(66.10) Corollary. *If T is normal then $\sigma(T)$ is a spectral set for T.*

Proof. If $f \in C(t; \sigma(T))$ then $f(T)$ is also normal, and therefore normaloid (58.3). ∎

The converse of (66.10) is in general false (66.28), but it is true when H is finite-dimensional (66.21). {The latter remark provides a striking 'involution-free' characterization of normal matrices.}

The involution in $\mathscr{L}(H)$ produces analogies between operators and complex numbers. For example, the fixed points under $\lambda \mapsto \lambda^*$ ($\lambda \in \mathbb{C}$) are the real numbers; the fixed points under $T \mapsto T^*$ ($T \in \mathscr{L}(H)$) are the self-adjoint operators; and the spectrum of a self-adjoint operator is real (57.24). Thus there is an analogy between self-adjointness and reality. But an operator with real spectrum need not be self-adjoint (66.29), so there is a certain flaw in the analogy; so to speak, the spectrum of T is in general not a sufficiently large set for its reality to imply self-adjointness of T. The theory of spectral sets perfects the analogy: T is self-adjoint iff $\mathbb{R}$ is a spectral set for T (this is proved below). Thus, the set of self-adjoint operators is precisely the set of all operators that admit $\mathbb{R}$ as a spectral set; the analogy pairs the set of all self-adjoint operators with the subset $\mathbb{R}$ of $\mathfrak{S}$. It is convenient to establish first another such pairing:

(66.11) Theorem. *The following conditions on T are equivalent:* (a) *T is unitary (i.e., $T^*T = TT^* = I$);* (b) *the unit circle $\mathbb{T} = \{\lambda \in \mathbb{C} : \lambda^*\lambda = 1\}$ is a spectral set for T.*

Proof. (a) implies (b): Since T is normal, $\sigma(T)$ is a spectral set for T (66.10); in view of (66.7), it will suffice to show that $\sigma(T) \subset \mathbb{T}$. Note that $\|T\|^2 = \|T^*T\| = 1$. If $\lambda \in \sigma(T)$ then $|\lambda| \le \|T\| = 1$; since also $\lambda^{-1} \in \sigma(T^{-1}) = \sigma(T^*)$, we have $|\lambda|^{-1} \le \|T^*\| = 1$, thus $|\lambda| = 1$.

(b) implies (a): Part of the hypothesis is that $\sigma(T) \subset \mathbb{T}$, so T is invertible. The rational form $f = 1/t$ belongs to $C(t; \mathbb{T})$, therefore

$$\|T^{-1}\| = \|f(T)\| \le \|f\|_{\mathbb{T}} = 1.$$

On the other hand, the monomial $g = t$ also belongs to $C(t; \mathbb{T})$, so

$$\|T\| = \|g(T)\| \le \|g\|_{\mathbb{T}} = 1.$$

It follows that T is unitary. {Proof: For all $x \in H$,

$$\|Tx\| \le \|x\| = \|T^{-1}(Tx)\| \le \|Tx\|,$$

thus $\|Tx\| = \|x\|$; squaring this yields $(T^*Tx|x) = (x|x)$ for all x, whence $T^*T = I$—therefore also $TT^* = I$ by the invertibility of T.} ∎

The analogous characterization of self-adjointness can be inferred from (66.11) via the following elementary lemma (cf. 66.30):

(66.12) Lemma. *If τ is a spectral set for T and if $f \in C(t; \tau)$, then $f(\tau)$ is a spectral set for $f(T)$.*

Proof. From $\tau \supset \sigma(T)$ and the spectral mapping formula (53.6), we have

$$\sigma(f(T)) = f(\sigma(T)) \subset f(\tau),$$

therefore $\mathbb{C}(t; f(\tau)) \subset \mathbb{C}(t; \sigma(f(T)))$. If $g \in \mathbb{C}(t; f(\tau))$, it follows from (54.8) that the composite function $g \circ f$ is rational, $g \circ f \in \mathbb{C}(t; \sigma(T))$, and $(g \circ f)(T) = g(f(T))$. By hypothesis, g has no poles in $f(\tau)$, therefore $g \circ f$ can have no poles in τ, thus $g \circ f \in \mathbb{C}(t; \tau)$; since τ is a spectral set for T, we have

(1) $$\|g(f(T))\| = \|(g \circ f)(T)\| \le \|g \circ f\|_\tau.$$

Moreover,

(2) $$\|g \circ f\|_\tau = \sup \{|g(f(\lambda))| : \lambda \in \tau\}$$
$$= \sup \{|g(\mu)| : \mu \in f(\tau)\} = \|g\|_{f(\tau)};$$

from (1) and (2) we infer the desired inequality $\|g(f(T))\| \le \|g\|_{f(\tau)}$. ∎

(66.13) Theorem. *The following conditions on T are equivalent: (a) $T^* = T$;* (b) $\mathbb{R}$ *is a spectral set for T.*

Proof. (a) implies (b): If $T^* = T$ then $\mathbb{R} \supset \sigma(T)$ (57.24) and $\sigma(T)$ is a spectral set for T (66.10), therefore so is $\mathbb{R}$ (66.7).

(b) implies (a): Consider the rational form $f = (t - i)(t + i)^{-1}$, where $i^2 = -1$. As a function on $\mathfrak{S}$, f maps $\mathbb{R}$ bijectively onto $\mathbb{T} - \{1\}$, and $f(\infty) = 1$. The only pole of f is i, thus $f \in \mathbb{C}(t; \mathbb{R})$; since $\mathbb{R}$ is a spectral set for T, it follows from the lemma that $f(\mathbb{R}) = \mathbb{T} - \{1\}$ is a spectral set for $f(T)$, therefore $f(T)$ is unitary (66.11). Writing $U = f(T)$, we have

(i) $$U = (T - iI)(T + iI)^{-1} \text{ is unitary.}$$

Since $\sigma(U) = \sigma(f(T)) = f(\sigma(T)) \subset f(\mathbb{R}) = \mathbb{T} - \{1\}$, we have $1 \notin \sigma(U)$, therefore

(ii) $$I - U \text{ is invertible.}$$

From (i) and (ii), an elementary computation yields

$$T = i(I + U)(I - U)^{-1},$$

and another then yields $T^* = T$. {The unitary operator U is called the *Cayley transform* of the self-adjoint operator T.} ∎

Suppose $\sigma(T) \subset \sigma \subset \mathfrak{S}$ and consider the functional calculus

(*) $$f \mapsto f(T) \qquad (f \in \mathbb{C}(t; \sigma))$$

described in (66.4). The calculus (*) is defined on an algebra of rational *forms*. Each $f \in \mathbb{C}(t; \sigma)$ may also be viewed as a function $\mathfrak{S} \to \mathfrak{S}$; but collectively, these functions do not form an algebra of functions on $\mathfrak{S}$, because there is trouble with the algebraic operations at poles. {The exceptional case that $\sigma = \mathfrak{S}$ is trouble-free and interest-free: $\mathbb{C}(t; \mathfrak{S}) = \mathbb{C}$.} Why not switch to the algebra of functions $\mathscr{Q}(\sigma)$ (66.2)? The proposal is to define a functional calculus on $\mathscr{Q}(\sigma)$ by the prescription

(**) $$f|\sigma \mapsto f(T) \qquad (f \in \mathbb{C}(t; \sigma));$$

the problem is, is this well-defined? That is, if $f, g \in \mathbb{C}(t; \sigma)$ and if $f|\sigma = g|\sigma$, does it follow that $f(T) = g(T)$? In other words, does $\|f - g\|_\sigma = 0$ imply that $\|f(T) - g(T)\| = 0$? The definition of spectral set is made to order to provide the affirmative answer; formally stated:

(66.14) **Proposition.** *If σ is a spectral set for T, then the correspondence*

$$f|\sigma \mapsto f(T) \qquad (f \in \mathbb{C}(t; \sigma))$$

defines a unital algebra homomorphism $\mathcal{Q}(\sigma) \to \mathcal{L}(H)$.

Of particular interest is the case that σ is a *closed* spectral set for T: then $\mathcal{Q}(\sigma)$ is a subalgebra of the Banach algebra $\mathscr{C}(\sigma)$ (66.2). Moreover, the definition of spectral set means that

$$\|f(T)\| \le \|f|\sigma\|_{\infty}$$

for all $f \in \mathbb{C}(t; \sigma)$, thus the homomorphism $\mathcal{Q}(\sigma) \to \mathcal{L}(H)$ of (66.14) is continuous for the respective norms; this opens the door to extending the homomorphism to a Banach algebra of functions—namely, the closure of $\mathcal{Q}(\sigma)$ in $\mathscr{C}(\sigma)$—by straightforward limiting arguments. This we proceed to do.

(66.15) **Definition.** Let σ be a closed subset of $\mathfrak{S}$. The closure of $\mathcal{Q}(\sigma)$ in $\mathscr{C}(\sigma)$, with respect to the sup norm, is denoted $\mathscr{A}(\sigma)$; a function in $\mathscr{A}(\sigma)$ is said to be σ-*analytic*. Thus, $f \in \mathscr{C}(\sigma)$ is σ-analytic iff there exists a sequence $f_n \in \mathbb{C}(t; \sigma)$ such that $\|f_n - f\|_{\sigma} \to 0$. {The letter f is being overworked—standing for rational forms, rational functions, and functions with domain σ—but the unappealing alternative is to invent separate castes of notation for every use.}

(66.16) **Theorem.** *If $\sigma \subset \mathfrak{S}$ is a closed spectral set for T and if $\mathscr{A}(\sigma)$ is the Banach algebra of all σ-analytic functions, then there exists a unique continuous algebra homomorphism*

$$\Phi : \mathscr{A}(\sigma) \to \mathscr{L}(H)$$

such that $\Phi(f|\sigma) = f(T)$ for all $f \in \mathbb{C}(t; \sigma)$. Moreover, $\|\Phi(f)\| \le \|f\|_{\infty}$ for all $f \in \mathscr{A}(\sigma)$.

Proof. Uniqueness is obvious from the density of $\mathcal{Q}(\sigma)$ in $\mathscr{A}(\sigma)$. The proof of the existence of Φ is a straightforward 'extension by continuity' argument, based on the observation that if $f \in \mathscr{A}(\sigma)$, $f_n \in \mathbb{C}(t; \sigma)$, and $\|f_n - f\|_{\sigma} \to 0$, then $\|f_m(T) - f_n(T)\| \le \|f_m - f_n\|_{\sigma} \to 0$ as $m, n \to \infty$. ∎

If $\sigma \subset \mathbb{C}$ then $t \in \mathbb{C}(t; \sigma)$ and (66.16) can be formulated as follows:

(66.17) **Corollary.** *If $\sigma \subset \mathbb{C}$ is a compact spectral set for T, then there exists a unique continuous algebra homomorphism*

$$\Phi : \mathscr{A}(\sigma) \to \mathscr{L}(H)$$

such that $\Phi(1) = I$ and $\Phi(u) = T$, where u is the identity function $u(\lambda) \equiv \lambda$ on σ. Moreover, $\|\Phi(f)\| \le \|f\|_{\infty}$ for all $f \in \mathscr{A}(\sigma)$.

Proof. The existence of Φ is covered by (66.16). If also Ψ has the same properties, evidently $\Psi(f|\sigma) = \Phi(f|\sigma)$ for all $f \in \mathbb{C}(t; \sigma)$; since $\mathcal{Q}(\sigma)$ is dense in $\mathscr{A}(\sigma)$, it results from continuity that $\Psi = \Phi$. ∎

(66.18) Definition. Notation as in (66.16). If $f \in \mathscr{A}(\sigma)$, the operator $\Phi(f)$ is denoted $f^{\sigma}(T)$ (or briefly $f(T)$, with the dependence on σ tacitly understood).

In particular, if T is normal we may take $\sigma = \sigma(T)$ (66.10), and it is clear that the continuous functional calculus $\mathscr{C}(\sigma(T)) \to \mathscr{L}(H)$ given by the spectral theorem (65.6) is an extension of the $\sigma(T)$-analytic functional calculus $\mathscr{A}(\sigma(T)) \to \mathscr{L}(H)$ given by (66.17). It can easily happen that $\mathscr{A}(\sigma(T)) \subset \mathscr{C}(\sigma(T))$ properly (cf. 66.32):

(66.19) Proposition. *If σ is a closed subset of $\mathfrak{S}$ and if $f \in \mathscr{A}(\sigma)$, then f is analytic at every interior point of σ.*

Proof. If λ is an interior point of σ, then int (σ) is a neighborhood of λ on which f is the uniform limit of (rational) analytic functions. ∎

In particular, if σ has nonempty interior and if $u(\lambda) \equiv \lambda$ is the identity function on σ, then the complex conjugate function $u^*(\lambda) = \lambda^*$ does not belong to $\mathscr{A}(\sigma)$. Thus, while the continuous functional calculus for a normal operator T encompasses T^*, in general the $\sigma(T)$-analytic functional calculus does not. This 'defect' is precisely what makes the theory of spectral sets applicable to nonnormal operators.

If σ is a proper, closed subset of $\mathfrak{S}$, then the subalgebra $\mathscr{A}(\sigma)$ of $\mathscr{C}(\sigma)$ separates the points of σ. {Proof: If σ excludes ∞ (i.e., if $\sigma \subset \mathbb{C}$), then $t \in \mathbb{C}(t; \sigma)$; if σ excludes some complex number c, then $(t - c1)^{-1} \in \mathbb{C}(t; \sigma)$.} Also, $\mathscr{A}(\sigma)$ contains the constant functions. By the Weierstrass-Stone theorem, $\mathscr{A}(\sigma)$ can be a $*$-subalgebra of $\mathscr{C}(\sigma)$ only if $\mathscr{A}(\sigma) = \mathscr{C}(\sigma)$; this phenomenon has a name:

(66.20) Definition. A closed subset σ of $\mathfrak{S}$ is said to be *thin* if $\mathscr{A}(\sigma) = \mathscr{C}(\sigma)$.

If a closed subset of $\mathfrak{S}$ is thin, then its interior is empty (by the remark following (66.19)), but the converse is false; various results about thinness are gathered in (66.32). The main reason for mentioning thin sets is to point out that nonnormal operators can be treated by the present theory only by considering 'thick' sets:

(66.21) Theorem. *If T has a thin spectral set, then T is normal.*

Proof. Suppose σ is a thin spectral set for T; in particular, σ is a closed, proper subset of $\mathfrak{S}$ that contains $\sigma(T)$. We restrict attention to the case that σ is bounded, i.e., σ is a compact subset of $\mathbb{C}$; the unbounded case then follows easily using (66.31) and (66.12). By (66.17), there exists an algebra homomorphism

$$\Phi: \mathscr{C}(\sigma) \to \mathscr{L}(H)$$

such that $\Phi(1) = I$, $\Phi(u) = T$, and $\|\Phi(f)\| \leq \|f\|_{\infty}$ for all $f \in \mathscr{C}(\sigma)$. The range of Φ is a commutative subalgebra of $\mathscr{L}(H)$, so it will clearly suffice to show that Φ is a $*$-homomorphism.

More generally, let us show that if A is a C^*-algebra with unity and $\varphi: A \to \mathscr{L}(H)$ is a unital algebra homomorphism such that $\|\varphi(a)\| \leq \|a\|$ for all $a \in A$, then φ is a $*$-homomorphism. {By the Gel'fand-Naĭmark theorem (62.1), one could even replace $\mathscr{L}(H)$ by an arbitrary C^*-algebra with unity. Incidentally, the converse was proved in (60.1).} Assuming $a \in A$, $a^* = a$, it will suffice to show that $\varphi(a)$ is also self-adjoint. Since $\sigma(a)$ is real (58.4), we may define

$$v = (a - i1)(a + i1)^{-1}$$

(v is the 'Cayley transform' of a). Elementary algebra yields $v^*v = vv^* = 1$, thus v is unitary. Then $\|v\| = \|v^{-1}\| = 1$. {Proof: $\|v\|^2 = \|v^*v\| = 1$ and $v^{-1} = v^*$.} By the hypothesis on φ, we have

$$\|\varphi(v)\| \le \|v\| = 1, \qquad \|\varphi(v)^{-1}\| = \|\varphi(v^{-1})\| \le \|v^{-1}\| = 1;$$

thus, setting $V = \varphi(v)$, we have $\|V\| \le 1$ and $\|V^{-1}\| \le 1$. It follows that V is unitary (see the proof of (66.11)). Elementary algebra shows that $1 - v$ is invertible and $a = i(1 + v)(1 - v)^{-1}$, therefore

$$\varphi(a) = i(I + V)(I - V)^{-1};$$

since $V^* = V^{-1}$, further elementary algebra then shows that $\varphi(a)^* = \varphi(a)$. ∎

If T is normal then $\sigma(T)$ is a spectral set for T (66.10); this is not a satisfying example since, for a normal operator, the spectral theory described in Section 65 does even more. If T is nonnormal, then $\sigma(T)$ need not be a spectral set for T (66.29); but a trivial spectral set is $\mathbb{C}$. Does every operator T have a *bounded* spectral set? The affirmative answer is due to von Neumann: the closed disc $\{\lambda : |\lambda| \le \|T\|\}$ is always a spectral set for T. The rest of the section is devoted to a sketch of the proof (by an obvious normalization, one need only consider the case that T is a *contraction*, i.e., $\|T\| \le 1$). Following B. Sz.-Nagy, we base the proof on a general lemma on contractions:

(66.22) Lemma. *If $\|T\| \le 1$ then there exist a Hilbert space K containing H and a unitary operator U on K, such that, writing P for the projection operator on K with range H, one has $T^n x = PU^n x$ for all $x \in H$ and for all nonnegative integers n.*

The lemma is a simplified special case of a far-reaching theorem of Sz.-Nagy [cf. **138**]. A maximally elementary proof of the lemma has been given by J. J. Schäffer; to save space, we omit the proof and cite Schäffer's original paper [**127**].

(66.23) Notation. We write $\Delta_1 = \{\lambda \in \mathbb{C} : |\lambda| \le 1\}$, the closed unit disc in the complex plane.

(66.24) Lemma. *If $\|T\| \le 1$ and f is any complex polynomial, then $\|f(T)\| \le \|f\|_{\Delta_1}$.*

Proof. Adopt the notation of (66.22). For every $x \in H$, clearly $f(T)x = Pf(U)x$, therefore

$$\|f(T)x\| \le \|f(U)x\| \le \|f(U)\|\,\|x\|;$$

this shows that

(*) $\|f(T)\| \le \|f(U)\|$

(the two norms are calculated in the Banach algebras $\mathscr{L}(H)$ and $\mathscr{L}(K)$, respectively). On the other hand, since the unit circle is a spectral set for U (66.11), so is Δ_1 (66.7), thus

(**)
$$\|f(U)\| \leq \|f\|_{\Delta_1}.$$

The lemma is immediate from (*) and (**). ∎

(66.25) Lemma. *If $\|T\| \leq 1$ and $f \in \mathbb{C}(t; \Delta_1)$, then there exists a sequence of complex polynomials f_n such that $\|f_n(T) - f(T)\| \to 0$.*

Proof. Observe that $f(T)$ exists. {Proof: $r(T) \leq \|T\| \leq 1$, therefore $\sigma(T) \subset \Delta_1$.} Explicitly, if $f = p/q$ in reduced form, then $f(T) = p(T)q(T)^{-1}$ (53.5).

If f is already a polynomial, there is nothing to prove. Otherwise $g = 1/q$ is a nonconstant rational function whose poles are exterior to Δ_1; if a is the distance from the origin to the nearest pole of g, then $a > 1$ and g has a Taylor expansion

$$g(\lambda) \equiv \sum_{k=0}^{\infty} c_k \lambda^k,$$

valid for $|\lambda| < a$. For $n = 1, 2, 3, \ldots$ let g_n be the polynomial defined by

$$g_n(\lambda) \equiv \sum_{k=0}^{n} c_k \lambda^k;$$

since $a > 1$ we have $g_n(\lambda) \to g(\lambda)$ uniformly on Δ_1, that is,

(1)
$$\|g_n - g\|_{\Delta_1} \to 0.$$

Form the operators $g_n(T)$ and $g(T) = q(T)^{-1}$. We assert that

(*)
$$\|g_n(T) - g(T)\| \to 0.$$

At any rate, citing (66.24) we have

$$\|g_m(T) - g_n(T)\| = \|(g_m - g_n)(T)\| \leq \|g_m - g_n\|_{\Delta_1} \to 0$$

as $m, n \to \infty$, hence by (40.23) there exists $S \in \mathscr{L}(H)$ with

(2)
$$\|g_n(T) - S\| \to 0;$$

we are to show that $S = g(T)$. Since

$$g_n(\lambda) \to g(\lambda) = 1/q(\lambda)$$

uniformly on Δ_1, we have

(3)
$$\|g_n q - 1\|_{\Delta_1} \to 0;$$

citing (66.24) again,

$$\|g_n(T)q(T) - I\| \leq \|(g_n q - 1)(T)\| \leq \|g_n q - 1\|_{\Delta_1},$$

therefore (3) yields

(4)
$$\|g_n(T)q(T) - I\| \to 0.$$

From (2) and (4) we see that $Sq(T) = I$, therefore $S = q(T)^{-1} = g(T)$. This completes the proof of (*).

It follows from (*) that $\|p(T)g_n(T) - p(T)g(T)\| \to 0$; thus, writing $f_n = pg_n$, we have $f_n \in \mathbb{C}[t]$ and $\|f_n(T) - f(T)\| \to 0$. ∎

Von Neumann's fundamental theorem on spectral sets follows easily:

(66.26) Theorem. $\|T\| \leq 1$ iff Δ_1 is a spectral set for T.

Proof. {Note the analogy: in the 'calculus' of spectral sets, the closed unit ball $\{T : \|T\| \leq 1\}$ in $\mathscr{L}(H)$ corresponds to the closed unit disc $\Delta_1 = \{\lambda : |\lambda| \leq 1\}$ in $\mathbb{C}$.}

If Δ_1 is a spectral set for T, then $\|T\| \leq 1$ results at once from the fact that $t \in \mathbb{C}(t; \Delta_1)$.

Conversely, suppose $\|T\| \leq 1$. Assuming $f \in \mathbb{C}(t)$ and $\|f\|_{\Delta_1} \leq 1$, it is to be shown that $\|f(T)\| \leq 1$ (66.6). Let f_n be a sequence of polynomials such that $\|f_n(T) - f(T)\| \to 0$ (66.25). By (16.4) we have

$$\lim_{n \to \infty} \|f_n(T)\| = \|f(T)\|$$

and

$$\lim_{n \to \infty} \|f_n\|_{\Delta_1} = \|f\|_{\Delta_1} \leq 1;$$

since $\|f_n(T)\| \leq \|f_n\|_{\Delta_1}$ for all n (66.24), passage to the limit yields $\|f(T)\| \leq 1$. ∎

(66.27) Corollary. *The closed disc* $\{\lambda : |\lambda| \leq \|T\|\}$ *is a spectral set for T (for every $T \in \mathscr{L}(H)$).*

Exercises

(66.28) Let H be a Hilbert space with an orthonormal basis e_n ($n = 1, 2, 3, \ldots$) and let $U \in \mathscr{L}(H)$ be the unique operator on H such that $Ue_n = e_{n+1}$ for all n (U is called a *unilateral shift operator*). Then $\sigma(U)$ is a spectral set for U.

(66.29) With notation as in (66.28), let R be the operator on H such that $Re_n = (1/n)e_n$ for all n, and let $T = UR$. Then $\sigma(T) = \{0\}$ (an operator with this property is said to be *quasinilpotent*); thus $\sigma(T)$ is real but T is not self-adjoint, therefore $\sigma(T)$ is not a spectral set for T.

(66.30) If $\tau \subset \mathfrak{S}$ is a closed spectral set for T and if $f \in \mathscr{A}(\tau)$, then (i) $\sigma(f^\tau(T)) = f(\sigma(T))$, and (ii) $f(\tau)$ is a (closed) spectral set for $f^\tau(T)$. (iii) If, moreover, $g \in \mathscr{A}(f(\tau))$, then $g \circ f \in \mathscr{A}(\tau)$ and

$$(g \circ f)^\tau(T) = g^{f(\tau)}(f^\tau(T));$$

briefly, $(g \circ f)(T) = g(f(T))$ whenever the right side is defined.

(66.31) (i) A rational form $f \in \mathbb{C}(t)$ defines a bijection $f : \mathfrak{S} \to \mathfrak{S}$ iff there exist complex numbers a, b, c, d such that $ad - bc \neq 0$ and

$$f = (at + b)/(ct + d).$$

{Such a rational function is called a *linear fractional transformation* (briefly, LFT).}

(ii) The mapping inverse to an LFT is also an LFT.

(iii) Let f be an LFT and let σ be a closed subset of $\mathfrak{S}$; σ is thin iff $f(\sigma)$ is thin.

(iv) If σ is a proper, closed subset of $\mathfrak{S}$, then there exists an LFT f such that $f(\sigma)$ is a compact subset of $\mathbb{C}$.

(66.32)   Let σ be a nonempty compact subset of $\mathbb{C}$. Results of the following type are surveyed in **[146]**.
   (i) If σ is thin, then its interior is empty.
   (ii) The converse of (i) fails.
   (iii) The unit circle $\mathbb{T} = \{\lambda : |\lambda| = 1\}$ is thin (Weierstrass approximation theorem).
   (iv) If σ has planar Lebesgue measure zero, then σ is thin (Hartogs-Rosenthal theorem).
   (v) If $\mathbb{C} - \sigma$ has finitely many components, then $\mathscr{A}(\sigma)$ is the set of all $f \in \mathscr{C}(\sigma)$ such that f is analytic on the interior of σ (Mergelyan's theorem). In particular, if σ has empty interior and $\mathbb{C} - \sigma$ has finitely many components, then σ is thin.
   (vi) If the minimal boundary of $\mathscr{A}(\sigma)$ is σ, then σ is thin (Bishop's theorem).

(66.33) $T \geq 0$ iff $[0, \infty)$ is a spectral set for T.

(66.34)   Suppose $T \in \mathscr{L}(H)$, $\alpha \in \mathbb{C}$, and $\beta > 0$.
   (i) The closed disc $\{\lambda \in \mathbb{C} : |\lambda - \alpha| \leq \beta\}$ is a spectral set for T iff $\|T - \alpha I\| \leq \beta$.
   (ii) The set $\{\lambda \in \mathbb{C} : |\lambda - \alpha| \geq \beta\}$ is a spectral set for T iff $\alpha \notin \sigma(T)$ and $\|(T - \alpha I)^{-1}\| \leq 1/\beta$.

(66.35)   For every $T \in \mathscr{L}(H)$, one has $\sigma(T) = \bigcap \tau$, where τ varies over all closed spectral sets of T.

(66.36)   Let $T \in \mathscr{L}(H)$ and let $f = (t - 1)/(t + 1)$ (cf. 66.31). The following conditions on T are equivalent: (a) $f(T)$ exists (i.e., $-1 \notin \sigma(T)$) and $\|f(T)\| \leq 1$; (b) $\operatorname{Re} T \geq 0$ (where $\operatorname{Re} T = \frac{1}{2}(T + T^*)$); (c) the set $\{\lambda \in \mathbb{C} : \operatorname{Re} \lambda \geq 0\}$ is a spectral set for T.

67. Irreducible representations.

The concept of reducibility arises in the theory of group representations (cf. Section 69) and perhaps even more naturally in the theory of C^*-algebras (67.24), and is a major theme in both theories. The best way to begin is to plunge right in with the principal definition:

(67.1) Definition.   Let H be a nonzero Hilbert space, let $\mathscr{S}$ be a *-subset of $\mathscr{L}(H)$ (i.e., $S \in \mathscr{S}$ implies $S^* \in \mathscr{S}$), and let N be a closed linear subspace of H. From (43.11) it is clear that N is invariant under every $S \in \mathscr{S}$ iff N reduces every $S \in \mathscr{S}$; if there exists such an N other than the trivial subspaces $\{\theta\}$ and H, then the set $\mathscr{S}$ is said to be *reducible*, and otherwise it is said to be *irreducible*.
   An extremely useful criterion for irreducibility lies very near the surface:

(67.2) Theorem.   *Let H be a Hilbert space and let $\mathscr{S}$ be a *-subset of $\mathscr{L}(H)$. The following conditions on $\mathscr{S}$ are equivalent:*
   (a) $\mathscr{S}$ is irreducible;

(b) 0 *and I are the only projections in $\mathscr{S}'$ (the commutant of $\mathscr{S}$ in $\mathscr{L}(H)$);*
(c) $\mathscr{S}' = \{\lambda I : \lambda \in \mathbb{C}\}$.

Proof. The equivalence of (a) and (b) is immediate from (43.11), and it is obvious that (c) implies (b).

(a) implies (c): Suppose $\mathscr{S}$ is irreducible. We assert that if $T \in \mathscr{S}'$ then either $T = 0$ or T is injective. Indeed, suppose $T \in \mathscr{S}'$ and N is the null space of T. For all $S \in \mathscr{S}$ we have $TS = ST$, therefore

$$T(S(N)) = S(T(N)) = S\{\theta\} = \{\theta\}.$$

This shows that $S(N) \subset N$ for all $S \in \mathscr{S}$; since $\mathscr{S}$ is an irreducible $*$-subset of $\mathscr{L}(H)$, we conclude that either $N = H$ or $N = \{\theta\}$, in other words, either $T = 0$ or T is injective. It follows trivially that $\mathscr{S}'$ has no divisors of zero: if $T_1, T_2 \in \mathscr{S}'$ and $T_1 T_2 = 0$, then either $T_1 = 0$ or $T_2 = 0$.

Write $\mathscr{A} = \mathscr{S}'$; $\mathscr{A}$ is a norm-closed $*$-subalgebra of $\mathscr{L}(H)$ containing I (49.11). We are to show that every element of $\mathscr{A}$ is a scalar multiple of I; it clearly suffices to show that this is true of every self-adjoint element of $\mathscr{A}$. Suppose $T \in \mathscr{A}$, $T^* = T$. The spectral theorem (65.1) is applicable to the normal element T of the C^*-algebra $\mathscr{A}$; in particular, $\mathscr{C}(\sigma(T))$ may be identified with a $*$-subalgebra of $\mathscr{A}$. A fortiori, $\mathscr{C}(\sigma(T))$ has no divisors of zero, therefore $\sigma(T)$ must be a singleton; thus $\mathscr{C}(\sigma(T))$ is one-dimensional and T is, indeed, a scalar multiple of I. ∎

(67.3) Corollary. *If $\mathscr{S}$ is a commutative, irreducible $*$-subset of $\mathscr{L}(H)$, then H is necessarily one-dimensional.*

Proof. {Note that the trivial space $H = \{\theta\}$ is excluded by the definitions (67.1).} Commutativity means that $\mathscr{S} \subset \mathscr{S}'$; irreducibility means that $\mathscr{S}' = \{\lambda I : \lambda \in \mathbb{C}\}$ (67.2); therefore $\mathscr{S}$ is a set of scalar multiples of I. Then every closed linear subspace of H reduces $\mathscr{S}$. In particular, if N is a one-dimensional subspace of H, then N reduces $\mathscr{S}$; since $\mathscr{S}$ is irreducible, $N = H$. ∎

The principal results of the section will be cast in terms of sets $\mathscr{S}$ that arise from $*$-representations of $*$-algebras:

(67.4) Definition. Let A be a $*$-algebra (43.6), let H be a nonzero Hilbert space, and let $A \to \mathscr{L}(H)$ be a $*$-representation of A on H (60.2), denoted, say, $a \mapsto T_a$ $(a \in A)$. The $*$-representation is said to be *reducible* [*irreducible*] if the set of operators $\{T_a : a \in A\}$ is reducible [irreducible] in the sense of (67.1).

The best result is obtained when A is a C^*-algebra with unity: one knows that A has a faithful $*$-representation, by the Gel'fand-Naĭmark theorem (62.1); it need not have a faithful irreducible representation (cf. 67.3); but one can hope that A has sufficiently many irreducible representations. This hope is fulfilled (67.27). The proof depends on a careful analysis of the states of A. To accommodate unitless algebras, it is convenient (and a little simpler) to cast the early part of the theory in terms of positive sesquilinear forms.

(67.5)   Let φ be a positive sesquilinear form (briefly, a PSF) on a complex vector space A (41.9). A Hilbert space H_φ may be derived from φ as follows. Let

$$N_\varphi = \{x \in A : \varphi(x, x) = 0\};$$

as noted in the proof of (61.11),

$$N_\varphi = \{x \in A : \varphi(x, y) = 0 \quad \text{for all } y \in A\},$$

and in particular N_φ is a linear subspace of A. Let $A_\varphi = A/N_\varphi$ be the quotient vector space and write $x_\varphi = x + N_\varphi$ $(x \in A)$; thus $x \mapsto x_\varphi$ is the canonical linear mapping $A \to A_\varphi$. As shown in the proof of (61.11), the formula

$$(x_\varphi | y_\varphi) = \varphi(x, y)$$

defines an inner product on A_φ, thus A_φ is a pre-Hilbert space; its completion is a Hilbert space H_φ (41.2), called the Hilbert space *derived from* φ. We regard A_φ as a dense linear subspace of H_φ.

(67.6) **Definition.**   Let φ and ψ be Hermitian sesquilinear forms on a complex vector space A (cf. 41.14). We write $\psi \le \varphi$ in case $\psi(x, x) \le \varphi(x, x)$ for all $x \in A$. {This is a partial ordering of Hermitian sesquilinear forms, the essential point being that if $\psi \le \varphi$ and $\varphi \le \psi$, then $\psi = \varphi$ by the polarization identity.}

(67.7) **Lemma.**   *Let A be a dense linear subspace of a Hilbert space H and suppose ψ is a PSF on A such that $\psi(x, x) \le (x|x)$ for all $x \in A$. Then there exists a unique operator $T \in \mathscr{L}(H)$ such that $\psi(x, y) = (Tx|y)$ for all $x, y \in A$; moreover, one has $0 \le T \le I$.*

Proof.   By the Cauchy-Schwarz inequality (41.14),

(*) $|\psi(x, y)|^2 \le \psi(x, x)\psi(y, y) \le (x|x)(y|y) = \|x\|^2\|y\|^2$

for all $x, y \in A$.

We first extend ψ to a PSF on H with preservation of (*), as follows. It is clear from (*) that the mapping $\psi: A \times A \to \mathbb{C}$ has the following continuity properties: $\psi(x, y) \to 0$ if $x \to \theta$ and $\|y\|$ remains bounded; or if $\|x\|$ remains bounded and $y \to \theta$. It then follows from the identity

$$\psi(x, y) - \psi(x', y') = \psi(x - x', y) + \psi(x', y - y')$$

that ψ is continuous on $A \times A$. {Alternatively, one could cite (18.2).} Since A is dense in H, the extension of ψ to $H \times H$ 'by continuity' is routine (cf. 24.13), and the existence of T is a straightforward application of the Fréchet-Riesz theorem (cf. 43.9).   ∎

(67.8) **Proposition.**   *Let φ be a PSF on a complex vector space A, and let H_φ be the Hilbert space derived from φ (67.5).*
    (i) *For each $T \in \mathscr{L}(H_\varphi)$ with $0 \le T \le I$, the formula*

$$\varphi_T(x, y) = (Tx_\varphi | y_\varphi) \qquad (x, y \in A)$$

defines a PSF on A such that $\varphi_T \leq \varphi$.

(ii) *The correspondence $T \mapsto \varphi_T$ described in* (i) *is a bijection onto the set of all PSF's ψ on A such that $\psi \leq \varphi$.*

(iii) *Moreover, the correspondence preserves convex combinations.*

Proof. (i) is obvious (for the meaning of $0 \leq T \leq I$, see (57.28)).

(ii) Suppose ψ is a PSF on A with $\psi \leq \varphi$. Clearly $N_\varphi \subset N_\psi$, therefore

(*) $x_\varphi \mapsto x_\psi \qquad (x \in A)$

is a well-defined linear mapping $A_\varphi \to A_\psi$; moreover,

$$\|x_\psi\|^2 = \psi(x, x) \leq \varphi(x, x) = \|x_\varphi\|^2$$

shows that the mapping (*) is continuous for the respective norms. The formula

$$\psi'(x_\varphi, y_\varphi) = (x_\psi | y_\psi) = \psi(x, y) \qquad (x, y \in A)$$

defines a PSF ψ' on A_φ, and the inequalities $\psi(x, x) \leq \varphi(x, x)$ $(x \in A)$ may be written

$$\psi'(x_\varphi, x_\varphi) \leq (x_\varphi | x_\varphi) \qquad (x \in A).$$

Citing the lemma, there exists $T \in \mathscr{L}(H_\varphi)$ with $0 \leq T \leq I$, such that

$$\psi'(x_\varphi, y_\varphi) = (Tx_\varphi | y_\varphi)$$

for all $x, y \in A$; in other words, $\psi = \varphi_T$. If also $\psi = \varphi_S$, $S \in \mathscr{L}(H_\varphi)$, then

$$(Tx_\varphi | y_\varphi) = \psi(x, y) = (Sx_\varphi | y_\varphi)$$

for all $x, y \in A$; since A_φ is dense in H_φ, we infer from continuity that $T = S$.

(iii) The set

$$\mathscr{K} = \{T \in \mathscr{L}(H_\varphi) : 0 \leq T \leq I\}$$

(called the *positive unit ball* of $\mathscr{L}(H_\varphi)$) is obviously convex; so is the set of all PSF's ψ on A such that $\psi \leq \varphi$. If $S, T \in \mathscr{K}$ and $0 \leq \lambda \leq 1$, and if $R = \lambda S + (1 - \lambda)T$, clearly

$$\varphi_R = \lambda \varphi_S + (1 - \lambda)\varphi_T.$$

{The virtue of (iii) is that the bijection $T \mapsto \varphi_T$ pairs the extremal points of these two convex sets.} ∎

We now specialize to PSF's defined on *-algebras. The following definition is designed to lead to *-representations:

(67.9) Definition. A PSF φ on a *-algebra A is said to be *adjunctive* if

$$\varphi(xy, z) = \varphi(y, x^*z)$$

for all $x, y, z \in A$. This is, in a sense, an extension of the notion of state:

(67.10) Proposition. *Let A be a *-algebra.*

(i) *If f is a state on A, then the formula*

$$\varphi_f(x, y) = f(y^*x) \qquad (x, y \in A)$$

defines an adjunctive PSF φ_f on A.

(ii) *If A has a unity element 1 and if φ is an adjunctive PSF on A, then the formula*

$$f_\varphi(x) = \varphi(x, 1) \qquad (x \in A)$$

defines a state f_φ on A. Moreover, the correspondences $f \mapsto \varphi_f$ and $\varphi \mapsto f_\varphi$ are mutually inverse bijections between the set of all states f and the set of all adjunctive PSF's φ; both of these sets are convex, and the correspondences preserve convex combinations.

Proof. There is nothing deeper here than the associativity of multiplication in A. ∎

(67.11) Definition. If f is a state on a $*$-algebra A and if $\varphi = \varphi_f$ is the adjunctive PSF on A derived from f (67.10), the notations N_φ, x_φ, A_φ, H_φ of (67.5) are replaced by N_f, x_f, A_f, H_f; thus

$$N_f = \{x \in A : f(x^*x) = 0\},$$

$x_f = x + N_f$ $(x \in A)$, the inner product on $A_f = A/N_f$ is defined by the formula

$$(x_f \,|\, y_f) = f(y^*x) \qquad (x, y \in A),$$

and H_f is the Hilbert space completion of A_f.

The further we can get with adjunctive PSF's, the longer the assumption of a unity element can be deferred. {This is worth doing because many important $*$-algebras come into the world without unity elements and, although a unity element is easily conferred, the extendibility of states to the unitification is not automatic (cf. 67.29).}

(67.12) Lemma. *If φ is an adjunctive PSF on a $*$-algebra A, and if*

$$N_\varphi = \{x \in A : \varphi(x, x) = 0\},$$

then N_φ is a left ideal of A.

Proof. Suppose $x \in N_\varphi$ and $y \in A$. From (67.5) we know that $\varphi(x, z) = 0$ for all $z \in A$; it follows that

$$\varphi(yx, z) = \varphi(x, y^*z) = 0$$

for all $z \in A$, therefore $yx \in N_\varphi$. ∎

(67.13) Definition. Let φ be an adjunctive PSF on a $*$-algebra A. It follows from (67.12) that, for each $a \in A$, the correspondence

$$x_\varphi \to (ax)_\varphi$$

is a well-defined linear mapping in the quotient vector space $A_\varphi = A/N_\varphi$; we denote it by T_a. Thus, for each $a \in A$, the linear mapping $T_a \colon A_\varphi \to A_\varphi$ is defined by the formula

$$T_a x_\varphi = (ax)_\varphi \qquad (x \in A).$$

{One can write $T_a{}^\varphi$ when it is necessary to indicate the dependence of T_a on φ.}

(67.14) Proposition. *Let φ be an adjunctive PSF on a $*$-algebra A. The correspondence $a \mapsto T_a$ described in (67.13) is an algebra homomorphism of A into the algebra of all linear mappings on the inner product space A_φ, such that*

(*) $$(T_a x_\varphi | y_\varphi) = (x_\varphi | T_{a*} y_\varphi)$$

for all $a, x, y \in A$.

When A has a unity element 1, the homomorphism is unital, i.e., T_1 is the identity mapping on A_φ; moreover,

$$\varphi(x, y) = (T_x 1_\varphi | T_y 1_\varphi)$$

for all $x, y \in A$.

Proof. If $a, b \in A$ then

$$T_{ab} x_\varphi = ((ab)x)_\varphi = (a(bx))_\varphi = T_a(bx)_\varphi = T_a(T_b x_\varphi) = (T_a T_b) x_\varphi$$

for all $x \in A$, therefore $T_{ab} = T_a T_b$. Similarly $T_{a+b} = T_a + T_b$ and $T_{\lambda a} = \lambda T_a$ for all $\lambda \in \mathbb{C}$.

The 'adjoint relation' (*) is the relation $\varphi(ax, y) = \varphi(x, a^*y)$ in disguise.

If A has a unit element 1, it is obvious that $T_1 = I$. Also, $T_x 1_\varphi = (x1)_\varphi = x_\varphi$, thus

$$(T_x 1_\varphi | T_y 1_\varphi) = (x_\varphi | y_\varphi) = \varphi(x, y)$$

for all $x, y \in A$. ∎

Suppose φ is an adjunctive PSF on a $*$-algebra A. According to (67.8), the PSF's ψ on A such that $\psi \leq \varphi$ are the forms $\psi = \varphi_T$, where $T \in \mathscr{L}(H_\varphi)$ and $0 \leq T \leq I$. Which of these forms ψ are adjunctive? This is answered in the next proposition:

(67.15) Proposition. *Let φ be an adjunctive PSF on a $*$-algebra A, and let ψ be a PSF on A such that $\psi \leq \varphi$; say $\psi = \varphi_T$, where $T \in \mathscr{L}(H_\varphi)$ and $0 \leq T \leq I$ (67.8). In order that ψ be adjunctive, it is necessary and sufficient that, in the notation of (67.14),*

$$(TT_a x_\varphi | y_\varphi) = (Tx_\varphi | T_{a*} y_\varphi)$$

for all $a, x, y \in A$.

Proof. The displayed relation is

$$(T(ax)_\varphi | y_\varphi) = (Tx_\varphi | (a^*y)_\varphi),$$

i.e., $\psi(ax, y) = \psi(x, a^*y)$. ∎

To fully exploit the last two propositions, we wish to pass from the algebra homomorphism $a \mapsto T_a$ of (67.14) to a $*$-representation $A \to \mathscr{L}(H_\varphi)$. This will require that each of the linear mappings $T_a: A_\varphi \to A_\varphi$ be continuous and hence continuously extendible to H_φ. What conditions does this impose on φ? Simply that, for each $a \in A$, there exist a constant $K_a \geq 0$ such that

$$\|T_a x_\varphi\|^2 \leq K_a \|x_\varphi\|^2$$

for all $x \in A$, that is,

$$\varphi(ax, ax) \leq K_a \varphi(x, x)$$

for all $x \in A$. This prompts the following definition:

(67.16) Definition. A PSF φ on a $*$-algebra A is said to be *admissible* if (i) φ is adjunctive, and (ii) for each $a \in A$ there exists a constant $K_a \geq 0$ such that $\varphi(ax, ax) \leq K_a \varphi(x, x)$ for all $x \in A$.

(67.17) If φ is an admissible PSF on a $*$-algebra A then, by the above remarks, for each $a \in A$ the formula $T_a x_\varphi = (ax)_\varphi$ defines a continuous linear mapping on the inner product space A_φ; the unique continuous extension of T_a to the completion H_φ will also be denoted by T_a. Thus $T_a \in \mathscr{L}(H_\varphi)$ for all $a \in A$. The preceding two propositions may now be sharpened:

(67.18) Theorem. *If φ is an admissible PSF on a $*$-algebra A, then there exists a (unique) $*$-representation $a \mapsto T_a$ of A on H_φ such that*

$$\varphi(ax, y) = (T_a x_\varphi | y_\varphi)$$

for all $a, x, y \in A$. The adjunctive PSF's ψ on A such that $\psi \leq \varphi$ are precisely the forms $\psi = \varphi_T$, where $T \in \mathscr{L}(H_\varphi)$, $0 \leq T \leq I$, and $TT_a = T_a T$ for all $a \in A$ (briefly, T is in the positive unit ball of $\{T_a : a \in A\}'$, the commutant of the $*$-subalgebra $\{T_a : a \in A\}$ of $\mathscr{L}(H_\varphi)$).

Proof. The relation (*) of (67.14) may now be written

$$(x_\varphi | T_a^* y_\varphi) = (x_\varphi | T_{a*} y_\varphi)$$

for all $a, x, y \in A$; since A_φ is dense in H_φ, we infer from continuity that $T_a^* = T_{a*}$. Thus $a \mapsto T_a$ is a $*$-representation of A on H_φ (unital, when A has a unity element), and

$$(T_a x_\varphi | y_\varphi) = ((ax)_\varphi | y_\varphi) = \varphi(ax, y)$$

for all $a, x, y \in A$. The uniqueness assertion is clear from continuity and the density of A_φ in H_φ.

Since $T_{a*} = T_a^*$, the relation of (67.15) may be written

$$(TT_a x_\varphi | y_\varphi) = (T_a T x_\varphi | y_\varphi)$$

for all $a, x, y \in A$; this proves the final assertion of the theorem. ∎

(67.19) Definition. An admissible PSF φ on a $*$-algebra A is said to be *irreducible* if $H_\varphi \neq \{\theta\}$ (i.e., $\varphi \neq 0$) and if, in the notation of (67.17), the $*$-subset $\{T_a : a \in A\}$ of $\mathscr{L}(H_\varphi)$ is irreducible in the sense of (67.1). A valuable characterization:

(67.20) Theorem. *Let φ be a nonzero, admissible PSF on a $*$-algebra A. The following conditions on φ are equivalent:*
(a) φ is irreducible in the sense of (67.19);

(b) *the only adjunctive PSF's ψ on A such that $\psi \leq \varphi$ are the forms $\psi = \alpha\varphi$,*
$0 \leq \alpha \leq 1$.

Proof. This is immediate from (67.18) and (67.2). ∎

Let us bring the parallel discussion of states up to date:

(67.21) Definition. Let A be a $*$-algebra and let f, g be states on A. We
write $g \leq f$ in case $f - g$ is a state, i.e., $g(x^*x) \leq f(x^*x)$ for all $x \in A$.
{Writing φ_f, φ_g as in (67.10), we have $g \leq f$ iff $\varphi_g \leq \varphi_f$ in the sense of (67.6).}

A state f on A is said to be *admissible* if the adjunctive PSF φ_f is admissible
in the sense of (67.16). {This means that, for each $a \in A$, there exists a constant
$K_a \geq 0$ such that $f(x^*a^*ax) \leq K_a f(x^*x)$ for all $x \in A$.}

An admissible state f on A is said to be *irreducible* if φ_f is irreducible in the
sense of (67.19).

If A is a $*$-algebra with unity and if f is a state on A, then the states g on A
such that $g \leq f$ are in bijective correspondence with the set of all adjunctive
PSF's ψ such that $\psi \leq \varphi_f$ (67.10). Immediate consequences of this observation
are the following variants of (67.18) and (67.20):

(67.22) Theorem. *Let A be a $*$-algebra with unity element 1, and let f be
an admissible state on A.*

(i) *There exists a unique $*$-representation $a \mapsto T_a$ of A on H_f such that
$\{T_a 1_f : a \in A\}$ is a dense linear subspace of H_f and*

$$f(x) = (T_x 1_f | 1_f)$$

for all $x \in A$.

(ii) *A state g on A satisfies $g \leq f$ iff there exists an operator $T \in \mathscr{L}(H_f)$
such that $0 \leq T \leq I$, $TT_a = T_a T$ for all $a \in A$, and $g(x) = (TT_x 1_f | 1_f)$ for all
$x \in A$.*

(iii) *Assuming $f \neq 0$, the following conditions on f are equivalent:* (a) *f is
irreducible;* (b) *the only states g on A such that $g \leq f$ are the states $g = \alpha f$,
$0 \leq \alpha \leq 1$.*

(67.23) Definition. Let A be a $*$-algebra with unity element 1. We denote
by Ω the set (possibly empty) of all normalized admissible states on A, i.e.,
the set of all admissible states f such that $f(1) = 1$. Note that Ω is convex.

Let A be a $*$-algebra with unity element 1 and let f be a state on A. By the
Cauchy-Schwarz inequality,

$$|f(x)|^2 = |f(1^*x)|^2 \leq f(1^*1)f(x^*x) = f(1)f(x^*x)$$

for all $x \in A$; it follows that $f = 0$ iff $f(1) = 0$. Thus if f is a nonzero state on
A, then $f(1) > 0$ and $f(1)^{-1}f$ is a normalized state. It follows that, in the
notation of (67.23), Ω is nonempty iff A has at least one nonzero admissible
state. We have arrived at the central result of the section:

(67.24) Theorem. *Let A be a $*$-algebra with unity element 1, assume that
A has nonzero admissible states, and let Ω be the (nonempty, convex) set of all*

normalized admissible states on A. *The following conditions on* $f \in \Omega$ *are equivalent:*

(a) f *is an extremal point of* Ω;

(b) f *is irreducible;*

(c) *if* g *is any state on* A *such that* $g \le f$, *then* $g = \alpha f$ *with* $0 \le \alpha \le 1$.

Proof. The equivalence of (b) and (c) was proved in (67.22). The equivalence of (a) and (c) turns out to be a proposition in elementary linear algebra.

(c) implies (a): Suppose $f = \lambda g + (1 - \lambda)h$, where $g, h \in \Omega$ and $0 < \lambda < 1$; it is to be shown that $g = h = f$. Obviously $\lambda g \le f$, therefore $\lambda g = \alpha f$ with $0 \le \alpha \le 1$; since $g(1) = f(1) = 1$, we conclude that $\lambda = \alpha$, whence $g = f$ and therefore also $h = f$.

(a) implies (c): Let g be a state such that $g \le f$. Then $f - g$ is also a state. If $g = 0$ or $g = f$, there is nothing to prove. Otherwise, by the remarks following (67.23), we have $g(1) > 0$ and $(f - g)(1) > 0$, thus $0 < g(1) < f(1) = 1$. Write $\alpha = g(1)$ and define

$$f_1 = \alpha^{-1}g, \qquad f_2 = (1 - \alpha)^{-1}(f - g);$$

clearly f_1, f_2 are normalized states, $0 < \alpha < 1$, and

$$f = \alpha f_1 + (1 - \alpha)f_2.$$

To exploit the assumption that f is an extremal point of Ω, it must be observed that f_1, f_2 are in Ω, i.e., that g and $f - g$ are admissible; a slightly more general result is proved in the lemma (67.25) below. Granted this, we have $f = f_1 = f_2$ by extremality; in particular, $g = \alpha f_1 = \alpha f$ as desired. ∎

(67.25) **Lemma.** *If* φ *and* ψ *are adjunctive PSF's on a* *-algebra* A, *such that* $\psi \le \varphi$, *and if* φ *is admissible, then* ψ *is also admissible.*

Proof. Quote (67.18): writing $a \mapsto T_a$ for the *-representation of A derived from φ, there exists $T \in \mathcal{L}(H_\varphi)$ such that $0 \le T \le I$, $TT_a = T_a T$ for all $a \in A$, and

$$\psi(x, y) = (Tx_\varphi | y_\varphi)$$

for all $x, y \in A$. Let $\mathcal{B} = \{T_a : a \in A\}'$, the commutant of the set of all T_a in $\mathcal{L}(H_\varphi)$. Then $\mathcal{B}$ is a C^*-algebra with unity, $T \in \mathcal{B}$, and $T \ge 0$ (cf. (57.26), (58.5)). Write $T = S^2$ with $S \in \mathcal{B}$, $S \ge 0$ (61.5). Then also $ST_a = T_a S$ for all $a \in A$. Fix $a \in A$ and write $K_a = \|T_a^* T_a\|$; for all $x \in A$, we have

$$\psi(ax, ax) = (T(ax)_\varphi | (ax)_\varphi) = (S^2 T_a x_\varphi | T_a x_\varphi) = (T_a^* T_a S x_\varphi | S x_\varphi)$$

$$\le K_a(S x_\varphi | S x_\varphi) = K_a(S^2 x_\varphi | x_\varphi) = K_a(T x_\varphi | x_\varphi) = K_a \psi(x, x),$$

thus ψ is admissible. ∎

This is a pretty theory; it is time to show that the theory is nonvacuous, and even important:

(67.26) Lemma. *If A is a C^*-algebra with unity, then every state on A is admissible.*

Proof. This is shown in the proof of (61.11). ∎

(67.27) Theorem. (Gel'fand-Naĭmark) *Every C^*-algebra with unity has a complete set of irreducible $*$-representations.*

Proof. If A is a C^*-algebra with unity and if $a, b \in A$, $a \neq b$, the assertion is that there exist a Hilbert space H and an irreducible $*$-representation $x \mapsto T_x$ of A on H, such that $T_a \neq T_b$.

The set of all normalized states on A is nonempty (61.10); since all states of A are admissible (67.26), it is consistent with (67.23) to denote this set by Ω. {In (61.14) it was denoted Σ.} As shown in (61.15), Ω is a nonempty, weak* compact, convex subset of the dual space A' of A. By the Kreĭn-Mil'man theorem (36.9), Ω is the weak* closure of conv Ω_{ep}, where Ω_{ep} is the set of all extremal points of Ω.

Let $c = a - b$. Since $c \neq 0$, there exists $f \in \Omega_{ep}$ with $f(c^*c) > 0$. {Proof: If, on the contrary, the weak* continuous linear mapping $g \mapsto g(c^*c)$ $(g \in A')$ vanishes on Ω_{ep}, then it also vanishes on conv Ω_{ep} and hence on its weak* closure Ω; this means that $g(c^*c) = 0$ for every $g \in \Omega$, a contradiction to (61.10).}

Let $x \mapsto T_x$ be the unital $*$-representation derived from f as in (67.22) (or in (61.11)). In particular,

$$0 < f(c^*c) = (T_{c^*c}1_f | 1_f) = (T_c^* T_c 1_f | 1_f) = \|T_c 1_f\|^2,$$

therefore $T_c \neq 0$, i.e., $T_a \neq T_b$. Finally, the representation is irreducible by (67.24). ∎

Exercises

(67.28) If A is a $*$-algebra with unity, H is a nonzero Hilbert space, and $a \mapsto T_a$ is an irreducible $*$-representation of A on H, then either (i) the representation is unital, or (ii) H is one-dimensional and $T_a = 0$ for all $a \in A$.

(67.29) (i) If f is a state on a $*$-algebra A, then f is *Hermitian* in the sense that $f(x^*y) = f(y^*x)^*$ for all $x, y \in A$ (cf. 41.9).

(ii) If f is a state on a $*$-algebra A with unity, then f is self-adjoint (61.23).

(iii) Let A be a $*$-algebra without unity, let A_1 be its unitification (cf. the proof of (58.7)), and let f be a state on A. In order that f be extendible to a normalized state on A_1, it is necessary and sufficient that (1) f be self-adjoint, and that (2) there exist a constant $K > 0$ such that $|f(x)|^2 \leq Kf(x^*x)$ for all $x \in A$.

*(67.30) If A is a Banach $*$-algebra with continuous involution, then every state f on A is admissible; explicitly, $f(x^*a^*ax) \leq \|a^*a\| f(x^*x)$ for all $a, x \in A$.

*(67.31) If A is a Banach $*$-algebra, then every continuous state f on A is admissible; explicitly, $f(x^*a^*ax) \leq r(a^*a)f(x^*x)$ for all $a, x \in A$, where r denotes spectral radius.

68. Von Neumann algebras. Let H be a Hilbert space.

(68.1) Definition. A *von Neumann algebra* on H is a $*$-subalgebra $\mathscr{A}$ of $\mathscr{L}(H)$ such that $\mathscr{A} = \mathscr{A}''$ (the bicommutant of $\mathscr{A}$ in $\mathscr{L}(H)$ (49.11)).

For a full dress account of the theory of von Neumann algebras, the reader should consult the treatise of J. Dixmier [37], as well as the books by S. Sakai [124], J. T. Schwartz [130], and D. M. Topping [140]. Our objective in the present section is strictly limited: to introduce the reader to the principal locally convex topologies on $\mathscr{L}(H)$ that occur in the theory and to indicate their role in the concept of von Neumann algebra.

(68.2) Definition. The *norm* (or 'uniform') topology on $\mathscr{L}(H)$ is the topology derived from the norm

$$T \mapsto \|T\| = \sup\{\|Tx\| : x \in H, \|x\| \le 1\}.$$

{This is the 'topology of uniform convergence on the closed unit ball of H.'} We denote this topology by τ_n.

(68.3) Relative to the above norm, $\mathscr{L}(H)$ is a C^*-algebra (43.7) and every von Neumann algebra on H is a closed $*$-subalgebra of $\mathscr{L}(H)$ (49.11); thus every von Neumann algebra is a C^*-algebra.

(68.4) Definition. For each $x \in H$, the function

$$p_x(T) = \|Tx\| \qquad (T \in \mathscr{L}(H))$$

is a seminorm on $\mathscr{L}(H)$; the topology on $\mathscr{L}(H)$ generated by the set of all such seminorms p_x ($x \in H$) is called the *strong operator topology* (or 'strong topology'). We denote this topology by τ_s.

(68.5) Equipped with the strong operator topology τ_s, $\mathscr{L}(H)$ is a locally convex TVS (37.17); it is obvious from (37.21) that this topology is separated. A seminorm p on $\mathscr{L}(H)$ is strongly continuous iff there exists a finite set of vectors $x_1, \ldots, x_n$ in H such that $p \le p_{x_1} + \cdots + p_{x_n}$, i.e.,

$$p(T) \le \|Tx_1\| + \cdots + \|Tx_n\|$$

for all $T \in \mathscr{L}(H)$ (cf. (37.15), (37.17)); alternatively (37.23), p is strongly continuous iff there exists a finite set of vectors $x_1, \ldots, x_n$ in H such that

$$p(T) \le \max\{\|Tx_1\|, \ldots, \|Tx_n\|\}$$

for all $T \in \mathscr{L}(H)$. A basic neighborhood of 0 in $\mathscr{L}(H)$ for the strong operator topology is given by

$$\{T : \|Tx_i\| \le 1 \quad \text{for } i = 1, \ldots, n\},$$

where $x_1, \ldots, x_n$ is a finite set of vectors in H (cf. 37.9). A net of operators $T_j \in \mathscr{L}(H)$ converges strongly to $T \in \mathscr{L}(H)$ iff $p_x(T_j - T) \to 0$ for each $x \in H$ (37.29), i.e., $\|T_j x - Tx\| \to 0$ for each $x \in H$; this means that, for each $x \in H$, $T_j x \to Tx$ for the norm topology on H. {Thus the strong operator topology on $\mathscr{L}(H)$ is the 'topology of pointwise convergence on H.'} Obviously $\tau_n \supset \tau_s$. Moreover, if $T_j \to T$ strongly, then, for each $S \in \mathscr{L}(H)$, $T_j S \to TS$ strongly and $ST_j \to ST$ strongly; it follows that if $\mathscr{S}$ is any subset of $\mathscr{L}(H)$, then $\mathscr{S}'$, the commutant of $\mathscr{S}$ in $\mathscr{L}(H)$, is a strongly closed subalgebra of $\mathscr{L}(H)$ containing the identity operator I (49.11). In particular, every von Neumann algebra on H is a strongly closed $*$-subalgebra of $\mathscr{L}(H)$ containing I (cf. 68.21). Further facts about the strong topology are noted in the exercises.

(68.6) Definition. For each pair of vectors $x, y \in H$, the function

$$\omega_{x,y}(T) = (Tx|y) \qquad (T \in \mathscr{L}(H))$$

is a linear form on $\mathscr{L}(H)$; the initial topology on $\mathscr{L}(H)$ for the set of all such linear forms $\omega_{x,y}$ $(x, y \in H)$ is called the *weak operator topology* (or 'weak topology'). We denote this topology by τ_w. {This is to be distinguished from the (rarely used) weak topology that accrues to $\mathscr{L}(H)$ from its Banach space structure (44.1).}

(68.7) Equipped with the weak operator topology τ_w, $\mathscr{L}(H)$ is a separated, locally convex TVS (33.9); a basic neighborhood of 0 for τ_w is given by

$$\{T : |(Tx_i|y_i)| \le 1 \quad \text{for } i = 1, \ldots, n\},$$

where $x_1, \ldots, x_n, y_1, \ldots, y_n$ are fixed vectors in H. By (37.6), τ_w is the topology generated by the family of seminorms

$$p_{x,y}(T) = |(Tx|y)| \qquad (T \in \mathscr{L}(H)).$$

A net of operators $T_j \in \mathscr{L}(H)$ converges weakly to $T \in \mathscr{L}(H)$ iff $p_{x,y}(T_j - T) \to 0$ for each pair $x, y \in H$, i.e.,

$$(T_j x|y) \to (Tx|y)$$

for each pair $x, y \in H$. In view of the Fréchet-Riesz theorem (42.12) this means that, for each $x \in H$, $T_j x \to Tx$ for the weak topology on H regarded as a Banach space (44.1). Obviously $\tau_u \supset \tau_s \supset \tau_w$. Moreover, if $T_j \to T$ weakly, then, for each $S \in \mathscr{L}(H)$, $T_j S \to TS$ weakly and $ST_j \to ST$ weakly. {For example, $(ST_j x|y) = (T_j x|S^*y) \to (Tx|S^*y) = (STx|y)$ for each pair $x, y \in H$.} It follows that if $\mathscr{S}$ is any subset of $\mathscr{L}(H)$, then $\mathscr{S}'$ is a weakly closed subalgebra of $\mathscr{L}(H)$ containing I (49.11). In particular:

(68.8) Proposition. *Every von Neumann algebra on H is a weakly closed $*$-subalgebra of $\mathscr{L}(H)$ containing the identity operator.*

The converse is true (von Neumann's theorem on bicommutants): If $\mathscr{A}$ is a weakly closed $*$-subalgebra of $\mathscr{L}(H)$ containing I, then $\mathscr{A} = \mathscr{A}''$. Thus, if $\mathscr{A}$ is a $*$-subalgebra of $\mathscr{L}(H)$ containing I, then $\mathscr{A} = \mathscr{A}''$ iff $\mathscr{A}$ is weakly closed. This proposition (proved in (68.21)) is the reason von Neumann algebras are also called 'W^*-algebras.' {Von Neumann, who invented them in 1929 [107], called them 'rings of operators.'}

The weakly (strongly) continuous linear forms are readily described:

(68.9) Proposition. *Let $\mathcal{M}$ be a linear subspace of $\mathcal{L}(H)$ and let f be a linear form on $\mathcal{M}$. The following conditions on f are equivalent:*

(a) *f is weakly continuous;*
(b) *f is strongly continuous;*
(c) *there exist vectors $x_1, \ldots, x_n, y_1, \ldots, y_n$ in H such that*

$$f(T) = \sum_{i=1}^{n} (Tx_i | y_i)$$

for all $T \in \mathcal{M}$, i.e., f is the restriction of $\sum_{i=1}^{n} \omega_{x_i, y_i}$ to $\mathcal{M}$.

Proof. In statements (a) and (b), it is assumed that $\mathcal{M}$ bears the relative weak and relative strong operator topologies, respectively.

(a) implies (b): This is immediate from $\tau_s \supset \tau_w$.

(b) implies (c): By the Hahn-Banach theorem (34.8) we can suppose without loss of generality that $\mathcal{M} = \mathcal{L}(H)$. Since, by hypothesis, the set

$$\{T \in \mathcal{L}(H) : |f(T)| \leq 1\}$$

is a strong neighborhood of 0, there exist vectors $x_1, \ldots, x_n$ in H such that

$$(*) \qquad \{T : \|Tx_i\| \leq 1 \quad \text{for } i = 1, \ldots, n\} \subset \{T : |f(T)| \leq 1\}$$

(cf. (68.5)). Note that if $Tx_i = \theta$ for all i then $f(T) = 0$. {Proof: For each $\varepsilon > 0$ we have $(\varepsilon T)x_i = \varepsilon(Tx_i) = \theta$ for all i, therefore $|f(\varepsilon T)| \leq 1$ by $(*)$; thus $|f(T)| \leq \varepsilon^{-1}$ for all $\varepsilon > 0$, whence $f(T) = 0$.} Better yet,

$$(**) \qquad\qquad |f(T)| \leq \max \{\|Tx_1\|, \ldots, \|Tx_n\|\}$$

for all $T \in \mathcal{L}(H)$. {If $Tx_i = \theta$ for all i then $f(T) = 0$ and the inequality holds trivially. If $\max \{\|Tx_i\| : i = 1, \ldots, n\} = a > 0$, then $\|(a^{-1}T)x_i\| \leq 1$ for all i; citing $(*)$, we have $|f(a^{-1}T)| \leq 1$, i.e., $|f(T)| \leq a$ as desired.}

Let $K = H \oplus \cdots \oplus H$ be the orthogonal sum of n copies of H (cf. 60.5). The set

$$M = \{(Tx_1, \ldots, Tx_n) : T \in \mathcal{L}(H)\}$$

is a linear subspace of K. If $(Tx_1, \ldots, Tx_n) = \theta$, i.e., if $Tx_i = \theta$ for all i, then $f(T) = 0$ as shown above; it follows that the formula

$$\tilde{f}(Tx_1, \ldots, Tx_n) = f(T)$$

defines a linear form $\tilde{f} \colon M \to \mathbb{C}$. Moreover, it results from $(**)$ that

$$|\tilde{f}(Tx_1, \ldots, Tx_n)| = |f(T)| \leq \max \{\|Tx_i\| : i = 1, \ldots, n\}$$

$$\leq \left(\sum_{i=1}^{n} \|Tx_i\|^2\right)^{1/2} = \|(Tx_1, \ldots, Tx_n)\|$$

for all $T \in \mathcal{L}(H)$; this shows that $\tilde{f}$ is a continuous linear form on M. Extend $\tilde{f}$ by continuity to the closure N of M. By the Fréchet-Riesz theorem (42.12) there exists a vector $y = (y_1, \ldots, y_n)$ in N such that

$$\tilde{f}(z) = (z | y) = \sum_{i=1}^{n} (z_i | y_i)$$

for all $z = (z_1, \ldots, z_n)$ in N. In particular, for all $T \in \mathscr{L}(H)$ we have

$$f(T) = \tilde{f}(Tx_1, \ldots, Tx_n) = \sum_{i=1}^{n} (Tx_i | y_i),$$

thus $f = \omega_{x_1, y_1} + \cdots + \omega_{x_n, y_n}$ as desired.

(c) implies (a): This is immediate from the fact that the weak topology is defined as the coarsest topology on $\mathscr{L}(H)$ for which all of the linear forms $\omega_{x,y}$ are continuous. ∎

(**68.10**)   Two other standard topologies on $\mathscr{L}(H)$ remain to be described: the 'ultrastrong' and the 'ultraweak' topologies. It is technically convenient to introduce an auxiliary Hilbert space K, namely,

$$K = H \oplus H \oplus H \oplus \cdots,$$

the orthogonal sum of $\aleph_0$ copies of H (cf. 60.5). Thus, the elements of K are the sequences $x = (x_i)$, with $x_i \in H$ for all i, such that $\sum_{i=1}^{\infty} \|x_i\|^2 < \infty$; the operations in K are coordinatewise, and the inner product of $x = (x_i)$ and $y = (y_i)$ is given by

$$(x|y) = \sum_{i=1}^{\infty} (x_i | y_i).$$

Each operator $T \in \mathscr{L}(H)$ defines an operator $\tilde{T} \in \mathscr{L}(K)$ via the formula

$$\tilde{T}x = (Tx_i) \qquad (x = (x_i) \in K);$$

it is straightforward to show that $T \mapsto \tilde{T}$ is a unital $*$-monomorphism $\mathscr{L}(H) \to \mathscr{L}(K)$, and that $\|\tilde{T}\| = \|T\|$ for all $T \in \mathscr{L}(H)$. The operator $\tilde{T}$ is called the $\aleph_0$-fold *ampliation* of T; we write

$$\mathscr{L}(H)^{\sim} = \{\tilde{T} : T \in \mathscr{L}(H)\}.$$

(**68.11**) **Definition.**   Notation as in (68.10). For each $x = (x_i) \in K$, the function

$$\tilde{T} \mapsto \|\tilde{T}x\| \qquad (T \in \mathscr{L}(H))$$

is a seminorm on $\mathscr{L}(H)^{\sim}$; the topology τ on $\mathscr{L}(H)^{\sim}$ defined by the family of all such seminorms is the relative topology on $\mathscr{L}(H)^{\sim}$ induced by the strong operator topology of $\mathscr{L}(K)$ (cf. 68.4). The inverse image of τ under the bijection $T \mapsto \tilde{T}$ is called the *ultrastrong topology* on $\mathscr{L}(H)$, denoted τ_{us}. Thus, the ultrastrong topology on $\mathscr{L}(H)$ is the topology generated by the family of all seminorms

$$T \mapsto \left(\sum_{i=1}^{\infty} \|Tx_i\|^2\right)^{1/2} \qquad (T \in \mathscr{L}(H)),$$

where (x_i) is a sequence of vectors in H such that $\sum_{i=1}^{\infty} \|x_i\|^2 < \infty$.

(**68.12**)   The ultrastrong topology τ_{us} on $\mathscr{L}(H)$ is, by definition, the topology on $\mathscr{L}(H)$ that renders the bijection $T \mapsto \tilde{T}$ of $\mathscr{L}(H)$ onto $(\mathscr{L}(H)^{\sim}, \tau)$ a homeomorphism; thus τ_{us} is locally convex and separated (cf. 68.5). Since $T \mapsto \tilde{T}$ is norm-continuous, the relation $\tau_n \supset \tau_s$ for $\mathscr{L}(K)$ implies that

$\tau_n \supset \tau_{us}$ for $\mathscr{L}(H)$. Moreover, it follows from the strong continuity of $\tilde{T} \mapsto \tilde{S}\tilde{T}$ and $\tilde{T} \mapsto \tilde{T}\tilde{S}$ in $\mathscr{L}(K)$, that the mappings $T \mapsto ST$ and $T \mapsto TS$ are ultrastrongly continuous in $\mathscr{L}(H)$. Thus if $\mathscr{S}$ is any subset of $\mathscr{L}(H)$, then $\mathscr{S}'$ is an ultrastrongly closed subalgebra of $\mathscr{L}(H)$ containing I (49.11). In particular, every von Neumann algebra on H is an ultrastrongly closed $*$-subalgebra of $\mathscr{L}(H)$ containing I. Further facts about the ultrastrong topology are noted in the exercises. The following proposition gives a convenient neighborhood base at 0 for τ_{us}:

(68.13) Proposition. *A basic neighborhood of* $0 \in \mathscr{L}(H)$ *for the ultrastrong topology is given by*

$$\left\{ T \in \mathscr{L}(H) : \sum_{i=1}^{\infty} \|Tx_i\|^2 \leq 1 \right\},$$

where (x_i) *is a sequence of vectors in* H *such that* $\sum_{i=1}^{\infty} \|x_i\|^2 < \infty$.

Proof. With notation as in (68.11), a basic τ_{us}-neighborhood $\mathscr{V}$ of $0 \in \mathscr{L}(H)$ is defined by fixing $y_1 = (y_{1i}), \ldots, y_n = (y_{ni}) \in K$ and setting

$$\mathscr{V} = \{ T \in \mathscr{L}(H) : p_{y_k}(\tilde{T}) \leq 1 \quad \text{for } k = 1, \ldots, n \};$$

since $p_{y_k}(\tilde{T}) \leq 1$ iff $p_{y_k}(\tilde{T})^2 \leq 1$, this may be written

$$\mathscr{V} = \left\{ T : \sum_{i=1}^{\infty} \|Ty_{ki}\|^2 \leq 1 \quad \text{for } k = 1, \ldots, n \right\}.$$

Let $x = (x_i)$ be any enumeration of the doubly indexed sequence (y_{ki}); then

$$\sum_{i=1}^{\infty} \|x_i\|^2 = \sum_{k=1}^{n} \left(\sum_{i=1}^{\infty} \|y_{ki}\|^2 \right) < \infty$$

and

$$p_x(\tilde{T}) \geq \max \{ p_{y_1}(\tilde{T}), \ldots, p_{y_n}(\tilde{T}) \}$$

for all $T \in \mathscr{L}(H)$. It follows that

$$\{ T : p_x(\tilde{T}) \leq 1 \} \subset \mathscr{V},$$

thus the neighborhoods of the form

$$\left\{ T : \sum_{i=1}^{\infty} \|Tx_i\|^2 \leq 1 \right\}$$

are indeed basic. {Incidentally, 1 may be replaced by any $\varepsilon > 0$.} ∎

(68.14) Notation as in (68.10). For each pair of vectors $x = (x_i)$, $y = (y_i)$ in K, the formula

(*) $$f(T) = \sum_{i=1}^{\infty} (Tx_i | y_i) \qquad (T \in \mathscr{L}(H))$$

defines a linear form f on $\mathscr{L}(H)$. The set $\mathscr{E}$ of all such linear forms f is evidently a vector space. Consider the nondegenerate bilinear form

$$B : \mathscr{L}(H) \times \mathscr{E} \to \mathbb{C}$$

defined by

$$B(T, f) = f(T) \qquad (T \in \mathscr{L}(H), f \in \mathscr{E}).$$

The topology $\sigma(\mathscr{L}(H), \mathscr{E})$ with respect to the bilinear form B (38.6) is called the *ultraweak topology* on $\mathscr{L}(H)$, denoted τ_{uw}. From (38.7) we know that the ultraweakly continuous linear forms on $\mathscr{L}(H)$ are precisely the forms (*):

(68.15) Proposition.   *A linear form f on $\mathscr{L}(H)$ is ultraweakly continuous iff there exists a pair of sequences (x_i), (y_i) of vectors in H, with $\sum_{i=1}^{\infty} \|x_i\|^2 < \infty$ and $\sum_{i=1}^{\infty} \|y_i\|^2 < \infty$, such that*

$$f(T) = \sum_{i=1}^{\infty} (Tx_i | y_i)$$

for all $T \in \mathscr{L}(H)$.

In the notation of (68.10), the formula (*) in (68.14) may be written $f(T) = \omega_{x,y}(\tilde{T})$. Since the relative topology on $\mathscr{L}(H)^{\sim}$ induced by the weak operator topology of $\mathscr{L}(K)$ may be described as the initial topology for the set of all such linear forms $\tilde{T} \mapsto \omega_{x,y}(\tilde{T})$, we have at once:

(68.16) Proposition.   *With notation as in (68.10), let τ be the relative topology on $\mathscr{L}(H)^{\sim}$ induced by the weak operator topology of $\mathscr{L}(K)$. Then the ultraweak topology on $\mathscr{L}(H)$ is the topology that makes the bijection $T \mapsto \tilde{T}$ of $\mathscr{L}(H)$ onto $(\mathscr{L}(H)^{\sim}, \tau)$ a homeomorphism.*

(68.17)   All topologies considered here are on $\mathscr{L}(H)$. As noted earlier, $\tau_s \supset \tau_w$, and evidently $\tau_n \supset \tau_{us} \supset \tau_{uw}$. Also, it is clear that $\tau_{us} \supset \tau_s$ and $\tau_{uw} \supset \tau_w$ (consider finitely nonzero sequences of vectors). A diagrammatic résumé:

Topologies connected by a line are comparable, with 'higher' = 'finer.'

(68.18)   As in (68.10), we write K for the orthogonal sum of $\aleph_0$ copies of H. Preliminary to proving the von Neumann theorem on bicommutants, we discuss the analysis of an operator on K as a 'matrix' of operators on H. It will be useful to number the summands of K; that is, we write

$$K = H_1 \oplus H_2 \oplus H_3 \oplus \cdots,$$

where $H_i = H$ for $i = 1, 2, 3, \ldots$.

For each positive integer j, we define an isometric linear mapping $U_j: H_j \to K$ by the formula

$$U_j x_j = (\delta_{ij} x_j) \qquad (x_j \in H_j);$$

that is, $U_j x_j$ is the element of K with x_j in the jth coordinate and θ in all others. The adjoint operators $U_j^*: K \to H_j$ are easily described (cf. 43.2): if $x =$

$(x_i) \in K$ then $U_j^* x = x_j$ for all j. {Proof: For all $y_j \in H_j$, $(U_j^* x | y_j) = (x | U_j y_j)$ $= (x_j | y_j)$.} Thus $x = (U_i^* x)$ for all $x \in K$. In particular, it is clear that the intersection of the null spaces of the U_j^* is $\{\theta\}$.

Since each U_j is isometric, its range $U_j(H_j)$ is a closed linear subspace of K. The subspaces $U_j(H_j)$ are *total* in K, i.e., if $x \in K$ and $x \perp U_j(H_j)$ for all j (in other words, $U_j^* x = \theta$ for all j (43.14)), then $x = \theta$. It follows that the closed linear span of the subspaces $U_j(H_j)$ is K (42.17).

Suppose $P \in \mathscr{L}(K)$. For each ordered pair of positive integers (i, j), the mapping

$$x_j \mapsto U_i^* P U_j x_j \qquad (x_j \in H_j)$$

is an operator on H_j; we denote it by P_{ij}. Thus, associated with each $P \in \mathscr{L}(K)$, there is a 'matrix' (P_{ij}) of operators $P_{ij} \in \mathscr{L}(H)$. We leave to the reader the discovery of the various pleasant algebraic and topological properties of such matricial representations; the only property we shall need is as follows:

(68.19) Lemma. *With notation as in* (68.18), *if* $P \in \mathscr{L}(K)$ *and* $T \in \mathscr{L}(H)$ *then* $\tilde{T}P = P\tilde{T}$ *iff* $TP_{ij} = P_{ij}T$ *for all* i, j.

Proof. For every $y \in K$, the ith coordinate of y is $U_i^* y$, thus $y = (U_i^* y)$; in particular, for all $x \in K$,

$$\tilde{T}Px = \tilde{T}(U_i^* Px) = (TU_i^* Px), \qquad P\tilde{T}x = (U_i^* P\tilde{T}x)$$

(the parentheses indicate sequences, i.e., elements of K). Thus $\tilde{T}P = P\tilde{T}$ iff

(*) ' $TU_i^* Px = U_i^* P\tilde{T}x$

for all $x \in K$ and all i.

Assume (*) holds and fix a pair of indices (i, j). Note that, for all $x_j \in H_j$, one has $\tilde{T}U_j x_j = U_j(Tx_j)$; replacement of x in (*) by $U_j x_j$ yields

$$TU_i^* P U_j x_j = U_i^* P\tilde{T}U_j x_j = U_i^* P U_j(Tx_j),$$

that is,

$$TP_{ij}x_j = P_{ij}Tx_j$$

for all $x_j \in H_j$; this shows that $TP_{ij} = P_{ij}T$.

Conversely, assume $TP_{ij} = P_{ij}T$ for all i, j. Fix an index i; we are to show that (*) holds for all $x \in K$. For each index j and for all $x_j \in H_j$ we have, by hypothesis,

$$TU_i^* P U_j x_j = U_i^* P U_j Tx_j = U_i^* P\tilde{T}U_j x_j,$$

thus $TU_i^* P - U_i^* P\tilde{T} = 0$ on $U_j(H_j)$; since the closed linear span of the subspaces $U_j(H_j)$ is K, we conclude that $TU_i^* P - U_i^* P\tilde{T} = 0$, thus (*) holds for all $x \in K$. ∎

(68.20) Theorem. *Let* $\mathscr{A}$ *be a* ∗-*subalgebra of* $\mathscr{L}(H)$ *such that the only vector of* H *annihilated by every* $A \in \mathscr{A}$ *is* θ *(as is the case, e.g., if* $I \in \mathscr{A}$). *Then* $\mathscr{A}''$ *is the ultrastrong closure of* $\mathscr{A}$.

Proof. {The hypothesis on the ∗-algebra $\mathscr{A}$ is that the intersection of the null spaces of the operators in $\mathscr{A}$ is $\{\theta\}$; equivalently, the closed linear span of the ranges of the operators in $\mathscr{A}$ is H (cf. (43.14), (42.17)).}

Since $\mathscr{A}''$ is ultrastrongly closed (68.12), it contains the ultrastrong closure of $\mathscr{A}$. Conversely, assuming $B \in \mathscr{A}''$ it is to be shown that B is ultrastrongly adherent to $\mathscr{A}$. Thus, given a basic neighborhood $\mathscr{V}$ of B for the ultrastrong topology, we must show that $\mathscr{V} \cap \mathscr{A} \neq \varnothing$.

Let K be the orthogonal sum of $\aleph_0$ copies of H, with notations as in (68.10) and (68.18). By (68.13) we can suppose that, for a suitable vector $x = (x_i)$ in K,

$$\mathscr{V} = \{T \in \mathscr{L}(H) : \sum_{i=1}^{\infty} \|Tx_i - Bx_i\|^2 \leq 1\}$$

$$= \{T \in \mathscr{L}(H) : \|\tilde{T}x - \tilde{B}x\| \leq 1\};$$

thus, we seek an operator $A \in \mathscr{A}$ such that $\|\tilde{A}x - \tilde{B}x\| \leq 1$.

Let $M = \{\tilde{A}x : A \in \mathscr{A}\}$; M is a linear subspace of K, whose closure we denote by N. In essence, we are trying to approximate $\tilde{B}x$ by vectors in M: it will clearly suffice to show that $\tilde{B}x \in N$. This will be accomplished by showing (in reverse order) that (i) $x \in N$, and (ii) $\tilde{B}(N) \subset N$.

(ii) For all $A \in \mathscr{A}$, obviously $\tilde{A}(M) \subset M$, therefore $\tilde{A}(N) \subset N$ by continuity; since $\{\tilde{A} : A \in \mathscr{A}\}$ is a ∗-subalgebra of $\mathscr{L}(K)$, it follows that N reduces $\tilde{A}$ for all $A \in \mathscr{A}$, and, writing $P = P_N$ for the projection operator on K with range N, we have $\tilde{A}P = P\tilde{A}$ for all $A \in \mathscr{A}$ (43.11). If (P_{ij}) is the 'matrix' associated with P as in (68.18), we infer from (68.19) that $AP_{ij} = P_{ij}A$ for all $A \in \mathscr{A}$ and all i, j, thus $P_{ij} \in \mathscr{A}'$ for all i, j. Since $B \in \mathscr{A}''$, it follows that $BP_{ij} = P_{ij}B$ for all i, j; another application of (68.19) yields $\tilde{B}P = P\tilde{B}$, that is, N reduces $\tilde{B}$. In particular, $\tilde{B}(N) \subset N$.

(i) The hypothesis on $\mathscr{A}$ implies that $x \in N$. {Proof: Writing $y = Px - x$, we are to show that $y = \theta$. Say $y = (y_i)$. If $A \in \mathscr{A}$ then $\tilde{A}x \in M \subset N$, therefore $P\tilde{A}x = \tilde{A}x$; since $P\tilde{A} = \tilde{A}P$ it follows that $\tilde{A}(Px) = \tilde{A}x$, thus $\tilde{A}(Px - x) = \theta$, i.e., $\theta = \tilde{A}y = (Ay_i)$. In other words $Ay_i = \theta$ for all $A \in \mathscr{A}$ and for all i, whence $y_i = \theta$ for all i, by the hypothesis on $\mathscr{A}$.} ∎

(68.21) Corollary. *If $\mathscr{A}$ is a ∗-subalgebra of $\mathscr{L}(H)$ and if the only vector of H annihilated by every $A \in \mathscr{A}$ is θ (in particular, if $I \in \mathscr{A}$), then the following conditions on $\mathscr{A}$ are equivalent:*

   (a) *$\mathscr{A} = \mathscr{A}''$ (i.e., $\mathscr{A}$ is a von Neumann algebra on H);*

   (b) *$\mathscr{A}$ is weakly closed;*

   (b′) *$\mathscr{A}$ is ultraweakly closed;*

   (c) *$\mathscr{A}$ is strongly closed;*

   (c′) *$\mathscr{A}$ is ultrastrongly closed.*

Proof. (a) implies (b) by (68.8).

The implications (b) ⇒ (b′) ⇒ (c′) and (b) ⇒ (c) ⇒ (c′) follow from the inclusion relations in (68.17).

Finally, (c′) implies (a) by the theorem (68.20). ∎

Exercises

(68.22) (i) If H is infinite-dimensional, all inclusions described in (68.17) are proper; (ii) if H is finite-dimensional, all five topologies coincide.

(68.23) A convex subset of $\mathscr{L}(H)$ is weakly closed iff it is strongly closed.

(68.24) (i) Let $\mathscr{M}$ be a linear subspace of $\mathscr{L}(H)$. The following conditions on a linear form $f\colon \mathscr{M} \to \mathbb{C}$ are equivalent: (a) f is ultraweakly continuous; (b) f is ultrastrongly continuous; (c) there exists a pair of sequences (x_i), (y_i) of vectors in H, with $\sum_{i=1}^{\infty} \|x_i\|^2 < \infty$ and $\sum_{i=1}^{\infty} \|y_i\|^2 < \infty$, such that $f(T) = \sum_{i=1}^{\infty} (Tx_i | y_i)$ for all $T \in \mathscr{M}$.

(ii) A convex subset of $\mathscr{L}(H)$ is ultraweakly closed iff it is ultrastrongly closed.

*(68.25) If $\mathscr{A}$ and $\mathscr{B}$ are von Neumann algebras and $\varphi\colon \mathscr{A} \to \mathscr{B}$ is a *-isomorphism, then φ is ultrastrongly bicontinuous and ultraweakly bicontinuous.

(68.26) Let $\mathscr{B} = \{T \in \mathscr{L}(H) : \|T\| \leq 1\}$.

(i) The strong topology induces a uniform structure on $\mathscr{L}(H)$ (11.10); $\mathscr{B}$ is complete for the relative uniform structure.

(ii) The ultrastrong and strong topologies coincide on $\mathscr{B}$ (hence on any norm-bounded subset of $\mathscr{L}(H)$); if H is separable, then $\mathscr{B}$ is metrizable for this topology and has a countable base for open sets.

(iii) The ultraweak and weak topologies coincide on $\mathscr{B}$ (hence on any norm-bounded subset of $\mathscr{L}(H)$); for this topology, $\mathscr{B}$ is compact (and metrizable, if H is separable).

(iv) If $\mathscr{U}(H)$ is the group of all unitary operators on H, then the weak, strong, ultraweak, and ultrastrong topologies all coincide on $\mathscr{U}(H)$. Equipped with this topology, $\mathscr{U}(H)$ is a topological group.

*(68.27) (J. Dixmier)   Every von Neumann algebra is the dual of a Banach space.

Explicitly, suppose $\mathscr{M}$ is a von Neumann algebra and let $\mathscr{M}_*$ be the vector space of all linear forms $f\colon \mathscr{M} \to \mathbb{C}$ that are continuous for the (relative) ultraweak topology on $\mathscr{M}$. Equip $\mathscr{M}_*$ with the norm

$$\|f\| = \sup \{|f(T)| : T \in \mathscr{M}, \|T\| \leq 1\}.$$

Then $\mathscr{M}_*$ is a Banach space, whose dual is isometrically isomorphic to $\mathscr{M}$; the bounded linear form on $\mathscr{M}_*$ corresponding to $T \in \mathscr{M}$ is $f \mapsto f(T)$. Regarding $\mathscr{M}$ as the dual space of $\mathscr{M}_*$, the weak* topology on $\mathscr{M}$ coincides with the (relative) ultraweak topology.

*(68.28) (S. Sakai)   A C^*-algebra is *-isomorphic to a von Neumann algebra if and only if it is the dual of some Banach space.

*69. Group representations.

Let G be a locally compact topological group, notated multiplicatively, with neutral element e. We write dt for a left-invariant Haar measure on G (cf. [10, p. 262]).

(69.1) Definition.   A *unitary representation* of G on a Hilbert space H is a group homomorphism $t \mapsto U_t$ of G into the group $\mathscr{U}(H)$ of unitary operators on H. Equipped with the strong operator topology, $\mathscr{U}(H)$ is a topological group (68.26); a unitary representation $G \to \mathscr{U}(H)$ is said to be *continuous* if it is continuous for the given topology on G and the strong operator topology on $\mathscr{U}(H)$. {Thus, a unitary representation $t \mapsto U_t$ is continuous iff $t \to s$ implies $U_t \to U_s$ strongly; equivalently (5.1), $t \to e$ implies $U_t \to I$ strongly.} A unitary representation $t \mapsto U_t$ of G on H is said to be *irreducible* if the $*$-subset $\{U_t : t \in G\}$ of $\mathscr{L}(H)$ is irreducible in the sense of (67.1); equivalently (67.2), $\{U_t : t \in G\}' = \{\lambda I : \lambda \in \mathbb{C}\}$.

Our objective in this section is to prove the fundamental completeness theorem of Gel'fand and Raĭkov:

Every locally compact group G has a complete set of irreducible, continuous unitary representations; i.e., if $s \in G$, $s \neq e$, then there exists an irreducible, continuous unitary representation $t \mapsto U_t$ of G such that $U_s \neq I$.

The strategy of the proof is to associate with G a C^*-algebra $\mathscr{A}$, to which the reduction theory of Section 67 is applied: the complete set of irreducible $*$-representations of $\mathscr{A}$ provided by (67.27) leads to the desired complete set of unitary representations of G.

The construction of the C^*-algebra $\mathscr{A}$ is based on the Lebesgue spaces $L^1(G)$, $L^2(G)$ derived from the Haar measure dt (see Section 39): (1) $L^1(G)$ is shown to have the structure of a Banach $*$-algebra (with multiplication induced by the convolution of functions); (2) it is shown that $L^1(G)$ has a faithful $*$-representation $u \mapsto T_u$ on the Hilbert space $L^2(G)$ (the 'left regular representation,' also induced by the convolution product); (3) $\mathscr{A}$ is defined to be the norm-closure of the $*$-algebra $\{T_u + \lambda I : u \in L^1(G), \lambda \in \mathbb{C}\}$.

The two main technical problems are (i) the construction of the C^*-algebra $\mathscr{A}$ described above, and (ii) the proof that the continuous unitary representations of G are in natural one-to-one correspondence with the $*$-representations of $L^1(G)$ that are nondegenerate in the sense of the following definition:

(69.2) Definition. Let A be a (complex) $*$-algebra and let H be a Hilbert space. A $*$-representation $\varphi: A \to \mathscr{L}(H)$ is said to be *nondegenerate* if the only vector of H annihilated by every $\varphi(a)$ $(a \in A)$ is the vector θ. {In view of (43.14) and (42.17), this means that the set of all vectors $\varphi(a)x$ $(a \in A, x \in H)$ is total in H—equivalently, its closed linear span is H.}

(69.3) First we sketch the necessary measure-theoretic constructions. The particular style of measure theory is not crucial; for the sake of definiteness and brevity, the discussion is based on [10, Ch. 9]. The exposition in [10] is limited to real scalars, so the first task is to make the minor adjustments that are needed to accommodate complex scalars.

The class of *Baire sets* of G is the σ-ring generated by the compact G_δ's; a subset of G is *locally Baire* if its intersection with every Baire set is a Baire set [cf. **10**, p. 35]. A complex-valued function $f: G \to \mathbb{C}$ is said to be a *Baire function* (or to be *Baire measurable*) if $\operatorname{Re} f$ and $\operatorname{Im} f$ are Baire functions [cf. **10**, p. 39, Exer. 1]. In the class of all Baire functions, the relation '$f = g$ a.e.' (with respect to the left Haar measure dt) is an equivalence relation; the equivalence class of f is denoted $[f]$.

For $1 \le p < \infty$, $\mathscr{L}^p(G)$ denotes the vector space of all Baire functions $f: G \to \mathbb{C}$ such that $|f|^p$ is integrable (with respect to Haar measure), and $L^p(G)$ is the set of all equivalence classes $u = [f]$, $f \in \mathscr{L}^p(G)$; as shown in (39.9), $L^p(G)$ is a Banach space relative to the linear operations

$$[f] + [g] = [f + g], \qquad \lambda[f] = [\lambda f]$$

and the norm

$$\|[f]\|_p = \left(\int |f(t)|^p \, dt \right)^{1/p}.$$

Only the cases $p = 1$ and $p = 2$ will be needed for the applications in the present section. In particular, $L^2(G)$ is a Hilbert space, with inner products defined by

$$(u|v) = \int f(t)g(t)^* \, dt$$

for $u = [f]$, $v = [g]$ in $L^2(G)$ (39.10).

(69.4) If f and g are Baire functions on G then the set of all $s \in G$, such that the function

$$t \mapsto f(st)g(t^{-1}) \qquad (t \in G)$$

is integrable, is the complement of a Baire set [10, Sect. 88]; the function

$$s \mapsto \int f(st)g(t^{-1}) \, dt = \int f(t)g(t^{-1}s) \, dt,$$

defined for all such s, is called the *convolution* of f and g (in that order), denoted $f * g$ [10, p. 277].

(69.5) Assume $1 \le p < \infty$. If $f \in \mathscr{L}^1(G)$ and $g \in \mathscr{L}^p(G)$, then there exists a function $h \in \mathscr{L}^p(G)$ such that $h = f * g$ a.e. and $\|h\|_p \le \|f\|_1 \|g\|_p$ [10, p. 295, Exer. 3]; and if $h' = f' * g'$ a.e., where $f = f'$ a.e. and $g = g'$ a.e., then $h = h'$ a.e. Thus, writing $u = [f]$, $v = [g]$, $w = [h]$, the formula $uv = w$ is a well-defined law of composition

$$L^1(G) \times L^p(G) \to L^p(G).$$

The following relations are readily verified:

(1) $\|uv\|_p \le \|u\|_1 \|v\|_p$

(2) $u(v_1 + v_2) = uv_1 + uv_2$

(3) $u(\lambda v) = (\lambda u)v = \lambda(uv)$

(4) $(u_1 + u_2)v = u_1 v + u_2 v$

(5) $(u_1 u_2)v = u_1(u_2 v)$

where $u, u_1, u_2 \in L^1(G)$, $v, v_1, v_2 \in L^p(G)$, and $\lambda \in \mathbb{C}$. Nothing deeper than Fubini's theorem and the left invariance of the Haar measure figures in these computations [cf. **10**, Sect. 86]. In particular: $L^1(G)$ *is a Banach algebra with respect to the multiplication* $(u_1, u_2) \mapsto u_1 u_2$ *defined via convolution* [cf. **10**, p. 288, Exer. 4]. In general, $L^1(G)$ does not have a unity element (69.28); the discussion in this section circumvents the issue.

The *left regular representation* of $L^1(G)$ on $L^p(G)$ is defined as follows. For each $u \in L^1(G)$ we may define

$$T_u: L^p(G) \to L^p(G)$$

by the formula $T_u v = uv$ $(v \in L^p(G))$. From (1)–(3) it is clear that T_u belongs to the Banach algebra $\mathscr{L}(L^p(G))$ (40.23) and that $\|T_u\| \leq \|u\|_1$; in view of (3)–(5), *the mapping* $u \mapsto T_u$ *is a continuous algebra homomorphism* $L^1(G) \to \mathscr{L}(L^p(G))$. This mapping is in fact a monomorphism, as will be shown below (69.17) using 'approximate identity' techniques.

(69.6) If $f \in \mathscr{L}^1(G)$, the *adjoint* of f is the function $\tilde{f}$ defined by the formula

$$\tilde{f}(t) = \Delta(t)f(t^{-1})^* \qquad (t \in G),$$

where Δ is the modular function of G [**10**, Sect. 77]; it follows that $\tilde{f} \in \mathscr{L}^1(G)$ and

$$\int \tilde{f}(t)\, dt = \left(\int f(t)\, dt \right)^*$$

[**10**, Sect. 82], and in particular $\|\tilde{f}\|_1 = \|f\|_1$. If $f_1 = f_2$ a.e. then $\tilde{f}_1 = \tilde{f}_2$ a.e. [**10**, p. 271, Cor. of Th. 2]; thus if $u = [f] \in L^1(G)$, we may define $\tilde{u} = [\tilde{f}]$. The mapping $u \mapsto \tilde{u}$ $(u \in L^1(G))$ has the following properties:

(6) $$(\tilde{u})^{\sim} = u$$

(7) $$(u_1 + u_2)^{\sim} = \tilde{u}_1 + \tilde{u}_2$$

(8) $$(\lambda u)^{\sim} = \lambda^* \tilde{u}$$

(9) $$(u_1 u_2)^{\sim} = \tilde{u}_2 \tilde{u}_1$$

(10) $$\|\tilde{u}\|_1 = \|u\|_1$$

for all $u, u_1, u_2 \in L^1(G)$ and $\lambda \in \mathbb{C}$. Summarizing:

(69.7) Proposition. $L^1(G)$ *is a Banach* $*$-*algebra, with involution* $u \mapsto \tilde{u}$ *defined as above; moreover,* $\|\tilde{u}\|_1 = \|u\|_1$ *for all* $u \in L^1(G)$.

Preliminary to the main business of the section—representations—there remain two technical topics requiring attention: translates (69.9) and approximate identities (69.13). To avoid long digressions on principally measure-theoretic questions, most of the details (generally routine) will be omitted. {The treatise of E. Hewitt and K. A. Ross [**70**] is recommended as a universal covering reference.} The sketch to be given is intended to be full enough to be persuasive, and the only genuinely sticky measure-theoretic point is met head-on in the proof of (69.20).

(69.8) If $f: G \to \mathbb{C}$ and $s \in G$, the left and right *translates* of f by s are the functions f_s and f^s defined by the formulas

$$f_s(t) = f(s^{-1}t), \qquad f^s(t) = f(ts).$$

If $f \in \mathscr{L}^1(G)$ then also $f_s, f^s \in \mathscr{L}^1(G)$ and

$$\int f_s(t)\, dt = \int f(t)\, dt, \qquad \int f^s(t)\, dt = \Delta(s) \int f(t)\, dt$$

[**10**, Sect. 79]. If $u = [f] \in L^1(G)$ and $s \in G$, one defines

$$u_s = [f_s], \qquad u^s = [f^s].$$

A sampling of the straightforward algebraic properties of these notations:

(11) $(u + v)_s = u_s + v_s, \qquad (u + v)^s = u^s + v^s$

(12) $(\lambda u)_s = \lambda u_s, \qquad (\lambda u)^s = \lambda u^s$

(13) $u_{st} = (u_t)_s, \qquad u^{st} = (u^t)^s$

(14) $(uv)_s = u_s v, \qquad (uv)^s = u v^s$

(15) $\|u_s\|_1 = \|u\|_1, \qquad \|u^s\|_1 = \Delta(s)\|u\|_1$

(16) $(\tilde{u})_s = \Delta(s^{-1})(u^s)^{\tilde{\;}}$

for all $u, v \in L^1(G)$, $s, t \in G$, and $\lambda \in \mathbb{C}$. More substantial is the following:

(69.9) Lemma.   *If $u \in L^1(G)$ then the mappings $s \mapsto u_s$ and $s \mapsto u^s$ are continuous; i.e., given any $\varepsilon > 0$, there exists a neighborhood V of e such that $\|u_s - u\|_1 \leq \varepsilon$ and $\|u^s - u\|_1 \leq \varepsilon$ for all $s \in V$.*

(69.10) Definition.   An *approximate identity* in a Banach algebra A is a net $(a_j)_{j \in J}$ in A (i.e., a family of elements of A indexed by an increasingly directed set J) such that (1) $\|a_j\| = 1$ for all $j \in J$, and (2) for each $b \in A$, $\|a_j b - b\| \to 0$ and $\|b a_j - b\| \to 0$ in the sense of net convergence. {Explicitly, if $b \in A$ and $\varepsilon > 0$, there exists an index j_0 such that $\|a_j b - b\| \leq \varepsilon$ and $\|b a_j - b\| \leq \varepsilon$ for all $j \geq j_0$.}

An approximate identity is a valuable substitute for an outright unity element; our next objective is to construct an approximate identity for $L^1(G)$.

(69.11) Definition.   We denote by $\mathscr{F}$ the set of all functions $g \in \mathscr{L}^1(G)$ such that $g \geq 0$ (for convenience, everywhere on G), g is essentially bounded, and $\int g(t)\, dt = 1$. If A is any subset of G, we write $\mathscr{F}_A$ for the set of all $f \in \mathscr{F}$ such that $f = 0$ on $\complement A$. Note that $A \subset B$ implies $\mathscr{F}_A \subset \mathscr{F}_B$. If A has nonempty interior, $\mathscr{F}_A$ is assuredly nonempty:

(69.12) Lemma.   *If V is any neighborhood of e, then $\mathscr{F}_V$ contains a continuous function g with compact support, such that $\tilde{g} = g$.*

Proof.   Let W be a symmetric neighborhood of e with $W \subset V$, and let h be a nonzero continuous function with compact support, such that $h = 0$ on

$\mathfrak{C} W$ and $h \geq 0$. Replacing h by $h + \tilde{h}$, one can suppose that $\tilde{h} = h$. Let $g = \alpha^{-1}h$, where $\alpha = \int h(t) \, dt$. ∎

(69.13) Lemma. *Suppose $f \in \mathscr{L}^1(G)$. Given any $\varepsilon > 0$, there exists a neighborhood V of e such that $\|f * g - f\|_1 \leq \varepsilon$ for all $g \in \mathscr{F}_V$.*

Proof. Let $g \in \mathscr{F}$. It is straightforward to check that $f * g$ is everywhere defined on G, hence $f * g \in \mathscr{L}^1(G)$. {Also $|(f * g)(t)| \leq \|f\|_1\|g\|_\infty$ and $|(f * g)(s) - (f * g)(t)| \leq \|f_{ts^{-1}} - f\|_1\|g\|_\infty$ for all $s, t \in G$, thus $f * g$ is bounded and continuous.} Since $\int \tilde{g}(t) \, dt = (\int g(t) \, dt)^* = 1$,

$$(f * g)(s) - f(s) = \int f(st)g(t^{-1}) \, dt - f(s) \int \tilde{g}(t) \, dt$$

$$= \int [f(st) - \Delta(t)f(s)]g(t^{-1}) \, dt$$

for all $s \in G$; writing $F(s, t)$ for the integrand in the last integral, elementary algebra yields

(i) $\qquad F(s, t) = f(st)[1 - \Delta(t)]g(t^{-1}) + [f(st) - f(s)]\tilde{g}(t).$

It follows from (i) that, for each fixed $t \in G$,

$$\int |F(s, t)| \, ds \leq \|f^t\|_1|1 - \Delta(t)|g(t^{-1}) + \|f^t - f\|_1\tilde{g}(t)$$

$$= \Delta(t)\|f\|_1|1 - \Delta(t)|g(t^{-1}) + \|f^t - f\|_1\tilde{g}(t)$$

$$= (\|f\|_1|1 - \Delta(t)| + \|f^t - f\|_1)\tilde{g}(t).$$

Thus, defining $k(t) = \int |F(s, t)| \, ds$, k is a Baire function such that

(ii) $\qquad 0 \leq k(t) \leq (\|f\|_1|1 - \Delta(t)| + \|f^t - f\|_1)\tilde{g}(t)$

for all $t \in G$.

Given any $\varepsilon > 0$, let V be a symmetric neighborhood of e such that

(iii) $\qquad \|f\|_1|1 - \Delta(t)| \leq \varepsilon/2$ and $\|f^t - f\|_1 \leq \varepsilon/2$

for all $t \in V$. {This is possible by the continuity of Δ and (69.9).}

Suppose now that $g \in \mathscr{F}_V$. Since $\tilde{g} = 0$ outside $V^{-1} = V$, it results from (ii) and (iii) that

(iv) $\qquad 0 \leq k \leq \varepsilon\tilde{g}.$

From (iv) we see that $k \in \mathscr{L}^1(G)$ and

$$\int k(t) \, dt \leq \varepsilon \int \tilde{g}(t) \, dt = \varepsilon,$$

thus

$$\int \left(\int |F(s, t)| \, ds \right) dt \leq \varepsilon;$$

it follows from the Fubini-Tonelli theorems that

$$\int \left| \int F(s, t) \, dt \right| ds \leq \varepsilon.$$

Since $\int F(s, t) \, dt = (f * g)(s) - f(s)$, we thus have $\|f * g - f\|_1 \leq \varepsilon$. ∎

The preceding lemma can be symmetrized:

(69.14) Lemma. *Suppose $f \in \mathcal{L}^1(G)$. Given any $\varepsilon > 0$, there exists a neighborhood V of e such that $\|f * g - f\|_1 \leq \varepsilon$ and $\|g * f - f\|_1 \leq \varepsilon$ for all $g \in \mathcal{F}_V$.*

Proof. As noted in the proof of (69.13), $f * g \in \mathcal{L}^1(G)$ for every $g \in \mathcal{F}$. Applying (69.13) to f and to $\tilde{f}$, we may choose a neighborhood V of e such that

(*) $\|f * g - f\|_1 \leq \varepsilon$ and $\|\tilde{f} * g - \tilde{f}\|_1 \leq \varepsilon$

whenever $g \in \mathcal{F}_V$. We can suppose that V is compact and symmetric; then $g \in \mathcal{F}_V$ implies $\tilde{g} \in \mathcal{F}_V$. {The point is that $\tilde{g}(t) = \Delta(t)g(t^{-1})$ is essentially bounded because Δ is continuous and g vanishes outside the compact set $V^{-1} = V$.} Thus if $g \in \mathcal{F}_V$ we may replace g by $\tilde{g}$ in the second inequality of (*); the proof is concluded by the observation that $\tilde{f} * \tilde{g} = (g * f)^{\sim}$. ∎

(69.15) Definition. The set $\mathscr{V}$ of all neighborhoods of e in G is increasingly directed by defining $W \geq V$ in case $W \subset V$. Thus every family $(x_V)_{V \in \mathscr{V}}$ indexed by $\mathscr{V}$ may be regarded as a net; if the x_V lie in a topological space X and if the net converges to $x \in X$, we say that '$x_V \to x$ as $V \to e$.'

The key result on approximate identities is at hand:

(69.16) Proposition. *For each neighborhood V of e, choose $g_V \in \mathcal{F}_V$ (cf. 69.12) and define $u_V = [g_V]$. Then the family $(u_V)_{V \in \mathscr{V}}$ is an approximate identity in $L^1(G)$; i.e., for each $u \in L^1(G)$ we have $\|uu_V - u\|_1 \to 0$ and $\|u_V u - u\|_1 \to 0$ as $V \to e$.*

Proof. Note that $\|u_V\|_1 = \int g_V(t) \, dt = 1$ for all $V \in \mathscr{V}$. Let $u = [f] \in L^1(G)$. Given any $\varepsilon > 0$, choose V as in (69.14). If $W \geq V$ then $g_W \in \mathcal{F}_W \subset \mathcal{F}_V$, therefore $\|f * g_W - f\|_1 \leq \varepsilon$ and $\|g_W * f - f\|_1 \leq \varepsilon$. ∎

The preliminaries are over; we return to the main theme of representation theory:

(69.17) Lemma. *For each p, $1 \leq p < \infty$, the left regular representation $u \mapsto T_u$ of $L^1(G)$ on $L^p(G)$ (69.5) is faithful, i.e., $T_u = 0$ implies $u = 0$.*

Proof. Assuming $u = [f] \in L^1(G)$ and $uv = 0$ for all $v \in L^p(G)$, we are to show that $u = 0$. For each neighborhood V of e, let g_V be a continuous function with compact support such that $g_V \in \mathcal{F}_V$ (69.12). By hypothesis, $f * g = 0$ a.e. for every $g \in \mathcal{L}^p(G)$. In particular, $f * g_V = 0$ for all V; thus, writing $u_V = [g_V] \in L^1(G)$, we have $uu_V = 0$ for all V. Since $\|u\|_1 = \|u - uu_V\|_1 \to 0$ as $V \to e$ (69.16), we conclude that $u = 0$. ∎

Let us specialize to $p = 2$:

(69.18) Theorem. *The left regular representation $u \mapsto T_u$ of $L^1(G)$ on $L^2(G)$ is a faithful $*$-representation.*

Proof. In view of the lemma, the only point remaining to be settled is that $T_{\tilde{u}} = (T_u)^*$.

Let $u \in L^1(G)$ and $v_1, v_2 \in L^2(G)$. Say $u = [f]$, $v_1 = [g_1]$, $v_2 = [g_2]$. From the properties of the modular function Δ, we have

$$(T_{\tilde{u}} v_1 | v_2) = (\tilde{u} v_1 | v_2)$$

$$= \int \left(\int \tilde{f}(st) g_1(t^{-1})\, dt \right) g_2(s)^*\, ds$$

$$= \iint \Delta(st) f(t^{-1}s^{-1})^* g_1(t^{-1}) g_2(s)^*\, dt\, ds$$

$$= \int \Delta(s) \left(\int \Delta(t) f(t^{-1}s^{-1})^* g_1(t^{-1})\, dt \right) g_2(s)^*\, ds$$

$$= \int \Delta(s) \left(\int f(ts^{-1})^* g_1(t)\, dt \right) g_2(s)^*\, ds$$

$$= \int \left(\int f(ts)^* g_1(t)\, dt \right) g_2(s^{-1})^*\, ds$$

$$= \iint f(ts)^* g_1(t) g_2(s^{-1})^*\, dt\, ds,$$

whereas

$$(T_u^* v_1 | v_2) = (v_1 | T_u v_2) = (v_1 | u v_2)$$

$$= \int g_1(t) \left(\int f(ts) g_2(s^{-1})\, ds \right)^*\, dt$$

$$= \iint g_1(t) f(ts)^* g_2(s^{-1})^*\, ds\, dt,$$

thus $(T_{\tilde{u}} v_1 | v_2) = (T_u^* v_1 | v_2)$ by Fubini's theorem. ∎

We now turn to the study of general $*$-representations of $L^1(G)$ and their relation to the continuous unitary representations of G.

(69.19) Lemma. *If $\varphi \colon L^1(G) \to \mathscr{L}(H)$ is a $*$-representation of $L^1(G)$ on a Hilbert space H, then $\|\varphi(u)\| \le \|u\|_1$ for all $u \in L^1(G)$.*

Proof. The essential point is that $L^1(G)$ is a Banach $*$-algebra whose involution is norm-preserving. Writing r for spectral radius (cf. (58.3), (51.18)), we have

$$\|\varphi(u)\|^2 = \|\varphi(u)^* \varphi(u)\| = r(\varphi(u)^* \varphi(u)) = r(\varphi(\tilde{u}u))$$

$$\le r(\tilde{u}u) \le \|\tilde{u}u\|_1 \le \|\tilde{u}\|_1 \|u\|_1 = (\|u\|_1)^2$$

for all $u \in L^1(G)$. ∎

(69.20) Theorem. *Suppose $\varphi: L^1(G) \to \mathscr{L}(H)$ is a nondegenerate $*$-representation of $L^1(G)$ on a Hilbert space H. Let $(u_j)_{j \in J}$ be any approximate identity in $L^1(G)$ (cf. 69.16). Then:*

(1) For each $t \in G$, the net of operators $\varphi((u_j)_t)$ converges strongly to a unitary operator U_t on H; U_t depends only on φ and t, not on the particular approximate identity (u_j).

(2) $\varphi(u_t) = U_t \varphi(u)$ for all $u \in L^1(G)$ and $t \in G$.

(3) $t \mapsto U_t$ is a continuous unitary representation of G on H.

(4) For all $u = [f] \in L^1(G)$ and $x, y \in H$, one has

$$(\varphi(u)x \,|\, y) = \int f(t)(U_t x \,|\, y)\, dt.$$

Proof. By the lemma, φ is continuous for the norm topologies, and $\|\varphi\| \leq 1$. {The rest of the proof uses only the fact that φ is a continuous algebra homomorphism such that $\|\varphi\| \leq 1$ and such that the vectors $\varphi(u)x$ $(u \in L^1(G), x \in H)$ have closed linear span H.}

Fix $t \in G$ and let $T_j = \varphi((u_j)_t)$; it is to be shown that the net (T_j) is strongly convergent. Since

$$\|T_j\| \leq \|(u_j)_t\|_1 = \|u_j\|_1 = 1$$

for all j, it suffices to show that $T_j y$ is convergent for all y in some dense linear subspace of H (cf. 47.8)—equivalently, for all y in some total subset of H. By the nondegeneracy hypothesis, it suffices to consider $y = \varphi(u)x$, where $u \in L^1(G)$ and $x \in H$. Since $u = \lim_j (u_j u)$, and therefore

$$u_t = \lim_j (u_j u)_t = \lim_j ((u_j)_t u),$$

we have

$$\varphi(u_t) = \lim_j \varphi((u_j)_t u) = \lim_j [\varphi((u_j)_t)\varphi(u)] = \lim_j [T_j \varphi(u)],$$

that is, $\|T_j \varphi(u) - \varphi(u_t)\| \to 0$; in particular,

$$\|T_j \varphi(u)x - \varphi(u_t)x\| \to 0.$$

Let U_t be the unique operator on H such that $\varphi((u_j)_t) \to U_t$ strongly. The foregoing computation shows that

$$U_t \varphi(u)x = \varphi(u_t)x$$

for all $u \in L^1(G)$ and $x \in H$, thus

(i) $U_t \varphi(u) = \varphi(u_t)$ for all $u \in L^1(G)$.

Since φ is nondegenerate, it is clear from (i) that U_t is independent of the particular approximate identity (u_j).

Routine computations based on (i) show that

(ii) $U_e = I$

(iii) $U_{st} = U_s U_t$.

Also, it is clear from the construction that $\|U_t\| \leq 1$ for all $t \in G$, hence also $\|U_t{}^{-1}\| = \|U_{t^{-1}}\| \leq 1$, thus the operators U_t are unitary.

Summarizing, $t \mapsto U_t$ is a unitary representation of G on H. Moreover, if $t \to e$ then $U_t \to I$ strongly. {Proof: Suppose $t \to e$. For each $u \in L^1(G)$, $u_t \to u$ by (69.9), therefore $\varphi(u) = \lim \varphi(u_t) = \lim U_t\varphi(u)$ (in norm) by (i). Thus $U_t\varphi(u)x \to \varphi(u)x$ for all $u \in L^1(G)$ and $x \in H$; since the vectors $\varphi(u)x$ are total and $\|U_t\| = 1$ for all t, we infer that $U_t \to I$ strongly.} Thus $t \mapsto U_t$ is a continuous unitary representation of G on H. This completes the proof of (1)–(3).

(4) Fix $x, y \in H$. We are to show that

(*)
$$(\varphi(u)x \mid y) = \int f(t)(U_tx \mid y) \, dt$$

for all $u = [f] \in L^1(G)$. We can suppose $\|y\| \leq 1$. Note that the integral on the right side is defined for every $f \in \mathcal{L}^1(G)$. {Proof: The function $t \mapsto (U_tx \mid y)$ is locally Baire measurable [10, p. 182, Exer. 13] and bounded by $\|x\| \|y\|$, therefore the function $t \mapsto f(t)(U_tx \mid y)$ is an integrable Baire function.} Moreover,

$$\left| \int f(t)(U_tx \mid y) \, dt \right| \leq \left(\int |f(t)| \, dt \right) \|x\| \|y\| = \|u\|_1 \|x\| \|y\|$$

for all $u = [f] \in L^1(G)$. Thus, both sides of (*) are continuous linear functions of u.

Let $u = [f]$, with $f = \chi_E$ the characteristic function of a Baire set E of finite Haar measure. Since $L^1(G)$ is the closed linear span of such elements, it is enough to show that (*) holds for u. Given any $\varepsilon > 0$, it suffices to show that

(**)
$$\left| (\varphi(u)x \mid y) - \int f(t)(U_tx \mid y) \, dt \right| \leq \varepsilon \int f(t) \, dt.$$

In view of conclusion (1) (with $t = e$) and (69.14), there exists a neighborhood V of e such that

$$\|\varphi([g])x - x\| \leq \varepsilon$$

for all $g \in \mathscr{F}_V$.

We first verify (**) assuming that $E^{-1}E \subset V$ (so the speak, E is 'small of order V'). If E has Haar measure zero, then (**) reduces to the triviality $|0 - 0| \leq 0$. Supposing that E has Haar measure $\alpha > 0$, let

$$g = \alpha^{-1}f.$$

Then $g \geq 0$, g is bounded, $g \in \mathscr{L}^1(G)$, and $\int g(s) \, ds = 1$; thus $g \in \mathscr{F}$. Moreover, if $t \in E$ then $g_{t^{-1}} \in \mathscr{F}_V$. {Proof: Let $t \in E$. Obviously $g_{t^{-1}} \in \mathscr{F}$. If $s \notin V$ then $s \notin E^{-1}E$, $(Es) \cap E = \varnothing$, $ts \notin E$, $f(ts) = 0$, $g(ts) = 0$, $g_{t^{-1}}(s) = 0$; thus $g_{t^{-1}}$ vanishes outside V.} Thus, if $t \in E$ then it follows from the choice of V that

$$\|\varphi([g_{t^{-1}}])x - x\| \leq \varepsilon,$$

therefore, by (2),

$$\|U_{t^{-1}}\varphi([g])x - x\| \leq \varepsilon;$$

since U_t is unitary, we conclude that

(†) $$\|\varphi([g])x - U_t x\| \leq \varepsilon \quad \text{for all } t \in E.$$

From $f = \alpha g$ and $\int g(t)\, dt = 1$ we see that

$$(\varphi(u)x\,|\,y) - \int f(t)(U_t x\,|\,y)\, dt = \alpha \int [(\varphi([g])x\,|\,y) - (U_t x\,|\,y)]g(t)\, dt;$$

since $g = 0$ outside E, whereas (†) holds for all $t \in E$, it follows that (recall that $\|y\| \leq 1$)

$$\left|(\varphi(u)x\,|\,y) - \int f(t)(U_t x\,|\,y)\, dt\right| \leq \alpha\varepsilon \int_E g(t)\, dt = \alpha\varepsilon = \varepsilon \int f(t)\, dt,$$

thus (**) is verified in the special case that $f = \chi_E$ with $E^{-1}E \subset V$.

Now suppose $f = \chi_E$ with E any Baire set of finite measure. Let W be a neighborhood of e such that $W^{-1}W \subset V$; we can suppose W is a Baire set. Let t_n be a sequence of elements of G such that

$$E \subset \bigcup_{n=1}^{\infty} t_n W.$$

{Construction: Cover E with a sequence of compact sets, then cover each of the compact sets with a finite number of left translates of W.} Defining $F_n = E \cap (t_n W)$, we have

$$E = \bigcup_{n=1}^{\infty} F_n,$$

where F_n is a sequence of Baire sets such that

$$F_n^{-1}F_n \subset (W^{-1}t_n^{-1})(t_n W) = W^{-1}W \subset V$$

for all n. Disjointify: setting $E_1 = F_1$ and

$$E_n = F_n - (F_1 \cup \cdots \cup F_{n-1})$$

for $n > 1$, we have

$$E = \bigcup_{n=1}^{\infty} E_n,$$

where the E_n are mutually disjoint Baire sets such that $E_n^{-1}E_n \subset V$ for all n. Let $f_n = \chi_{E_n}$ and let

$$h_n = \sum_{k=1}^{n} f_k = \chi_{E_1 \cup \ldots \cup E_n}.$$

As shown in the preceding paragraph, (**) holds for each $u_k = [f_k]$, therefore by linearity we have

$$\left|(\varphi([h_n])x\,|\,y) - \int h_n(t)(U_t x\,|\,y)\, dt\right| = \left|\sum_{k=1}^{n} [(\varphi(u_k)x\,|\,y) - \int f_k(t)(U_t x\,|\,y)\, dt]\right|$$

$$\leq \sum_{k=1}^{n} \varepsilon \int f_k(t)\, dt = \varepsilon \int h_n(t)\, dt;$$

since $h_n \uparrow f$ pointwise, we have $\|[h_n] - [f]\|_1 \to 0$, thus passage to the limit yields (**) for $u = [f]$. ∎

Conversely, every continuous unitary representation of G induces a non-degenerate $*$-representation of $L^1(G)$:

(69.21) Theorem. *Suppose $t \mapsto U_t$ is a continuous unitary representation of G on a Hilbert space H. Then:*
  (1) *There exists a unique $*$-representation $\varphi: L^1(G) \to \mathscr{L}(H)$ such that*

(*) $$(\varphi(u)x \mid y) = \int f(t)(U_t x \mid y)\, dt$$

for all $u = [f] \in L^1(G)$ and $x, y \in H$. Moreover,
  (2) φ *is nondegenerate.*
  (3) $\varphi(u_s) = U_s \varphi(u)$ *for all $u \in L^1(G)$ and $s \in G$.*
  (4) *If $(u_j)_{j \in J}$ is any approximate identity in $L^1(G)$ (cf. 69.16), then, for each $t \in G$, $\varphi((u_j)_t) \to U_t$ strongly.*

  Proof. (1) The uniqueness assertion is obvious, since the right side of (*) depends only on the given unitary representation.
  Suppose $u \in L^1(G)$, say $u = [f]$. For each pair of vectors x, y in H, define

$$B_u(x, y) = \int f(t)(U_t x \mid y)\, dt$$

(it is shown in the proof of (69.20) that the integral exists). Clearly B_u is a sesquilinear form on H such that

$$|B_u(x, y)| \le \|u\|_1 \|x\| \|y\|$$

for all $x, y \in H$; let $\varphi(u)$ be the operator on H such that $B_u(x, y) = (\varphi(u)x \mid y)$ for all $x, y \in H$ (43.9).
  We have thus defined a mapping $\varphi: L^1(G) \to \mathscr{L}(H)$ satisfying (*). It is routine to check that φ is linear; moreover, it follows from $|(\varphi(u)x \mid y| \le \|u\|_1 \|x\| \|y\|$ that $\|\varphi(u)\| \le \|u\|_1$, thus φ is continuous for the norm topologies.
  Next we note that $\varphi(\tilde{u}) = \varphi(u)^*$ $(u \in L^1(G))$. Indeed, if $u = [f]$ and $x, y \in H$ then

$$(\varphi(u)^* x \mid y) = (x \mid \varphi(u)y) = (\varphi(u)y \mid x)^*$$

$$= \left(\int f(t)(U_t y \mid x)\, dt \right)^*$$

$$= \int f(t)^*(x \mid U_t y)\, dt$$

$$= \int \Delta(t) f(t^{-1})^* (x \mid U_{t^{-1}} y)\, dt$$

$$= \int \tilde{f}(t)(U_t x \mid y)\, dt = (\varphi(\tilde{u})x \mid y).$$

To complete the proof of (1), we must show that φ is multiplicative, that is, $\varphi(uv) = \varphi(u)\varphi(v)$ $(u, v \in L^1(G))$. Suppose $u = [f]$, $v = [g]$, and $uv = [h]$, and let $x, y \in H$. Since $h = f * g$ a.e., we have (by Fubini's theorem and the left invariance of Haar measure)

$$(\varphi(uv)x \mid y) = \int h(t)(U_t x \mid y) \, dt$$

$$= \int \left(\int f(s)g(s^{-1}t) \, ds \right)(U_t x \mid y) \, dt$$

$$= \int f(s)\left(\int g(s^{-1}t)(U_t x \mid y) \, dt \right) ds$$

$$= \int f(s)\left(\int g(t)(U_{st} x \mid y) \, dt \right) ds$$

$$= \int f(s)\left(\int g(t)(U_t x \mid U_s^* y) \, dt \right) ds$$

$$= \int f(s)(\varphi(v)x \mid U_s^* y) \, ds$$

$$= \int f(s)(U_s \varphi(v)x \mid y) \, ds$$

$$= (\varphi(u)\varphi(v)x \mid y).$$

(2) Let $x \in H$, $\|x\| = 1$; we seek $u \in L^1(G)$ such that $\varphi(u)x \neq \theta$. Since $(U_e x \mid x) = (Ix \mid x) = 1$ and the representation $t \mapsto U_t$ is continuous, there exists a neighborhood V of e such that

$$|(U_t x \mid x) - 1| \leq 1/2 \qquad \text{for all } t \in V.$$

We can suppose that V is a Baire set (e.g., a compact G_δ). Choose any $f \in \mathscr{F}_V$ (cf. 69.12) and let $u = [f]$. Then

$$(\varphi(u)x \mid x) - 1 = \int f(t)(U_t x \mid x) \, dt - \int f(t) \, dt$$

$$= \int_V f(t)[(U_t x \mid x) - 1] \, dt,$$

thus

$$|(\varphi(u)x \mid x) - 1| \leq \int_V f(t)|(U_t x \mid x) - 1| \, dt \leq \tfrac{1}{2} \int_V f(t) \, dt = \tfrac{1}{2},$$

whence $(\varphi(u)x \mid x) \neq 0$.

(3) From the formula (*), we have

$$(\varphi(u_s)x \mid y) = \int f_s(t)(U_t x \mid y) \, dt$$

$$= \int f(s^{-1}t)(U_t x \mid y) \, dt$$

$$= \int f(t)(U_{st} x \mid y) \, dt$$

$$= \int f(t)(U_t x \mid U_s^* y) \, dt$$

$$= (\varphi(u)x \mid U_s^* y) = (U_s \varphi(u)x \mid y)$$

for all $u = [f] \in L^1(G)$, $s \in G$ and $x, y \in H$.

(4) Let $(u_j)_{j \in J}$ be an approximate identity in $L^1(G)$, and let $t \in G$. By (3), we have

$$\varphi((u_j)_t) = U_t \varphi(u_j)$$

for all j; on the other hand, $\varphi(u_j) \to I$ strongly by (69.20) applied to φ, thus $\varphi((u_j)_t) \to U_t I = U_t$ strongly. ∎

Let us combine the preceding two theorems:

(69.22) Theorem. *The correspondences*

$$\varphi \mapsto U_{(\cdot)}$$

$$U_{(\cdot)} \mapsto \varphi$$

of Theorems (69.20) and (69.21) are mutually inverse bijections between the set of all nondegenerate ∗-representations φ of $L^1(G)$ and the set of all continuous unitary representations $t \mapsto U_t$ of G.

Proof. If $\varphi : L^1(G) \to \mathscr{L}(H)$ is a nondegenerate ∗-representation and if $U_{(\cdot)}$ is the continuous unitary representation of G derived from φ via (69.20), then it results from (4) of (69.20) that φ is the ∗-representation of $L^1(G)$ derived from $U_{(\cdot)}$ via (69.21).

Conversely, if $t \mapsto U_t$ is a continuous unitary representation of G and if φ is the ∗-representation of $L^1(G)$ derived from $U_{(\cdot)}$ via (69.21), then it results from (4) of (69.21) that $U_{(\cdot)}$ is the unitary representation of G derived from φ via (69.20). ∎

The pairing in the foregoing theorem reduces the study of unitary representations of G to the study of ∗-representations of $L^1(G)$. Under this pairing, irreducibility is preserved:

(69.23) Theorem. *If φ and $U_{(\cdot)}$ correspond as in (69.22), then*

$$\{\varphi(u) : u \in L^1(G)\}' = \{U_t : t \in G\}'$$

(the commutants are calculated in $\mathscr{L}(H)$, where H is the Hilbert space on which the representations act). In particular, φ is irreducible iff $U_{(\cdot)}$ is irreducible.

Proof. If $T \in \mathscr{L}(H)$ commutes with every U_t, then, for all $u = [f] \in L^1(G)$ and $x, y \in H$,

$$(T\varphi(u)x \mid y) = (\varphi(u)x \mid T^*y) = \int f(t)(U_t x \mid T^* y)\, dt = \int f(t)(TU_t x \mid y)\, dt$$

$$= \int f(t)(U_t Tx \mid y)\, dt = (\varphi(u)Tx \mid y),$$

thus T commutes with every $\varphi(u)$.

Conversely, suppose $T \in \mathscr{L}(H)$ commutes with every $\varphi(u)$. Fix $t \in G$ and let $(u_j)_{j \in J}$ be an approximate identity in $L^1(G)$. For all $j \in J$,

$$TU_t\varphi(u_j) = T\varphi((u_j)_t) = \varphi((u_j)_t)T = U_t\varphi(u_j)T;$$

since $\varphi(u_j) \to I$ strongly, it follows that $TU_t = U_t T$.

The final assertion of the theorem is immediate from the definitions (cf. (67.2), (67.4), (69.1)). ∎

(69.24) Definition. Let $u \mapsto T_u$ be the (faithful) left regular representation of $L^1(G)$ on the Hilbert space $L^2(G)$ (69.18). The set of operators

$$\{T_u + \lambda I : u \in L^1(G), \lambda \in \mathbb{C}\}$$

is a $*$-subalgebra of $\mathscr{L}(L^2(G))$ containing the identity operator I; we write $\mathscr{A}$ for its closure with respect to the norm topology of $\mathscr{L}(L^2(G))$. Thus $\mathscr{A}$ is a C^*-algebra with unity element I (43.7).

(69.25) Lemma. *Notation as in (69.24). Let $\Phi : \mathscr{A} \to \mathscr{L}(H)$ be a unital $*$-representation of $\mathscr{A}$ on a nonzero Hilbert space H, and let φ be the $*$-representation of $L^1(G)$ on H obtained by composing Φ with the left regular representation, i.e.,*

$$\varphi(u) = \Phi(T_u) \qquad (u \in L^1(G)).$$

Then

$$\{\varphi(u) : u \in L^1(G)\}' = \{\Phi(A) : A \in \mathscr{A}\}';$$

in particular, φ is irreducible iff Φ is irreducible.

Proof. The following conditions on $T \in \mathscr{L}(H)$ are equivalent: T commutes with $\varphi(u) = \Phi(T_u)$ for all u; T commutes with $\varphi(u) + \lambda I = \Phi(T_u + \lambda I)$ for all u and λ (recall that Φ is unital); $T_u + \lambda I \in \Phi^{-1}(\{T\}')$ for all u and λ; $\mathscr{A} \subset \Phi^{-1}(\{T\}')$ (recall that Φ is norm-continuous); $\Phi(\mathscr{A}) \subset \{T\}'$; T commutes with $\Phi(A)$ for all $A \in \mathscr{A}$. ∎

(69.26) Theorem. (Gel'fand-Raĭkov) *Every locally compact group G has a complete system of irreducible, continuous unitary representations.*

Proof. Let $s \in G$, $s \neq e$. Choose any $v \in L^1(G)$ such that $v_s \neq v$. {E.g., let $v = [f]$, where f is a continuous function with compact support such that $f(s^{-1}) \neq f(e)$.}

Set $w = v_s - v$. Since $w \neq 0$ and the left regular representation $u \mapsto T_u$ of $L^1(G)$ on $L^2(G)$ is faithful (69.18), we have $T_w \neq 0$. Therefore, by the Gel'fand-Naĭmark completeness theorem (67.27), applied to the C^*-algebra $\mathscr{A}$ of (69.24), there exists an irreducible unital $*$-representation

$$\Phi : \mathscr{A} \to \mathscr{L}(H),$$

on a suitable Hilbert space H, such that $\Phi(T_w) \neq 0$. Let $\varphi : L^1(G) \to \mathscr{L}(H)$ be the composed $*$-representation, i.e.,

$$\varphi(u) = \Phi(T_u) \qquad (u \in L^1(G)).$$

Since φ is irreducible (69.25) and not identically zero, it must be nondegenerate; the continuous unitary representation $U_{(\cdot)}$ of G paired with φ via (69.22) is also irreducible (69.23). Finally,

$$\varphi(v_s) - \varphi(v) = \varphi(v_s - v) = \varphi(w) = \Phi(T_w) \neq 0,$$

thus $U_s\varphi(v) = \varphi(v_s) \neq \varphi(v)$, whence $U_s \neq I$. ∎

Exercises

(69.27)   Prove (69.9).

(69.28)   The Banach algebra $L^1(G)$ has a unity element iff G is discrete.

(69.29)   What is the unitary representation of G associated with the left regular representation $L^1(G) \to \mathscr{L}(L^2(G))$?

(69.30)   Let G be a locally compact group, let H be a Hilbert space, and suppose $\varphi: L^1(G) \to \mathscr{L}(H)$ is a continuous algebra homomorphism such that (i) $\|\varphi\| \leq 1$, and (ii) H is the closed linear span of the vectors $\varphi(u)x$ ($u \in L^1(G)$, $x \in H$). Then φ is a $*$-representation of $L^1(G)$.

*70. The character group of an LCA group.

Let us specialize the results of the preceding section to locally compact *abelian* groups (briefly, LCA groups). Assume for the rest of the section that G is an LCA group.

The first observation is that the Banach algebra $L^1(G)$ is commutative. {This is immediate from the invariance of the Haar integral under the substitution $t \mapsto t^{-1}$ [cf. **10**, p. 259, Exer. 2].} It follows that if φ is a $*$-representation of $L^1(G)$ on a Hilbert space H, then the range of φ is a commutative $*$-subalgebra of $\mathscr{L}(H)$. Also, if $t \mapsto U_t$ is a unitary representation of G on H, then $\{U_t : t \in G\}$ is a commutative $*$-subset of $\mathscr{L}(H)$.

The foregoing remarks are decisive for reducibility. If $\mathscr{S}$ is an irreducible, commutative $*$-subset of $\mathscr{L}(H)$, then H must be one-dimensional (67.3). Identifying the one-dimensional Hilbert space with $\mathbb{C}$—and hence its unitary group with the circle group $\mathbb{T} = \{\lambda : |\lambda| = 1\}$—we have the following result:

(70.1) Proposition.   (i) *The irreducible, continuous unitary representations of G are the continuous homomorphisms $\alpha: G \to \mathbb{T}$.*

(ii) *The irreducible, nondegenerate $*$-representations of $L^1(G)$ are the $*$-epimorphisms $\varphi: L^1(G) \to \mathbb{C}$.*

(70.2) Definition.   A continuous homomorphism $\alpha: G \to \mathbb{T}$ is called a *continuous character* of G; the set of all continuous characters of G is denoted $\hat{G}$. Relative to the pointwise product of characters (i.e., $(\alpha\beta)(t) = \alpha(t)\beta(t)$), $\hat{G}$ is an abelian group, called the *character group* of G.

(70.3)   In view of (69.22), the continuous characters $G \to \mathbb{T}$ correspond bijectively to the $*$-epimorphisms $L^1(G) \to \mathbb{C}$. Explicitly, if α is the character corresponding to the $*$-epimorphism $\varphi: L^1(G) \to \mathbb{C}$, then

$$\varphi(u) = \int f(t)\alpha(t)\, dt \qquad (u = [f] \in L^1(G));$$

moreover, if u is any element of $L^1(G)$ such that $\varphi(u) \neq 0$, then

$$\alpha(t) = \varphi(u_t)/\varphi(u) \qquad (t \in G).$$

By the Gel'fand-Raïkov theorem (69.26), the character group $\hat{G}$ is a separating set of functions on G.

Condition (ii) of (70.1) can be sharpened: an algebra epimorphism $\varphi: L^1(G) \to \mathbb{C}$ is automatically a $*$-epimorphism. This property of $L^1(G)$, known as *symmetry*, is proved in the next proposition.

(70.4) Lemma. *If A is a Banach algebra and $\varphi: A \to \mathbb{C}$ is an algebra epimorphism, then φ is continuous and $\|\varphi\| \leq 1$.*

Proof. {It is not assumed that A has a unity element; and if A does have a unity element, it is not assumed to have norm one.}

Let $A_1 = A \oplus \mathbb{C}$ be the unitification of A constructed in (49.5). In particular, every $x \in A_1$ has a unique representation $x = a + \lambda u$ ($a \in A$, $\lambda \in \mathbb{C}$), and $\|x\| = \|a\| + |\lambda|$. Define $\psi: A_1 \to \mathbb{C}$ by the formula

$$\psi(a + \lambda u) = \varphi(a) + \lambda \qquad (a \in A, \lambda \in \mathbb{C}).$$

Evidently ψ is an algebra epimorphism that extends φ. Since A is isometrically embedded in A_1, it will suffice to show that ψ is continuous and $\|\psi\| = 1$.

Let M be the kernel of ψ. Since the quotient algebra A/M is isomorphic to the field $\mathbb{C}$, M is a maximal ideal; by the proof of (52.3), M is closed in A_1. Since ψ is a linear form, we conclude that ψ is continuous (22.2).

If $x \in A_1$ and $\|x\| = 1$, then

$$|\psi(x)|^n = |\psi(x^n)| \leq \|\psi\| \|x^n\| \leq \|\psi\|$$

for every positive integer n, therefore $|\psi(x)| \leq 1$; this shows that $\|\psi\| \leq 1$. On the other hand, it follows from $\psi(u) = 1 = \|u\|$ that $\|\psi\| \geq 1$. ∎

(70.5) Proposition. *If $\varphi: L^1(G) \to \mathbb{C}$ is an algebra epimorphism, then (i) φ is continuous, (ii) $\|\varphi\| \leq 1$, and (iii) $\varphi(\tilde{u}) = \varphi(u)^*$ for all $u \in L^1(G)$.*

Proof. (i) and (ii) follow from the lemma, and (iii) is then a special case of (69.30).

{The proof does not require that G be abelian—but we are not asserting that an arbitrary locally compact group G admits such a mapping φ, other than the trivial mapping $\varphi(u) = \int f(t)\,dt$ ($u = [f] \in L^1(G)$).} ∎

(70.6) Definition. We denote by $\mathscr{X}$ the set of all algebra epimorphisms $\varphi: L^1(G) \to \mathbb{C}$.

Summarizing: The sets $\hat{G}$ and $\mathscr{X}$ are in one-one correspondence. Explicitly, paired elements $\alpha \in \hat{G}$ and $\varphi \in \mathscr{X}$ are linked by the formula

$$\varphi(u) = \int f(t)\alpha(t)\,dt \qquad (u = [f] \in L^1(G)).$$

On the other hand, $\mathscr{X}$ is in one-one correspondence with the set $\mathscr{M}$ of all modular maximal ideals of $L^1(G)$ (52.19), the bijection $\mathscr{X} \to \mathscr{M}$ being defined by

$$\varphi \mapsto \ker \varphi.$$

Thus the sets $\hat{G}$ and $\mathscr{M}$ are also in one-one correspondence:

(70.7) Definition. If $M \in \mathcal{M}$ and if $\varphi_M \colon L^1(G) \to \mathbb{C}$ is the algebra epimorphism with kernel M, we write α_M for the continuous character of G corresponding to φ_M as in (70.3).

As noted in (52.19), $\mathcal{X}$ is a weak* locally compact subset of the dual space of $L^1(G)$; and $\mathcal{M}$ thereby acquires a locally compact topology. The Gel'fand transformation $L^1(G) \to \mathscr{C}^0(\mathcal{M})$ can be described explicitly in terms of (70.7):

(70.8) Proposition. *If* $u = [f] \in L^1(G)$ *and* $\hat{u}$ *denotes the Gel'fand transform of* u, *then*

$$\hat{u}(M) = \int f(t)\alpha_M(t)\, dt$$

for all $M \in \mathcal{M}$.

Proof. For fixed M, both of the mappings

$$u \mapsto \hat{u}(M) \quad \text{and} \quad u \mapsto \int f(t)\alpha_M(t)\, dt$$

are epimorphisms $L^1(G) \to \mathbb{C}$ with kernel M, hence they are identical. ∎

If $f \in \mathscr{L}^1(G)$, the functions on $\hat{G}$ defined by

$$\alpha \mapsto \int f(t)\alpha(t)\, dt \quad \text{and} \quad \alpha \mapsto \int f(t)\alpha(t)^*\, dt$$

are called the *Fourier cotransform* and *Fourier transform* of f [**26**, Ch. II, §1, No. 2]; in view of (70.8) they are, respectively, the Gel'fand transforms of $u = [f]$ and $v = [g]$, where $g(t) = f(t^{-1})$ $(t \in G)$. In this way, the theory of the Fourier transformation for an LCA group may be founded on the theory of the Gel'fand transformation, a stunning fulfillment of the idea of Gel'fand sketched in Section 0.

(70.9) Concluding apéritif: The topology on $\hat{G}$ induced by $\mathcal{X}$ is compatible with the group structure of $\hat{G}$. Thus $\hat{G}$ is itself an LCA group, and one can form $(\hat{G})^\wedge$. In fact, $(\hat{G})^\wedge = G$ (Pontrjagin duality theorem).

Hints, Notes, and References

(1.3), *(1.4)* The field $\mathbb{Q}$ of rational numbers may be given a topology that makes it a topological ring but not a topological field [20, Ch. III, §6, Exer. 20e].

(2.21) A theorem of R. Ellis [44, Th. 2].

(2.22) Cf. (6.8).

(4.10) Cf. [20, Ch. III, §2, Props. 13, 14], [35, p. 53, (12.11.2)].

(4.11) Cf. [20, Ch. III, §2, Exer. 16].

(4.13) For normed spaces, this result appears in a paper of F. Hausdorff [69, p. 302, Satz V].

(4.14) Hint: (4.9).

(5.3) Uniformity in topological groups was formulated by A. Weil, who also invented uniform structures [144]. For topological vector spaces, the idea of uniformity goes back to J. von Neumann [108].

(5.4) For the general theory of uniform structures, see [19, Ch. II].

(5.5)–*(5.20)* Cf. [20, Ch. III, §3, No. 1].

(5.21) Cf. [20, Ch. III, §3, No. 2].

(5.22) For the characterization of uniformizable spaces via continuous real functions, see [21, Ch. IX, §1, Th. 2].

(5.23) Cf. [19, Ch. II, §1, Prop. 3].

(5.24) Cf. [19, Ch. II, §4, Th. 1].

(5.25) An example is given in (10.8).

(5.26) Cf. [85, p. 159, Cor. and p. 172, Prob. Y], [76, p. 68, Th. 1].

(6.2) Cf. [14], [21, Ch. IX, §1, Prop. 2].

(6.3) See G. Birkhoff [14] and S. Kakutani [80]. See also the historical footnote on page 13 of Weil's monograph [144].

(6.6) The proof is given in (10.5).

(6.7) [144, p. 16].

(7.1) Cf. [19, Ch. II].

(7.10) Cf. [20, Ch. III, §3, Prop. 4], [76, p. 68, Th. 3].

(7.12) [34, p. 775].

(7.15) [20, Ch. III, §3, Th. 1].

(9.1) [21, Ch. IX, §3, Prop. 4].

(9.2) Better yet, see (9.4).

(9.3) Cf. (12.3).

(9.4) [20, Ch. III, §3, Exer. 9].

(10.2) [20, Ch. III, §3, Exer. 3].

(10.5) [88, Th. (1.5)]; cf. [21, Ch. IX, §3, Exers. 1, 2].

(10.8) Cf. [20, Ch. III, §3, Exer. 4].

(10.10) [88, Th. (2.3)].

(10.11) [58, p. 46, Th. 6].

(10.12) [88, Th. (1.3)].

(11.1) The concept of topological vector space apparently goes back to M. Fréchet [50, pp. 201–204]. The definition of a TVS in terms of axioms on the neighborhoods of zero (cf. Section 17) is due to von Neumann [108]. For historical remarks on the subject, see the paper of J. V. Wehausen [143].

(11.10) The notion of completeness in a general TVS was formulated by von Neumann [108]; see the note for (5.3).

(12.3) Cf. [24, Ch. IV, §4, Exer. 10b], [86, p. 195, Prob. 20D], [92, p. 124], [126, p. 192, Exer. 11].

(*12.4*)   See (13.9).

(*13.1*)   [8, p. 35].

(*13.3*)   This result, at least as it pertains to spaces of type (*F*), is attributed to S. Mazur by Banach [8, p. 232]. The proof given here is modeled on an argument of R. Arens [2, Th. 5]. Cf. [23, Ch. I, §3, Exer. 1].

(*13.6*)   The reference to Fréchet's paper [49] is given in Banach's book [8, p. 232]. Fréchet spaces (modern usage) are treated in the books of N. Bourbaki [25, p. 29], T. Husain [75], J. L. Kelley and I. Namioka [86], G. Köthe [92], and H. H. Schaefer [126].

(*13.7*)   The solution to this problem, posed in Banach's book [8, p. 232], is contained in a paper of V. L. Klee, Jr. [88, Th. (2.6)]. Klee's more general result is noted in (10.10).

(*14.1*)   [8, p. 10].

(*14.6*)   [8, p. 50, Th. 11].

(*15.10*)   Cf. [41, Part I, pp. 329–330].

(*15.12*)   [8, p. 9].

(*15.13*)   Cf. [82, p. 155, Lemma 3].

(*16.2*)   A basic exposition of normed spaces is the 1932 paper of F. Hausdorff [69], who called them 'linear metric spaces'; a principal aim of the exposition is to free certain arguments of superfluous completeness hypotheses.

(*16.11*)   See [69, p. 301, IV].

(*16.13*)   Cf. (13.8).

(*16.14*)   Minkowski's inequality is a special case of (39.4) for a measure space with finitely many (equally weighted) points.

(*17.12*)   [23, Ch. I, §1, Prop. 5]. The neighborhood axioms for a TVS were given by J. von Neumann [108, p. 4, Def. 2b].

(*17.14*)   Cf. (17.15), (33.17).

(*18.2*)   [23, Ch. I, §1, Prop. 6].

(*19.1*)   [19, Ch. I, §2, Prop. 4].

(*19.3*)   [23, Ch. I, §1, Prop. 15].

(*20.2*)   [23, Ch. I, §1, Def. 5].

(*20.10*)   Cf. [147].

(*20.11*)   An example is given in [66, §15]. See also [135, pp. 21–22].

(*21.10*)   [23, Ch. II, §6, No. 1].

(*21.17*)   Cf. [131, p. 289, Lemme].

(*21.18*)   Hint: Consider $f = g_0 - ih_0, f' = g_0' - ih_0'$ (cf. 21.13).

(*22.1*)   [23, Ch. I, §2, Prop. 2].

(*23.1*)   Cf. [23, Ch. I, §2, Th. 2]. The original reference is a paper of A. Tychonoff [141, p. 769]. For the special case of a finite-dimensional normed space, see F. Hausdorff [69, p. 298]. For a proof of the elementary fact that a complete subspace of a separated uniform space is closed, see [19, Ch. II, §3, Prop. 8].

(*23.7*)   This is proved in [23, Ch. I, §2, Prop. 3] under the hypothesis that *E* is separated. That this hypothesis can be omitted is noted in [35, pp. 60–61, Th. (12.13.2)].

(*23.10*)   This result goes back to F. Riesz [117, p. 78, Hilfs. 5], who proved it for normed spaces. See also Banach's book [8, p. 84, Th. 8]; the historical note on p. 236 of Banach's book refers to the footnote on p. 151 of his book. The proof given here follows that given by R. E. Edwards [43, p. 65], who attributes the proof to J.-P. Serre. The general result is given in Bourbaki [23, Ch. I, §2, Th. 3]; I am unable to trace the result further back.

(24.1)   [**19**, Ch. II, §3, No. 7].

(24.2)   [**19**, Ch. II, §3, Th. 2].

(24.3)   [**19**, Ch. II, §3, Th. 3].

(24.6)–(24.9)   [**20**, Ch. III, §3, No. 4].

(24.10)–(24.11)   [**20**, Ch. III, §3, Th. 2].

(24.13)   [**20**, Ch. III, §6, Th. 1].

(24.14)   [**23**, Ch. I, §1, No. 5].

(24.15)–(24.16)   [**19**, Ch. II, §4, No. 2].

(24.17)   [**20**, Ch. III, §3, Th. 1].

(24.18)   See [**22**, Ch. X, §3, Exer. 16]. The original reference is a paper of J. Dieudonné [**34**]. See also [**85**, pp. 211–212, Prob. Q].

(24.19)   See (24.11).

(25.1)   A general reference on convexity is the book of F. A. Valentine [**142**]. The present section draws most heavily on the exposition of Bourbaki [**23**, Ch. II, §1].

(25.17)–(25.22)   Cf. V. L. Klee, Jr. [**87**, Part I, (4.1)–(4.6)] and N. Bourbaki [**23**, Ch. II, §1, No. 6].

(25.27)   [**23**, Ch. II, §6, No. 2].

(25.28)   Cf. [**23**, Ch. II, §6, No. 2], [**92**, p. 164].

(25.30)   Cf. [**23**, Ch. II, §1, Exer. 8], [**142**, p. 14].

(26.2)   In this form the definition is due to J. von Neumann [**108**, p. 7, Def. 5].

(26.4)   The original reference is the paper of A. Kolmogoroff [**90**].

(26.5)   The notion of sequential convergence is definable (though generally useless) in any topological space: $t_n \to t$ means that for any neighborhood V of t, there exists an index n such that $t_k \in V$ for all $k \geq n$.

For historical remarks concerning the notion of boundedness, see the paper of J. V. Wehausen [**143**, p. 159, footnote], where the formulation of boundedness in (26.5) is attributed to Banach.

(26.7)   Indeed, no neighborhood of zero in (S) is bounded (cf. (15.8), (26.5)).

(26.8)   Hint: (26.5).

(27.1)–(27.14)   [**23**, Ch. II, §1, No. 4].

(27.15)   [**23**, Ch. II, §1, Cor. 2 of Prop. 15].

(28.1)   I am unable to trace fully the history of this result. The proof given here is taken from N. Bourbaki [**23**, Ch. II, §3, Th. 1]. For normed spaces, the result goes back to S. Mazur [**100**, p. 73, Satz 1]. Relevant references in between are the papers of J. Dieudonné [**32**] and V. L. Klee, Jr. [**87**]. Proofs of the result given here may also be found in the books of R. E. Edwards [**43**, p. 117, Th. 2.2.1], G. Köthe [**92**, p. 191], and A. P. Robertson and W. Robertson [**120**, p. 27, Th. 2].

(28.6)   The proof given here is the one given in Banach's book [**8**, p. 27, Th. 1], except that Zorn's lemma replaces a transfinite induction. For historical remarks, see also [**92**, p. 193].

(28.8)   The complex case is proved in a paper of H. F. Bohnenblust and A. Sobczyk [**16**].

(28.11)   [**8**, p. 29, Cor.].

(29.1)   [**74**, p. 257].

(29.3)   The systematic study of amenability of semigroups begins with the papers of M. M. Day ([**29**], [**30**]) and J. Dixmier [**36**]. A general reference is the book of F. P. Greenleaf [**63**]. These references may be consulted for the earlier history of the subject.

(29.5)   The proof is modeled on the argument given in Banach's book [**8**,

pp. 30–31]. A quite different proof has been given by J. Dixmier [**36**, Th. 2]. Still another proof can be derived from the Markov-Kakutani fixed point theorem [cf. **23**, Ch. II, Appendix, pp. 114–115].

(*29.6*)   A slightly different proof is given in Banach's book [**8**, p. 34].

(*29.8*)   Cf. Section 21.

(*29.9*)   If T is a semigroup with operation $(x, y) \mapsto x + y$, the semigroup T' opposite to T is defined to be the set T equipped with the operation $(x, y) \mapsto y + x$.

(*29.10*)   This was proved by M. M. Day [**30**, p. 529, Cor. 2].

(*29.11*)   In view of (29.10), the problem is to show that there exist groups that possess no left-invariant mean. Examples are given in the papers of M. M. Day [**29**, p. 290, No. 7] and J. Dixmier [**36**, Th. 4]; notably, the free group with two generators is not amenable.

(*29.12*)   As noted by Dixmier [**36**, Th. 2], the amenability of a compact group derives immediately from its Haar measure. Banach showed that a generalized limit could be used to prove the existence of Haar measure for metrizable locally compact groups (see Banach's Appendix to the book of S. Saks [**125**, p. 314]).

(*29.13*)   Cf. M. M. Day [**29**, Th. 6] and J. Dixmier [**36**, Ths. 2, 3].

(*29.14*)   By the Hahn-Banach theorem (28.6) there exists a linear form f such that $f(x) \le p(x)$ for all $x \in (m)$, but f need not satisfy the invariance property (i) of (29.6) [cf. **103**, p. 244, Exer. s].

(*29.15*)   For the additive group of integers, and the additive group of real numbers, the theorem is due to B. Sz.-Nagy [**137**]. For general amenable groups, the result is noted in the papers of Day [**29**, Th. 8], Dixmier [**36**, Th. 6], and M. Nakamura and Z. Takeda [**106**]. For a consideration of the similarity question independently of amenability, see R. V. Kadison [**79**]. A general reference on amenability in topological groups is the book of F. P. Greenleaf [**63**].

(*29.16*)   [**148**, p. 271, Exer. 11].

(*30.3*)   Cf. [**23**, Ch. II, §3, Defs. 1, 2].

(*30.6*)   [**23**, Ch. II, §3, Prop. 1].

(*30.12*)   [**23**, Ch. II, §3, Cor. of Prop. 3].

(*31.1*)–(*31.5*)   [**23**, Ch. II, §1, No. 5].

(*32.2*)   Cf. [**132**, p. 940, Lemma 2.2].

(*32.3*)   Cf. [**23**, Ch. II, §3, Prop. 5].

(*32.4*)   Cf. [**23**, Ch. II, §3, Prop. 6]. The prototype theorem is due to M. Krein [**93**].

(*32.5*)   If f is a linear form on the Banach space $\mathscr{B} = \mathscr{B}_{\mathbb{R}}(T)$, T any nonempty set (16.15), then the following conditions on f are equivalent: (a) inf $x \le f(x) \le$ sup x for all $x \in \mathscr{B}$; (b) $f(1) = 1$, and $x \ge 0$ implies $f(x) \ge 0$. Such a linear form satisfies the condition $|f(x)| \le \|x\|_\infty$ and is therefore continuous.

(*33.1*)–(*33.11*)   [**23**, Ch. II, §2]. The concept of local convexity is due to von Neumann [**108**].

(*33.12*)   [**23**, Ch. II, §3, Cor. 3 of Prop. 4].

(*33.17*)   [**23**, Ch. II, §2, Prop. 1 and Cor.], [**25**, §3, Nos. 14, 15], [**43**, pp. 429–430].

(*33.18*)   [**23**, Ch. II, §2, No. 4], [**25**, §3, No. 15], [**86**, p. 148, Prob. C].

(*34.1*)–(*34.10*)   [**23**, Ch. II, §3, No. 3]. For the property of uniform spaces cited in the proof of (34.1), see [**19**, Ch. II, §4, Prop. 4]. For normed spaces, (34.3) goes back to S. Mazur [**100**, p. 80, Satz 3].

(*34.13*)   An example of a locally convex TVS that admits no nonzero continuous linear forms: E any vector space over $\mathbb{K}$, with $\varnothing$ and E as the only open sets. See also [**23**, Ch. II, §3, Exer. 14].

(*35.1*)   [**23**, Ch. II, §4, Prop. 1].

(*35.3*)   [**23**, Ch. II, §4, Prop. 2]. The general topological result cited in the proof (a uniform space is precompact iff it is totally bounded) is proved in [**19**, Ch. II, §4, Th. 3].

(*35.6*)   [**23**, Ch. II, §4, Prop. 2].

(*35.7*)   Cf. [**23**, Ch. II, §4, Cor. of Prop. 2]. The prototype result is due to S. Mazur [**99**].

(*35.8*)   Cf. [**114**, Th. 9.3]. Hint: (25.30).

(*36.3*)   The term 'face' is appropriated from lectures of P. R. Halmos (1952). Other terms in use: 'supporting set' [**84**]; 'support' [**86**, p. 130]; 'extremal subset' [**149**, p. 362]. A linear variety V such that $V \cap A$ is a face of A is called a 'support variety' [**23**, Ch. II, §4, No. 2] or a 'manifold of support' [**105**, §3, No. 9, III].

(*36.8*)   [**23**, Ch. II, §4, Prop. 3].

(*36.9*)   Originally proved by M. Krein and D. Milman [**94**] for a weak* compact convex set in the dual of a Banach space. The proof given here is due to J. L. Kelley [**84**]; cf. [**23**, Ch. II, §4, Th. 1], [**86**, p. 131].

(*36.10*)   [**23**, Ch. II, §4, Prop. 4].

(*36.12*)   [**105**, §3, No. 9].

(*36.13*)   [**114**, Th. 4.4]. Cf. (41.17).

(*36.14*)   The term 'uniformly convex' is due to J. A. Clarkson [**28**]. Cf. [**31**, p. 112].

(*37.1*)–(*37.4*)   Cf. [**108**, Th. 25], [**23**, Ch. II, §5, Prop. 3].

(*37.5*)   [**23**, Ch. II, §5, No. 4].

(*37.17*)   Cf. [**23**, Ch. II, §5, No. 4, Remark].

(*37.19*)   Cf. [**108**, Th. 26], [**23**, Ch. II, §5, Prop. 4].

(*37.21*)   [**23**, Ch. II, §5, Prop. 5].

(*37.22*)   [**23**, Ch. II, §5, Th. 1 and §6, Th. 1].

(*37.23*)   Cf. [**35**, p. 66, (12.14.12)].

(*37.25*)–(*37.26*)   [**23**, Ch. II, §5, Prop. 7]. Cf. (39.7).

(*37.27*)   [**23**, Ch. II, §5, No. 4, Remark].

(*37.28*)   Cf. [**24**, Ch. IV, §3, Def. 1]. Hints: (i) Cf. (26.6); (ii) cf. (37.17) and (37.21); (iii) note that $\{x\} \in \mathfrak{S}$ for all $x \in E$.

(*37.30*)   [**26**, Ch. I, §4, No. 1], [**43**, pp. 429–430].

(*38.1*)–(*38.17*)   [**33**], [**24**, Ch. IV, §1].

(*39.4*)   Cf. [**119**, p. 43].

(*39.6*)   Cf. [**119**, p. 58].

(*40.7*)   [**69**, p. 299, II].

(*40.10*)   The result goes back to H. Hahn [**64**, p. 218, Satz V]. See also S. Banach [**6**, p. 213], [**8**, p. 55, Th. 3].

(*40.26*)   The equivalence of (a) and (b) is elementary. The proof that (c) implies (a) rests on the uniform boundedness principle (47.3).

(*41.5*)–(*41.6*)   The original reference is a paper of P. Jordan and J. von Neumann [**78**]. For recent expositions see the books of A. Wilansky [**148**, p. 124, Th. 1], A. N. Kolmogorov and S. V. Fomin [**91**, pp. 160–162], and M. Schechter [**129**, p. 244, Th. 1.1].

(*41.17*)   (iii) Cf. [**12**]. A more general result is proved in a paper of V. L. Klee, Jr. [**89**, proof of Th. 2.1].

(iv) Cf. (36.11).

(v) Cf. (36.13).

(42.7) In this form the result appears in B. Sz.-Nagy's monograph [**136**, p. 8]; the essential details of the proof go back to F. Riesz [**118**].

(42.12) For separable Hilbert spaces, this result goes back to M. Fréchet [**48**, p. 439, Th.]. The proof given here is that of F. Riesz [**118**].

(43.12) [**119**, p. 399].

(43.13) Hint: (43.12).

(44.2) The result goes back to H. Hahn [**64**, p. 217, Satz III] (strictly speaking, Hahn considered only closed linear subspaces of Banach spaces, but, via the technique of completion, this limitation is inessential). See also S. Banach [**6**, p. 213, Th. 2], [**8**, p. 55, Th. 2].

(44.3)–(44.4) Cf. [**95**, p. 574, Th. 12′], [**33**, p. 124, Th. 17].

(44.8) Cf. [**31**, p. 39, Lemma 1].

(44.11) (iii) Cf. [**31**, p. 39, Lemma 1].

(44.12) The original references are N. Bourbaki [**18**, Cor. of Th. 1], L. Alaoglu [**1**] and J. Dieudonné [**33**, p. 128, Th. 22]. The first reference is an announcement of results for which detailed proofs are given in the third.

(44.13) Hint: (11.7), (16.7), (44.4), and (40.10). See S. Banach [**6**, p. 215, Th. 4], [**8**, p. 57, Lemme].

(44.14) This result goes back to F. Riesz. Cf. [**119**, pp. 78 and 211].

(44.15) Hint: (44.12) and (40.10). Cf. [**8**, p. 185, Th. 9].

(44.16) The original reference is [**95**, p. 564, Th. 5]. Cf. [**41**, Part I, pp. 427–429], [**24**, Ch. IV, §2, Th. 5], [**86**, p. 177].

(45.3) [**33**, p. 137, Th. 35], [**24**, Ch. IV, §5, Prop. 5].

(45.4) In the generality given here, the theorem is announced in a paper of N. Bourbaki [**18**, Th. 3] and the proof is given in a paper of J. Dieudonné [**33**, p. 130, Th. 24]. For the case of real scalars, a variant of the result appears in a paper of H. H. Goldstine [**62**, p. 128, Th. IV]. For separable spaces, the theorem goes back to Banach's book [**8**, p. 123, Th. 3 and p. 189, Th. 13], with 'weak compactness' stated in sequential form. A proof that the closed unit ball of a reflexive Banach space is weakly (sequentially) compact appears in a paper of V. Gantmakher and V. Šmulian [**51**, p. 92, Th. 1]. {The debate between weak compactness and weak sequential compactness is ended by a theorem of W. F. Eberlein [**42**]: a norm-bounded, weakly closed subset of a Banach space is weakly compact iff it is weakly sequentially compact (cf. [**31**, p. 51, Th. 3], [**92**, p. 316], [**149**, p. 141]).} The evolution of the theorem mirrors the gradual ascendency of the general topological language.

(45.5) Necessity: A routine application of (44.3), (44.4).

Sufficiency: Note that E is a Banach space (16.12). Assuming $\varphi \in E''$, we seek $x \in E$ with $\varphi = x''$. By (44.4) and the reflexivity of E/M, there exists $z \in E$ such that $\varphi = z''$ on $M^{\perp}$. Then $\varphi_0 = \varphi - z''$ vanishes on $M^{\perp}$, hence induces a bounded linear form on $E'/M^{\perp}$; then (44.3) and the reflexivity of M yield $y \in M$ such that $\varphi_0 = y''$ on E'. Thus $\varphi = (y + z)''$.

(45.7) Only if: Obvious.

If: Infer from the reflexivity of E' that $E_0^{\perp} = \{0\}$, then apply (44.4) to the closed linear subspace E_0 of E''.

(45.8) The equivalence of (b) and (c) is immediate from (45.7).

(a) implies (c): Obviously $(E_0)^{\perp} = \{0\}$, thus (44.3) yields $(E_0)' = E'''$.

(c) implies (a): If $\Phi \in (E_0')'$ and $\Phi = 0$ on E_0, one sees from the reflexivity of E' that $\Phi = 0$, thus E_0 is dense in E'' by the Hahn-Banach theorem (33.12).

(45.9) [**102**]. Cf. [**111**], [**33**, p. 139].

(46.1)   Cf. N. Bourbaki [**21**, Ch. IX, §5].

(46.3)   The proof is taken from the book of E. Hewitt and K. Stromberg [**71**, p. 68]; the argument goes back to R. Baire [**4**, p. 65].

(46.4)–(46.5)   [**21**, Ch. IX, §5, No. 3].

(46.7)   See S. Banach and H. Steinhaus [**9**, p. 53, Lemme 2].

(46.9)   (i) [**21**, Ch. IX, §5, Prop. 3].

(iii) follows at once from (i) and (ii), via the one-point compactification. For a unified proof of (iii) and (46.3), see [**21**, Ch. IX, §5, Th. 1].

(46.11)   [**21**, Ch. IX, §5, Prop. 5].

(47.1)   See S. Banach and H. Steinhaus [**9**, p. 53, Lemme 3]; also S. Banach [**8**, p. 80, Th. 5]. For earlier historical indications, and an elegant proof based directly on completeness and avoiding the notion of category, see F. Hausdorff [**69**, p. 304, VII].

(47.2)   The special case of a pointwise convergent sequence of continuous linear forms on a Banach space goes back to Banach's 1920 thesis [cf. **5**, p. 157, Th. 5].

(47.3)   The argument goes back to S. Banach [**7**, p. 231, Lemme 4].

(47.6)   Hint: Let E be the linear subspace of (m) consisting of all finitely nonzero sequences, and, for each positive integer n, let f_n be the linear form on E such that $f_n(x) = n\lambda_n$ for $x = (\lambda_n)$.

(47.7)   Cf. the proof of (47.5).

(47.8)   See S. Banach and H. Steinhaus [**9**, p. 53, Lemme 4]; also S. Banach [**8**, p. 79, Th. 3].

(48.1)   See S. Banach [**8**, p. 38, item (1) in the proof of Th. 3]. The proof given here is taken from Bourbaki [**23**, Ch. I, §3, Th. 1]. For Banach spaces, the result goes back to J. Schauder [**128**, p. 6, Satz 2]. For a systematic study of theorems of this type, see the monograph of T. Husain [**75**].

(48.4)   See S. Banach [**8**, p. 41, Th. 5].

(48.5)   See S. Banach [**8**, p. 41, Th. 6].

(48.6)   See S. Banach [**8**, p. 41, Th. 7].

(48.7)   [**23**, Ch. I, §3, Cor. 4 of Th. 1].

(48.8)   [**23**, Ch. I, §3, Cor. 3 of Th. 1].

(49.1)   For a historical note on the terminology, see the book of E. Hille [**72**, p. 12].

(49.14)   R. Arens [**2**]. Cf. (13.3).

(50.9)   If y is a left inverse for x, then so is $y + (1 - xy)$.

(50.10)   Cf. [**77**, p. 7].

(50.11)   (iii) Cf. (48.12).

(iv) See (57.19), (57.20).

(50.12)   [**26**, Ch. I, §2, Prop. 5]. Hint: (50.5), (50.7).

(50.14)   (ii) Cf. (50.12).

(51.8)   The original references are I. M. Gel'fand [**52**, Satz 3] and S. Mazur [**101**]. For an elementary proof avoiding complex function theory, see [**116**, p. 38, Th. 1.7.1 and p. 40].

(51.13)   Hint: The closed linear span of the vectors $x_0, x_1, x_2, \ldots$ is invariant under every operator in $\mathscr{B}$.

(51.14)   (i) Cf. (51.13).

(ii) See (55.2).

(51.15)   [**116**, p. 40, Th. 1.7.6].

(51.16)   Cf. [**26**, Ch. I, §1, Prop. 1].

(51.17)   Cf. (49.12).

(51.19)   (i)–(iv) [**77**, p. 9, Th. 1].

(v) [**116**, p. 57, Th. 2.3.5].

(52.1)   For the general theory of commutative Banach algebras, see the books of I. M. Gel'fand, D. A. Raĭkov, and G. E. Šilov [**57**], M. A. Naĭmark [**105**, Ch. III], and C. E. Rickart [**116**, Ch. III].

(52.13)   [**52**, Satz 10].

(52.14)   [**116**, p. 43, Cor. 2.1.4].

(52.15)   Hint (Chinese remainder theorem): Note that $M_j \not\supset \bigcap_{i \neq j} M_i$.

(52.16)   Hint: A character of A is determined by its value at x. Cf. [**35**, p. 287, (15.3.6)].

(52.17)   Hint: Drop down to the full subalgebra $\{x, y\}''$.

(52.18)   The notations are taken from the book of J. Dieudonné [**35**, pp. 273–274].

(ii) Concerning complete regularity, see N. Bourbaki [**21**, Ch. IX, §1, Th. 2 and Def. 4]. For an application of the Gel'fand theory to general topology, see Section 64.

(52.19)   [**116**, p. 119, Th. 3.1.20].

(53.6)   Cf. [**35**, p. 278, (15.2.3.2)].

(53.10)   Cf. the books of N. Bourbaki [**26**, Ch. I, §4, Th. 3], J. Dieudonné [**35**, Ch. XV, §2, Exers. 11, 12], N. Dunford and J. T. Schwartz [**41**, Part I, pp. 568–570], E. Hille [**73**, p. 300, Th. 9.4.2], E. Hille and R. S. Phillips [**74**, p. 168, Th. 5.2.5], E. R. Lorch [**97**, p. 105, Th. 7-1], C. E. Rickart [**116**, p. 157, Th. 3.5.1], M. Schechter [**129**, p. 143], and A. E. Taylor [**139**, p. 290, Th. 5.6-A].

(55.1)   [**52**, Satz 8'].

(55.3)   (i) Cf. (49.7).

(ii) [**116**, p. 10, Th. 1.4.1].

(55.6)   Cf. (51.18).

(56.1)   For a historical note on the concept of TD Z, see the book of C. E Rickart [**116**, p. 27].

(56.3)   Cf. [**74**, p. 129, Th. 4.11.1].

(56.5)   Cf. [**26**, Ch. I, §2, Prop. 6], [**74**, p. 130, Th. 4.11.2], [**116**, p. 33, Th. 1.6.12].

(56.6)   Cf. [**26**, Ch. I, §2, Cor. of Prop. 6], [**74**, p. 130, Cor. 1].

(56.7)   Cf. [**116**, p. 185, Th. 4.1.9].

(56.8)   (ii) [**116**, p. 21, Lemma 1.5.2].

(56.9)   Hint: If x is singular, then its minimal polynomial $p(t)$ is divisible by t.

(56.10)   Cf. [**26**, Ch. I, §2, Prop. 4], [**35**, Ch. XV, §2, Exer. 3c], [**116**, p. 22, Th. 1.5.4].

(56.12)   Immediate from (56.5).

(57.2)   [**67**, p. 37].

(57.9)–(57.12)   Cf. [**26**, Ch. I, §2, Exer. 12], [**35**, Ch. XV, §2, Exer. 8].

(57.13)   The classical nomenclature is applicable to a linear mapping $T: D \to E$, where E is a normed space, D is a dense linear subspace of E, and T is not necessarily continuous (cf. [**61**, p. 71, Def. II.5.2], [**135**, p. 128, Def. 4.1] [**139**, p. 264]). On the other hand, the definitions in Section 57 are framed in the setting of the algebra $\mathscr{L}(E)$. This makes for some discrepancies in usage. In the following 'remarks on concordance,' the notations are decorated with circumflexes to distinguish them from the notations introduced in Section 57. With T as described above, the 'resolvent set' $\hat{\rho}(T)$ is defined to be the set of all complex

numbers λ such that $T - \lambda I$ is bounded below and its range is dense in E. (If $D = E$ and T is continuous—so that $\rho(T)$ is also defined, in terms of invertibility in $\mathscr{L}(E)$—then $\hat{\rho}(T) \supset \rho(T)$; if, in addition, E is a Banach space, it is elementary that $\hat{\rho}(T) = \rho(T)$ (40.28).) The 'spectrum' $\hat{\sigma}(T)$ is defined to be $\mathbb{C} - \hat{\rho}(T)$. (Thus $\hat{\sigma}(T) \subseteq \sigma(T)$ in the parenthetical special case, with equality when E is a Banach space.) The definitions of $\hat{\sigma}_p(T)$, $\hat{\sigma}_{\text{com}}(T)$, $\hat{\sigma}_{\text{ap}}(T)$, and $\hat{\sigma}_r(T)$ run the same as those in Section 57. The 'continuous spectrum' $\hat{\sigma}_c(T)$ is defined to be the set

$$\hat{\sigma}_{\text{ap}}(T) - \{\hat{\sigma}_p(T) \cup \hat{\sigma}_{\text{com}}(T)\};$$

this is in fact the same as the set

$$\hat{\sigma}(T) - \{\hat{\sigma}_p(T) \cup \hat{\sigma}_{\text{com}}(T)\},$$

because if $\lambda \in \hat{\sigma}(T)$ and $T - \lambda I$ has dense range, then $T - \lambda I$ must fail to be bounded below, i.e., $\lambda \in \hat{\sigma}_{\text{ap}}(T)$. Thus

$$\hat{\sigma}(T) = \hat{\sigma}_p(T) \cup \hat{\sigma}_c(T) \cup \hat{\sigma}_r(T),$$

the terms on the right being mutually disjoint; the desire to imitate this formula is the motivation for the definition of $\sigma_c(T)$ given in (57.13). In the special case that E is a Banach space, $D = E$ and T is continuous, all is harmonious (57.34).

(*57.15*) [**61**, p. 62, Lemma II.4.1].

(*57.16*) [**61**, p. 63, Th. II.4.4, (ii)], [**139**, p. 234, Ths. 4.7–B, C].

(*57.17*) For the equivalence of (a) and (b), see [**61**, p. 73, Th. II.6.3].

(*57.18*) [**61**, p. 63, Th. II.4.4, (i)], [**139**, p. 233, Th. 4.7-A].

(*57.19*) For the equivalence of (a) and (b), cf. [**139**, p. 251, Exer. 6].

(*57.29*) [**67**, p. 225, Solution 58].

(*57.30*) Hint: (57.10), (57.29).

(*57.31*) Hint: (57.11).

(*57.32*) [**61**, p. 62, Lemma II.4.2].

(*57.33*) [**67**, p. 227, Solution 62].

(*57.34*) (b) implies (c): (40.28).

(c) implies (a): (48.1).

(*57.35*) [**67**, p. 40, Prob. 64].

(*57.37*) (i) [**67**, p. 110, Prob. 166]; this result is known as the Toeplitz-Hausdorff theorem.

(ii) [**67**, p. 111, Prob. 169].

(iii) Cf. [**67**, p. 112, Prob. 171], [**135**, p. 327, Th. 8.14].

(*57.38*) [**66**, §29, Th. 2].

(*58.1*) Standard references for C^*-algebras are the books of J. Dixmier [**38**], M. A. Naĭmark [**105**], C. E. Rickart [**116**], and S. Sakai [**124**]; the Bourbaki term is 'algèbre stellaire' [**26**, Ch. I, §6, Def. 3].

(*58.7*) [**83**, Appendix], [**116**, p. 186, Lemma 4.1.13], [**38**, p. 7, Prop. 1.3.8], [**124**, p. 2, Prop. 1.1.7], [**26**, Ch. I, §6, Prop. 2].

(*58.9*) Cf. [**116**, p. 188, Th. 4.1.19].

(*59.2*) The original reference is [**54**]. See also [**26**, Ch. I, §6, Th. 1], [**38**, p. 9, Th. 1.4.1], [**57**, §8, Th. 1], [**105**, §16, Th. 2], [**116**, p. 190, Th. 4.2.2], [**124**, p. 4, Th. 1.2.1].

(*60.7*) Hint: Cayley transform (see the proof of (66.21)).

(*60.9*) Hint: (58.7).

(*60.10*) [**38**, p. 17, Prop. 1.8.2], [**105**, §24, Th. 6], [**116**, p. 249, Th. 4.9.2].

(*60.11*) (i) [**38**, p. 16, Prop. 1.8.1], [**105**, §24, Th. 3], [**116**, p. 241, Th. 4.8.5].

(ii) [**38**, p. 18, Cor. 1.8.3].

(*60.13*)   An easy variation on the proof of (60.1). More generally, see [116, p. 188, Th. 4.1.20].

(*61.8*)   [38, p. 12, Prop. 1.6.1], [105, §24, Th. 4], [116, p. 243, Th. 4.8.9], [124, p. 8, Th. 1.4.4].

(*61.9*)   A theorem of H. F. Bohnenblust and S. Karlin [15, p. 226, Th. 10]. Cf. [37, p. 50, Lemme 5], [38, p. 25, Prop. 2.1.9], [116, p. 247, Th. 4.8.16], [124, p. 9, Prop. 1.5.2].

(*61.11*)   [38, p. 32, Prop. 2.4.4], [104, §6, Th. 3], [105, §17, Th. 2], [116, p. 201, Th. 4.3.7].

(*61.13*)   [104, §12, No. 2], [105, §23, No. 2]. See also (62.4).

(*61.14*)   The most commonly used term is 'numerical range' [cf. 17], but this has a drawback. If $T \in \mathcal{L}(H)$, H a Hilbert space, then $\Sigma_{\mathcal{L}(H)}(T)$ is the closure of what has been called the numerical range for decades, namely the set $W(T) = \{(Tx|x) : x \in H, \|x\| = 1\}$ (57.37). A compromise (that didn't catch on) is to call $\Sigma(a)$ the 'closed numerical range' of a [13]. The term 'numerical status' is coined in desperation.

(*61.19*)   [38, p. 31, Prop. 2.4.1], [104, §6, Th. 3], [105, §17, Th. 2], [116, p. 206, Th. 4.4.3].

(*61.20*)   (i) Hint: (60.11).

(ii) [15, p. 228, Th. 12], [13, p. 501, Th. 3].

(*61.21*)   (i) Hint: (62.1), (61.20), (57.37). Cf. [17, p. 88, Th. 1].

(ii) Hint: Kreĭn-Mil′man theorem (36.9). Cf. [13, p. 501, Cor. 1].

(iii) Cf. [17, p. 54, Th. 14].

(*62.1*)   The original reference is [54]. See also [38, p. 39, Th. 2.6.1], [104, §12, Cor. of Th. 3], [105, §24, Th. 5], [116, p. 244, Th. 4.8.11], [124, p. 41, Th. 1.16.6].

(*62.2*)   Cf. (58.7).

(*62.3*)   Hint: If A is represented as an algebra of operators on H, then A_n may be represented as an algebra of operators on the orthogonal sum of n copies of H (cf. [37, Ch. I, §2, No. 3], [105, §5, No. 15]).

(*62.4*)   In addition to the references for (62.1), see [83, Ch. 12].

(*63.4*)   Cf. (51.17).

(*65.1*)   Cf. [26, Ch. I, §6, Prop. 5], [116, p. 241, Th. 4.8.7].

(*65.3*)   Cf. [26, p. 69], [116, p. 241, Th. 4.8.7], [11, p. 111, Th. 39].

(*65.4*)   [26, Ch. I, §6, Cor. 2 of Prop. 6].

(*65.5*)   [26, Ch. I, §6, Prop. 6].

(*65.7*)   Cf. [35, p. 305, (15.4.15)].

(*65.9*)   Worked out in the proof of (66.21). Cf. (60.7).

(*65.10*)   [81, p. 207, Cor. 3.3].

(*65.11*)   Cf. [66, §41, Th. 2], [67, p. 306], [123, p. 300, Th. 12.16].

(*65.12*)   [68].

(*65.13*)   (i) Proved in (61.5).

(ii) If $a^* = a$ and $\|a\| \le 1$, then $u = a + i(1 - a^2)^{1/2}$ is unitary and $u + u^* = 2a$.

(*66.28*)   Hint: $\sigma(U) = \Delta_1$ [cf. 67, p. 230, Solution 67].

(*66.29*)   Hint: Calculate $\|T^n\|$ (cf. 55.1).

(*66.30*)   (i) [47, p. 369, (i)].

(ii), (iii) [110, p. 266, (3.4)].

(*66.32*)   (i) [110, p. 279, (B)].

(ii) [146, p. 77, Example].

(iv) [146, p. 78, Th. 7.8].

(v) [**146**, pp. 77–78, Ths. 7.6, 7.7], [**122**, p. 390, Exer. 1].

(vi) [**146**, p. 78, Th. 7.9], [**112**, p. 122].

(*66.34*)   [**110**, p. 274].

(*66.35*)   [**110**, p. 276].

(*66.36*)   [**110**, p. 276].

(*67.9*)   Cf. [**35**, p. 312, (15.6.3)].

(*67.16*)   Cf. [**116**, p. 213], [**35**, p. 314, (15.6.9)].

(*67.18*)   Cf. [**35**, p. 315, (15.6.10)], [**105**, §19, Th. 1], [**116**, p. 215, Th. 4.5.4].

(*67.20*)   Cf. [**38**, p. 37, Prop. 2.5.4], [**105**, §19, Th. 2].

(*67.24*)   [**105**, §19, Th. 2 and §19, No. 4, I], [**116**, p. 223, Th. 4.6.4].

(*67.27*)   The original reference is [**55**]. See also [**38**, p. 41, Th.], [**57**, §49, Th. 4], [**104**, §10, Th. 2], [**105**, §19, Th. 3], [**116**, p. 225, Th. 4.6.7].

(*67.29*)   [**96**, p. 96], [**105**, §10, No. 2, IV], [**116**, p. 218].

(*67.30*)   Cf. [**83**, Th. 11.2], [**116**, p. 214, Th. 4.5.2].

(*67.31*)   Cf. [**83**, Th. 11.3], [**116**, p. 214, Th. 4.5.2].

(*68.20*)   This lovely proof goes back to von Neumann ([**107**, Satz 5], [**109**]), who called the ultrastrong topology the 'strongest topology.' See also [**59**, p. 99, Prop. 22].

(*68.21*)   Cf. [**37**, p. 41, Th. 2].

(*68.22*)   (i) [**37**, p. 34].

(ii) Cf. Tihonov's theorem (23.3).

(*68.23*)   Cf. [**37**, p. 38, Th. 1]. Hint: (34.3) and (68.9).

(*68.24*)   (i) [**37**, p. 35, Lemme 2]. Hint: (68.9).

(ii) Cf. (68.23).

(*68.25*)   [**37**, p. 54, Cor. 1].

(*68.26*)   (i) [**37**, p. 30].

(ii) [**37**, p. 34].

(iii) [**37**, p. 32].

(iv) Hint: If U and V are unitary operators and x is any vector, then
$$\| Ux - Vx \|^2 = 2\|x\|^2 - 2\,\mathrm{Re}\,(Ux\,|\,Vx).$$

(*68.27*)   [**37**, p. 38, Th. 1].

(*68.28*)   [**124**, p. 41, Th. 1.16.7].

(*69.20*)   The proof is modeled on an argument of D. A. Raĭkov [**115**, §5, Th. 4]. See also [**26**, Ch. II, §1, Prop. 1], [**38**, p. 253, Prop. 13.3.4], [**57**, §21, Th. 2], [**70**, Vol. I, p. 338, Th. 22.7], [**96**, p. 128, Th. 32C], [**105**, §29, Th. 1], [**121**, p. 7, Th. 1.2.2], [**133**, p. 263, Th. 10.1'].

(*69.26*)   The original reference is [**56**]. See also [**38**, p. 262, Cor. 13.6.6], [**70**, Vol. I, p. 343, Th. 22.12], [**105**, §29, Th. 2].

(*69.27*)   Cf. [**70**, Vol. I, p. 285, Th. 20.4], [**96**, p. 118, Th. 30C], [**105**, §28, No. 2, I], [**133**, p. 199, Scholium 7.1].

(*69.28*)   Cf. [**96**, p. 123, Th. 31D], [**105**, §28, No. 1].

(*69.30*)   In the proof of (69.20), the hypothesis that φ is involution-preserving is used only to show that φ is continuous and $\|\varphi\| \leq 1$. But the right side of the formula in (4) defines a ∗-representation by (69.21).

(*70.9*)   Cf. [**26**, Ch. II, §1, Th. 2], [**57**, §26, Th. 1], [**70**, Vol. I, p. 378, Th. 24.8] [**96**, p. 151, 37D], [**105**, §31, Th. 5], [**113**, Sect. 40, Th. 52], [**115**, §12, Cor.]

Bibliography

[1] ALAOGLU, L.: Weak topologies in normed linear spaces. *Ann. of Math.* (2) **41**, 252–267 (1940).

[2] ARENS, R.: Linear topological division algebras. *Bull. Amer. Math. Soc.* **53**, 623–630 (1947).

[3] BACHMAN, G. and NARICI, L.: *Functional analysis.* New York: Academic 1966.

[4] BAIRE, R.: Sur les fonctions de variables réelles. *Ann. Mat. Pura Appl.* (3) **3**, 1–123 (1899).

[5] BANACH, S.: Sur les opérations dans les ensembles abstraits et leur application aux équations intégrales. *Fund. Math.* **3**, 133–181 (1922).

[6] BANACH, S.: Sur les fonctionelles linéaires. *Studia Math.* **1**, 211–216 (1929).

[7] BANACH, S.: Sur les fonctionelles linéaires. II. *Studia Math.* **1**, 223–239 (1929).

[8] BANACH, S.: *Théorie des opérations linéaires.* Warsaw 1932. Reprinted New York: Chelsea 1955.

[9] BANACH, S. and STEINHAUS, H.: Sur le principe de la condensation de singularités. *Fund. Math.* **9**, 50–61 (1927).

[10] BERBERIAN, S. K.: *Measure and integration.* New York: Macmillan 1965. Reprinted New York: Chelsea 1970.

[11] BERBERIAN, S. K.: *Notes on spectral theory.* Princeton, N.J.: Van Nostrand 1966.

[12] BERBERIAN, S. K.: Compact convex sets in inner product spaces. *Amer. Math. Monthly* **74**, 702–705 (1967).

[13] BERBERIAN, S. K. and ORLAND, G. H.: On the closure of the numerical range of an operator. *Proc. Amer. Math. Soc.* **18**, 499–503 (1967).

[14] BIRKHOFF, G.: A note on topological groups. *Compositio Math.* **3**, 427–430 (1936).

[15] BOHNENBLUST, H. F. and KARLIN, S.: Geometrical properties of the unit sphere of Banach algebras. *Ann. of Math.* (2) **62**, 217–229 (1955).

[16] BOHNENBLUST, H. F. and SOBCZYK, A.: Extensions of functionals on complex linear spaces. *Bull. Amer. Math. Soc.* **44**, 91–93 (1938).

[17] BONSALL, F. F. and DUNCAN, J.: *Numerical ranges of operators on normed spaces and of elements of normed algebras.* London Math. Soc. Lecture Note Series No. 2. Cambridge: Cambridge University Press 1971.

[18] BOURBAKI, N.: Sur les espaces de Banach. *C. R. Acad. Sci. Paris* **206**, 1701–1704 (1938).

[19] BOURBAKI, N.: *Topologie générale. Chs. I, II.* Fasc. II of *Éléments de mathématique,* Third edition. Paris: Hermann 1961.

[20] BOURBAKI, N.: *Topologie générale. Chs. III, IV.* Fasc. III of *Éléments de mathématique,* Third edition. Paris: Hermann 1960.

[21] BOURBAKI, N.: *Topologie générale. Ch. IX.* Fasc. VIII of *Éléments de mathématique,* Second edition. Paris: Hermann 1958.

[22] BOURBAKI, N.: *Topologie générale. Ch. X.* Fasc. X of *Éléments de mathématique,* Second edition. Paris: Hermann 1961.

[23] BOURBAKI, N.: *Espaces vectoriels topologiques. Chs. I, II.* Fasc. XV of *Éléments de mathématique.* Paris: Hermann 1953.

[24] BOURBAKI, N.: *Espaces vectoriels topologiques. Chs. III–V.* Fasc. XVIII of *Éléments de mathématique.* Paris: Hermann 1955.

[25] BOURBAKI, N.: *Espaces vectoriels topologiques. Fascicule de résultats.* Fasc. XIX of *Éléments de mathématique.* Paris: Hermann 1955.

[26] BOURBAKI, N.: *Théories spectrales. Chs. I, II.* Fasc. XXXII of *Éléments de mathématique.* Paris: Hermann 1967.

[27] BROWN, A. L. and PAGE, A.: *Elements of functional analysis.* London: Van Nostrand Reinhold 1970.

[28] CLARKSON, J. A.: Uniformly convex spaces. *Trans. Amer. Math. Soc.* **40**, 396–414 (1936).

[29] DAY, M. M.: Means for the bounded functions and ergodicity of the bounded representations of semi-groups. *Trans. Amer. Math. Soc.* **69**, 276–291 (1950).

[30] DAY, M. M.: Amenable semigroups. *Illinois J. Math.* **1**, 509–544 (1957).

[31] DAY, M. M.: *Normed linear spaces.* Second corrected printing. Berlin: Springer-Verlag 1962.

[32] DIEUDONNÉ, J.: Sur le théorème de Hahn-Banach. *Rev. Scientifique* **79**, 642–643 (1941).

[33] DIEUDONNÉ, J.: La dualité dans les espaces vectoriels topologiques. *Ann. Sci. École Norm. Sup.* **59**, 107–139 (1942).

[34] DIEUDONNÉ, J.: Sur la complétion des groupes topologiques. *C. R. Acad. Sci. Paris* **218**, 774–776 (1944).

[35] DIEUDONNÉ, J.: *Éléments d'analyse. II.* Paris: Gauthier-Villars 1968.

[36] DIXMIER, J.: Les moyennes invariantes dans les semi-groupes et leurs applications. *Acta Sci. Math. (Szeged)* **12**, 213–227 (1950).

[37] DIXMIER, J.: *Les algèbres d'opérateurs dans l'espace hilbertien (Algèbres de von Neumann).* Second edition. Paris: Gauthier-Villars 1969.

[38] DIXMIER, J.: *Les C*-algèbres et leurs représentations.* Second edition. Paris: Gauthier-Villars 1969.

[39] DOUGLAS, R. G.: *Banach algebra techniques in operator theory.* New York: Academic 1972.

[40] DOUGLAS, R. G. and PEARCY, C.: On the spectral theorem for normal operators. *Proc. Cambridge Philos. Soc.* **68**, 393–400 (1970).

[41] DUNFORD, N. and SCHWARTZ, J. T. (with the assistance of W. G. BADE and R. G. BARTLE): *Linear operators. Part I: General theory. Part II: Spectral theory. Part III: Spectral operators.* New York: Interscience 1958, 1963, 1971.

[42] EBERLEIN, W. F.: Weak compactness in Banach spaces. I. *Proc. Nat. Acad. Sci. U.S.A.* **33**, 51–53 (1947).

[43] EDWARDS, R. E.: *Functional analysis. Theory and applications.* New York: Holt, Rinehart and Winston 1965.

[44] ELLIS, R.: Locally compact transformation groups. *Duke Math. J.* **24**, 119–125 (1957).

[45] EPSTEIN, B.: *Linear functional analysis. Introduction to Lebesgue integration and infinite-dimensional problems.* Philadelphia: Saunders 1970.

[46] FOIAȘ, C.: Sur certains théorèmes de J. von Neumann concernant les ensembles spectraux. *Acta Sci. Math. (Szeged)* **18**, 15–20 (1957).

[47] FOIAȘ, C.: Some applications of spectral sets. I. Harmonic-spectral measure (Romanian). *Stud. Cerc. Mat.* **10**, 365–401 (1959). (English translation in *Amer. Math. Soc. Translations, Series 2, Vol.* 61, pp. 25–62. Providence, R.I.: American Mathematical Society 1967.)

[48] FRÉCHET, M.: Sur les opérations linéaires. *Trans. Amer. Math. Soc.* **8**, 433–446 (1907).

[49] FRÉCHET, M.: Les espaces abstraits topologiquement affines. *Acta Math.* **47**, 25–52 (1926).

[50] FRÉCHET, M.: *Les espaces abstraits et leurs théorie considérée comme introduction a l'analyse générale.* Paris: Gauthier-Villars 1928.

[51] GANTMAKHER, V. and ŠMULIAN, V.: Sur les espaces linéaires dont la sphère unitaire est faiblement compacte. *C. R. (Doklady) Acad. Sci. U.R.S.S.* **17**, 91–94 (1937).

[52] GEL'FAND, I. M.: Normierte Ringe. *Mat. Sb. N.S.* **9 (51)**, 3–24 (1941).

[53] GEL'FAND, I. M.: Über absolut konvergente trigonometrische Reihen und Integrale. *Mat. Sb. N.S.* **9 (51)**, 51–66 (1941).

[54] GEL'FAND, I. M. and NAĬMARK, M. A.: On the embedding of normed rings into the ring of operators in Hilbert space (Russian). *Mat. Sb. N.S.* **12 (54)**, 197–213 (1943).

[55] GEL'FAND, I. M. and NAĬMARK, M. A.: Rings with involution and their representations (Russian). *Izvest. Akad. Nauk SSSR Ser. Mat.* **12**, 445–480 (1948).

[56] GEL'FAND, I. M. and RAĬKOV, D. A.: Continuous unitary representations of locally bicompact groups (Russian). *Mat. Sb. N.S.* **13 (55)**, 301–316 (1943).

[57] GELFAND, I., RAIKOV, D., and SHILOV, G.: *Commutative normed rings.* Translated from the Russian. New York: Chelsea 1964.

[58] GODEMENT, R.: Mémoire sur la théorie des caractères dans les groupes localement compacts unimodulaires. *J. Math. Pures Appl.* **30**, 1–110 (1951).

[59] GODEMENT, R.: Sur la théorie des représentations unitaires. *Ann. of Math.* (2) **53**, 68–124 (1951).

[60] GOFFMAN, C. and PEDRICK, G.: *First course in functional analysis.* Englewood Cliffs, N.J.: Prentice-Hall 1965.

[61] GOLDBERG, S.: *Unbounded linear operators: Theory and applications.* New York: McGraw-Hill 1966.

[62] GOLDSTINE, H. H.: Weakly complete Banach spaces. *Duke Math. J.* **4**, 125–131 (1938).

[63] GREENLEAF, F. P.: *Invariant means on topological groups and their applications.* New York: Van Nostrand Reinhold 1969.

[64] HAHN, H.: Über lineare Gleichungen in linearen Räumen. *J. Reine Angew. Math.* **157**, 214–229 (1927).

[65] HALMOS, P. R.: *Measure theory.* New York: Van Nostrand 1950.

[66] HALMOS, P. R.: *Introduction to Hilbert space and the theory of spectral multiplicity.* New York: Chelsea 1951.

[67] HALMOS, P. R.: *A Hilbert space problem book.* Princeton, N.J.: Van Nostrand 1967.

[68] HALMOS, P. R.: Continuous functions of Hermitian operators. *Proc. Amer. Math. Soc.* **31**, 130–132 (1972).

[69] HAUSDORFF, F.: Zur Theorie der linearen metrischen Räume. *J. Reine Angew. Math.* **167**, 294–311 (1932).

[70] HEWITT, E. and ROSS, K. A.: *Abstract harmonic analysis. Vols. I, II.* New York/Heidelberg/Berlin: Springer-Verlag 1963, 1970.

[71] HEWITT, E. and STROMBERG, K.: *Real and abstract analysis.* New York: Springer-Verlag 1965.

[72] HILLE, E.: *Functional analysis and semi-groups.* Amer. Math. Soc. Colloq. Publ. Vol. 31. New York: American Mathematical Society 1948.

[73] HILLE, E.: *Methods in classical and functional analysis.* Reading, Mass.: Addison-Wesley 1972.

[74] HILLE, E. and PHILLIPS, R. S.: *Functional analysis and semi-groups.* Amer. Math. Soc. Colloq. Publ. Vol. 31, Revised edition. Providence, R.I.: American Mathematical Society 1957.

[75] HUSAIN, T.: *The open mapping and closed graph theorems in topological vector spaces.* Oxford: Clarendon 1965.

[76] HUSAIN, T.: *Introduction to topological groups.* Philadelphia: Saunders 1966.

[77] JACOBSON, N.: *Structure of rings.* Amer. Math. Soc. Colloq. Publ. Vol. 37, Revised edition. Providence, R.I.: American Mathematical Society 1964.

[78] JORDAN, P. and VON NEUMANN, J.: On inner products in linear, metric spaces. *Ann. of Math.* (2) **36**, 719–723 (1935).

[79] KADISON, R. V.: On the orthogonalization of operator representations. *Amer. J. Math.* **77**, 600–620 (1955).

[80] KAKUTANI, S.: Über die Metrisation der topologischen Gruppen. *Proc. Imp. Acad. Tokyo* **12**, 82–84 (1936).

[81] KANTOROVITZ, S.: Classification of operators by means of their operational calculus. *Trans. Amer. Math. Soc.* **115**, 194–224 (1965).

[82] KAPLANSKY, I.: Topological rings. *Amer. J. Math.* **69**, 153–183 (1947).

[83] KAPLANSKY, I.: *Topological algebra.* University of Chicago Mimeographed Lecture Notes, Chicago, Ill., 1952 (with an Appendix added October, 1955). Reprinted (without the 1955 appendix) as Notas de Matemática No. 16. Rio de Janeiro: Instituto de Matemática Pura e Aplicada 1959.

[84] KELLEY, J. L.: Note on a theorem of Krein and Milman. *J. Osaka Inst. Sci. Tech. Part I* **3**, 1–2 (1951).

[85] KELLEY, J. L.: *General topology.* New York: Van Nostrand 1955.

[86] KELLEY, J. L. and NAMIOKA, I.: *Linear topological spaces.* Princeton, N.J.: Van Nostrand 1963.

[87] KLEE, V. L., JR.: Convex sets in linear spaces. *Duke Math. J.* **18**, 443–466 (1951); Part II, ibid. **18**, 875–883 (1951); Part III, ibid. **20**, 105–111 (1953).

[88] KLEE, V. L., JR.: Invariant metrics in groups (solution of a problem of Banach). *Proc. Amer. Math. Soc.* **3**, 484–487 (1952).

[89] KLEE, V. L., JR.: Extremal structure of convex sets. II. *Math. Z.* **69**, 90–104 (1958).

[90] KOLMOGOROFF, A. [KOLMOGOROV, A. N.]: Zur Normierbarkeit eines allgemeinen topologischen linearen Raumes. *Studia Math.* **5**, 29–33 (1934).

[91] KOLMOGOROV, A. N. and FOMIN, S. V.: *Introductory real analysis.* Revised English edition, translated from the Russian and edited by R. A. Silverman. Englewood Cliffs, N.J.: Prentice-Hall 1970.

[92] KÖTHE, G.: *Topologische lineare Räume. I.* Berlin/Göttingen/Heidelberg: Springer-Verlag 1960.

[93] KREIN, M.: On positive additive functionals in linear normed spaces. *Comm. Inst. Sci. Math. Méc. Univ. Kharkoff* [*Zapiski Inst. Mat. Mech.*] (4) **14**, 227–237 (1937).

[94] KREIN, M. and MILMAN, D.: On extreme points of regularly convex sets. *Studia Math.* **9**, 133–138 (1940).

[95] KREIN, M. and ŠMULIAN, V.: On regularly convex sets in the space conjugate to a Banach space. *Ann. of Math.* (2) **41**, 556–583 (1940).

[96] LOOMIS, L. H.: *An introduction to abstract harmonic analysis.* New York: Van Nostrand 1953.

[97] LORCH, E. R.: *Spectral theory.* New York: Oxford University Press 1962.

[98] MADDOX, I. J.: *Elements of functional analysis.* Cambridge: Cambridge University Press 1970.

[99] MAZUR, S.: Über die kleinste konvexe Menge die eine gegebene kompakte Menge enthält. *Studia Math.* **2**, 7–9 (1930).

[100] Mazur, S.: Über konvexe Mengen in linearen normierten Räumen. *Studia Math.* **4**, 70–84 (1933).

[101] MAZUR, S.: Sur les anneaux linéaires. *C. R. Acad. Sci. Paris* **207**, 1025–1027 (1938).

[102] MILMAN, D. P.: On some criteria for the regularity of spaces of the type (B). *Dokl. Akad. Nauk SSSR (N.S.)* **20**, 243–246 (1938).

[103] MUNROE, M. E.: *Measure and integration.* Second edition. Reading, Mass.: Addison-Wesley 1971.

[104] NAĬMARK, M. A.: *Rings with involution.* Amer. Math. Soc. Translation No. 25 (English translation of an article originally appearing in Russian [*Uspehi Mat. Nauk (N.S.)* **3**, no. 5 (27), 52–145 (1948)]). New York: American Mathematical Society 1950.

[105] NAIMARK, M. A.: *Normed rings.* Revised English edition. Translated from the Russian by Leo F. Boron. Groningen: Noordhoff 1964.

[106] NAKAMURA, M. and TAKEDA, Z.: Group representation and Banach limit. *Tôhoku Math. J.* (2) **3**, 132–135 (1951).

[107] NEUMANN, J. VON: Zur Algebra der Funktionaloperatoren und Theorie der normalen Operatoren. *Math. Ann.* **102**, 370–427 (1929).

[108] NEUMANN, J. VON: On complete topological spaces. *Trans. Amer. Math. Soc.* **37**, 1–20 (1935).

[109] NEUMANN, J. VON: On a certain topology for rings of operators. *Ann. of Math.* (2) **37**, 111–115 (1936).

[110] NEUMANN, J. VON: Eine Spektraltheorie für allgemeine Operatoren eines unitären Raumes. *Math. Nachr.* **4**, 258–281 (1950/51).

[111] PETTIS, B. J.: A proof that every uniformly convex space is reflexive. *Duke Math. J.* **5**, 249–253 (1939).

[112] PHELPS, R. R.: *Lectures on Choquet's theorem.* Princeton, N.J.: Van Nostrand 1966.

[113] PONTRYAGIN, L. S.: *Topological groups.* Second edition. Translated from the Russian by Arlen Brown. New York: Gordon and Breach 1966.

[114] PRICE, G. B.: On the extreme points of convex sets. *Duke Math. J.* **3**, 56–67 (1937).

[115] RAĬKOV, D. A.: Harmonic analysis on commutative groups with a Haar measure and the theory of characters (Russian). *Trudy Mat. Inst. Steklov* **14** (1945). (German translation in *Sowjetische Arbeiten zur Funktionalanalysis*, pp. 11–87. Berlin: Verlag Kultur u. Fortschritt 1954.)

[116] RICKART, C. E.: *General theory of Banach algebras.* Princeton, N.J.: Van Nostrand 1960.

[117] RIESZ, F.: Über lineare Funktionalgleichungen. *Acta Math.* **41**, 71–98 (1918).

[118] RIESZ, F.: Zur Theorie des Hilbertschen Raumes. *Acta Sci. Math. (Szeged)* **7**, 34–38 (1934).

[119] RIESZ, F. and SZ.-NAGY, B.: *Leçons d'analyse fonctionelle.* Budapest: Akadémiai Kiadó 1952.

[120] ROBERTSON, A. P. and ROBERTSON, W.: *Topological vector spaces.* Cambridge: Cambridge University Press 1964.

[121] RUDIN, W.: *Fourier analysis on groups.* New York: Interscience 1962.

[122] RUDIN, W.: *Real and complex analysis.* New York: McGraw-Hill 1966.

[123] RUDIN, W.: *Functional analysis.* New York: McGraw-Hill 1973.

[124] SAKAI, S.: *C*-algebras and W*-algebras.* New York/Heidelberg/Berlin: Springer-Verlag 1971.

[125] SAKS, S.: *Theory of the integral.* Second revised edition. Warsaw 1937. Reprinted New York: Hafner.

[126] SCHAEFER, H. H.: *Topological vector spaces.* New York/Heidelberg/Berlin: Springer-Verlag 1971.

[127] SCHÄFFER, J. J.: On unitary dilations of contractions. *Proc. Amer. Math. Soc.* **6**, 322 (1955).

[128] SCHAUDER, J.: Über die Umkehrung linearer stetiger Funktionaloperationen. *Studia Math.* **2**, 1–6 (1930).

[129] SCHECHTER, M.: *Principles of functional analysis.* New York: Academic 1971.

[130] SCHWARTZ, J. T.: *W*-algebras.* New York: Gordon and Breach 1967.

[131] SCHWARTZ, L.: *Analyse. Topologie générale et analyse fonctionnelle.* Paris: Hermann 1970.

[132] SEGAL, I. E.: Postulates for general quantum mechanics. *Ann. of Math.* (2) **48**, 930–948 (1947).

[133] SEGAL, I. E. and KUNZE, R. A.: *Integrals and operators.* New York: McGraw-Hill 1968.

[134] SIMMONS, G. F.: *Introduction to topology and modern analysis.* New York: McGraw-Hill 1963.

[135] STONE, M. H.: *Linear transformations in Hilbert space and their applications to analysis.* Amer. Math. Soc. Colloq. Publ. Vol. XV. New York: American Mathematical Society 1932.

[136] SZ.-NAGY, B.: *Spektraldarstellung linearer Transformationen des Hilbertschen Raumes.* Berlin: Springer-Verlag 1942.

[137] SZ.-NAGY, B.: On uniformly bounded linear transformations in Hilbert space. *Acta Sci. Math.* (*Szeged*) **11**, 152–157 (1946/48).

[138] SZ.-NAGY, B. *Prolongements des transformations de l'espace de Hilbert qui sortent de cet espace.* Appendix to F. Riesz and B. Sz.-Nagy's *Leçons d'analyse fonctionnelle.* Budapest: Akadémiai Kiadó 1955.

[139] TAYLOR, A. E.: *Introduction to functional analysis.* New York: Wiley 1958.

[140] TOPPING, D. M.: *Lectures on von Neumann algebras.* London/New York: Van Nostrand Reinhold 1971.

[141] TYCHONOFF, A. [TIHONOV, A.]: Ein Fixpunktsatz. *Math. Ann.* **111**, 767–776 (1935).

[142] VALENTINE, F. A.: *Convex sets.* New York: McGraw-Hill 1964.

[143] WEHAUSEN, J. V.: Transformations in linear topological spaces. *Duke Math. J.* **4**, 157–169 (1938).

[144] WEIL, A.: *Sur les espaces a structure uniforme et sur la topologie générale.* Paris: Hermann 1938.

[145] WEIL, A.: *L'intégration dans les groupes topologiques et ses applications.* Paris: Hermann 1940.

[146] WERMER, J.: Banach algebras and analytic functions. *Adv. Math.* **1**, 51–102 (1961).

[147] WHITLEY, R.: Projecting m onto c_0. *Amer. Math. Monthly* **73**, 285–286 (1966).

[148] WILANSKY, A.: *Functional analysis.* New York: Blaisdell 1964.

[149] YOSIDA, K.: *Functional analysis.* Berlin/Göttingen/Heidelberg: Springer-Verlag 1965.

[150] ZAANEN, A. C.: *Linear analysis. Measure and integral, Banach and Hilbert space, linear integral equations.* Amsterdam: North-Holland 1953.

INDEX